14th International Conference on Nanomaterials - Research and Application (NANOCON 2022)

Brno, Czech Republic
19 – 21 October 2022

ISBN: 978-1-7138-7044-9

Printed from e-media with permission by:

Curran Associates, Inc.
57 Morehouse Lane
Red Hook, NY 12571

CURRAN ASSOCIATES INC.
proceedings.com

Some format issues inherent in the e-media version may also appear in this print version.

Tanger Ltd
Keltickova 62
710 00 Slezka Ostrava
Czech Republic

www.tanger.cz

Additional copies of this publication are available from:

Curran Associates, Inc.
57 Morehouse Lane
Red Hook, NY 12571 USA
Phone: 845-758-0400
Fax: 845-758-2633
Email: curran@proceedings.com
Web: www.proceedings.com

TANGER Ltd.
The Czech Society for New Materials and Technologies
Regional Centre of Advanced Technologies and Materials

2022 NANOCON®

14th International Conference on Nanomaterials - Research & Application

CONFERENCE PROCEEDINGS

October 19 - 21, 2022
Brno, Czech Republic, EU

ISSN 2694-930X

NANOCON 2022 **Conference Proceedings**

Different Authors

October 19-21, 2022 Orea Congress Hotel Brno, Czech Republic, EU

Issued by: TANGER Ltd., Keltickova 62, 710 00 Ostrava, Czech Republic, EU

Edition: 1st Edition, 2023

SESSION A - NANOMATERIALS FOR ELECTRONIC, MAGNETIC AND OPTIC APPLICATIONS. CARBON NANOSTRUCTURES, QUANTUM DOTS

SESSION B - INDUSTRIAL AND ENVIRONMENTAL APPLICATIONS OF NANOMATERIALS

SESSION C - BIONANOTECHNOLOGY, NANOMATERIALS IN MEDICINE

SESSION D - MONITORING AND TOXICITY OF NANOMATERIALS

SESSION E - ADVANCED METHODS OF PREPARATION AND CHARACTERIZATION OF NANOMATERIALS

SESSION A
NANOMATERIALS FOR ELECTRONIC, MAGNETIC AND OPTIC APPLICATIONS. CARBON NANOSTRUCTURES, QUANTUM DOTS

Chairmen

Prof. Dr. Eduard HULICIUS	Institute of Physics of the CAS, Prague, Czech Republic, EU
Assoc. Prof. Dr. Martin KALBAC	J. Heyrovsky Institute of Physical Chemistry of the CAS, Prague, Czech Republic, EU

INVESTIGATION OF NUCLEATION MECHANISMS OF NANOCRYSTALLINE PYROCHLORE PHOSPHORS EMITTING IN INFRARED

[1]Ivo BARTOŇ, [1]Soňa KAMRÁDKOVÁ, [1]Jana PROBOŠTOVÁ, [2]Luminita PREDOANA, [2]Jeanina PANDELE-CUSU, [1]Jan MRÁZEK

[1]Institute of Photonics and Electronics of the Czech Academy of Sciences, Prague, Czech Republic, EU, barton@ufe.cz

[2]"Ilie Murgulescu" Institute of Physical Chemistry, Bucharest, Romania, EU, lpredoana@icf.ro

https://doi.org/10.37904/nanocon.2022.4573

Abstract

High-power infrared lasers are the heart of modern equipment in telecommunications, tracking and navigation systems, etc. The increasing power of lasers places high demands on the materials used as active laser media, which must exhibit high thermal stability and luminescence efficiency. Rare earth-doped yttrium titanium oxides, with the general formula $(RE_xY_{1-x})_2Ti_2O_7$, represent a perspective class of materials for their phenomenal optical properties. Because the optical properties are highly sensitive to the structure and uniformity of the nanocrystals constituting the material, the knowledge of crystallization kinetics is necessary to prepare the nanocrystalline materials with tailored properties. We present a versatile sol-gel approach to nanocrystalline $Y_2Ti_2O_7$ and $(Er_{0,05}Y_{0,95})_2Ti_2O_7$. We studied the nucleation process and crystallization mechanism of $Y_2Ti_2O_7$ and $(Er_{0.05}Y_{0.95})_2Ti_2O_7$ from the amorphous xerogels. The crystallization temperatures of $Y_2Ti_2O_7$ and $(Er_{0,05}Y_{0,95})_2Ti_2O_7$ were 792.5 ± 0.9 and 789.9 ± 0.5 °C, respectively. Based on calculated Avrami parameters the formation of $Y_2Ti_2O_7$ from amorphous xerogel was driven by homogenous site-saturated nucleation. The introduction of Er^{3+} ions into host lattice of $Y_2Ti_2O_7$ changed the crystallization kinetics causing the formation of $(Er_{0.05}Y_{0.95})_2Ti_2O_7$ occurred in a manner of homogenous nucleation with a constant nucleation rate. The different crystallization kinetics caused the nanocrystals of $(Er_{0.05}Y_{0.95})_2Ti_2O_7$ to be larger and exhibited broader nanocrystals size distribution than the nanocrystals of undoped $Y_2Ti_2O_7$. The results provide fundamental information about nucleation and growth properties and crystal structure of investigated luminophores and give necessary information for preparing nanocrystalline powders with tailored properties for high-power photonic devices.

Keywords: Nucleation, crystallization, pyrochlore, phosphors, $(Er_{0.05}Y_{0.95})_2Ti_2O_7$

1. INTRODUCTION

Lanthanide titanium oxides, crystallizing in a pyrochlore structure with a general formula $A_2B_2O_7$ [1], have attracted attention in recent material research for their magnetic and optical properties. The spin arrangement of rare earths (RE) in the pyrochlore lattice provides distinctive magnetic properties allowing the preparation of spin-glass and spin-ice compounds [2,3]. Although the $RE_2B_2O_7$ pyrochlores are optically inactive, the yttrium ions in the pyrochlore structure break the spin interactions and prevent the non-radiative transitions between RE ions [4,5]. Consequently, the RE-doped $Y_2Ti_2O_7$ exhibit effective luminescence properties and they have been intensively studied as luminophores for high-power photonic applications [6,7]. The final optical properties of RE-doped $Y_2Ti_2O_7$ strongly depend on the local arrangement and regular distribution of RE ions inside the crystal lattice. The local clustering of RE ions supports the non-radiative transitions and reduces the luminescence efficiency. The structural defects and broad nanocrystal size distribution can contribute to the scattering of the light and increase the optical losses [8,9]. Therefore, the processing of RE-doped $Y_2Ti_2O_7$ must be carefully tailored to keep the regular statistical distribution of RE inside the pyrochlore lattice and to prepare highly homogenous luminophores or coatings [6,7].

Several approaches have been used for the synthesis of pyrochlores. In addition to standard high-temperature sintering ceramic methods [1,2], the "bottom-up" sol-gel methods have been intensively studied to prepare broad set of luminescent powders [6,10] and active optical waveguides [7,11]. The "bottom-up" methods benefit from targeted thermally induced crystallization of amorphous materials. However, the processing temperature must be carefully controlled to prevent the side-formation of parasitic phases or unwanted nanocrystal overgrowth. Knowledge of the crystallization kinetics of pyrochlores from amorphous precursors opens up possibilities for extending the synthesis of RE-doped $Y_2Ti_2O_7$ and preparing the luminophores with tailored structural properties.

This study presents a versatile sol-gel approach to nanocrystalline $Y_2Ti_2O_7$ and $(Er_{0.05}Y_{0.95})_2Ti_2O_7$. We evaluated the structural properties of prepared nanocrystals. We study the nucleation process and crystallization mechanism of $Y_2Ti_2O_7$ and $(Er_{0.05}Y_{0.95})_2Ti_2O_7$ from the amorphous xerogels. The acquired knowledge is necessary for the preparation of nanocrystalline powders with tailored properties for high-power photonic devices.

2. EXPERIMENTAL

The samples were prepared by a sol-gel method followed by thermal treatment of the xerogels. To prepare the sols, a total of 5 g titanium(IV)butoxide (Fluka, Purum) was dissolved in 250 ml of anhydrous ethanol (Sigma-Aldrich, Spectranal grade), after which a total of 5.63 g of yttrium(III) nitrate hexahydrate (Aldrich, 99.8%), or a total of 5.35 g yttrium(III) nitrate hexahydrate (Aldrich, 99.8%) and 0.325 g of erbium(III) nitrate pentahydrate (Aldrich, 99.9%) were dispersed in the solution. The mixtures were stirred at ambient temperature to form the transparent solutions, which were refluxed at 80 °C for 24 h and then allowed to cool. The sols were dried on a rotary evaporator (R100, Buchi) to form the xerogels.

Differential thermal analysis (DTA) and thermal gravimetry (TG) measurements were performed using a Mettler Toledo TGA/SDTA 851^e apparatus. The samples were analyzed in alumina crucibles at heating rates ranging from 10 to 40 $°C \cdot min^{-1}$ under an oxygen flow of 50 $ml \cdot min^{-1}$.

X-ray powder diffraction (XRD) patterns were collected using a Bruker D8 Discover diffractometer in the Bragg-Brentano reflecting geometry. The copper tube was operated at voltage of 40 kV and current of 40 mA providing Cu-K$\alpha 1$ radiation (λ=1.540596 Å).

Scanning electron microscopy (SEM) images were obtained with a TESCAN Lyra 3 XMU device. A thin carbon layer was sputtered onto the samples to prevent sample charging. To evaluate the nanocrystal size distribution, we marked the nanocrystal boundaries in the SEM images and we used Gwyddion 2.55 data visualization and analysis software.

3. RESULTS AND DISCUSSION

The thermal processing of xerogels represents a complex process consisting of a broad set of physical and chemical processes. It usually includes several steps involving the evaporation of volatile substances, burning and decomposition of organic compounds and inorganic ligands, and the crystallization process itself. A key factor for the successful synthesis of pyrochlores with tailored structural properties is a perfect knowledge of the physicochemical processes occurring during thermal treatment. **Figure 1a** shows the general thermal analysis curves recorded for the heat rate of 40 $°C \cdot min^{-1}$. DTA curve shows broad endothermic peak below 300 °C and two exothermic peaks around 350 °C and 843 °C. The broad endothermic peaks below 600 °C and the first exothermic peak around 350 °C were accompanied by regular weight loss. The minor weight loss accompanied the second exothermic peak at 820 °C. According to the chemical analysis of the compounds liberated during the thermal treatment of pyrochlores, the broad endothermic peak below 300 °C can be attributed to the liberation of water, vaporization, and carbonization of the adsorbed alcohols and esters [12]. The exothermic peak around 350 °C was attributed to the decomposition of nitrates and the condensation of -OH groups presented in the xerogels [13,14]. The remaining nitrates act as complexing agents that stabilized Y^{3+} and RE^{3+} ions inside the amorphous titanium dioxide matrix preventing the crystallization of titanium dioxide [14]. Once the nitrates are burn-out the crystallization of pyrochlores takes place. The burning-out of the nitrates was attributed to mass loss at 820 °C and the minor exothermic peak at 843 °C corresponded to the crystallization of $Y_2Ti_2O_7$.

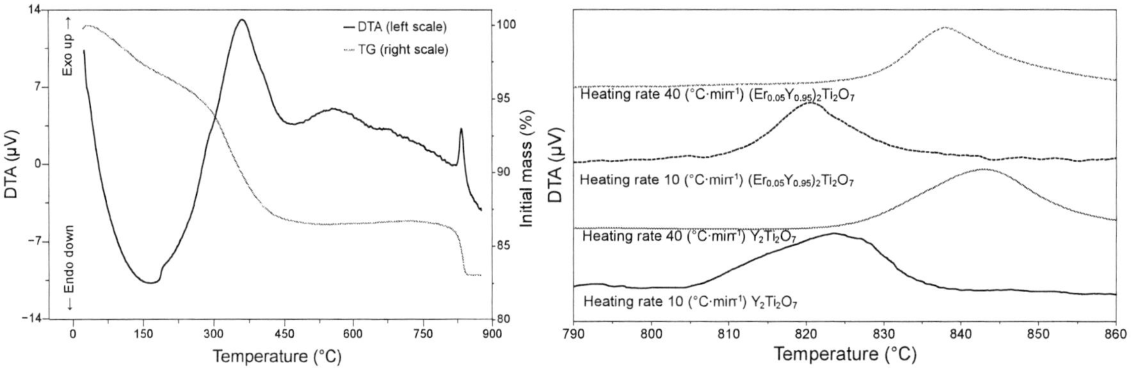

Figure 1 Representative results of thermal analysis. a) Differential thermal analysis and thermal gravimetry records for $Y_2Ti_2O_7$ powder recorded for heating rate of 40 $°C{\cdot}min^{-1}$. b) Zoom on the DTA curves demonstrating the shift of the crystallization peaks with different heating rates

The crystallization is a dynamic process depending on the overheating or overcooling of the thermally treated materials. The shape and position of the crystallization peak depend on the heating rate of the analyzed samples. **Figure 1b** shows the shifts in the positions of representative crystallization peaks in detail. The crystallization of $Y_2Ti_2O_7$ occurs at 824.6 °C for the heating rate of 10 $°C{\cdot}min^{-1}$. With increasing heating rates, the positions of the crystallization peaks shift to higher temperatures and the peaks became better pronounced. Compared with $Y_2Ti_2O_7$, the crystallization peaks of $(Er_{0,05}Y_{0,95})_2Ti_2O_7$ shifted to lower temperatures.

The shifts in the positions of crystallization peaks have been used in many theoretical and experimental models to evaluate fundamental crystallization parameters. The crystallization temperature of nanocrystalline materials, t_c, can be calculated using the empirical Lasocka's equation [15]:

$$t_p = t_c + A_c \cdot \ln(\alpha) \qquad (1)$$

where t_p represents the position of the crystallization peak (°C), α is the heating rate ($°C{\cdot}min^{-1}$), and A_c is the experimental constant. **Figure 2a** shows the experimental positions of the crystallization peaks for various heating rates and corresponding linear fits. The y-intercept representing t_c was calculated from a linear fit of α equal to 1 $°C{\cdot}min^{-1}$. The crystallization temperatures of $Y_2Ti_2O_7$ and $(Er_{0,05}Y_{0,95})_2Ti_2O_7$ were 792.5 ± 0.9 and 789.9 ± 0.5 °C, respectively. These values were in line with the crystallization temperatures of pyrochlores prepared by Pechini's methods, e.g. 817.7 °C for $Er_2Ti_2O_7$ [16], 821 °C for $Y_2Ti_2O_7$ [17], and ~800 °C for Er-doped $Y_2Ti_2O_7$ [18].

XRD records of the samples heat-treated at 1000 °C are shown in **Figure 2b**. The pure pyrochlore cubic phase of $Y_2Ti_2O_7$ was formed and the crystal lattice parameter *a* of $Y_2Ti_2O_7$ was 10.0963 Å. This value matches the value of 10.0976 Å reported for single-crystal sample [17]. The XRD pattern of $(Er_{0,05}Y_{0,95})_2Ti_2O_7$ exhibited isostructural diffraction lines confirming the formation of the pyrochlore structural type. The samples were free of side-formed oxides, such as TiO_2, Y_2O_3, or $Er_2Ti_2O_7$, and the diffraction lines did not show the twining, proving the regular statistical distribution of Er^{3+} ions inside the pyrochlore crystal lattice. The positions of the diffraction lines were shifted to longer angles providing the crystal lattice parameter *a* = 10.1071 Å.

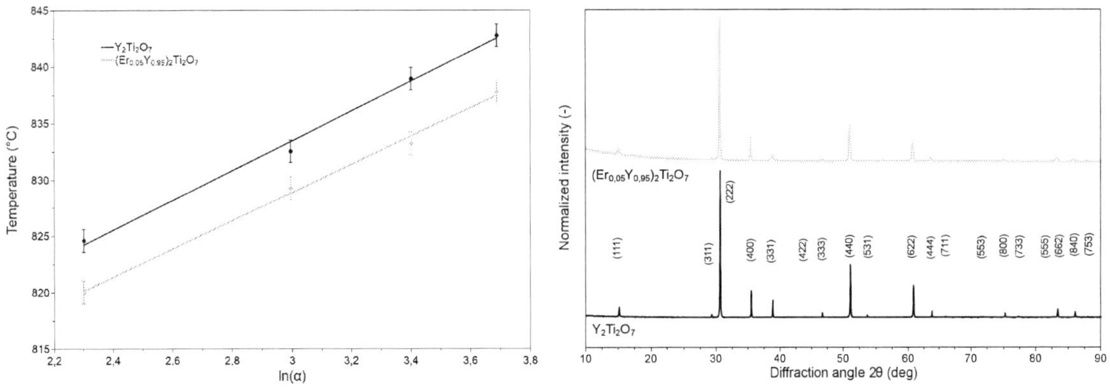

Figure 2 a) Linear fits of Lasocka's equation approximating the crystallization temperature of the compounds. b) Diffractograms of the prepared compounds heat-treated at 1000 °C

To determine the crystallization kinetics, we used Johnson, Mehl, and Avrami model (JMA). In the JMA approximation, the relationship between the crystallization rate, $d\psi/d\tau$, and volume fraction of crystallized material, ψ, at time, τ, can be expressed in the general form of the JMA kinetic equation:

$$\frac{d\psi}{d\tau} = k \cdot \eta \cdot (1-\psi) \cdot [-ln(1-\psi)]^{1-\frac{1}{\eta}} \quad (2)$$

where η is the Avrami parameter that depends on the nucleation and growth mechanism, *k* is the rate constant, and volume fraction of crystallized material, ψ, is also known as the relative crystallinity of the sample crystallized at time τ (min). A representative course of relative crystallinity is shown in **Figure 3a**.

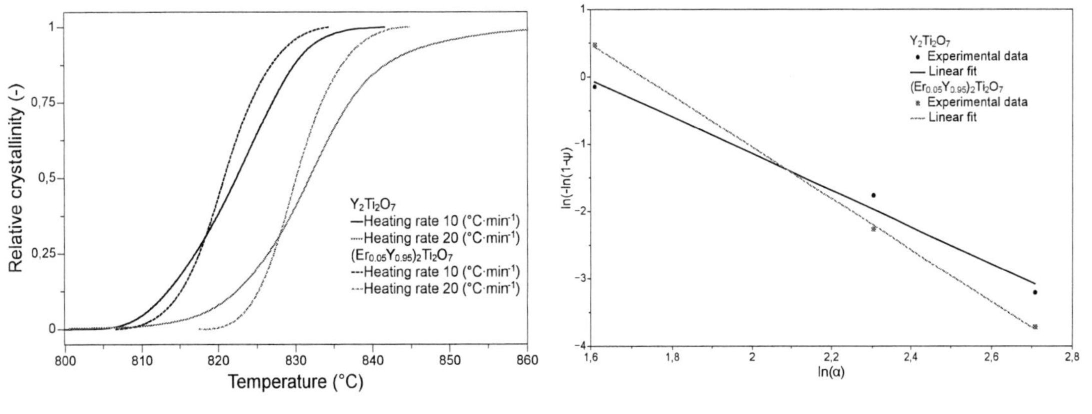

Figure 3 a) Variation of the relative crystallinity with heating rate for particular compounds. b) Logarithmic plot of the dependence of the term *-ln(1-Ψ)* on the heating rate for the evaluation of Avrami parameters for the non-isothermal crystallization kinetics approach.

The positions of the curves shifted to higher temperatures with increasing heating rates. The curves' slopes increased with increasing dimensionality of the nucleation process and the crystal growth [19]. For linear heating rates the JMA kinetic equation can be expressed by the Ozawa's non-isothermal modification [20]:

$$-\eta = \frac{d\{ln[-ln(1-\psi)]\}}{d\,ln\,a}\bigg|_{t_a} \quad (3)$$

where the Avrami parameter η, is determined from the slope of the plot of ln[-ln(1-ψ)] vs. ln α. The relative crystallinity, ψ, is calculated as the ratio of the partial integration area of the crystallization peak at temperature t_a to the total area of the crystallization peak. **Figure 3b** shows the corresponding data and fits allowing evaluation of the Avrami parameters, which were 2.9 ± 0.3 and 4.1 ± 0.2 for $Y_2Ti_2O_7$ and $(Er_{0.05}Y_{0.95})_2Ti_2O_7$, respectively. The values of the Avrami parameter are characteristic of particular nucleation mechanism and mass transfer causing the nanocrystal growth [16,21]. Each dimension of the crystal growth or formation of nuclei contributes a unit to the total Avrami parameter. Therefore, the values corresponding to homogenous nucleation with a constant nucleation rate are greater by a unit than those corresponding to site-saturated nucleation. Because the real crystallization process usually combines several mechanisms, the Avrami parameters slightly differ from the predicted theoretical integers. The value of η calculated for $Y_2Ti_2O_7$ matched the value of 2.8 reported for $Y_2Ti_2O_7$ prepared by Pechini's method [17]. The excellent agreement of these results suggested that the crystallization mechanism of $Y_2Ti_2O_7$ from the amorphous "soft chemistry" precursors is identical and the pyrochlores crystallize in a manner of homogenous 3-D site-saturated nucleation. However, the introduction of Er^{3+} ions significantly increased the value of the Avrami parameter up to 4. Such a high value suggested that the crystallization of $(Er_{0.05}Y_{0.95})_2Ti_2O_7$ runs in a manner of homogenous nucleation with a constant nucleation rate.

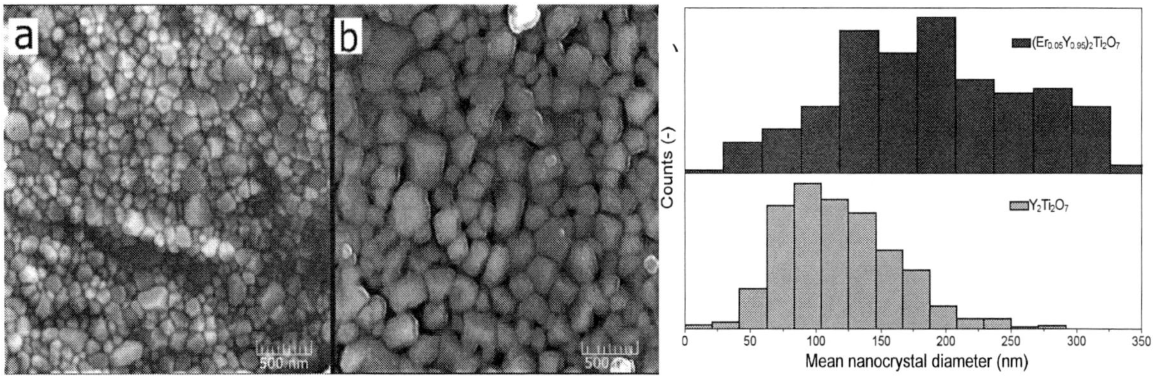

Figure 4 SEM image of the compounds heat-treated at 1000 °C. a) Sample of $Y_2Ti_2O_7$. b) Sample of $(Er_{0.05}Y_{0.95})_2Ti_2O_7$. c) Histogram summarizing the nanocrystal size distribution

The different crystallization kinetics have a significant effect on the nanocrystalline structure of the prepared materials. To verify the proposed crystallization mechanism, the powders heat-treated at 1000 °C for 30 minutes were visualized by SEM and the images are shown in **Figure 4**. The powder of $Y_2Ti_2O_7$ was composed of well-grown nanocrystals with a regular shape without any preferential size orientation. **Figure 4c** shows that the nanocrystal diameter ranged from 50 to 200 nm with a mean diameter of 100 nm. Uniform and narrow nanocrystal size distribution is typical for sited-saturated nucleation. Nuclei are formed in the entire volume of the amorphous material at once in a very narrow temperature interval. The prolongated thermal treatment promotes the recrystallization of the material, causing the nanocrystal size distribution to broaden and grow. In the case of $(Er_{0.05}Y_{0.95})_2Ti_2O_7$, the observed variability of the nanocrystal size and shape was higher than for $Y_2Ti_2O_7$. The nanocrystals were larger, the smaller nanocrystals filled the spaces between larger one and the nanocrystal diameter ranged from 25 to 325 nm with a mean frequency of 175 nm. A broad nanocrystal size distribution is typical for homogenous nucleation with a constant nucleation rate. The nuclei formed at the beginning of the crystallization process regularly grew into larger nanocrystals. Simultaneously, the nucleation of new, smaller nanocrystals occurs. The final materials consist of nanocrystals with broad size distribution.

The presented Avrami parameters and nanocrystal size analyses supported the conclusion that the introduction of RE into the host matrix of $Y_2Ti_2O_7$ changed the crystallization mechanism and the crystallization process of $(Er_{0.05}Y_{0.95})_2Ti_2O_7$ occurred in the manner of homogenous nucleation with a constant nucleation rate. The homogenous nucleation with constant nucleation rate can reduce the associated energy consumption due to the lower nucleation energy barrier [22]. However, the broad nanocrystal size distribution can increase the Rayleigh scattering. This phenomenon can reduce the optical quality of prepared materials and limit their application in photonics [23,24].

4. CONCLUSION

We demonstrated a sol-gel approach to nanocrystalline $Y_2Ti_2O_7$ and $(Er_{0.05}Y_{0.95})_2Ti_2O_7$. The crystallization temperatures of $Y_2Ti_2O_7$ and $(Er_{0.05}Y_{0.95})_2Ti_2O_7$ were 792.5 ± 0.9 and 759.9 ± 0.5 °C, respectively. The Er^{3+} ions statistically substituted the Y^{3+} ions inside the pyrochlore structural lattice causing the increase of the crystal lattice parameter *a* up to 10.1071 Å. We used the non-isothermal crystallization kinetics approach and image analysis of SEM to evaluate the crystallization mechanisms. The crystallization of $Y_2Ti_2O_7$ occurred in the manner of site-saturated nucleation. The introduction of Er^{3+} ions into the host lattice of $Y_2Ti_2O_7$ changed the nucleation mechanism and $(Er_{0.05}Y_{0.95})_2Ti_2O_7$ crystallized in the manner of homogenous nucleation with a constant nucleation rate. The acquired knowledge is necessary to prepare nanocrystalline powders with tailored properties for high-power photonic devices.

ACKNOWLEDGEMENTS

This research was financially supported by the Czech Science Foundation number 22-14200S and the mobility project of the Czech Academy of Sciences number AR-20-01.

REFERENCES

[1] GLERUP, M., NIELSEN, O. F., POULSEN, F. W. The structural transformation from the pyrochlore structure, A(2)B(2)O(7), to the fluorite structure, AO(2), studied by Raman spectroscopy and defect chemistry modeling. *Journal of Solid State Chemistry*. 2001, vol. 160, no. 1, pp. 25-32.

[2] GARDNER, J. S., GINGRAS, M.J., GREEDAN, J.E. Magnetic pyrochlore oxides. *Reviews of Modern Physics*. 2010, vol. 82, no. 1, pp. 53-107.

[3] VLASKOVA, K., PROSCHEK, P., DIVIS, M., Duc, L.E., COLMAN, R.H., KLICPERA, M. Magnetic properties and crystal field splitting of the rare-earth pyrochlore $Er_2Ir_2O_7$. *Physical Review B*. 2020, vol. 102, vol. 5.

[4] JENOUVRIER, P., BOCCARDI, G., FICK, J., JURDYC, A. M., LANGLET, M. Up-conversion emission in rare earth-doped $Y_2Ti_2O_7$ sol-gel thin films. *Journal of Luminescence*. 2005, vol. 113, no. 3-4, pp. 291-300.

[5] SINGH, B. P., PARCHUR, A. K., SINGH, R. K., ANSARI, A. A., SINGH, P., RAI, S. B. Structural and up-conversion properties of Er^{3+} and Yb^{3+} co-doped $Y_2Ti_2O_7$ phosphors. *Physical Chemistry Chemical Physics* [online]. 2013, vol. 15, no. 10, pp. 3480-3489.

[6] GUO, Y., WANG, D., ZHAO, X., WANG, F. Fabrication, microstructure and upconversion luminescence of Yb^{3+}/Ln(3+) (Ln=Ho, Er, Tm) co-doped $Y_2Ti_2O_7$ ceramics. *Materials Research Bulletin*. 2016, vol. 73, pp. 84-89.

[7] VYTYKACOVA, S., MRAZEK, J., PUCHY, V., DZUNDA, R., SKALA, R., PETERKA, P., KASIK, I. Sol-gel route to highly transparent $(Ho_{0.05}Y_{0.95})(2)Ti_2O_7$ thin films for active optical components operating at 2 mu m. *Optical Materials Express*. 2018, vol. 78, pp. 415-420.

[8] ALMEIDA, R. M., MARQUES, A. C., PORTAL, S. Glassy and nanocrystalline photonic materials and structures by sol-gel. *Optical Materials*. 2005, vol. 27, no. 11, pp. 1718-1725.

[9] MATEJEC, V., MRAZEK, J., HAYER, M., KASIK, I., PETERKA, P., KANKA, J., HONZATKO, P., BERKOVA, D. Microstructure fibers for gas detection. *Materials Science & Engineering C-Biomimetic and Supramolecular Systems.* 2006, vol. 26, no. 2-3, SI, pp. 317-321.

[10] JENOUVRIER, P., FICK, J., AUDIER, M., LANGLET, M. Microstructure and photoluminescence properties of sol-gel $Y_{2-x}Er_xTi2O_7$ thin films. *Optical Materials.* 2004, vol. 27, no. 2, pp. 131-137.

[11] JENOUVRIER, P., LANGLET, M., RIMET, R., FICK, J. Influence of crystallisation on the photoluminescence properties of $Y_{2-x}Er_xTi_2O_7$ sol-gel thin films. *Applied Physics a-Materials Science & Processing.* 2003, vol. 77, no. 5, pp. 687-692.

[12] STARUKH, G., TOSCANI, S., BOURSICOT, S., SPANHEL, L. Photoactivity of sol-gel derived nitridated ZnxTiyOz-films. *Zeitschrift Fur Physikalische Chemie-International Journal Of Research In Physical Chemistry & Chemical Physics.* 2007, vol. 221, no. 3, pp. 349-360.

[13] BALTRUSAITIS, J., SCHUTTLEFIELD, J., JENSEN, J.H., GRASSIAN, V.H. FTIR spectroscopy combined with quantum chemical calculations to investigate adsorbed nitrate on aluminium oxide surfaces in the presence and absence of co-adsorbed water. *Physical Chemistry Chemical Physics.* 2007, vol. 9, no. 36, pp. 4970-4980.

[14] MRAZEK, J., POTEL, M., BURSIK, J., MRACEK, A., KALLISTOVA, A., JONASOVA, S., BOHACEK, J., KASIK, I. Sol-gel synthesis and crystallization kinetics of dysprosium-titanate Dy2Ti2O7 for photonic applications. *Materials Chemistry and Physics.* 2015, vol. 168, pp. 159-167.

[15] LASOCKA, M. Effect of scanning rate on glass-transition temperature of splat-cooled Te-85-Ge-15. *Materials Science and Engineering.* 1976, vol. 23, no. 2-3, pp. 173-177.

[16] MILICEVIC, B., KUZMAN, S., POROBIC, S. J., MARINOVIC-CINCOVIC, M., DRAMICANIN, M. D. Non-isothermal crystallization kinetics of the heavy-group lanthanide dititanates. *Optical Materials.* 2017, vol. 74, pp. 86-92.

[17] MILICEVIC, B., MARINOVIC-CINCOVIC, M., DRAMICANIN, M. D. Non-isothermal crystallization kinetics of Y(2)Ti(2)O7. *Powder Technology.* 2017, vol. 310, pp. 67-73.

[18] TING, C.C., CHIU, Y.S., CHANG, C.W., CHUANG, L.C. Visible and infrared luminescence properties of Er^{3+}-doped $Y_2Ti_2O_7$ nanocrystals. *Journal of Solid State Chemistry.* 2011, vol. 184, no. 3, pp. 563-571.

[19] VAZQUEZ, J., GONZALEZ-PALMA, R., VILLARES, P., JIMENEZ-GARAY, R. Theoretical study on the glass-crystal transformation and deduction of its kinetic parameters by DSC, using non-isothermal regime. *Physica B-Condensed Matter.* 2003, vol. 336, no. 3-4, pp. 297-307.

[20] MALEK, J., MITSUHASHI, T. Testing method for the Johnson-Mehl-Avrami equation in kinetic analysis of crystallization processes. *Journal of the American Ceramic Society.* 2000, vol. 83, no. 8, pp. 2103-2105.

[21] PEREZ-MAQUEDA, L. A., CRIADO, J.M., J. MALEK, J. Combined kinetic analysis for crystallization kinetics of non-crystalline solids. *Journal of Non-Crystalline Solids.* 2003, vol. 320, no. 1-3, pp. 84-91.

[22] YASNO, J. P., CONCONI, S., VISINTIN, A., SUAREZ, G. Non-isothermal reaction mechanism and kinetic analysis for the synthesis of monoclinic lithium zirconate (m-Li2ZrO3) during solid-state reaction. *Journal of Analytical Science and Technology.* 2021, vol. 12, no. 1.

[23] ALMEIDA, R. M., MARQUES, A.C., CHIASERA, A., CHIAPPINI, A., FERRARI, M. Rare-earth doped photonic crystal microcavities prepared by sol-gel. *Journal of Non-Crystalline Solids.* 2007, vol. 353, no. 5, pp. 490-493.

[24] KASIK, I., MRAZEK, J., PODRAZKY, O., SEIDL, M., AUBRECHT, J., TOBISKA, P., POSPISILOVA, M., MATEJEC, V., KOVACS, B., MARKOVICS, A., SZILI, M. Fiber-optic detection of chlorine in water. *Sensors and Actuators B-Chemical.* 2009, vol. 139, no. 1, pp. 139-142.

XPS CHARACTERIZATION OF NIOBIUM-OXIDE NANOCOLUMN ARRAYS VIA SELF-ORGANIZED ANODIZING IN SELENIC ACID

Maria BENDOVA, Kirill KAMNEV, Alexander MOZALEV

CEITEC - Central European Institute of Technology, Brno University of Technology, Brno, Czech Republic, EU
maria.bendova@ceitec.vutbr.cz, alexander.mozalev@ceitec.vutbr.cz

https://doi.org/10.37904/nanocon.2022.4574

Abstract

As previously reported, arrays of Nb_2O_5 nanocolumns on substrates may be synthesized by anodizing Al-on-Nb bilayers in oxalic, sulfuric, malonic, or phosphoric acids. Here, for the first time, we employed a new anodizing electrolyte - selenic acid (H_2SeO_4) due to a unique morphology of porous anodic alumina (PAA) formed in this electrolyte featuring extremely narrow pores and therefore, the minimal possible porosity (~3%) among the PAA films prepared in the usual solutions. Such a morphology may result in distinctive structural, physical, and chemical properties of the niobium-oxide nanoarrays complemented by the specific dielectric, optical, electro-optical, catalytic, and antibacterial properties in case the selenium species are incorporated into the column structure during the anodizing. The columns were prepared from the magnetron-sputtered Al/Nb bilayers by anodizing at 12 or 40 V, followed by re-anodizing to 80 or 120 V, respectively, and selective dissolution of the alumina overlayer. Detailed XPS characterization confirmed that selenium species is present in the column material. The fitting of the narrow-scan Se 3d spectra, complicated by overlapping with Nb 4s peaks, shows that selenium is incorporated in the form of selenate (SeO_4^{2-}) and selenide (Se^{2-}) anions, likely creating a core-shell structure. Quantitative analysis reveals that about 1.5-3.0 % of O^{2-} anions in the oxide structure are replaced by selenate or selenide anions. Further investigation is in progress to understand the formation-structure-morphology relationship and explore the functional properties of the arrays.

Keywords: Anodizing, porous anodic alumina, niobium pentoxide, selenium

1. INTRODUCTION

Nanostructuring is beneficial or crucial for many applications of metal oxides due to an increased surface-to-volume ratio and a shorter distance to the surface [1,2]. One-dimensional nanostructures (nanowires, nanorods, or nanocolumns), anchored to the substrate, can be synthesized in various ways, including sol-gel synthesis and other wet-chemical methods, physical and chemical vapor deposition, and electrodeposition [1,3]. An alternative approach, the so-called PAA-assisted anodizing [4], may be applied to a thin film of aluminum deposited on a layer of another valve metal (i.e. Nb, Ta, W, Ti, Hf, or Zr). Initially, the Al layer is converted into porous anodic alumina (PAA) by anodizing in an acidic electrolyte, immediately followed by self-organized growth of the underlying metal oxide locally into the barrier layer of the PAA film beneath the alumina pores. This results in the formation of short nanoprotrusions of the underlying metal oxide, mixed to some extent with alumina. Further, re-anodizing to a higher voltage may be carried out to prolong the oxide protrusions within the alumina pores. In this way, metal-oxide nanocolumns, upright standing on a substrate, attaining the morphology of the initially formed PAA layer may be created.

Niobium pentoxide (Nb_2O_5) is a wide-bandgap *n*-type semiconductor [5]. It is a non-toxic metal oxide with excellent chemical stability and corrosion resistance [5], suitable for applications including photocatalysis, gas sensing, electrochemical energy storage, electrochromic devices, or photoanodes. The PAA-assisted anodizing has been used to synthesize arrays of Nb_2O_5 nanocolumns of various dimensions [4], with diameters ranging between 30 and 150 nm, depending on the electrochemical and electrical conditions applied during the Al anodizing. So far, various electrolytes have been tested, including oxalic, sulfuric, malonic, or phosphoric acids [4]. However, anodizing in *selenic acid* (H_2SeO_4), which offers a unique morphology of PAA with extremely narrow pores and very low porosity of ~3% [6], may result in arrays of very thin Nb_2O_5 nanocolumns possessing specific structural, physical, or chemical properties. In addition, if selenium species are incorporated into the column material, altering the dielectric, optical, catalytic, or antibacterial properties may be achieved.

In the present work, we have anodized and re-anodized the Al-on-Nb bilayers in selenic acid to synthesize arrays of Se-doped Nb_2O_5 nanocolumns of various morphologies. The chemical composition of the films, focusing on the presence of selenium species, has been examined by X-ray photoelectron spectroscopy (XPS).

2. EXPERIMENTAL PART

2.1. Sample preparation

Thin Al-on-Nb bilayer film was used as a precursor for anodizing. It was prepared on oxidized Si wafers by magnetron sputtering of a 140-nm thick Nb layer followed by a 1-µm thick Al layer from the Nb (99.95%) and Al (99.999%) targets, respectively. Pieces of 1 cm × 1 cm cut from the wafers were anodized and re-anodized in 1.5 M H_2SeO_4 aqueous solution at room temperature. The anodizing was performed galvanostatically at 0.5 or 5 $mA \cdot cm^{-2}$, followed by a short potentiostatic hold at 12 or 40 V, respectively, to anodize Nb through the barrier layer. For selected samples, potentiodynamic re-anodizing was carried out at 10 $V \cdot s^{-1}$ to 80 or 120 V, respectively, followed by a short voltage stabilizing period. To obtain PAA-free samples, the PAA overlayer was selectively dissolved away in Cr_2O_3/H_3PO_4-based etchant [7]. Field emission SEM was employed to examine the morphology of the films.

2.2. XPS characterization

XPS analysis was carried out for the PAA-free films on a Kratos Axis Ultra DLD spectrometer using a monochromatic Al Kα source. The X-ray emission power was 150 W with a 15 kV accelerating voltage focused to a spot of 300 µm × 700 µm. The emitted electrons were detected at fixed pass energies of 160 eV for the survey spectra and 20 eV for the high-resolution spectra. The Kratos charge neutralizer system was used for all specimens.

The spectra were analyzed using CasaXPS. GL(30) profiles were used for all components except the metallic core lines of Nb 3d, for which asymmetric profiles in the form of LA(1.2,5,12) were applied. A standard Shirley background was used in all fitted spectra. Spectra from all samples have been charge corrected to give the adventitious C 1s spectral component (C−C, C−H) binding energy of 284.8 eV. The deconvolution of C 1s spectra was performed as described in previous works [4].

3. RESULTS AND DISCUSSION

3.1. Film morphology

The PAA-assisted anodizing of Nb bilayers in selenic acid at 12 and 40 V was expected to result in arrays of Nb_2O_5 nanocolumns of two dissimilar diameters. **Figures 1a,c** shows the top-view and cross-fractional SEM images of the PAA-free 40V columns. Protrusions of presumably Nb_2O_5 are observed, about 70 nm long and 20 nm in diameter. They are anchored to the Nb metal below through oxide bases and are separated from each other by the remainings of Al and possibly Nb metals. Increasing the voltage during the re-anodizing to

120 V results in the lengthening and widening of the protrusions to 260 and 30 nm, respectively (**Figures 1b,d**). The nanocolumns are still connected to metal oxide bases below them and are separated by the residues of the metals, evidently of a smaller amount than in the case of the anodized 40V sample. The nanocolumns obtained by anodizing at 12 V followed by re-anodizing to 80 V (i.e., 12+80V columns, SEM images not shown) are thinner and shorter than the 40+120V columns, as expected, having a diameter of 12 nm and a length of 120 nm. However, they are not separated by metal residues. They are standing on a continuous layer of metal oxide residing between the columns and the unoxidized Nb on the substrate. In the following, we will focus on the wider, 40V and 40+120V nanocolumns.

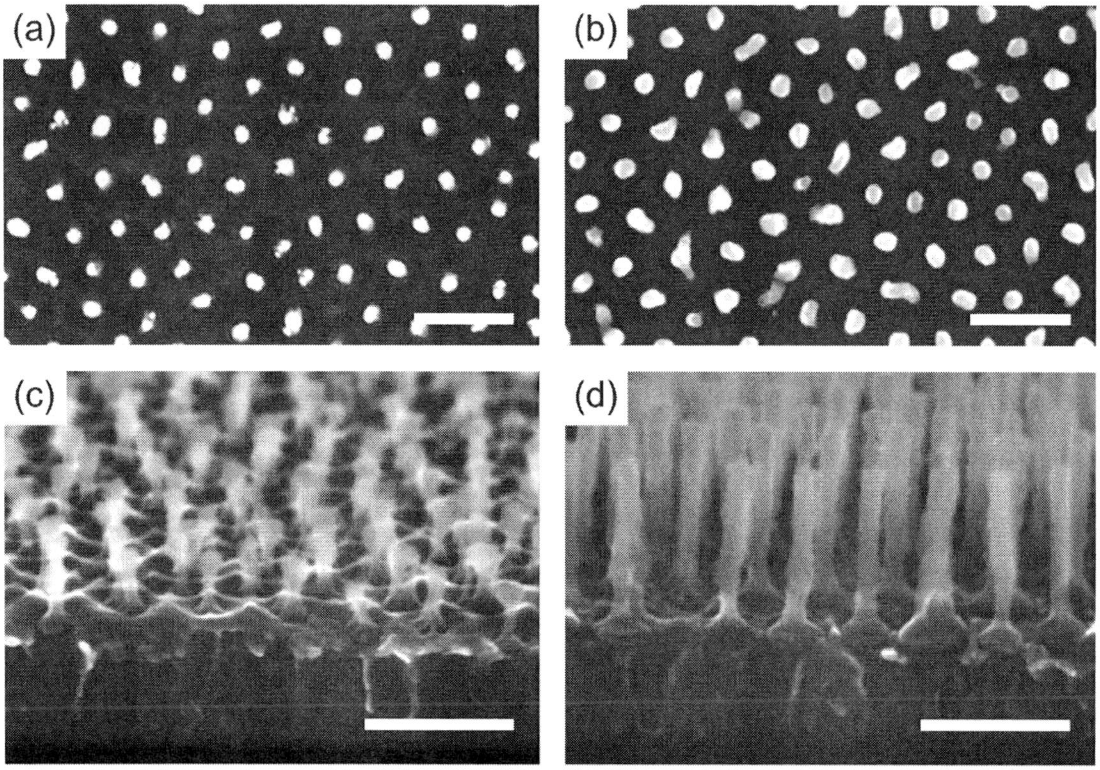

Figure 1 (a,b) Top-view and (c,d) cross-fractional SEM images of the PAA-free (a,c) 40V (anodized) and (b,d) 40+120V (re-anodized) nanocolumns. The scale bar is 200 nm.

Based on the SEM observation, schematic images of the formation of the 40V and 40+120V films were developed (**Figure 2**). Major features of the Nb_2O_5 nanocolumns formed in this way in the selenic acid are their smaller diameter (compared with that of the columns formed at similar electrical conditions in oxalic acid [4]) and their separation from each other by the residual Al and Nb metals. These characteristics may influence the possible utilization of the nanocolumn arrays, in addition to the expected presence of selenium species in the columns, as investigated by XPS.

3.2. Film composition

Examination of the chemical composition and bonding states in the surface layer of the PAA-free 40V and 40+120V films was performed by XPS. The presence of C, Nb, O, Al, Se, P, and Cr was identified in the survey spectra of the samples; the last two elements were present due to the PAA dissolution in the Cr_2O_3/H_3PO_4-based etchant. Narrow-scan C 1s, Nb 3d, O 1s, Al 2p, and Se 3d spectra were collected to analyze the bonding states. The experimental and fitted Nb 3d, Al 2p, and Se 3d spectra are shown in **Figure 3**.

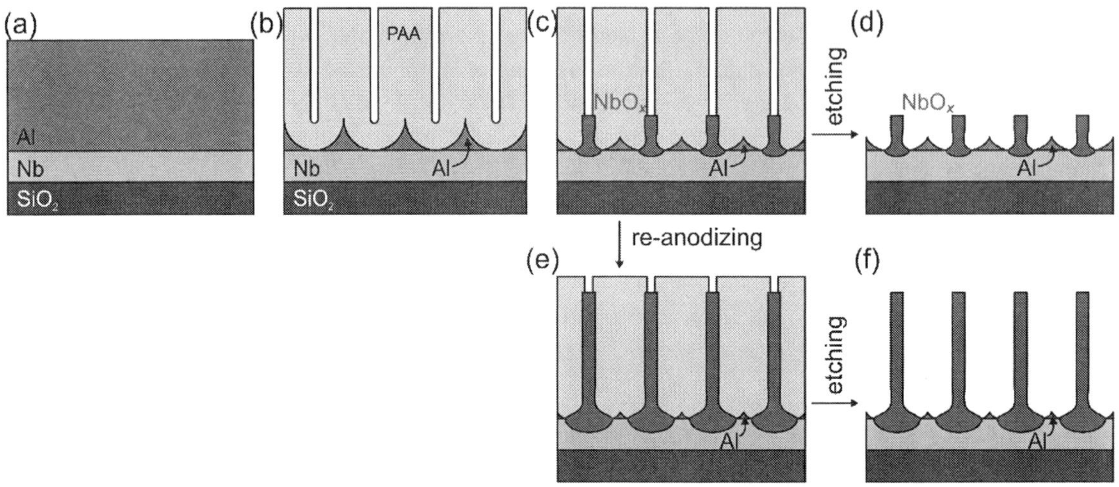

Figure 2 Schematic images of the processes for forming the 40V and 40+120V nanocolumn arrays: (a) the sputter-deposition of an Al/Nb bilayer onto an oxidized Si wafer, (b) the anodizing of the Al layer to form a PAA film, (c) the anodizing of the Nb underlayer into the barrier layer (the anodized, 40V film), (d) possible etching of the PAA overlayer to obtain the PAA-free 40V film, (e) the re-anodizing of Nb layer to 120 V (the re-anodized, 40+120V film), (f) possible etching of the PAA overlayer to obtain the PAA-free 40+120V sample

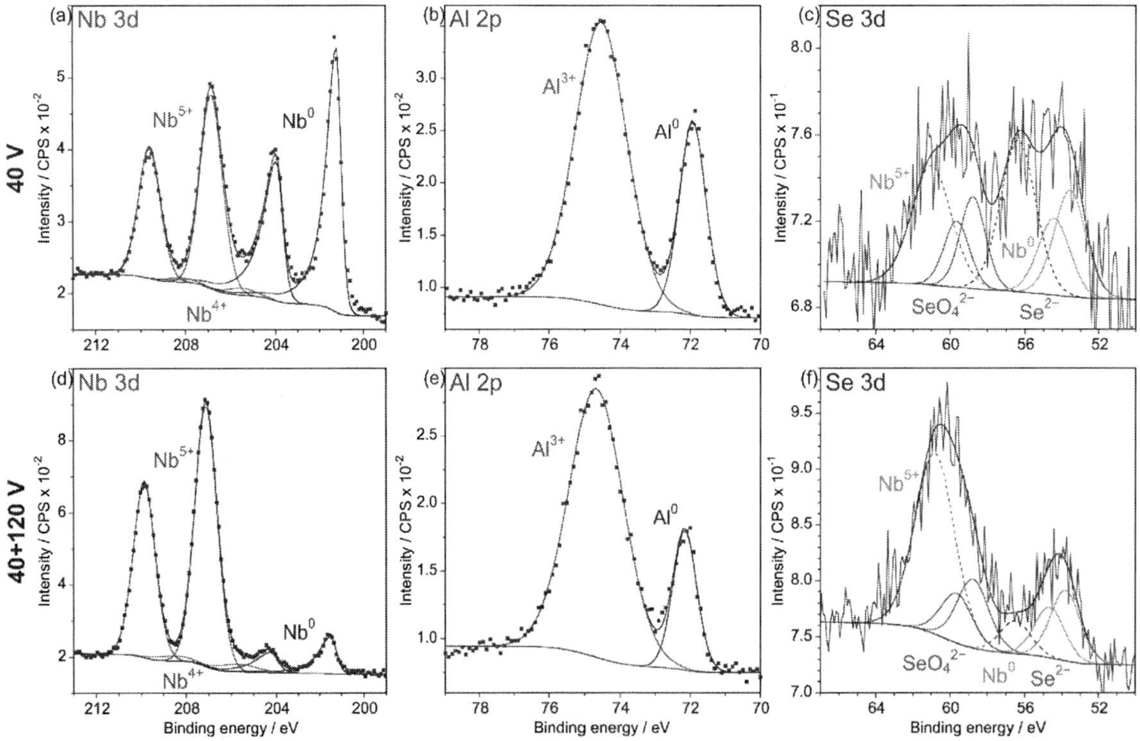

Figure 3 Experimental and fitted XPS spectra of the surface of the PAA-free (a-c) 40V and (d-f) 40+120V anodic films. (a,d) Nb 3d, (b,e) Al 2p, and (c,f) Se 3d high-resolution spectra are shown.

The *Nb 3d spectra* of the two samples (**Figures 3a,d**) are both fitted with three doublets of appropriately constrained peaks (Nb $3d_{5/2}$ and $3d_{3/2}$ with fixed peak separation energy of 2.75 eV and with fixed intensity ratio of 3:2 [4]). The highest binding energy (BE) doublet is due to the presence of Nb^{5+} cations, whereas the lowest BE doublet is associated with metallic Nb^0 [4], separating the column bases. The relative amount of the metallic Nb^0 is much smaller in the 40+120V anodic film than in the 40V sample (12 % vs. 53 % of all Nb species), which is in agreement with the SEM observations (**Figure 1**). The third, lowest-intensity doublet, shifted to −1.35 eV from the Nb^{5+} component, is attributed to the presence of Nb^{4+} cations in the film [4]. The ratio of Nb^{5+}:Nb^{4+} cations is 97:3 and 95:5 for the 40V and 40+120V films, respectively, indicating the presence of a substoichiometric Nb_2O_5 in the columns. The presence of Nb^{4+} cations (and lower-oxidation-state metal cations in general) at the surface of re-anodized PAA-assisted films seems to be a common feature [4,7].

The *Al 2p spectra* (**Figures 3b,e**) are fitted with two single peaks: a high-intensity peak associated with Al_2O_3 and a low-intensity peak associated with metallic Al. The metallic Al and a portion of the alumina seen in the spectra are attributed to Al metal residues (covered with native oxide) situated around the columns and over the metallic Nb residues. These metallic contributions can be subtracted from both the Nb 3d and Al 2p spectra to obtain an approximate amount of Nb and Al present only on the column surface. For Al, also the corresponding native oxide was taken into account and subtracted [8]. The result is 65 and 61 at.% of Al for the 40V and 40+120V films, respectively (providing at.%(Al+Nb) = 100 %). Similarly larger amounts of Al in the outer layer of PAA-assisted oxide columns of about 75 at.% have been reported for N-doped TiO_2 surfaces [7]. These results mean that the outmost column material is composed of mixed alumina-niobia, whereas the column interior is alumina-free Nb_2O_5 [4].

Figures 3c,f shows the measured and fitted *Se 3d spectra* for the two anodic films. The spectra are noisy due to the low content of Se, which, in addition to overlapping with Nb 4s peaks, complicates the deconvolution of the Se 3d spectra. When comparing the spectra of the various samples (including the 12V and 12+80V films, not shown), fitting them, and verifying the peak positions with the literature [9,10], multiple components were well assigned: Nb 4s peak of Nb^{5+} cations (~60.9 eV), Nb 4s peak of Nb^0 atoms (~56.4 eV), Se 3d peaks of selenate anions (SeO_4^{2-}, Se $3d_{5/2}$ at ~58.8 eV), and Se 3d peaks of selenide anions (Se^{2-}, Se $3d_{5/2}$ at ~54.0 eV). For fitting the Nb 4s peaks, the area below the peaks was considered proportional to the area of the corresponding Nb 3d peaks and the area of the Nb^0 peak was constrained as a multiple of the Nb^{5+} peak to agree with the corresponding Nb 3d spectrum. The Se 3d peaks were fitted as doublets with a constrained spin-orbit splitting of 0.86 eV [10] and the ratio of the Se $3d_{3/2}$ and $3d_{5/2}$ components was fixed at 0.72 [10].

The ratio of selenate:selenide is found 39:61 and 51:49 for the 40V and 40+120V films, respectively. The amount of Se at the column surface was calculated using the area of the Se 3d components and after subtracting the metallic contributions from the Al 2p and Nb 3d spectra. 5.3 and 3.2 at.% of Se were obtained for the 40V and 40+120V samples, respectively (provided at.%(Al+Nb+Se) = 100 %).

The chemical composition of the surface layer of the columns in the 40V and 40+120V anodic films can thus be expressed as a mixture of Al_2O_3 and substoichiometric Nb_2O_5, with some percentage of the O^{2-} anions in both oxides replaced by SeO_4^{2-} and Se^{2-} anions. There are about 65 at.% of Al_2O_3 and 3.0 and 1.8 % of O^{2-} anions replaced by selenium species in the 40V and 40+120V samples, respectively.

Selenate anions in the columns originate from the anodizing electrolyte. They move within the growing oxide under the high electric field, with a slower velocity than the O^{2-} anions, and are incorporated into the oxide structure, similarly to other electrolyte anions during the PAA-assisted anodizing [11]. The presence of selenide anions is less obvious and may be explained by the reduction of SeO_4^{2-} anions at the cathode, their further diffusion inside the electrolyte, and incorporation into the growing anodic oxide in a similar way as in the case of the selenate anions. Most probably, the anions move more easily within the surface layer of the columns (which is mixed with alumina [4]). Therefore they form a kind of core-shell structure. The shell contains the Se species and alumina, while the core is free from them.

4. CONCLUSIONS

The Al-on-Nb bilayers were anodized via the PAA-assisted anodizing in selenic-acid electrolyte for the first time. Varying the electrical conditions during the oxide growth resulted in the arrays of thinner and wider nanocolumns of niobium oxide, with diameters of 12 and 30 nm, respectively. XPS analysis of the wider columns revealed that their surface layer consists of substantial amounts of alumina (about 65 at.%) and substoichiometric Nb_2O_5. In both oxides, ~2.5 % of O^{2-} anions are replaced by selenate and selenide anions. Probably, the columns grow as a core-shell structure with the shell being alumina- and Se-species-rich and the core being free of them.

The presence of selenate and selenide anions in the column material may impact their functional properties, such as optical, catalytic, and antibacterial. Moreover, the self-formed heterojunctions may lead to even more advantages. Such promising properties of the anodic films prepared here may be investigated in future works.

ACKNOWLEDGEMENTS

The research leading to these results was supported by GACR grant no. 20-25486S. We acknowledge the use of CzechNanoLab Research Infrastructure supported by MEYS CR (LM2018110).

REFERENCES

[1] DEVAN, R.S., PATIL, R.A., LIN, J.-H., MA, Y.-R. One-Dimensional Metal-Oxide Nanostructures: Recent Developments in Synthesis, Characterization, and Applications. *Advanced Functional Materials.* 2012, vol. 22, pp. 3326-3370.

[2] OSTERLOH, F.E. Inorganic nanostructures for photoelectrochemical and photocatalytic water splitting. *Chemical Society Reviews.* 2013, vol. 42, pp. 2294-2320.

[3] LEE, W., PARK, S.-J. Porous Anodic Aluminum Oxide: Anodization and Templated Synthesis of Functional Nanostructures. *Chemical Reviews.* 2014, vol. 114, pp. 7487-7556.

[4] MOZALEV, A., VÁZQUEZ, R.M., BITTENCOURT, C., COSSEMENT, D., GISPERT-GUIRADO, F., LLOBET, E., HABAZAKI, H. Formation-structure-properties of niobium-oxide nanocolumn arrays via self-organized anodization of sputter-deposited aluminum-on-niobium layers. *Journal of Materials Chemistry C.* 2014, vol. 2, pp. 4847-4860.

[5] ZHOU, C., SHI, R., WATERHOUSE, G.I.N., ZHANG, T. Recent advances in niobium-based semiconductors for solar hydrogen production. *Coordination Chemistry Reviews.* 2020, vol. 419, 213399.

[6] NASIR, M.E., PERUCH, S., VASILANTONAKIS, N., WARDLEY, W.P., DICKSON, W., WURTZ, G.A., ZAYATS, A.V. Tuning the effective plasma frequency of nanorod metamaterials from visible to telecom wavelengths. *Applied Physics Letters.* 2015, vol. 107, 121110.

[7] BENDOVA, M., KOLAR, J., MARIK, M., LEDNICKY, T., MOZALEV, A. Influence of nitrogen species on the porous-alumina-assisted growth of TiO_2 nanocolumn arrays. *Electrochimica Acta.* 2018, vol. 281, pp. 796-809.

[8] STROHMEIER, B.R. An ESCA method for determining the oxide thickness on aluminum alloys. *Surface and Interface Analysis.* 1990, vol. 15, pp. 51-56.

[9] BIESINGER, M.C. *X-ray Photoelectron Spectroscopy (XPS) Reference Pages, Selenium* [online]. 2009-2021 [viewed: 2022-05-16]. Available from: http://www.xpsfitting.com/search/label/Selenium

[10] WERTHEIM, G.K., DiSALVO, F.J., BUCHANAN, D.N.E. Site inequivalence in $Fe_{1+x}Nb_{3-x}Se_{10}$. *Physical Review B.* 1983, vol. 28, pp. 3335-3338.

[11] MOZALEV, A., BENDOVA, M., GISPERT-GUIRADO, F., LLOBET, E. Hafnium-Oxide 3-D Nanofilms via the Anodizing of Al/Hf Metal Layers. *Chemistry of Materials.* 2018, vol. 30, pp. 2694-2708.

TECHNOLOGICAL CHALLENGES IN THE FABRICATION OF MoS_2/DIAMOND HETEROSTRUCTURES

[1,2]Marian VARGA, [1]Michaela SOJKOVA, [1,3]Jana HRDA, [1]Peter HUTAR, [1,3]Saviz PARSA SAEB, [1]Gabriel VANKO, [1]Lenka PRIBUSOVA SLUSNA, [2]Lukas ONDIC, [2]Jan FAIT, [2]Alexander KROMKA, [1]Martin HULMAN

[1]Slovak Academy of Sciences, Institute of Electrical Engineering, Bratislava, Slovakia, EU

[2]Czech Academy of Sciences, Institute of Physics, Prague, Czech Republic, EU

[3]Slovak University of Technology, Faculty of Electrical Engineering and Information Technology, Bratislava, Slovakia, EU

https://doi.org/10.37904/nanocon.2022.4586

Abstract

Nowadays, 2D materials are one of the most studied classes of materials. In addition to the most famous graphene, progress has been achieved in studying and using fundamental properties of transition metal dichalcogenides (TMD). Complementary, diamond as a representative of 3D materials has gained a reputation as an extremely versatile material due to its extraordinary combination of physical/chemical/electrical/optical properties. Besides these particular forms of 2D and 3D materials, their heterostructures have become very attractive due to new phenomena and functions (bandgap engineering, enhanced charge transport, optical interaction, etc.). However, individual technological procedures are still minimally investigated and described. Here, we will demonstrate a proof-of-concept for the preparation of MoS_2/diamond heterostructures, where two different strategies were employed: a) growth of MoS_2 layers on diamond films, and b) growth of diamond films on Si/MoS_2 substrates. Considering the growth conditions for MoS_2 and diamond materials, heterostructures based on MoS_2 on diamond can be prepared more easily. Regardless of the diamond film's morphology, the grown MoS_2 layer simply copies the diamond surface and does not damage the diamond film. However, the heterostructure in the configuration of diamond on MoS_2 is a highly challenging task. It was found experimentally that the combination of deposition temperature and aggressive chemical-plasma environment during diamond growth places high demands on the resistance and stability of MoS_2 layers.

Keywords: Transition metal dichalcogenides, diamond, heterostructures, chemical vapor deposition

1. INTRODUCTION

Two-dimensional (2D) materials have been one of the most extensively studied classes of materials due to their unusual physical properties. Besides the most known graphene, in recent years, progress has been achieved in exploring and using the fundamental properties of the transition metal dichalcogenides (TMD). Depending on the combination of transition metal (e.g. Mo, Pt, W, Ni or Ti) and chalcogen (e.g. S, Se, or Te), the group of TMD materials has roughly more than 60 different members which exhibit various properties [1]. These properties make them highly attractive for fundamental studies of novel physical phenomena as well as for a wide range of applications. In addition, heterostructures of two or more atomically thin materials with a large variety of bandgaps and electron affinities may have properties suitable for electronic and photonic applications [2,3]. MoS_2 is recently one of the most studied TMDs composed of covalently bonded S-Mo-S sheets which are bound by weak van der Waals forces [4]. MoS_2 is a semiconductor with an indirect bandgap of about 1.3 eV, however, the monolayer MoS_2 is a direct gap semiconductor with a bandgap of 1.8 eV. This is, in particular, important for photovoltaic and photocatalytic applications due to its strong absorption in the solar spectral region. However, properly used fabrication methods and growth conditions are crucial aspects that tailor the crystallographic structure, material quality, number of layers, and thus the properties of the TMD material in a wide range [5].

Conversely, diamond as a representative of 3D materials has gained a reputation as an extremely versatile material due to its extraordinary combination of physical/chemical/electrical/optical properties. These fascinating properties make diamond an ideal material for a wide range of applications [6]. Although single-crystal diamond is a better candidate than polycrystalline diamond films, its production and machining of single-crystals are quite cumbersome [7]. In contrast, polycrystalline diamond thin films with desired properties (thickness, morphology, crystal size, etc.) can be easily deposited at low cost on a wide range of substrates and also over large areas. Similar to TMD materials, the diamond properties (bulk or surface) can be significantly controlled by the growth conditions and post-growth processing [8,9].

Besides the individual forms of 2D materials and/or 3D diamond crystals, their heterostructures have also attracted the research community's attention due to novel phenomena and functionalities at artificial hetero-interfaces, and the enhancement of advantageous properties that further expand the applicability fields [10]. For instance, a computational study of the thermal stability of the MoS_2/diamond structure suggests that the single-layer MoS_2 film on diamond can be used as a field-effect transistor or a heat-generating gate junction heterostructure up to the temperature of 550 K [11]. In addition, a successful horizontal or vertical growth of a MoS_2 layer on a thick well-faceted microcrystalline diamond film has already been demonstrated by our group [4]. However, CVD diamond growth on TMD materials is a challenging issue due to thermal expansion differences, crystallographic mismatches and the aggressiveness of CVD conditions. Here, we focused on the preparation of MoS_2/diamond heterostructures in two different configurations and highlighted the growth mechanism and technological obstacles.

2. MATERIALS AND METHODS

In this work, MoS_2 was chosen as a representative TMD material. The MoS_2 layers were prepared by sulfurization of sputtered Mo layers (4 nm thick) in a custom-designed one-zone CVD chamber (**Figure 1a**) at 800 °C for 30 min. The Mo-coated substrate was placed in the center of the tube furnace along with sulfur powder (3 g, purity of 99,999 %). Sulfurization took place at atmospheric pressure in a constant N_2 flow.

For the diamond growth, a pulsed MW plasma CVD system with a linear antenna arrangement (Roth&Rau AK400) was selected (**Figure 1b**). This CVD system reveals a homogeneous "cold" plasma distribution at a low working pressure (units of Pa) over a water-cooled graphite substrate holder of 30 × 20 cm^2. Two gas compositions of CH_4:CO_2:H_2 with flow rates of 5:20:50 sccm and 10:40:50 sccm were tested. Apart from the gas composition, the deposition parameters were the same for both depositions, i.e. deposition time t = 20 hours, working pressure p = 0.15 mbar, microwave power P = 2×1700 W (ON/OFF duty cycle = 6/3) and substrate holder temperature T = 400 °C. Prior to diamond growth, reference Si substrates were ultrasonically seeded in an aqueous suspension with detonation diamond powder (~5 nm). In the case of Si/MoS_2 samples, only half of the sample was nucleated, while the remaining area was kept as untreated. It should be noted that the samples were immersed only in an aqueous solution with diamond nanoparticles (without the application of ultrasound) to exclude possible mechanical damage to the MoS_2 layers already before the CVD process.

The surface morphology of the fabricated samples was analyzed using a field-emission scanning electron microscope (SEM) operating in secondary electron mode (MAIA3, Tescan Ltd.) or FEI Quanta 250 FEG-SEM. Raman spectra were acquired by Renishaw InVia Reflex Raman spectrometer equipped with He-Cd laser with an excitation wavelength of 442 nm or by a confocal Raman microscope (Alpha 300R, WiTec, Germany) using a 532 nm excitation laser wavelength.

Energy-dispersive X-ray spectra (EDX) were obtained using a dual-beam SEM combined with a high current focused ion beam - FEI Quanta 3D 200i equipped with EDS AMETEK (Oxford). EDX spectra were collected with an accelerating voltage of 10 kV, an emission current of 0.1 nA and a working distance of 10 mm.

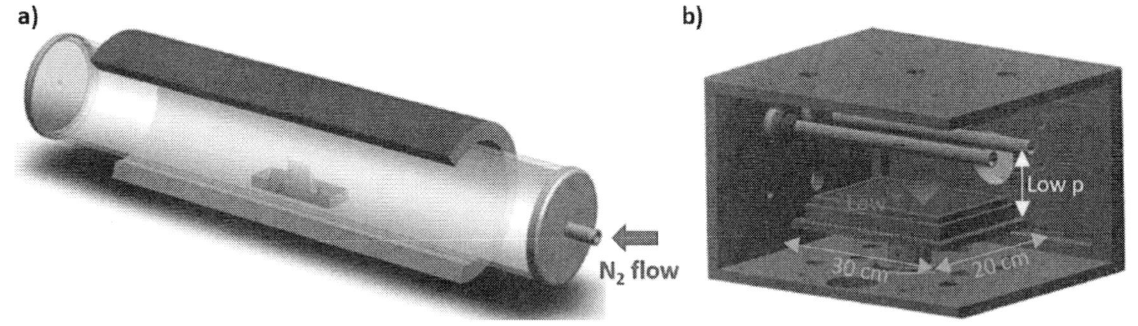

Figure 1 Schematic representation of the **a)** one-zone sulfurization method and **b)** pulsed MW plasma CVD system with a linear antenna arrangement (Roth&Rau AK400) [4,12].

3. RESULTS AND DISCUSSION

3.1. Growth and characterization of reference MoS_2 and diamond films

First, reference MoS_2 and diamond films were grown on Si substrates. **Figure 2** shows top-view SEM images and Raman spectra of MoS_2 film and diamond films deposited at two different gas compositions. The morphology of the MoS_2 film is characterized by a smooth surface with a small number of separated mounds.

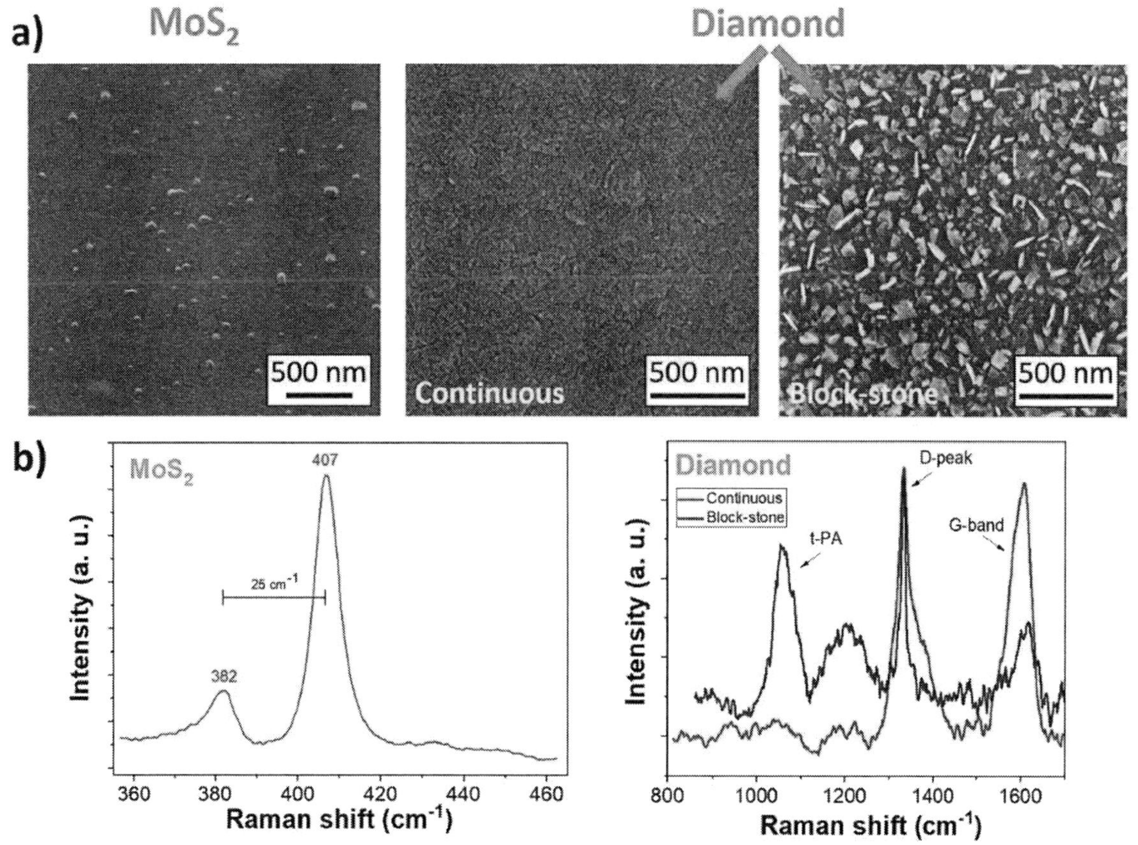

Figure 2 a) Top-view SEM images and **b)** Raman spectra of the deposited MoS_2 film and diamond films deposited in a linear antenna PE CVD system at two different gas compositions.

Raman measurements confirmed the MoS_2 character of the deposited film, with two dominant peaks in the spectrum corresponding to E^1_{2g} (382 cm^{-1}) and A^1_{1g} (407 cm^{-1}) Raman modes of MoS_2. In this case, the peak-to-peak (E^1_{2g} and A^1_{1g}) distance is more than 25 cm^{-1}, indicating that the film has more than 8 monolayers [13]. Based on results from the polarized Raman spectroscopy (not shown here), a horizontal alignment of the MoS_2 layer was confirmed [14]. The thickness of the MoS_2 film was about 18 nm, as measured by AFM.

Diamond films grown at two different gas compositions resulted in films with different morphologies, i.e. a continuous nanocrystalline film and a discontinuous layer consisting of individual well-faceted block-stone diamond crystals. The root-mean-square (RMS) surface roughness of the films was 7 nm and 40 nm, respectively. For both diamond films, a distinct peak (D-peak) centered at 1 332 cm^{-1} confirms the sp^3-bonded diamond character of the film. Moreover, the Raman spectrum of the continuous nanocrystalline diamond film also reveals the presence of a trans-polyacetylene chains (t-PA) at 1 150 cm^{-1} and a broad graphite band (G-band) at 1 450–1 650 cm^{-1} attributed to sp^2-bonded carbon. These non-diamond phases are usually localized at the diamond grain boundaries, which confirms the polycrystalline character of the diamond film.

3.2. Growth of MoS_2 on diamond

Figure 3a shows the surface morphologies of diamond films coated with MoS_2 layers. As found, the MoS_2 growth process does not damage the initial diamond films. The MoS_2 layers just copy the primary diamond surface morphology (the RMS roughness trend is similar to the reference diamond films). However, it should be noted that this is not at all the case when the diamond surface roughness is too high (e.g. in the case of a microcrystalline diamond, a shadowing effect may occur) or if the diamond film has a porous nature with a high-aspect-ratio. Due to the diamond surface roughness (RMS values are close to or higher than the MoS_2 thickness itself), it is not possible to evaluate the thickness and/or homogeneity of the grown MoS_2 film. Concerning the chemical structure, Raman measurements (**Figure 3b**) confirmed the successful conversion of Mo to MoS_2 layers on both types of diamond films [15].

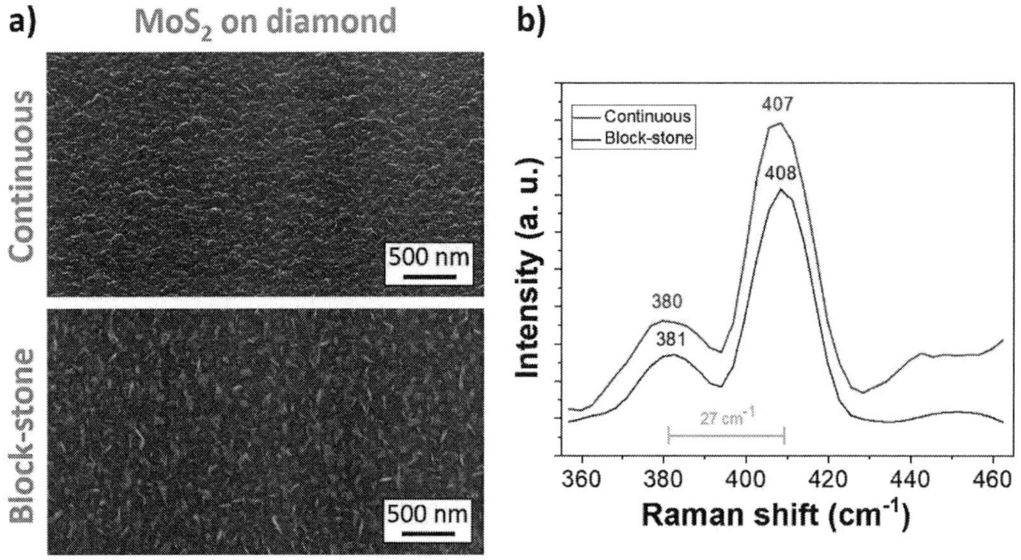

Figure 3 a) Top-view SEM images and **b)** Raman spectra of the fabricated heterostructure in the configuration MoS_2 on the continuous and block-stone diamond films.

3.3. Growth of diamond on MoS_2

The fabrication of MoS_2/diamond heterostructures was also carried out in the opposite way. To study the overall resistance of MoS_2 layers to the diamond CVD process, one-half of MoS_2/Si sample was nucleated, and the remaining area was left untreated. **Figure 4a** shows an optical image of the boundary area between the nucleated and non-nucleated regions after the diamond growth. The nucleated region is clearly overgrown with a continuous nanocrystalline diamond film (as confirmed by Raman measurements) and reveals the same morphology as the reference diamond film, **Figure 4b**-nucleated vs. **Figure 2a**. The non-nucleated region reveals a damaged-like surface with locally overgrown carbon-based deposits initialized by the spontaneous diamond nucleation. Raman measurements (**Figure 4c**) confirmed that the MoS_2 layer on the non-nucleated region survived the harsh plasma environment during the diamond growth. However, a significant broadening and blue-shift of the characteristic peaks was observed. The MoS_2 layer under the diamond film was not clearly recognized by Raman measurements, although some indications of MoS_2 peaks can be observed in the spectrum. The presence of MoS_2 in both regions was also studied by energy-dispersive X-ray spectroscopy (EDX) (**Figure 4d**). Although the K-lines of sulfur overlap strongly with the L-lines of molybdenum [16], the intensity of the molybdenum and sulfur lines is similar in both film regions (with and without diamond film). So, as a next step, a detailed characterization of the MoS_2 layer under the diamond film is required.

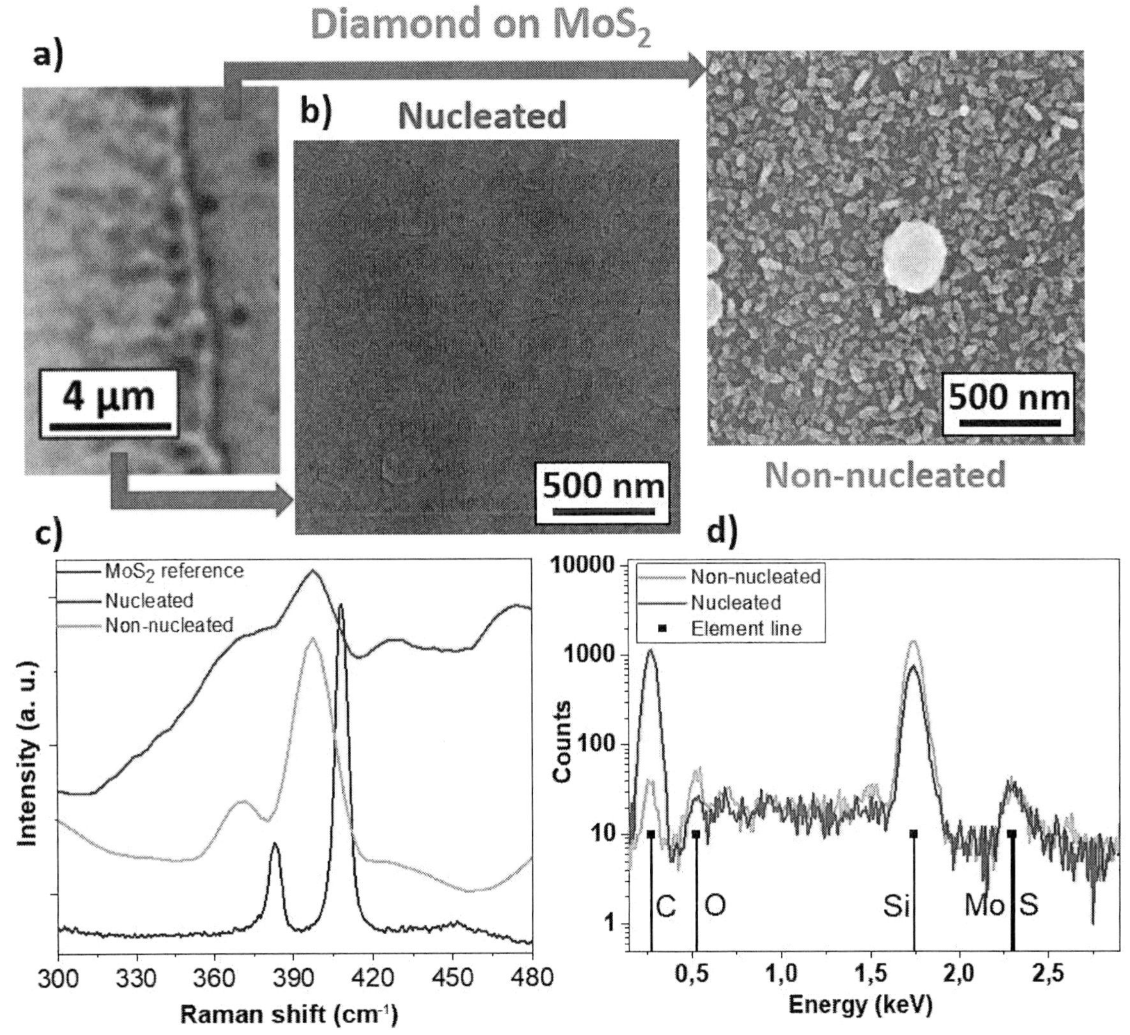

Figure 4 a) Optical image of the fabricated MoS_2/diamond heterostructure highlighted at the boundary of the nucleated and non-nucleated regions of the sample. **b)** Top-view SEM images, **c)** Raman spectra and **d)** EDX spectra of diamond films grown on nucleated and non-nucleated MoS_2 regions.

4. CONCLUSION

We investigated two technological concepts for the fabrication of MoS_2/diamond heterostructures. In the first concept, the diamond films were not damaged/modified with the MoS_2 growth process, and the formed MoS_2 layer well replicated the primary diamond morphology (continuous or porous). In the second concept, the MoS_2 film exposed to the diamond deposition conditions revealed some changes, as confirmed by Raman measurements. Some aspects of the diamond growth mechanism on the MoS_2 layer are still undiscovered and need deeper studies. Our preliminary results confirm technological progress in the fabrication of MoS_2/diamond heterostructures as a novel material platform attractive not only for fundamental research but also for industrially oriented applications.

ACKNOWLEDGEMENTS

This work was supported by project 19MRP0010 financed from the MoRePro program and funding from the Slovak Academy of Sciences, Slovak Grant Agency for Science VEGA 2/0059/21, bilateral GAAV project no. SAV-AV ČR 21-10 and project Lumina Queruntur of the Czech Academy of Sciences.

REFERENCES

[1] N. SHANTO, J.R. MCCARTHY, eds. Biomedical applications of graphene and 2D nanomaterials. *Elsevier*. Amsterdam, 2019.

[2] A.K. GEIM, I.V. GRIGORIEVA. Van der Waals heterostructures. *Nature*. 2013, vol. 499, pp. 419-425. Available from: https://doi.org/10.1038/nature12385.

[3] H. TERRONES, F. LÓPEZ-URÍAS, M. TERRONES. Novel hetero-layered materials with tunable direct band gaps by sandwiching different metal disulfides and diselenides. *Sci Rep*. 2013, vol. 3, pp. 1549. Available from: https://doi.org/10.1038/srep01549.

[4] M. SOJKOVÁ, P. SIFFALOVIC, O. BABCHENKO, G. VANKO, E. DOBROČKA, J. HAGARA, N. MRKYVKOVA, E. MAJKOVÁ, T. IŽÁK, A. KROMKA, M. HULMAN. Carbide-free one-zone sulfurization method grows thin MoS2 layers on polycrystalline CVD diamond. *Sci Rep*. 2019, vol. 9, 2001. Available from: https://doi.org/10.1038/s41598-018-38472-9.

[5] M. SAMADI, N. SARIKHANI, M. ZIRAK, H. ZHANG, H.-L. ZHANG, A.Z. MOSHFEGH. Group 6 transition metal dichalcogenide nanomaterials: synthesis, applications and future perspectives. *Nanoscale Horiz*. 2018, vol. 3, pp. 90-204. Available from: https://doi.org/10.1039/C7NH00137A.

[6] J. WILKS, E. WILKS. Properties and applications of diamond. paperback ed, Butterworth-Heinemann, Oxford, 1994.

[7] M. DOMONKOS, M. VARGA, L. ONDIČ, L. GAJDOŠOVÁ, A. KROMKA. Microsphere lithography for scalable polycrystalline diamond-based near-infrared photonic crystals fabrication. *Materials & Design*. 2018, vol. 139, pp. 363-371. Available from: https://doi.org/10.1016/j.matdes.2017.10.076.

[8] A.G. MONTAÑO-FIGUEROA, J.J. ALCANTAR-PEÑA, P. TIRADO, A. ABRAHAM, E. DE OBALDIA, O. AUCIELLO. Tailoring of polycrystalline diamond surfaces from hydrophilic to superhydrophobic via synergistic chemical plus micro-structuring processes. *Carbon*. 2018, vol. 139, pp. 361-368. Available from: https://doi.org/10.1016/j.carbon.2018.06.062.

[9] S. STEHLIK, M. VARGA, P. STENCLOVA, L. ONDIC, M. LEDINSKY, J. PANGRAC, O. VANEK, J. LIPOV, A. KROMKA, B. REZEK. Ultrathin Nanocrystalline Diamond Films with Silicon Vacancy Color Centers via Seeding by 2 nm Detonation Nanodiamonds. *ACS Appl. Mater. Interfaces*. 2017, vol. 9, pp. 38842-38853. Available from: https://doi.org/10.1021/acsami.7b14436.

[10] P. YU, Y.-H. CHU, R. RAMESH. Oxide interfaces: pathways to novel phenomena. *Materials Today*. 2012, vol. 15, pp. 320-327. Available from: https://doi.org/10.1016/S1369-7021(12)70137-2.

[11] A. GALASHEV, K. IVANICHKINA. Computer study of the structure and thermal stability of a monolayer MoS2 film on a diamond substrate. *Lett. Mater.* 2019, vol. 9, pp. 270-275. Available from: https://doi.org/10.22226/2410-3535-2019-3-270-275.

[12] M. VARGA, Š. POTOCKÝ, M. DOMONKOS, T. IŽÁK, O. BABČENKO, A. KROMKA. Great Variety of Man-Made Porous Diamond Structures: Pulsed Microwave Cold Plasma System with a Linear Antenna Arrangement, *ACS Omega.* 2019, vol. 4, pp. 8441-8450. Available from: https://doi.org/10.1021/acsomega.9b00323.

[13] G. SIEGEL, Y.P. VENKATA SUBBAIAH, M.C. PRESTGARD, A. Tiwari, Growth of centimeter-scale atomically thin MoS_2 films by pulsed laser deposition. *APL Materials.* 2015, vol. 3, 056103. Available from: https://doi.org/10.1063/1.4921580.

[14] M. HULMAN, M. SOJKOVÁ, K. VÉGSÖ, N. MRKYVKOVA, J. HAGARA, P. HUTÁR, P. KOTRUSZ, J. HUDEC, K. TOKÁR, E. MAJKOVA, P. SIFFALOVIC. Polarized Raman Reveals Alignment of Few-Layer MoS_2 Films. *J. Phys. Chem. C.* 2019, vol. 123, pp. 29468-29475. Available from: https://doi.org/10.1021/acs.jpcc.9b08708.

[15] K. GOŁASA, M. GRZESZCZYK, R. BOŻEK, P. LESZCZYŃSKI, A. WYSMOŁEK, M. POTEMSKI, A. BABIŃSKI. Resonant Raman scattering in MoS 2 -From bulk to monolayer. *Solid State Communications.* 2014, vol. 197, pp. 53-56. Available from: https://doi.org/10.1016/j.ssc.2014.08.009.

[16] E. NOLOT, S. CADOT, F. MARTIN, P. HÖNICKE, C. ZECH, B. BECKHOFF. In-line characterization of ultrathin transition metal dichalcogenides using X-ray fluorescence and X-ray photoelectron spectroscopy. *Spectrochimica Acta Part B: Atomic Spectroscopy.* 2020, vol. 166, 105788. Available from: https://doi.org/10.1016/j.sab.2020.105788.

PREPARATION AND CHARACTERIZATION OF CONDUCTING COMPOSITE FROM ANILINIUM-MONTMORILLONITE INTERCALATE

[1]Markéta DAVIDOVÁ, [1]Lenka KULHÁNKOVÁ, [2]Jonáš TOKARSKÝ, [1]Lenka ŘEHÁČKOVÁ, [1]Michal RITZ, [1]Vlastimil MATĚJKA

[1]*VSB - Technical University of Ostrava, Faculty of Materials Science and Technology, Ostrava, Czech Republic, EU*

[2]*VSB - Technical University of Ostrava, CEET, Nanotechnology Centre, Ostrava, Czech Republic, EU*

https://doi.org/10.37904/nanocon.2022.4581

Abstract

The main purpose of this work is to explore the possibilities of preparing conducting composite containing graphite or graphene from anilinium/montmorillonite intercalates. The use of only a precursor (without polymerization) is a fundamental difference compared to earlier research devoted to the preparation of graphite/graphene-silicate composites from polyaniline/montmorillonite intercalates. Therefore, the preparation procedure tested in this study represents another way to obtain an electrically conductive material, the electrically conductive component of which is formed *in situ*. Four different concentrations of anilinium sulfate solution were used to prepare these intercalates (0.2M; 0.4M; 0.6M; 0.8M). Montmorillonite was added to these solutions, and after 7 days of intercalation, the solutions were filtered and the resulting intercalates were dried. From each sample, one tablet was pressed, and the electrical conductivity of these tablets was measured before and after their calcination in argon atmosphere at 1300 °C. Electrical conductivity of the calcined samples crushed to powder was also measured. For the analysis of uncalcined samples, X-ray diffraction analysis was used. Calcined samples were analyzed using X-ray diffraction analysis and Raman spectroscopy. These analyses in combination with electrical conductivity measurements revealed that conducting composite can be prepared using the method described above.

Keywords: Graphite, montmorillonite, anilinium sulfate, electrical conductivity, calcination

1. INTRODUCTION

Composites containing graphite or graphene as conductive component are of interest to researchers mainly because of their high electrical conductivity [1,2]. The most common method of preparation is the use of ready-made graphite or graphene, which are commercially available, and their application to the selected matrix [1- 4]. However, graphite or graphene can also be prepared *in situ*, from a suitable precursor. The starting material in these cases can be, for example, SiC [5]. Phyllosilicate matrices intercalated with organic substances can also be used for this purpose [6,7]. Calcination of intercalates produces graphite or graphene in the resulting composite. In this work, the clay mineral montmorillonite (MMT) was used as a matrix. This most conventional layered silicate has a great ability to expand and accept other molecules into the interlayers to form an intercalate. The aim of this work was to use a simple organic molecule for the preparation of electrically conductive composite. Inspiration was taken from the article by Čapková et al. [7] in which the graphene was prepared from the polyaniline/MMT intercalate. The same precursor, aniline, was used in this work. However, it was only intercalated into the interlayers of MMT, the polymerization step was omitted here, and the preparation was thus generally easier. Taking into account that one of the reasons for the successful transformation of polyaniline into graphite and graphene was stated to be the arrangement of polyaniline chains on the surface of MMT, we were interested in the behavior of a similar intercalate in which anilinium polymerization was not performed.

2. MATERIALS AND METHODS

2.1. Materials and preparation of the samples

Na-montmorillonite (MMT) was purchased from Ankerpoort NV, The Netherlands. Basal spacing of the MMT is ~ 1.193 nm and the crystallochemical formula $(M_{0,5})(Al, Mg)_2(Si)_4O_{10}(OH)_2$ with a layer charge ~ 0.7 el. per unit cell. Aniline and sulfuric acid were used as received from Lach-Ner, Czech Republic. By addition of aniline to an aqueous sulfuric acid solution, solutions of anilinium sulfate were prepared. For the preparation of different concentrations of samples, different amounts of aniline were used. Then MMT was added to each sample and the suspensions were allowed to intercalate for 7 days. The final products were collected on filters by washing them with distilled water. Filter cakes were dried for 24 h at 40 °C. The prepared samples were denoted as 0.2An_MMT, 0.4An_MMT, 0.6An_MMT and 0.8An_MMT. To press the samples into square-shaped tablets, a table hydraulic press was used without using any binders or lubricants. The pressed tablets were calcined in a high-temperature tube resistance furnace (CLASIC CZ, spol. s.r.o., Czech Republic) equipped with a Pt-13% Rh/Pt thermocouple. In a tube, an inert atmosphere (>99.9999% Ar) was maintained under a constant overpressure of $1.06 \cdot 10^5$ Pa. The process included controlled heating and cooling within a temperature range of 25-1300 °C at a rate of 5 °C min^{-1}. The tablets were exposed to the target temperature of 1300 °C for 1 h. The Ar atmosphere was replaced four times during the process: at 900 °C, at 1100 °C, at the beginning and at the end of the 1 h interval at 1300 °C. The rate of each replacement was 5 $dm^3 \cdot min^{-1}$, and each replacement took 5 min. The internal volume of the tube was 2.26 dm^3.

2.2. Characterization methods

XRD analysis was performed on a Miniflex600 powder diffractometer (Rigaku, JP) with a DTex/Ultra 1D detector. The samples were measured in reflex mode, ranged from 5 to 80° 2Theta, with a step of 0.02° and a step time of 5° min^{-1}. During the measurement, a Co tube was used. Before measurement, the samples were ground to a fine powder. The Raman spectra were measured on the DXR SmartRaman dispersive Raman spectrometer (ThermoScientific, USA) with a CCD detector using a 180-degree measurement technique. The wavelength of the excitation laser was 780 nm, the grating was 400 lines mm^{-1}, the aperture 50 μm, the exposure time 1 s, and the number of exposures 250. An empty sample compartment was used for background measurement. The spectra were corrected for fluorescence (by a polynomial function of the 6th order).

2.3. Conductivity measurements

For DC conductivity measurement, we constructed a special measuring apparatus using DC POWER SUPPLY HY 3003 D-2, programmable DC POWER SUPPLY BK PRECISION 9120, pA-meter KEITHLEY 6487. The DC conductivity of the uncalcined tablets has been measured in two perpendicular directions, in the tablet plane ($\sigma_{\leftrightarrow}$; **Figure 1**) and in an orthogonal direction to the tablet plane ($\sigma_{\downarrow}$; **Figure 1**). Flat Cu electrodes were used for the measurements. The tablets of samples 0,2An_MMT and 0,4An_MMT after calcination were broken. Therefore, to measure the DC conductivity, these samples were first crushed into fine powder. In this case, the conductivity was denoted as σ_{pow}. The samples were then measured on a special measuring apparatus intended to measure powder samples. To estimate the effect that this crushing had on electrical conductivity, the electrical conductivity of samples 0.6An_MMT and 0.8An_MMT was measured first in tablets and then in crushed powder.

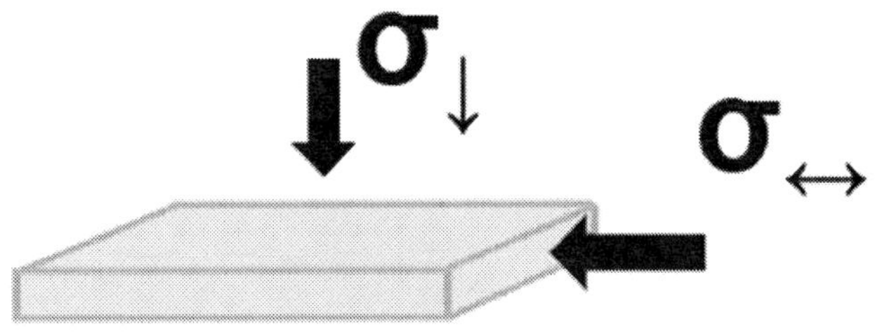

Figure 1 Directions of conductivities measured

3. RESULTS AND DISCUSSION

The profile of the (001) basal diffraction line of pure MMT and 0.2An_MMT XRD is shown in **Figure 2**. The shape of (001) diffraction line registered for pure MMT is of a diffuse character with a nonsharp maximum, which proves the disorder of the MMT particles. The intercalation of anilinium to interlayers of MMT caused the arrangement of MMT particles, which was manifested by a narrowing of the peak and an increase in its intensity. In all uncalcined samples, the presence of a peak at around 8.4° 2Theta proving intercalation in the MMT interlayers (**Figure 3**). However, in the XRD pattern of 0.8An_MMT (**Figure 3d**), additionally an intensive peak at around 6.5° 2Theta appears. This peak belongs to pure anilinium sulfate, which is confirmed by the XRD pattern of 0.8An_sulfate (**Figure 3e**). The comparison of these two patterns shows that the anilinium is located not only in the MMT interlayer space, but also on the surface of the MMT particles.

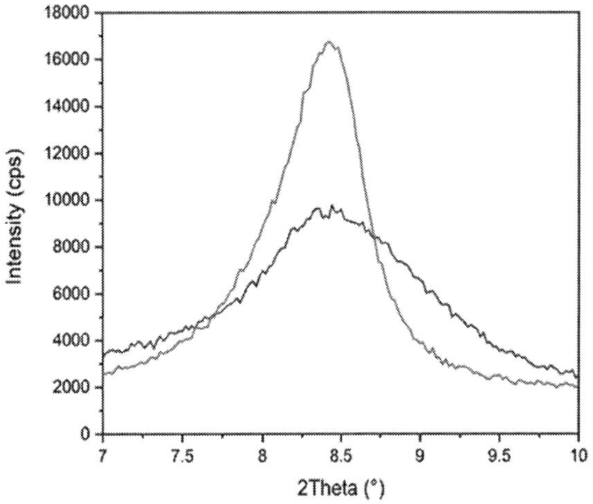

Figure 2 XRD patterns of pure MMT (black) and 0.2An_MMT (red)

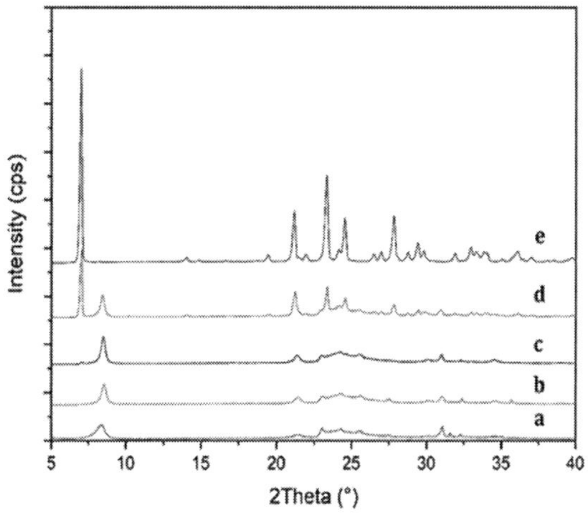

Figure 3 XRD patterns of uncalcined samples; a) 0.2An_MMT, b) 0.4An_MMT, c) 0.6An_MMT, d) 0.8An_MMT, e) 0.8An_sulfate

In the case of calcined samples (**Figure 4**), the diffraction line (001) is missing due to dehydroxylation of the MMT structure. Furthermore, the calcination of the intercalates led to the formation of cristobalite (PDF card No. 01-076-0941), mullite (PDF card No. 01-079-1457) and graphite (PDF card No. 03-065-6212).

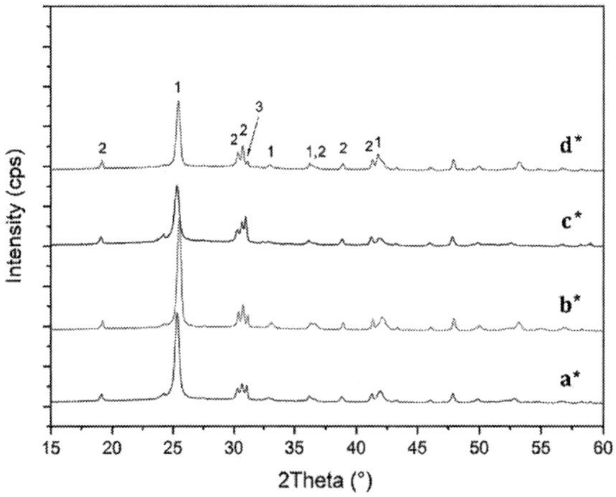

Figure 4 XRD patterns of calcined samples; a*) 0.2An_MMT, b*) 0.4An_MMT, c*) 0.6An_MMT, d*) 0.8An_MMT. 1 ... cristobalite, 2 ... mullite, 3 ... graphite

Raman spectra of the calcined samples are shown in **Figure 5.** The graphitic band confirming the orderliness of the carbon structure (~1590 cm^{-1}) and the disorder band (shows defects in the samples; ~1315 cm^{-1}) appear in all spectra. The fact that disorder bands are more intensive than graphitic bands show the predominance of defects in the carbon structure [7-9]. In the spectra of sample 0.8An_MMT the peak of the band around 1600 cm^{-1} is divided to 1606 and 1579 cm^{-1}. The peak of 1606 cm^{-1} can prove the presence of aromatic rings bonded to nitrogen [10]. Also, a strong band at 2613 cm^{-1} can be observed in the spectra of 0.8An_MMT. This band is attributed to the 2D overtone of graphite [11]. The highest intensity of this band can be seen in spectrum of the 0.8An_MMT originally containing anilinium not only in the MMT interlayer space.

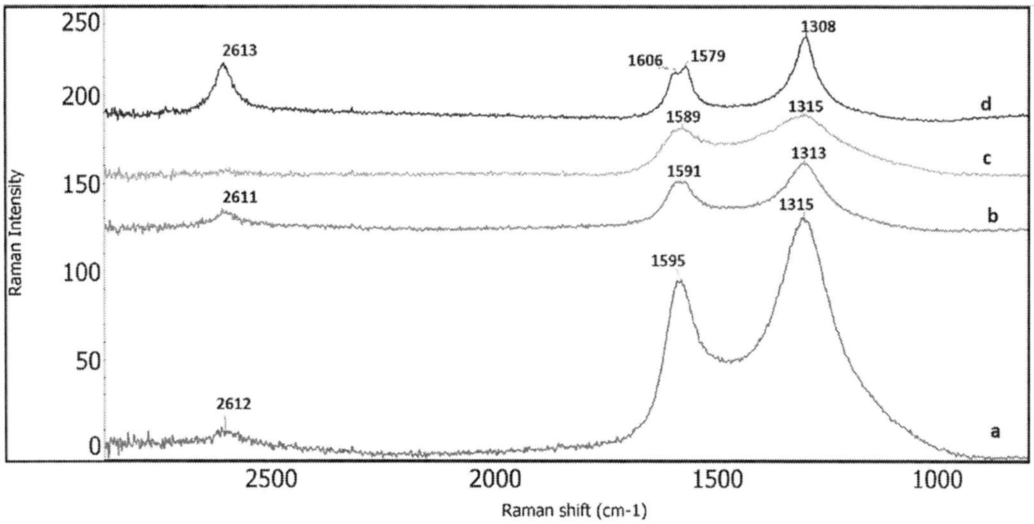

Figure 5 Raman spectra of calcined samples; a) 0.2An_MMT, b) 0.4An_MMT, c) 0.6An_MMT, d) 0.8An_MMT

The conductivities measured for the uncalcined samples (**Table 1**) showed very low values. This result was expected because in these samples, there is no ingredient that could provide electrical conductivity. Electrical conductivities of calcined samples are also listed in the **Table 1**. For these samples, a huge effect of calcination on conductivity can be seen. When comparing the electrical conductivities of samples measured in powder form and in tablet form, significantly higher σ_{pow} value can be observed for the 0.8An_MMT sample. This electrical conductivity is about twenty times higher in comparison with conductivity measurements of the 0.8An_MMT sample in tablet form. By pressing into tablets, the MMT particles were arranged in such a way that they had the function of an insulator [7]. Since the arrangement was disrupted when the sample was crushed, the flow of electric current was less restricted. This crushing together with the highest amount of anilinium and its location not only in the MMT interlayer space led to the highest electrical conductivity obtained for the 0.8An_MMT sample.

Table 1 Electrical conductivities of uncalcined and calcined samples

Sample	**uncalcined**		**calcined**		
	$\sigma_{\downarrow}$ ($S \cdot m^{-1}$)	**$\sigma_{\leftrightarrow}$ ($S \cdot m^{-1}$)**	**σ_{pow} ($S \cdot m^{-1}$)**	**$\sigma_{\downarrow}$ ($S \cdot m^{-1}$)**	**$\sigma_{\leftrightarrow}$ ($S \cdot m^{-1}$)**
0.2An_MMT	$4.009 \cdot 10^{-7}$	$1.018 \cdot 10^{-4}$	2.574		
0.4An_MMT	$1.632 \cdot 10^{-7}$	$3.145 \cdot 10^{-5}$	11.652		
0.6An_MMT	$1.304 \cdot 10^{-7}$	$2.601 \cdot 10^{-5}$	5.651	0.071	12.424
0.8An_MMT	$9.159 \cdot 10^{-8}$	$1.839 \cdot 10^{-5}$	53.210	0.069	2.642

4. CONCLUSION

In this work, an effort was made to synthesize conducting composite from An_MMT intercalate. Samples with four different concentrations of anilinium sulfate were prepared. Subsequently, the samples were calcined at 1300 °C. The XRD analysis of uncalcined samples revealed the anilinium intercalated into the MMT interlayers in all samples. In the case of the highest concentration (sample 0.8An_MMT), the anilinium was found also outside the MMT interlayer space. Raman spectroscopy and XRD analysis of calcined samples proved the presence of graphite. The formation of graphite was also confirmed by electrical conductivity, which in calcined samples was several orders of magnitude higher than in uncalcined samples. The highest conductivity correlates with the largest amount of anilinium and its presence outside the MMT interlayer. The graphite thus formed is probably located on the surface of the silicate grains, as was observed earlier [12], whereby contributes to the conductivity achieved. The question of the presence of graphene in the MMT interlayer and the number of layers of graphite formed will be the subject of further study. The present results confirm that the electrically conductive composite can be prepared from the intercalate without the need to polymerize the anilinium.

ACKNOWLEDGEMENTS

This work was supported by the Czech Republic Ministry of Education, Youth, and Sports project (SP2022/61).

REFERENCES

[1] RAMÍREZ, C., F.M. FIGUEIREDO, P. MIRANZO, P. POZA, M.I. OSENDI, Graphene nanoplatelet/silicon nitride composites with high electrical conductivity. *Carbon.* 2012, vol. 50, pp. 3607-3615.

[2] MUÑOZ-FERREIRO, C., A. MORALES-RODRÍGUEZ, T.C. ROJAS, E. JIMÉNEZ-PIQUÉ, C. LÓPEZ-PERNÍA, R. POYATO, A. GALLARDO-LÓPEZ. Microstructure, interfaces, and properties of 3YTZP ceramic composites with 10 and 20 vol% different graphene-based nanostructures as fillers. *J. Alloys Compd.* 2019, vol. 777, pp. 213-224.

[3] ŁUKASZEWICZ, M., P. GŁUCHOWSKI, B. CICHY, W. STRĘK. Persistent photoconductance in graphene ceramics. *Phys. Pro cedia.* 2015, vol. 76, pp. 155-159.

[4] BELMONTE, M., A. NISTAL, P. BOUTBIEN, B. ROMÁN-MANSO, M.I. OSENDI, P. MIRANZO. Toughened and strengthened silicon carbide ceramics by adding graphene-based fillers. *Scripta Mater.* 2016, vol. 113 pp. 127- 130.

[5] MIRANZO, P., C. RAMÍREZ, B. ROMÁN-MANSO, L. GARZÓN, H.R. GUTIÉRREZ, M. TERRONES, C. OCAL, M. I. OSENDI, M. BELMONTE. In situ processing of electrically conducting graphene/SiC nanocomposites. *J. Eur. Ceram. Soc.* 2013, vol. 33, pp. 1665-1674.

[6] C. XU, G. NING, X. ZHU, G. WANG, X. LIU, J. GAO, Q. ZHANG, W. QIAN, F. WEI. Synthesis of graphene from asphaltene molecules adsorbed on vermiculite layers. *Carbon.* 2013, vol. 62, pp. 213-221.

[7] ČAPKOVÁ, P., V. MATĚJKA, J. TOKARSKÝ, P. PEIKERTOVÁ, L. NEUWIRTHOVÁ, L. KULHÁNKOVÁ, J. BEŇO a V. STÝSKALA. Electrically conductive aluminosilicate/graphene nanocomposite. *Journal of the European Ceramic Society.* 2014, vol. 34, no. 12, pp. 3111-3117.

[8] CALIZO I., GHOSH S., BAO W., MIAO F., LAU C.N., BALANDIN A.A. Raman Nanometrology of Graphene: Temperature and Substrate Effects. *Solid State Communications.* 2009, vol. 149, pp. 1132-1135.

[9] ZHOU Z., GU H., WANG C., ZHANG J., LV M., HE R. Study on the Synthesis and Surface Enhanced Raman Spectroscopy of Graphene-Based Nanocomposites Decorated with Noble Metal Nanoparticles. *Colloids and Surfaces A.* 2013, vol. 430, pp. 103-109.

[10] M. BADAWI, H., W. FÖRNER a S. A. ALI. A comparative study of the infrared and Raman spectra of aniline and o-, m-, p-phenylenediamine isomers. *Spectrochimica Acta Part A: Molecular and Biomolecular Spectroscopy.* 2013, vol. 112, pp. 388-396.

[11] MALARD, L.M., M.A. PIMENTA, G. DRESSELHAUS a M.S. DRESSELHAUS. Raman spectroscopy in graphene. *Physics Reports.* 2009, vol. 473, pp. 51-87.

[12] TOKARSKÝ, J., K. MAMULOVÁ KUTLÁKOVÁ, P. PEIKERTOVÁ, L. ŘEHÁČKOVÁ, M. KORMUNDA, P. MATĚJKOVÁ, S. ŠTUDENTOVÁ a L. KULHÁNKOVÁ. Polypyrrole/montmorillonite and polypyrrole/ghassoul intercalates as a source of graphite and multi-layer graphene: Preparation of nanocomposites exhibiting strongly anisotropic electrical conductivity. *Materials Research Bulletin.* 2021, vol. 142, 111429.

NANOCERIA PREPARED BY ELECTRON BEAM EVAPORATION

[1]Naděžda PIZÚROVÁ, [2]Antonín HLAVÁČEK, [1,3]Zuzana KAVČIAKOVÁ, [1,3]Pavla ROUPCOVÁ, [1]Ivo KUBĚNA, [1]Jiří BURŠÍK, [4,5]Sergey Yu. SOKOVNIN

[1]Czech Academy of Sciences, Institute of Physics of Materials, Brno, Czech Republic, EU, pizurova@ipm.cz, kavciakova@ipm.cz, kubena@ipm.cz, bursik@ipm.cz

[2]Czech Academy of Sciences, Department of Bioanalytical Instrumentation, Institute of Analytical Chemistry, Brno, Czech Republic, EU, hlavacek@iach.cz

[3]CEITEC - Central European Institute of Technology, Brno University of Technology, Brno, Czech Republic; EU, roupcova@ipm.cz

[4]Institute of Electrophysics Ural Branch RAS, Yekaterinburg, Russia

[5]Ural Federal University, Yekaterinburg, Russia; sokovnin@iep.uran.ru

https://doi.org/10.37904/nanocon.2022.4585

Abstract

Cerium oxide nanoparticles (nanoceria) are currently one of the most investigated nanomaterials because of their attractive properties used in biomedical applications, catalysis, fuel cells, and many others. These attractive properties are connected with the Ce^{3+} and Ce^{4+} valency state ratio. In the nanoparticle form, cerium oxides contain a mixture of Ce^{3+} and Ce^{4+} on the nanoparticle surfaces. Switching between these two states requires oxygen vacancies. Therefore, nanoceria's inherent ability to act as an antioxidant in an environmentally-dependent manner and a "redox switch" to confer auto-regenerating capabilities by automatically shifting between Ce^{4+} and Ce^{3+} oxidation states is significantly affected by surface morphology. Regarding this demanded behavior, we aimed to characterize synthesized nanoparticle surface quality and its influence on the cerium oxidation states. The received results were used to evaluate the synthesis method's suitability for suggested utilization.

We used nanoparticles prepared by electron beam evaporation. This unique physical method includes nanoparticle creation through the fast cooling process followed by breaking radiation damaging nanoparticle surfaces to create surface off-stoichiometry. We prepared a sample containing clusters of a mixture of ultra-small nanoparticles and approximately 100 nm particles. X-ray diffraction confirmed the CeO_2 phase in both components. To extract the finest component, we used centrifugal size fractionation. We received 200 nm clusters of 2-10 nm nanoparticles. Nanoparticle shapes and facet types were analyzed using transmission electron microscopy methods. We found out most nanoparticles were formed with truncated octahedrons containing {1,1,1} and {1,0,0} facet types and truncated cuboctahedrons containing {1,1,1}, {1,0,0}, and additional {1,1,0} facets. No octahedron (without truncation) containing only {1,1,1} facets was observed. Nanoparticle shapes containing {1,1,0} and {1,0,0} are suitable for redox activity. Some amount of irregular shapes, beneficial for redox activity, was also observed. Spectroscopy methods confirmed Ce^{3+} content.

Keywords: Nanoceria, Ce^{3+} and Ce^{4+}, electron beam evaporation, facet types, HRTEM

1. INTRODUCTION

Cerium in compounds exists in two valency states, Ce^{3+} (Ce_2O_3) and Ce^{4+} (CeO_2). These valency types are associated with different properties usable for catalysis and biomedical applications, especially for radioprotection of healthy tissues during radiation therapy (Ce^{4+}) or for supporting radio-oncological treatment toxic effect (Ce^{3+}). The formation of oxygen vacancies on the CeO_2 surfaces leads to the reduction of the Ce^{4+}

ions to Ce^{3+} and vice versa. Such oxygen off-stoichiometry correlates with the redox activity of cerium oxide nanoparticles [1-3].

Cerium oxides formed with the CeO_2 fluorite lattice (**Figure 1a**) contain Ce^{4+}. In the nanoparticle form, CeO_2 contain also Ce^{3+} ions on the surfaces. With a nanoparticle size decrease, the number of surface oxygen vacancies and Ce^{3+} increases [4,5]. Therefore, the Ce^{3+}/Ce^{4+} ratio depends on the nanoparticle size. Concurrently, nanoparticle shapes determined with surface facet types can also contribute to this surface effect. Pérez-Bailac et al. [6] demonstrated {1,1,0} facets are the most beneficial. The stable nanoparticle shapes are formed with the surface facets of the lowest surface energy. The lowest surface energies for the CeO_2 phase were found for {1,1,1} and {1,0,0} facets depending on the terminal atomic layer. **Table 1** containing the calculated surface energy values shows favorable O-terminated facets in both cases [7]. Considering these values, octahedrons are preferred (**Figure 1b**). This shape is formed with only {1,1,1} facet types, unsuitable for vacancy creation. The second lowest energy was found for {1,0,0}. Therefore, the shapes formed with {1,1,1} and {1,0,0} are assumed to be the second most favorable nanoparticle forms (**Figure 1c**).

The consideration of the stable morphology, especially nanoparticle shapes and their surface crystallography, is required for redox efficiency and its time stability assessment. For this reason, we analyzed a sample prepared using electron beam evaporation and aged for two years. The high-resolution transmission electron microscopy (HRTEM) analyses, X-ray photoelectron spectroscopy (XPS), X-ray diffraction (XRD), and Electron energy-loss spectroscopy (EELS) were employed for morphological, phase, and oxidation state analyses. TEM samples were purified with centrifugal size fractionation to decrease nanoparticle clusters, receive better nanoparticle distribution, and make more facets visible.

Table 1 Surface energies for {1,1,1} and {1,0,0} according to the terminal atomic layer [7]

facet	Ce	O	facet	Ce	O
{1,1,1}	0.823 $eVÅ^{-2}$	0.084 $eVÅ^{-2}$	{1,0,0}	0.777 $eVÅ^{-2}$	0.455 $eVÅ^{-2}$

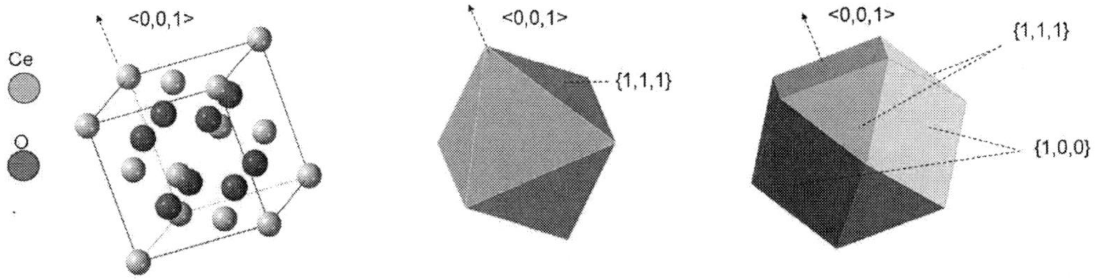

Figure 1 a) CeO2 lattice, b) octahedron, the most typical CeO2 nanoparticle morphology containing only {1,1,1} facet types, c) cuboctahedron containing {1,1,1} and {1,0,0} facet types

2. EXPERIMENTAL

2.1. Synthesis - Electron Beam Evaporation

The sample was produced through the evaporation of a solid target (CeO_2 micropowder) by a pulsed electron beam in a H_2 atmosphere using a NANOBEAM 2 device. This method, including the setup for CeO_2 synthesis, was published previously [8].

2.2. Centrifugal Size Fractionation

The sample of nanoparticles (25 mg) was dispersed in a mixture of sodium oleate (1 g), oleylamine (1 g), and octadec-1-ene (5 g). The mixture was supplemented with 5 mL of methanol and slowly heated to 300 °C under a protective N_2 atmosphere (methanol was evaporated at a lower temperature). The temperature of 300 °C was kept for 5 min. After cooling to laboratory temperature (~25 °C), the nanoparticles were precipitated by adding 10 mL of propanol and centrifuged (1500 g, 5 min). The pellet was dispersed in 5 mL of methanol and shortly sonicated in an ultrasonic bath (~20 s). After centrifugation (1500 g, 5 min), the pellet was dispersed in cyclohexane (1 mL) and shortly sonicated (~20 s). The cyclohexane dispersion was centrifuged (1500 g, 5 min), and the supernatant containing the CeO_2 nanoparticles was applied on a TEM grid.

2.3. Chemicals

Centrifugal size fractionation: Methyl alcohol (p.a.), isopropyl alcohol (p.a.), and cyclohexane (p.a.) were from Ing. Petr Švec - PENTA s.r.o. (www.pentachemicals.eu). Sodium oleate (82 %), oleylamine (70 %, technical grade), octadec-1-ene (90 %, technical grade) were from Sigma-Aldrich (www.sigmaaldrich.com).

2.4. Characterization Methods

A Titan Themis 60-300 cubed transmission electron microscope was used for morphological analyses. The micrographs were processed with the TIA (ThermoFisher), the DigitalMicrograph, and the JEMS (by P. Stadelmann) softwares. An Empyrean diffractometer (PanAnalytical) with Co $K\alpha_{1,2}$ was used for X-ray powder diffraction study. Electron energy-loss spectra were taken with a Gatan GIF with resolution of 0.9 eV. The X-ray photoelectron spectroscopy (XPS) (Kratos AXIS Supra) was carried out with a monochromatic Al Kα (1486.7 eV) excitation source to determine the chemical composition and bonding environment of the sample. The CrystalMaker software was used for crystal structure and nanoparticle shape simulations (drawing, plots).

3. RESULTS AND DISCUSSION

3.1. XRD

The synthesized sample was analyzed using X-ray diffraction (**Figure 2**) confirming only the CeO_2 phase. The sample contained two morphological components. The first one contained ≈ 5 nm CeO_2 nanoparticles (approximately 87 %, blue), the second one contained ≈ 150 nm crystallites (green).

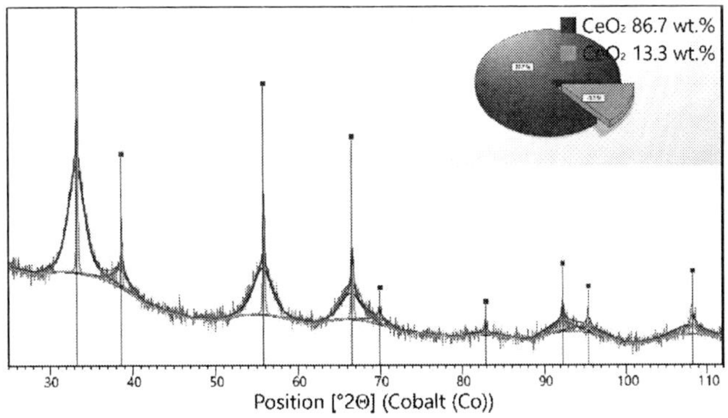

Figure 2 XRD patterns confirmed the CeO_2 phase and a high content of the nanocrystalline component (blue)

3.2. TEM

We purified the finest morphological component with centrifugal size fractionation to prepare TEM samples. We received approximately 100 nm nanoparticle clusters revealing individual nanoparticles at the cluster edges. TEM analysis revealed approximately 5 nm nanoparticles. We took 200 micrographs of 84 unduplicated nanoparticles at suitable orientations for facet type evaluation. The phase analyses using a Fast Fourier transform pattern evaluation confirmed the CeO_2 phase. We observed octahedron shapes (**Figure 3a)** in most cases (approximately 80 %). All of these octahedrons were truncated, so {1,0,0} facet types were observed in all nanoparticles. Truncated cuboctahedrons typically contained additional (1,1,0) facets (**Figure 3b**). Some amounts of irregular nanoparticles were also observed (**Figure 3c).**

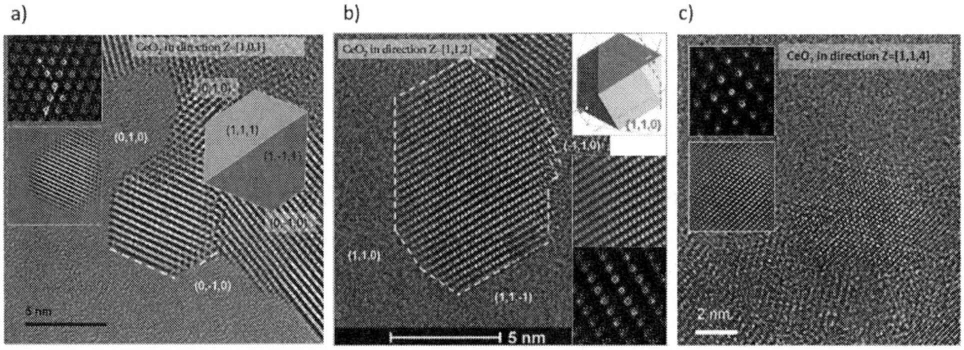

Figure 3 a) HRTEM image shows 5 nm truncated octahedron CeO2 nanoparticle in crystallographic direction Z= [1,0,1]; b) truncated cuboctahedron in the direction Z= [1,1,2] containing additional {1,1,0} facet types, and c) irregular nanoparticle in the direction Z = [1,1,4]. Evaluated Fast Fourier transform patterns and Inverse filtered Fast Fourier patterns are included in the corresponding images

3.3. XPS

The XPS measurement proved the sample purity. Only the Ce and O were detected (**Figure 4a**). The visible peaks at the binding energies of 882.19 eV and 916.54 eV corresponded to Ce^{4+} $3d_{5/2}$ and the 2nd satellite of Ce^{4+} $d_{3/2}$, respectively (**Figure 4b**). These energies indicated the CeO_2 phase content [9,10]. The identification of the Ce^{3+} $3d_{5/2}$ and Ce^{3+} $d_{3/2}$ peaks was unclear. However, satellites at the energies of 884.26 eV and 902.69 eV confirmed the presence of Ce^{3+}. The O 1s spectra in more detail are visible in **Figure 4c**. These results suggested Ce^{4+} in majority compare to the Ce^{3+} compounds.

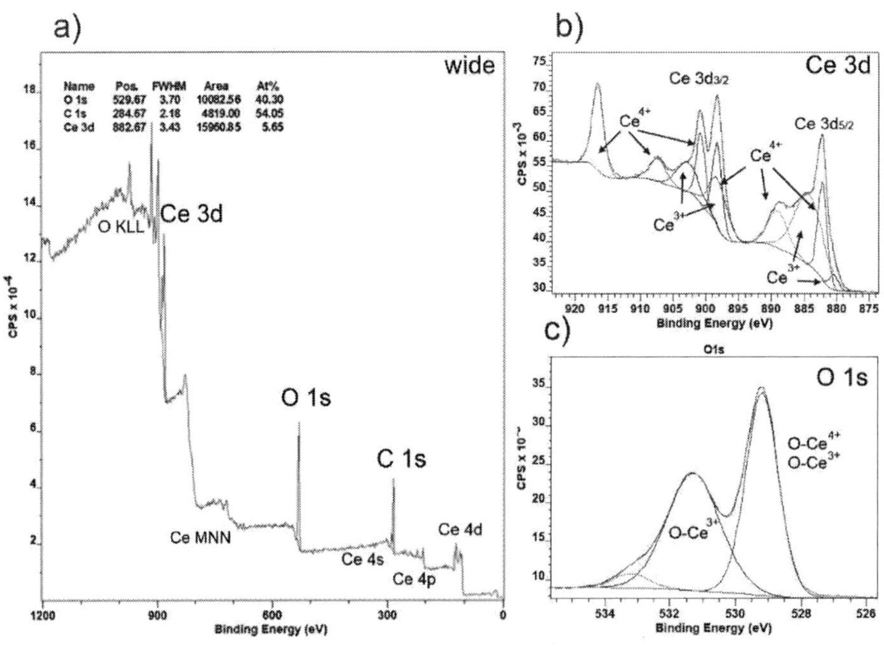

Figure 4 a) XPS wide spectra; b) Ce 3d; c) O1s spectra

3.4. EELS

EELS measurements were conducted in STEM mode and at an energy resolution of 0.9 eV. Data were taken from the thin areas of nanoparticle clusters near the surface. According to data received previously [11], we detected both oxidation states (**Figure 5**). The Ce^{3+} oxidation state was found as often as Ce^{4+} in the randomly chosen sites for analysis. These results indicate approximately equal representation of both states, contrary to XPS measurement results. This difference is intelligible because XPS measurements were conducted on the full sample surfaces, including large particles, in contrast to EELS data taken from the smallest nanoparticles at the cluster edge.

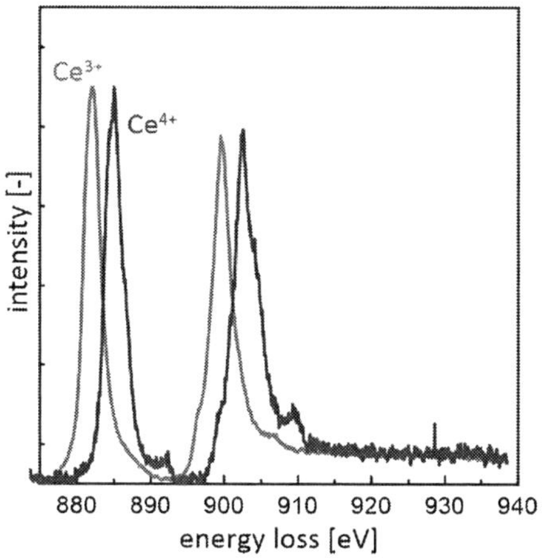

Figure 5 EELS measurements detect two oxidation states of cerium (Ce^{3+} and Ce^{4+})

4. CONCLUSION

This work confirmed the possible long-period redox activity of the samples prepared using electron beam evaporation. The results proved that nanoparticles formed with truncated shapes can be active in the redox processes for a long time. Nanoparticles maintained their morphology even after thermal treatment in the purification process. The phase analyses of many nanoparticles using FFT pattern evaluation confirmed only the CeO_2 phase. Therefore, the Ce^{3+} state is assumed to be in the nanoparticle surface layer formed with the off-stoichiometry CeO_{2-x} phase. The sample aging did not cause CeO_2 - Ce_2O_3 phase transformation even in the ultra-small nanoparticles.

ACKNOWLEDGEMENTS

The research has been supported by the Czech Science Foundation (Project 21-31852J) and by Russian Foundation for Basic Research (Project 20-58-26002). We acknowledge CzechNanoLab Research Infrastructure supported by MEYS CR (LM2018110).

REFERENCES

[1] DHALL, A., SELF, W. Cerium oxide nanoparticles: A Brief Review of Their Synthesis Methods and Biomedical Applications. *Antioxidants (Basel)*. 2018, vol. 7, no. 8, pp. 97.

[2] ULITKO, M.V., NAUMOVA, A.S., SULTANOVA, T.R., VAZIROV, R.A., AGDANTSEVA, E., OLSHVANG, O. Yu., SOKOVNIN, S. Yu. Investigation of the effect of cerium dioxide nanoparticles on the radiosensitivity of various cell types. *AIP Conference Proceedings*. 2020, vol. 2313, 080031.

[3] XU, C., QU, X. Cerium oxide nanoparticle: a remarkably versatile rare earth nanomaterial for biological applications. *NPG Asia Mater.* 2014, vol. 6, no. e90.

[4] SAIFI, M.A., SEAL, S., GODUGU, Ch. Nanoceria, the versatile nanoparticles: Promising biomedical applications. *Journal of Controlled Release*. 2021, vol. 338, pp.164-189.

[5] DESHPANDE, S., PATIL, S., KUCHIBHATLA, S. VNT, SEAL, S. Size dependency variation in lattice parameter and valency states in nanocrystalline cerium oxide. *Appl. Phys. Lett.* 2005, vol. 87, 133113.

[6] PÉREZ-BAILAC, P., LUSTEMBERG, P.G., GANDUGLIA-PIROVANO, M.V., Facet - dependent stability of near-surface oxygen vacancies and excess charge localization at CeO_2 surfaces. *Journal of Physics: Condensed Matter*. 2021, vol. 33, no. 50, 504003.

[7] DAN, X., WANG Chao, XU Xi, LIU Ya, CHENG Xiaowei, FRONZI, M., BI Lei, ZHAO, X. S. Improving the sinterability of CeO_2 by using plane-selective nanocubes. *Journal of the European Ceramic Society*. 2019, vol. 39, pp. 4429-4434.

[8] IL'VES, V.G., SOKOVNIN, S.Yu. Production and studies of properties of naanopowders on the basis of CeO_2. 2012. *Nanotechnol. Russ.* 2012, vol. 7, pp. 213 - 226.

[9] ELOIRDI, R., CAKIR, P., HUBER, F., SEIBERT, A., KONINGS, R., GOUDER, T. X-ray photoelectron spectroscopy study of the reduction and oxidation of uranium and cerium single oxide compared to (U-Ce) mixed oxide films. *Applied Surface Science*. 2018, vol. 457, pp. 566-571.

[10] MASLAKOV, K.I., TETERIN, Yu.A., POPEL, A.J., TETERIN, A.Yu., IVANOV, K.E., KALMYKOV, N., PETROV, V.G., PETROV, V.G., PETROV, P.K., FARNAN, I. XPS study of ion irradiated and unirradiated CeO_2 bulk and thin film samples. *Applied Surface Science*. 2018, vol. 448, pp. 154-162.

[11] GARVIE, L.A.J., BUSECK, P.R. Determination of Ce^{4+}/Ce^{3+} in electron beam damaged CeO_2 by electron energy-loss spectroscopy. *Journal of Physics and Chemistry of Solids*. 1999, vol. 60, pp. 1943-1947.

THE EFFECTS OF EMBEDDING MEDIUM AND SIZES ON THE OPTICAL ABSORTION IN VERTICALLY ALIGNED GAAS NANOWIRES ARRAY BASED PHOTOVOLTAIC SOLAR CELLS

[1]R.M. DE LA CRUZ, [2]C. KANYINDA-MALU, [1]J.E. MUÑOZ SANTIUSTE, [1]E. SALAS, [1]E. GARCÍA-TABARÉS, [1]M. TARDÍO, [1]S. ATHANASOPOULOS, [1]B. GALIANA

[1]*University Carlos III of Madrid, Leganés (Madrid), Spain, EU, rmc@fis.uc3m.es*

[2]*University Rey Juan Carlos, FCSJ, Madrid, Spain, EU, clement.kanyindamalu@urjc.es*

https://doi.org/10.37904/nanocon.2022.4580

Abstract

We theoretically investigate the effects of the surrounding media and sizes on the optical absorbance of a periodic square array of GaAs nanowires (NWs) as a candidate of photovoltaic solar cells. The filling media considered are: PMMA, Polycarbonate, Polystyrene and PVP, commonly used in the science and technology of semiconductors. The sizes considered are the NWs length and radius and the pitch array. The simulated system consists of a top layer of ITO, the active layer of GaAs NWs surrounded by an embedding medium and a substrate layer of Si. The system absorbance is modelled though the transfer matrix method; while the GaAs NWs with the embedding medium is considered as an effective medium modelled by Maxwell-Garnett theory. For s- and p-polarization, we find that the embedding medium effects on the absorbance spectra are scarcely perceptible compared with the size effects. Indeed, for longer and wider NWs and smaller pitches array, the absorbance values are greater around 600-1100 nm. These results are in good agreement with those reported in the literature by experiments and other sophisticated simulations in III-V semiconductor NWs.

Keywords: Absorption spectra, GaAs NWs array, Solar cells, Transfer matrix method, Maxwell-Garnett theory

1. INTRODUCTION

Nowadays, the research on gallium arsenide (GaAs) nanowires (NWs) is gaining an increasing interest due to their relevant applications in different fields, specifically in photovoltaics solar cells [1]. Indeed, the NWs material has exciting absorption characteristics because of their geometry [2], giving them exceptional light interaction features, not found in a bulk material, as consequence of their large aspect ratio. Increasing the absorption of light in GaAs NWs based solar cells is crucial to enhance the overall efficiency and reducing cost. Two main aspects to investigate in order to improve the light absorption in GaAs NWs can be the embedding medium of NWs and their sizes (NWs length and radius and pitch array).

Floris et al. [2] investigated the optical reflectance of InAs NWs for different filling media. In addition, several authors investigated experimentally and theoretically the sizes dependence on the optical spectra in semiconductor NWs array based solar cells [3,4]. These authors used finite difference time domain (FDTD) method, which takes a long time of calculation. However, we recently investigated the geometrical parameters dependence on the optical reflectance of GaAs NWs array simulated by means of the transfer matrix formalism and effective medium theory, where the computing time of calculation is less than that of FDTD method [5]. In that work, we only simulated the optical reflectance and we neglected the substrate and top layer effects on the reflectance.

Therefore, we extend the previous work in order to investigate the embedding medium and sizes effects on the absorbance of a GaAs NWs array based photovoltaic solar cells to gain insight about the system efficiency.

We also included the effect of the top layer and substrate. A better absorption entails a higher efficiency. We use the transfer matrix formalism to simulate the optical absorbance. To evaluate the dielectric function of the continuum medium (GaAs NWs+embedding material), we use the Maxwell-Garnett (M-G) theory [6].

2. INVESTIGATED SYSTEM AND THEORETICAL MODEL

Figure 1 depicts the scheme of the investigated system constituted by (1) a semi-infinite air space; (2) a layer of indium tin oxide (ITO), transparent in the VIS range and extensively used in solar cells technology, with dielectric function (ε_2) [7] and thickness d_2 = 100 nm; (3) the active layer based on effective medium of GaAs NWs embedded in insulator material, with thickness d_3 = h (NWs length) and dielectric function described later; (4) a layer of silicon (Si) as substrate, with dielectric constant ε_4 = 17.749 + 0.50381 *i* (value at 517 nm) [8] and thickness d_4 = 2000 nm and (5) a semi-infinite air space. The light is impinging from the air and is going out to the air, then $\varepsilon_1 = \varepsilon_5 = 1$. The election of the above thicknesses is taken with the objective of a better light trapping in the array and a better confinement of the GaAs NWs.

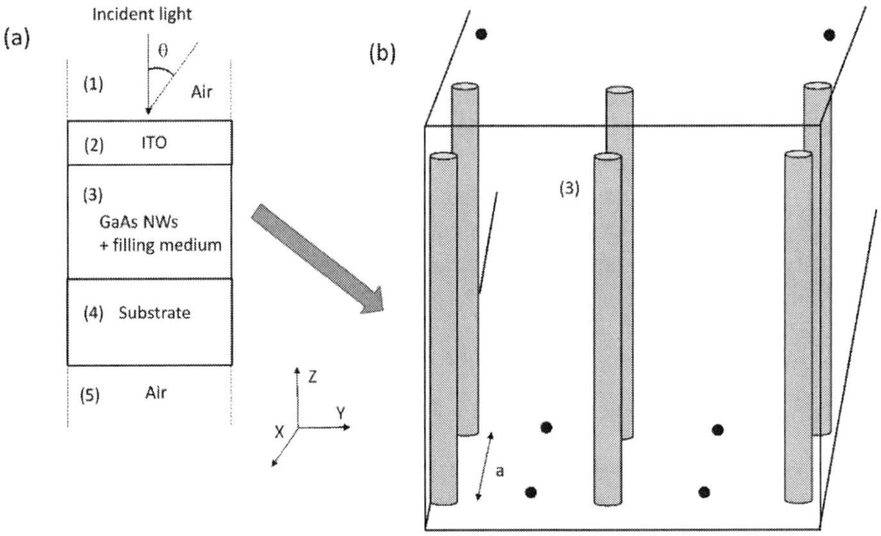

Figure 1 Scheme of the investigated system with the different layers (Figure 1a). Figure 1b is the zoom of the active layer (GaAs NWs+embedding medium). The black points represent the NWs array.

As embedding media, we chose PMMA, Polycarbonate, Polystyrene and PVP. Their dielectric function dependences on the wavelength are taken from ref. [8]. The array is characterized by cylinder radii ranging between R = 15 nm to 45 nm, length h = 500 nm and 1000 nm and pitch array a = 100 nm, 200 nm and 300 nm. These geometrical values yield filling factor ($f = \pi(R/a)^2$) belonging to the range between 0.018 to 0.196. The dielectric functions of the effective medium (GaAs embedded in insulator material) are characterized by $\varepsilon_3^{\parallel}(\omega)$ and $\varepsilon_3^{\perp}(\omega)$, being the effective dielectric functions in the directions parallel and perpendicular to the axis of cylindrical NWs, respectively. We use the M-G model under the assumption of a small inclusion density [6,9]; therefore

$$\varepsilon_3^{\parallel}(\omega) = f\,\varepsilon_{NW}(\omega) + (1 - f)\,\varepsilon_e \quad (1)$$

where f is the filling factor, ε_e is the embedding material dielectric function and $\varepsilon_{NW}(\omega)$ is the GaAs NWs dielectric function that we will describe later. On the other hand, $\varepsilon_3^{\perp}(\omega)$ is written as [9]

$$\varepsilon_3^{\perp}(\omega) = \varepsilon_e\,[(\varepsilon_{NW}(\omega)\,(1 + f) + \varepsilon_e\,(1 - f))/(\varepsilon_{NW}(\omega)\,(1 - f) + \varepsilon_e\,(1 + f))] \quad (2)$$

To describe $\varepsilon_{NW}(\omega)$, we use the formalism of Webb [10], where the exciton confinement energy is considered as a resonance. All parameters are extensively defined in ref. [10]. This formalism is appropriate for low-dimensional systems such as our NWs array.

With the knowledge of the dielectric permittivity of each layer, we use the transfer matrix method for our system constituted by five layers with four interfaces, similarly reported in ref. [11]. Then, the reflectance is $R^{(s,p)} = |r^{(s,p)}|^2$, where the superscripts s, p refer to s- and p- polarization, respectively; while the transmittance is $T^{(s,p)} = |t^{(s,p)}|^2$, being $r^{(s,p)}$ and $t^{(s,p)}$ the reflection and transmission coefficients, defined in ref. [11]. The absorbance of the stacked layer is defined as $A^{(s,p)} = 1 - R^{(s,p)} - T^{(s,p)}$. We implement the numerical simulations over the wavelength range 300 nm to 1100 nm, which cover the relevant parts of solar spectrum.

3. RESULTS AND DISCUSSION

To clarify the presentation of our results, we divide the discussion in two subsections, accounting for the effects of embedding material and the geometrical parameter sizes on the absorbance of GaAs NWs for s- and p-polarized light.

3.1. Embedding material effects on the absorbance of GaAs NWs

We show in **Figure 2a** the absorbance for the four investigated embedding media with fixed geometrical parameters (R = 25 nm, a = 200 nm and h = 500 nm) for s-polarized light at normal incidence. **Figure 2b** represents the comparison between s- and p-polarization at normal incidence when the insulator material is PMMA, as an example.

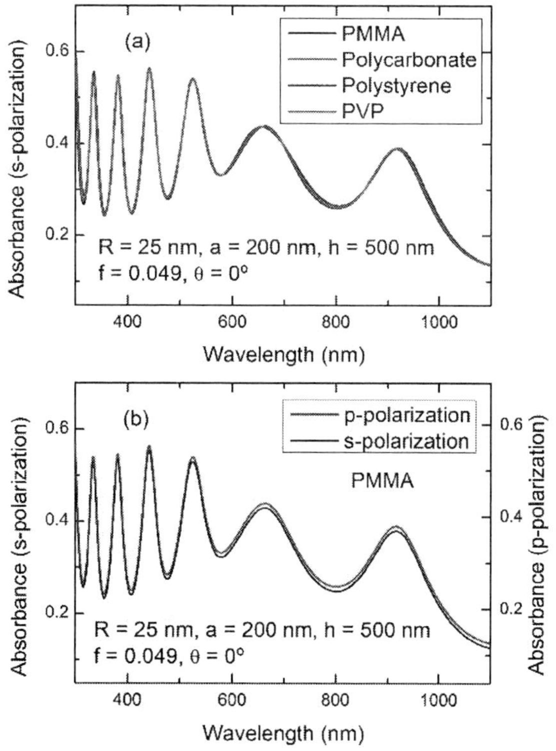

Figure 2 Absorbance for s-polarization of GaAs NWs surrounded by PMMA, Polycarbonate, Polystyrene and PVP at normal incidence (a). Absorbance for s- and p-polarization of GaAs NWs embedded in PMMA. In this figure, left and right axis are sligthly displaced to show both spectra (b).

We obtain an oscillatory behavior of the absorbance with the wavelength for all investigated embedding media for s- and p-polarization. This oscillatory behavior is typical in periodical systems [12] as a consequence of the interference effects; i.e., as light impinges on the NWs, photons can interact with NWs and undergo multiple scattering, yielding interference effects due to relative phase difference between reflected waves. The absorbance spectra for s- and p-polatization at normal incidence are identical. **Figure 2b** depicts the absorbance for both polarizations at normal incidence when the insulator material is PMMA, as an example. We previously obtained identical reflectance values for s- and p-polarization in GaAs NWs at normal incidence [5]. Sikdar et al. [11] also obtained reflectance spectra with identical reflection for both s- and p-polarized light at normal incidence for nanoparticles layers. On the other hand, the embedding medium effect in the absorbance is more significant in the range 600-800 nm, where there is a slight absorbance difference. The similar refractive indexes of the investigated polymers, appear to be the origin of the small embedding effect found in the absorption spectra of GaAs NWs. For the next section, we chose PMMA as filling medium in GaAs NWs to investigate the size effects.

3.2. Size effects on the absorbance of GaAs NWs

Figure 3 depicts the absorbance dependence on NWs length for s- and p-polarization at normal incidence with fixed R = 25 nm and a = 200 nm. The values of investigated length are h = 500 nm and 1000 nm.

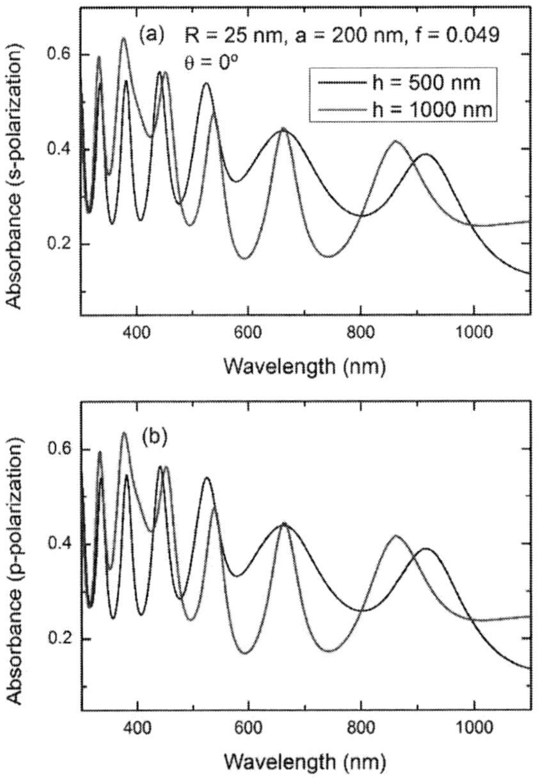

Figure 3 Absorbance for s-polarization (a) and p-polarization (b) of GaAs NWs embedded by PMMA with fixed R, a and f. The code of colors is the same for figures (a) and (b).

For both polarizations and lengths, we obtain again oscillatory behavior of the absorbance, typical of periodical systems. The absorbance is greater for h = 1000 nm in the range 300 nm to 500 nm. As the length changes, there is a difference in the phase factor; then, the absorbance values clearly depend on this geometrical parameter. Several authors showed experimentally and theoretically that the optical absorbance is greater for longer III-V semiconductor NWs [13]. Then, longer NWs benefit the light trapping in the array and consequently, the optical absorbance increases yielding a plausible gain efficiency in GaAs NWs photovoltaic solar cells.

We show in **Figure 4** the absorbance dependence on NWs radius for s- and p-polarization at normal incidence with fixed a = 200 nm and h = 500 nm. The investigated radii are R = 15 nm, 25 nm, 35 nm and 45 nm.

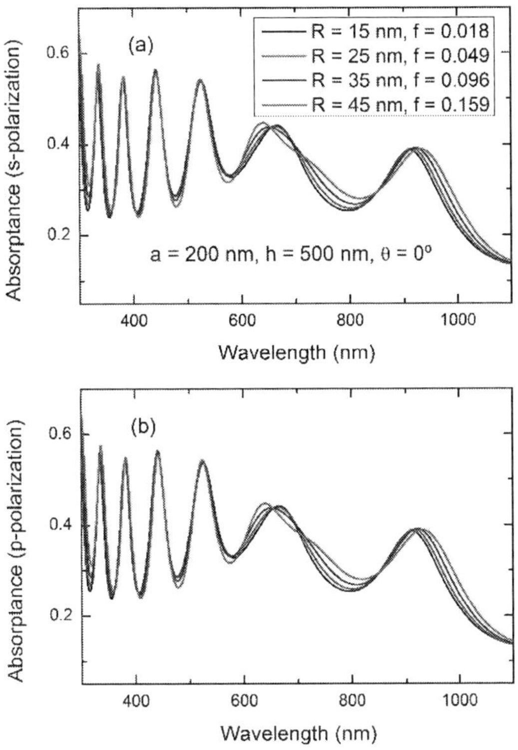

Figure 4 Absorbance for s-polarization (a) and p-polarization (b) of GaAs NWs embedded by PMMA with fixed a and h. The code of colors is the same for figures (a) and (b).

For increasing radius, our simulations show that the absorbance is slightly greater in the range 600 nm to 1100 nm. Ali et al. [14] theoretically showed that the absorption is greater for wider GaAs/GaSb NWs, in good agreement with our results. On the other hand, we obtain a red-shift of the absorbance modulations with increasing radius. Dhindsa et al. [3] previously reported this phenomenon for experimental and simulated reflectance by FDTD method in GaAs NWs array.

Although it is not shown here, for smaller pitch array, our simulations show greater absorbance values in the range 600 nm to 1100 nm, in good agreement with that reported by Dhindsa et al. [3] in GaAs NWs. A smaller pitch array entails a greater number of absorbance centers (greater filling factor), then the absorbance values should be greater.

4. CONCLUSION

We have calculated the embedding material and sizes effects on the optical absorbance of GaAs NWs array based photovoltaic solar cells. We simulated the absorbance spectra using the transfer matrix formalism and Maxwell-Garnett model. An oscillatory behavior of the absorbance for s- and p-polarized light, independently of the embedding medium and sizes, is found. We investigated as embedding media PMMA, Polycarbonate, Polystyrene and PVP, commonly used in the science and technology of semiconductors. We have found that for both polarizations, the embedding medium effects on the absorbance are scarcely perceptible compared with the sizes effects. Indeed, for longer and wider NWs and smaller pitches array, the absorbance values are greater in the range 600-1100 nm. These numerical results are in good agreement with those reported in the literature by experiments and sophisticated FDTD simulations in III-V semiconductor NWs.

ACKNOWLEDGEMENTS

This work was partially supported by Spanish MICINN under grant RTI 2018-101020-B-I00 and TECHNOFUSION III CM-S2018IEMAT-4437. This work has also been supported by Comunidad de Madrid under the agreement with UC3M in the line of Excellence of University Professors (EPUC3M14).

REFERENCES

[1] MARIANI, G., SCOFIELD, A.C., HUNG, C.-H., HUFFAKER, D.L. GaAs nanopillar-array solar cells employed in situ surface passivation. *Nat. Commun.* 2013, vol. 4, no. 1497, p. 7.

[2] FLORIS, F., FORNASARI, L., MARINI, A., BELLANI, V., BANFI, F., RODDARO, S., ERCOLANI, D., ROCCI, M., BELTRAM, F., CECCHINI, M., SORBA, L., ROSSELLA, F. Self-assembled InAs nanowires as optical reflectors. *Nanomaterials.* 2017, vol. 7, no. 400, p. 11.

[3] DHINDSA, N., CHIA, A., BOULANGER, J., KHODADAD, I., LAPIERRE, R., SAINI, S.S. Highly ordered vertical GaAs nanowire arrays with dry etching and their optical properties. *Nanotechnology.* 2014, vol. 25, 305303, p. 11.

[4] AZIZUR-RAHMANAND, K.M., LAPIERRE, R.R. Wavelength-selective absorptance in GaAs, InP and InAs nanowire arrays. *Nanotechnology.* 2015, vol. 26, 295202, p. 7.

[5] DE LA CRUZ, R.M., KANYINDA-MALU, C., MUÑOZ SANTIUSTE, J.E. Simulations of optical reflectance invertically aligned GaAs nanowires array: The effect of the geometrical structural parameter. *Physica B.* 2022, vol. 639, 41393, p. 7.

[6] MAXWELL-GARNETT, J.C. Colours in Metal Glasses and in Metallic Films. *Trans. Of the Royal Society.* 1904, vol. CCIII, pp 385-420.

[7] KORNYSHEV, A.A., MARINESCU, M., PAGET, J., URBAKH, M. Reflection of light by metal nanoparticles at electrodes. *Phys. Chem. Chem. Phys.* 2012, vol. 14, pp 1850-1859.

[8] RefractiveIndex.INFO. Refractive index database [online] 2022. [viewed 2022-07-30]. Available from: https://refractiveindex.info/?shelf=3dbook=plasticspage.

[9] STARKO-BOWES, R., ATKINSON, J., NEWMAN, H., HU, T., KALLOS, T., PALIKARAS, G., FEDOSEJEVS, R., PRAMANIK, S., JACOB, Z. Optical characterization of epsilon-near-zero pole, and hyperbolic response in nanowire metamaterials. *J. of the Optical Society of America B.* 2015, vol. 32, pp 2074-2080.

[10] WEBB, K.J., LUDWIG, A. Semiconductor quantum dot mixture as a lossless negative dielectric constant optical material. *Phys. Rev. B* 2008, vol. 78, 153303, p. 4.

[11] SIKDAR, D., KORNYSHEV, A.A. Theory of tailorable optical response of two-dimensional arrays of plasmonic nanoparticles at dielectric interfaces. *Scientific Reports.* 2016, 33712, p. 6.

[12] CHEN, Y, HÖHN, O., TUCHER, N., PISTOL, M.-E., ANTTU, N. Optical analysis of a III-V-nanowire-array-on-Si dual junction solar cell. *Optics Express.* 2017, vol. 25, A665, p. 15.

[13] DHINDSA, N., KOANDHANI, R., SAINIA, S.S. Length dependent optical characteristics analysis for semiconductor nanowires. *Nanotechnology.* 2020, vol. 31, 224001, p. 9.

[14] ALI, L.M., ABED, F.A. Numerical modeling of opto-electric characterization of GaAs/GaSb nanowire solar cells. *Optical and Quantum Electronics.* 2020, vol. 52, no. 154, p. 9.

EVALUATION OF PRINTED INTERDIGITAL ELECTRODES ARRAY FOR CHEMO-RESISTIVE GAS SENSOR APPLICATION

[1]Vojtěch POVOLNÝ, [1]Alexandr LAPOSA, [1]Jiří KROUTIL, [1]Jan VOVES, [2]Marina DAVYDOVA, [1]Pavel HAZDRA

[1]Faculty of Electrical Engineering, Czech Technical University in Prague, Prague, Czech Republic, EU, povolvoj@fel.cvut.cz

[2]Institute of Physics of the Czech Academy of Sciences, Prague, Czech Republic, EU

https://doi.org/10.37904/nanocon.2022.4588

Abstract

In this article, we present the design and preparation of a platform for a gas sensor application based on the ink print of silver nanoparticles on flexible substrates. The Inkjet printing allows localized deposition of ink at low temperatures on a large area and can be used for the deposition of both conductive and dielectric materials. The printed platform consists of four interdigital electrodes (IDE), and connector pads printed with silver ink and sintered with an intense pulse light (IPL). The dielectric ink printed around the IDE structures protects the connection pathways and defines the sensing area. The multilayer structure was printed on polyethylene terephthalate (PET) and polyimide (PI) substrates. Before printing, the plasma treatment of the surface of the polyimide is necessary. Three combinations of electrodes with various widths were designed and the structures were printed with three various resolutions to verify the printability on the used combination of substrate and ink. Bending tests, that show the flexibility of printed structures, were performed as well.

Keywords: Inkjet, gas sensor array, PET, Polyimide, IPL

1. INTRODUCTION

Printing technology has been increasingly popular in recent years. Due to its low manufacturing costs, printing technologies such as screen printing [1,2], micro-contact printing [3,4], and inkjet printing [5] have become increasingly popular. Inkjet printing has sparked the most interest among the techniques mentioned because the components produced using this technology have undeniable advantages over conventional ones in terms of the ease of production, on a wide range of substrates, the contactless deposition of the resulting motive without the use of a mask, low-temperature processing without the use of vacuum, and low cost. All-polymer transistors [5,6] polymer light emitting diode (PLED) [7,8], nanoparticle Micro Electro Mechanical Systems [9], and polymer capacitors [10] have all been made with inkjet printing.

Inkjet printing technology has grown in popularity as a technique for producing electronic components in a variety of academic and industrial fields during the last decade. The use of a drop-on-demand or on-demand material deposition technology is a key advantage for inkjet printing and allows the fully additive processing of electronic materials. Inkjet eliminates the requirement for a cleanroom, toxic chemical waste, and expensive photolithographic masks when compared to traditional electronic production processes [11].

In this work, we report the preparation of an inkjet printed sensor array on flexible substrates. We discuss the results obtained for printing with various resolutions, the conductivity of the printed paths according to the annealing method, and the results of a cyclic bend test to analyze the reliability of the printed structures.

2. MATERIALS AND METHODS

2.1. Sensor array design

The prepared platform was designed for the purpose of multiple sensor testing. It was printed with material inkjet printer Fujifilm Dimatix 2831 and cartridge DMC-11610 (16 nozzles, 10 pl drop volume) and the design of the platform was prepared with the CleWin5 layout editor (Wieweb software, Netherland). The dimensions of the array are 13.6 mm x 12 mm and the platform consists of two basic parts - conductive and insulating layers. The conductive layer is divided into two parts - conductive pathways including connector pads and four interdigital electrodes (IDE). The connector part is designed to be simply connected with a flexible tip or to be mounted into an Amphenol Clincher flex connector. The IDE structures have dimensions of approximately 2 mm x 2 mm with finger widths of 20 µm, 40 µm, and 60 µm and were printed with a combination of three various resolutions and three various widths of fingers. The printed structure and the layout of the platform with dimensions can be seen in **Figure 1**.

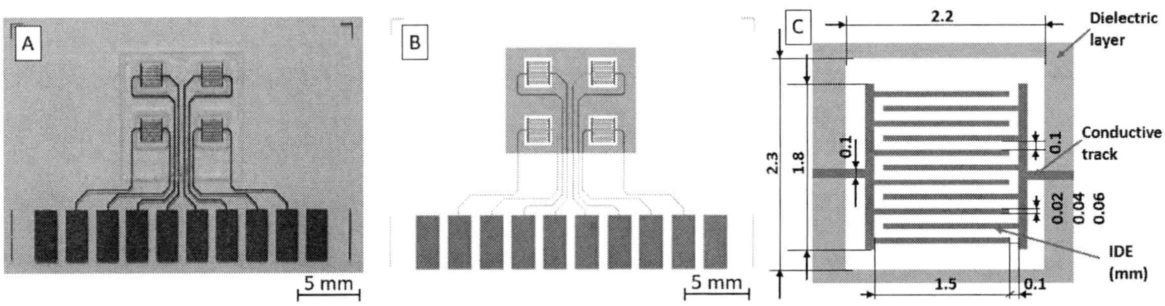

Figure 1 Platform of printed interdigital sensor array: A) Structure printed with silver ink and with the dielectric protection layer, B) Layout of the complete platform, C) Detail of the IDE with dimensions

2.2. Substrates

Two types of flexible materials were used as substrates for printing. The adopted polyethylene terephthalate (PET) substrate Mitsubishi Novele™ IJ-220 by Novacentrix (thickness 140 µm) is a printed electronics mesoporous substrate appropriate for low-cost and low-temperature applications where the flexible and transparent substrate is desired and specifically engineered for inkjet-compatible inks [12]. The second substrate used for printing is polyimide (PI) foil DuPont™ Kapton HN (thickness 75 µm). The parameters of the PET and polyimide materials are listed in **Table 1**. Polyimide can withstand higher temperatures than PET and can be used for the preparation of components in processes with temperatures of several hundred degrees. However, before printing, it is necessary to carry out the plasma treatment that will prevent bleeding of the ink by slightly etching the surface. In this research, we used SF_6 plasma treatment with a power of 200 W of ICP (Inductively Coupled Plasma) source and 30 W of RF (Radio Frequency) generator, respectively with Sentech ICP Etch System SI 500 (SENTECH Instruments GmbH, Germany).

2.3. Inks

The conductive layers were fabricated with the silver nanoparticle ink ANP Silverjet DGP-40LT-15C (solid content 35 %, resistivity 11 µΩ·cm, particle size <50 nm). During the printing, the platen temperature was set to 40 °C and the structures were printed with a resolution ranging from 635 dpi to 1693 dpi at 2 kHz jetting frequency and the print head was preheated to 35 °C. Printed structures were dried at room temperature for 24 h and then sintered with the high-intensity pulse light (IPL) system Xenon X-1100 by Xenon Corporation, USA (up to 9 J·cm^{-2} radiant energy per pulse, broad-spectrum light) by two flashes with the settings of 300 µs and 200 J.

Table 1 Typical properties of selected flexible substrates

Property	Polyethylene terephthalate (PET)	Polyester (PES)	Polyimide (PI)
Tensile strength (MPa)	60 - 85 *	10 - 123 *	72 - 231 *
Flexibility	Excellent	Excellent	Excellent
Dimensional stability	Good	Good	Excellent
Dielectric strength (kV/mm)	20 - 50 *	11 - 30 *	177 - 201 *
Max operating temperature (°C)	100 - 140 *	71 - 200 *	220 - 400 *
Chemical resistance	Very good	Excellent	Excellent
Moisture absorption (%)	0.10	0.20	0.30
Cost	Low	Low	High

* - According to type

The isolation protective layer was printed with dielectric ink Dycotec DM-INI-7003 by Dycotec Materials Ltd, UK (volume resistivity $1.4 \cdot 10^{14}$ Ω·cm, UV curing wavelength 380 nm, UV curing energy 500 mJ·cm^{-2} - 1000 mJ·cm^{-2}). The dielectric ink was printed with 15 nozzles at the resolution of 1016 dpi at 2 kHz, the print head was preheated to 32 °C and the substrate was heated to 50 °C. Immediately after printing, the layer was cured by UV radiation with a wavelength of 380 nm and energy of 1.7 W·cm^{-2} using a high-intensity UV LED.

2.4. Mechanical test

A bending endurance test was carried out for the verification of the flexibility and stability of the inkjet printed layers. For this purpose, an in-house made bend test apparatus was used, see **Figure 2**. This machine performs cyclic bending of a flexible substrate with a printed test structure, which puts mechanical stress on the entire structure. The bending radius (6 mm) and the distance (25 mm) are precisely specified according to the standard for mechanical bending tests IPC-TM-650 2.4.3.

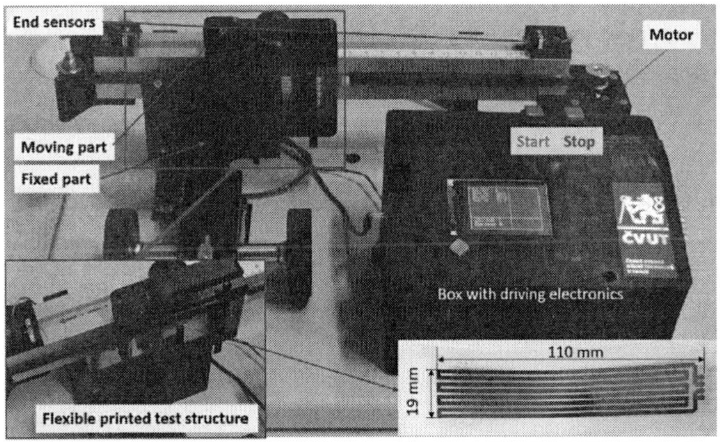

Figure 2 Setup for mechanical test of flexible printed circuit board

3. RESULTS AND DISCUSSIONS

3.1. PET

To ascertain the behavior of the silver ink on the PET substrate surface, a single nozzle line was prepared and printed using built-in control software. Optical characterization and measurement of printed structures were performed on a digital microscope Olympus BX 60. **Figure 3** shows silver lines printed at resolutions of 635 dpi, 725 dpi, 847 dpi, 1016 dpi, 1270 dpi, and 1693 dpi and also with measured widths. It can be seen that the lines printed with 635 dpi and 725 dpi are formed from separated drops and the lines are not continuous. For the line printed with 1693 dpi, we observed that the line is not uniform due to the deposition of a big amount of the material. The resolutions of 847 dpi, 1016 dpi, and 1270 dpi showed the most suitable shape of the lines and were chosen for further printing. The measured widths were varying from 30 μm to 33 μm. For the thickness evaluation of silver ink, a 200 µm wide and 2 mm long line was printed with a various number of layers at the resolution of 1016 dpi. The thickness of printed structures was measured with Stylus Type 3D Surface Profilometer (Taylor Hobson, UK). Mean heights with maximal deviation are listed in **Table 2**.

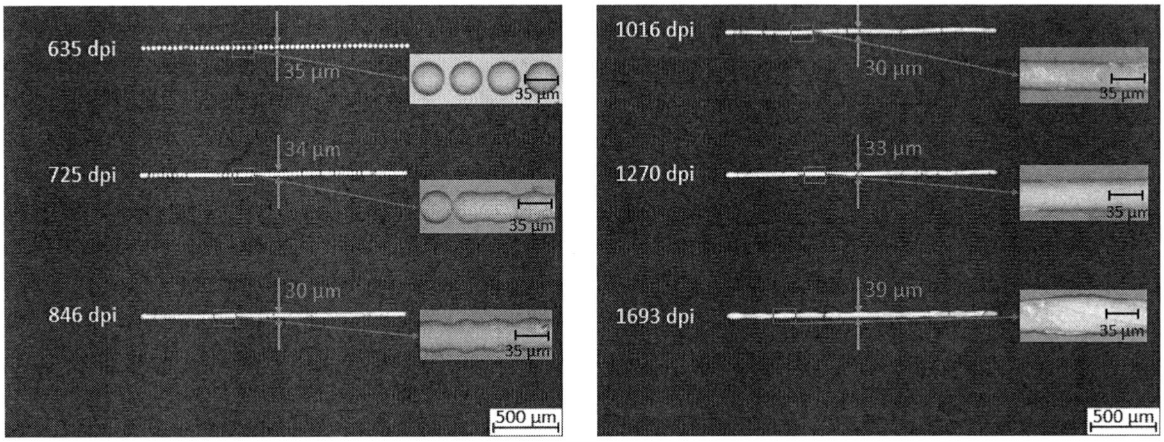

Figure 3 Resolution test of silver ink printed on PET substrate

Table 2 Thickness and maximal deviation for the various number of silver layers printed on PET

Number of layers	Thickness (µm)	Max deviation (µm)
1	0.45	0.19
2	0.88	0.23
3	1.37	0.25
4	1.73	0.26
5	2.13	0.31
10	3.93	0.34

Line width and shape

The interdigital electrodes were printed with two layers and resolutions of 847 dpi, 1016 dpi, and 1270. From the structures printed with the resolution of 847 dpi, we observed that the lines are not homogenous because of the greater drop spacing. Even if the structure was printed with two layers an incomplete connection may occur. This can make the platform unusable. For the structures printed with the resolution of 1016 dpi, the lines are homogenous and all patterns are completely connected. However, the edges of the IDE fingers are still not absolutely smooth. The best results of the printed structures with high smoothness and uniformity were achieved at the resolution of 1270 dpi. However, printing with higher resolution is more time-consuming. The printing time of conductive tracks and connection pads was 6 min, 7 min, and 9 min for the resolutions of 847 dpi, 1016 dpi, and 1270 dpi, respectively. For printing larger areas, it is important to take these parameters into account.

Sintering

Figure 4A shows the dependency of resistance on a various number of layers for different materials or sintering processes, respectively. From the plot can be seen that for structures printed with silver nanoparticle ink the values of the resistance are approximately 4 times higher compared to the calculated values of resistance for the specific resistivity of the bulk silver. This means that if we want to achieve these values with silver ink the thickness of the printed structure would be 4 times greater. **Figure 4A** also shows that we achieved better results than one can expect considering the values of resistance for the manufacturer´s specific resistivity. The reason for this phenomenon is caused by different sintering methods. The producer declares the specific resistivity of 11 µΩ·cm for structures sintered at 250 °C for 3 hours even though the recommended sintering temperature is 100 °C - 150 °C. The temperature of 250 °C is too high for a majority of substrates and can cause their damage. Therefore, it is unusable for printed electronics.

In our case, we use an intense pulse light system to sinter printed structures. The calculated resistivity of the structures is approximately 8.3 µΩ·cm, see the small plot in **Figure 4A**. To compare, **Figure 4B** shows the calculated resistance of silver ink sintered with a laboratory oven from the previous measurements. From this comparison, it can be seen that the resistance of the structures printed with five layers and sintered with the oven is 3 times higher than that of the IPL sintered structures.

When using IPL to sinter printed structures, the multiple flashes have the effect of decreasing the resistance, see **Figure 4B**. With each flash, the silver nanoparticles melt and gradually form a uniform layer, leading to an improvement in the conductivity of the layer.

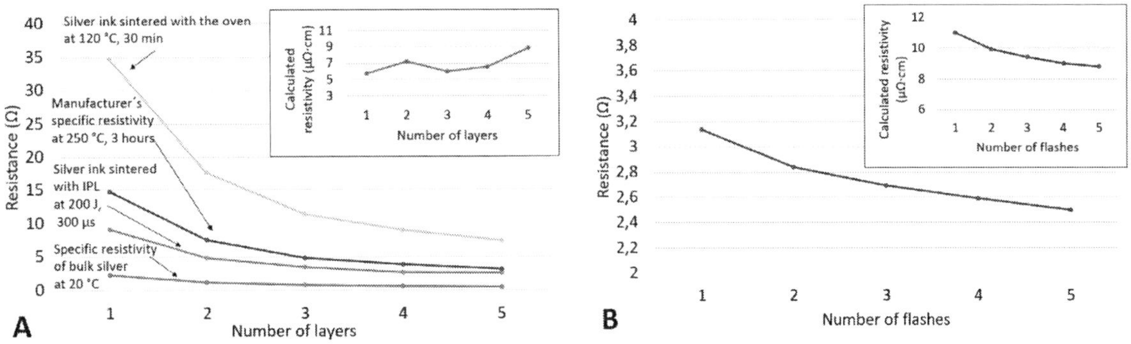

Figure 4 Dependency of the resistance on various number of layers for various value of the resistivity (left), dependency of resistance on various number of flashes (right)

3.2. Polyimide

Printing on polyimide is more demanding due to the necessity of the surface treatment. When no cleaning and surface activation has not been done the printed structure didn't create the desired shape due to the flooding of the ink. Therefore, the SF_6 plasma treatment was selected for the surface treatment of the substrate. The original process time of 30 seconds was gradually reduced to 5 seconds. This value was chosen as the most suitable in terms of ink behavior on the surface. The sensor array platform was printed with the same conditions as it was on the PET substrate. **Figure 5** shows the printed structure on the PI. The measured values of finger widths printed with two layers are summarized in **Table 3**.

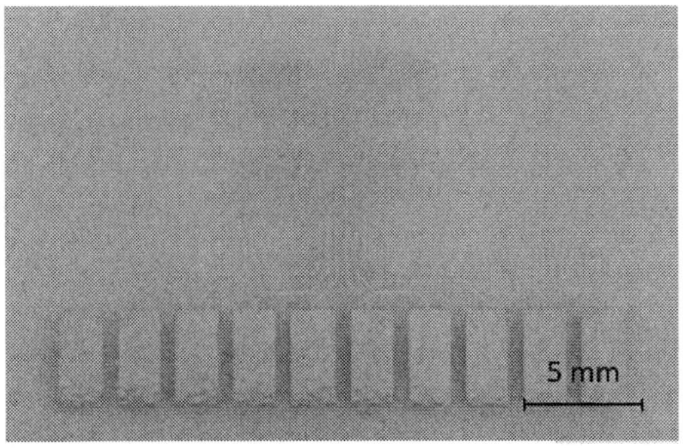

Figure 5 Interdigital sensor array printed on polyimide substrate

3.3. Mechanical test

The meander testing structure was printed on the PET substrate to verify the flexibility and stability of the printed electronics. The structure was tested with 1000 cycles which correspond to approximately 2 hours of test time. Before testing, the resistance of the structure was measured and compared with the value after testing. It was observed that the bending of the structure caused a relative change of the test structure of about 3 %. Furthermore, the printed structure has been shown to withstand the bending test without any destructive effect.

Table 3 Measured widths of printed fingers of IDE structures for each resolution

Designed width (μm)	Measured width on PET (μm)			Measured width on PI (μm)		
	847 dpi	1016 dpi	1270 dpi	847 dpi	1016 dpi	1270 dpi
20	26.5	25	21	48	25	38.1
40	58.5	37.5	29.5	51	51.6	54.3
60	66	55.5	42	72	65.7	70.6

4. CONCLUSION

In this work, the interdigital sensor array was fabricated on two flexible substrates by inkjet printing with various printing resolutions and various widths of IDE fingers. It was proved that for printing narrow lines (ca 40 μm) is better to use a higher resolution (1270 dpi), which is not so affected by the change in layout size of printed structures. On PET substrate, the thickness of printed layers was measured for a various number of layers. It was proved that the thickness increases linearly with a higher number of layers. Moreover, it was observed that with the use of the IPL sintering system we can obtain better resistivity (8.3 μΩ·cm measured after 2 x 200 J pulse energy and 300 μs each pulse length) than those indicated by the manufacturer (11 μΩ·cm measured after dry oven baking at 250 °C for 3 hours). By testing the flexural endurance of printed circuit boards, it was shown that printed silver structures can withstand at least 1000 cycles without any evidence of damage or delamination.

ACKNOWLEDGEMENTS

This work was supported by the Czech Science Foundation grant No. 22-04533S and by the Czech Technical University Student Grant No. SGS20/176/OHK3/3T/13.

REFERENCES

[1] F. GARNIER, R. HAJLAOUI, A. YASSAR, and P. SRIVASTAVA. “All-polymer field-effect transistor realized by printing techniques.” *Science (80-.).* 1994. Available from: https://doi.org/10.1126/science.265.5179.1684.

[2] Z. BAO, Y. FENG, A. DODABALAPUR, V. R. RAJU, and A. J. LOVINGER. “High-Performance Plastic Transistors Fabricated by Printing Techniques.” *Chem. Mater.* 1997. Available from: https://doi.org/10.1021/cm9701163.

[3] J. TATE *et al.* “Anodization and microcontact printing on electroless silver: Solution-based fabrication procedures for low-voltage electronic systems with organic active components.” *Langmuir.* 2000. Available from: https://doi.org/10.1021/la991646b.

[4] Z. BAO, J. A. ROGERS, and H. E. KATZ. “Printable organic and polymeric semiconducting materials and devices.” *Journal of Materials Chemistry.* 1999. Available from: https://doi.org/10.1039/a902652e.

[5] H. SIRRINGHAUS *et al.* “High-resolution inkjet printing of all-polymer transistor circuits.” *Science (80-.).* 2000. Available from: https://doi.org/10.1126/science.290.5499.2123.

[6] T. KAWASE, H. SIRRINGHAUS, R. H. FRIEND, and T. SHIMODA. “Inkjet printed via-hole interconnections and resistors for all-polymer transistor circuits.” *Adv. Mater..* 2001. Available from: https://doi.org/10.1002/1521-4095(200111)13:21<1601::AID-ADMA1601>3.0.CO;2-X.

[7] T. R. HEBNER, C. C. WU, D. MARCY, M. H. LU, and J. C. STURM. “Ink-jet printing of doped polymers for organic light emitting devices.” *Appl. Phys. Lett.* 1998. Available from: https://doi.org/10.1063/1.120807.

[8] Y. YANG, S. C. CHANG, J. BHARATHAN, and J. LIU. "Organic/polymeric electroluminescent devices processed by hybrid ink-jet printing." *J. Mater. Sci. Mater. Electron.* 2000. Available from: https://doi.org/10.1023/A:1008968511133.

[9] S. B. FULLER, E. J. WILHELM, and J. M. JACOBSON. "Ink-jet printed nanoparticle microelectromechanical systems." *J. Microelectromechanical Syst.*. 2002. Available from: https://doi.org/10.1109/84.982863.

[10] Y. LIU, T. CUI, and K. VARAHRAMYAN. "All-polymer capacitor fabricated with inkjet printing technique." *Solid. State. Electron.*. 2003. Available from: https://doi.org/10.1016/S0038-1101(03)00082-0.

[11] B. J. KANG, C. K. LEE, and J. H. OH. "All-inkjet-printed electrical components and circuit fabrication on a plastic substrate." *Microelectron. Eng.* 2012. Available from: https://doi.org/10.1016/j.mee.2012.03.032.

[12] B. ANDÒ and V. MARLETTA. "An All-InkJet Printed Bending Actuator with Embedded Sensing Feature and an Electromagnetic Driving Mechanism." *Actuators*. Jul. 2016, vol. 5, no. 3, p. 21. Available from: https://doi.org/10.3390/act5030021.

LOW-COST OPTICAL FIBERS MICROSCALE GRINDING AND POLISHING SYSTEM: TOWARDS OPEN-SOURCE DEVICES AND SYSTEMS

[1,2]Mohamed ABDELKADER, [1]Stanislav PETRIK

[1]Technical University of Liberec, Department of Advanced Materials, Institute for Nanomaterials, Advanced Technologies, and Innovation (CXI), Liberec, Czech Republic, EU, mohamed.fawzy@mena.vt.edu

[2]Technical University of Liberec, Department of Glass producing Machines and Robotics, Faculty of Mechanical Engineering, Liberec, Czech Republic, EU

https://doi.org/10.37904/nanocon.2022.4582

Abstract

Open-source software, devices, and systems allow other researchers and investigators to replicate, develop and enhance the reported systems in the literature. This paper introduces an open-source system for optical fiber grinding and polishing, which facilitates and speeds up the polishing and grinding process. The polishing system is based on an Arduino microcontroller board, an open-source development board that utilizes an easy programming language enabling almost everyone to develop and write their codes and projects. In addition, the reported device targets low-cost fabrication allowing more researchers to fabricate their grinding-polishing machine. The device consists of the control circuits, stepper motor, breadboard, and a 3D-printed desk, where the polishing and grinding films should be attached. The device is controlled via 3 buttons: start/shut down button, speed up button and speed down button.

Keywords: Low-cost, open source, grinding and polishing, optical fibers, optical sensors

1. INTRODUCTION

A recent trend in engineering is to share projects, findings, and accomplishments under open-source access. Open-source software can be defined as free software with public access to its source code so anyone can view, modify or edit its content [1]. Low-cost prototyping is associated with several open-source projects in the literature. For example, a group of researchers reported a low-cost open-source system to monitor inertia and pressure using an Arduino prototyping platform [2]. Several research groups reported low-cost open-source 3D printer prototypes [3-5].

Optical fibers are waveguides that can carry light signals for long distances and minimize power loss compared to conventional copper cables. In addition, optical fibers are immune to signal distortion and interference [6]. Optical fibers are manufactured from glass or polymers, glass optical fibers, and polymer optical fibers (POFs). Besides their conventional use in optical communications, they have been used extensively in various sensing applications [7-9]. Hence, there is an increasing demand to grind and polish glass optical fibers and POFs. Commercial optical fiber polishing machines can cost several thousands of dollars [10,11], forming a financial challenge for the research teams in the sensory field.

This study aims to introduce an open-source polishing and grinding system for optical fibers which is easy to replicate and use. The reported system automates the grinding and polishing process using a 3D-printed motorized stage. The system speeds up the grinding and polishing processes compared to using grinding and polishing films manually, as manual grinding and polishing require more time and can be less accurate on the optical fiber surface.

This paper introduces an open-source, low-cost Arduino-based optical fiber polishing stage that can use different paper polishing films to grind and polish glass and plastic optical fibers. Three push buttons can control the stage (on/off button, speed up button, and speed down button) and a graphical user interface (GUI) windows application. The paper includes the control circuit designs, and all the source codes can be downloaded from GitHub, as provided later.

2. MATERIALS AND METHODS

A- The grinding-polishing stage

The grinding polishing stage has several parts: Power supply (LONGWEI K3010D, LONGWEI Electric, China), D.C. stepper motor NEMA 17, L298N motor driver, Arduino Mega, and 3 push buttons. **Figure 1** shows the circuit design drawn by Fritzing software [12].

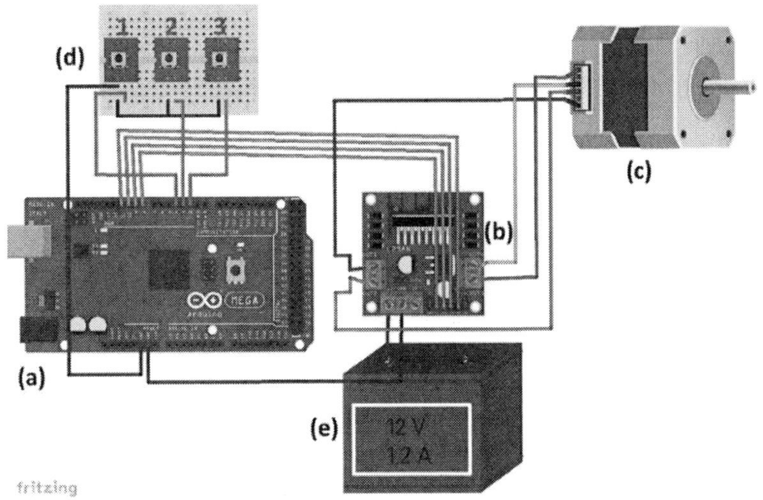

Figure 1 Grinding polishing stage electronic design: (a) Arduino Mega, (b) L298N motor driver, (c) NEMA17 stepper motor, (d) breadboard with control push buttons, and (e) power source

A 125 mm in diameter rounded desk was 3D designed using Autodesk Inventor Professional 2018 student's version [13] the designed desk is compatible with the stepper motor to fit into the motor's axe, Prusa i3 MK3S+ 3D printed was used to print the desk, the desk was printed using a 1.75 mm (Polylactic acid) PLA filament. **Figure 2** shows the desk's design. The desk's STL file is downloadable from the Thingiverse website [14].

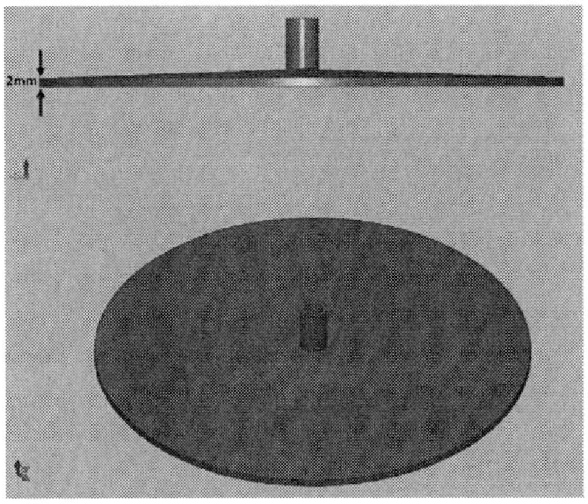

Figure 2 The polishing-grinding desk's 3D design

The 125 mm polishing and grinding films (OPTOKON [15] films: 3 µm and 5 µm diamond grinding films and Utlimas 5 final polishing films) were attached to the 3D printed desk using a double side tape.

The stage can be controlled via the 3 push buttons, as shown in **Figure 1d**. Button 1 is used to decrease the set starting default speed, button 2 is used to switch on/off the motor, and button 3 is used to increase the default speed. Setting the desired default speed is possible through editing the Arduino code, which can be downloaded from GitHub [16]. To set the motor's speed to a desired/default revolutions per minute (RPM) speed, the motor's step angle is applied in equation 1. NEMA 17 has a step angle of 1.8°, meaning it has 200 steps for each complete revolution.

$$Steps\ per\ revolution = \frac{360}{Step\ angle} \quad (1)$$

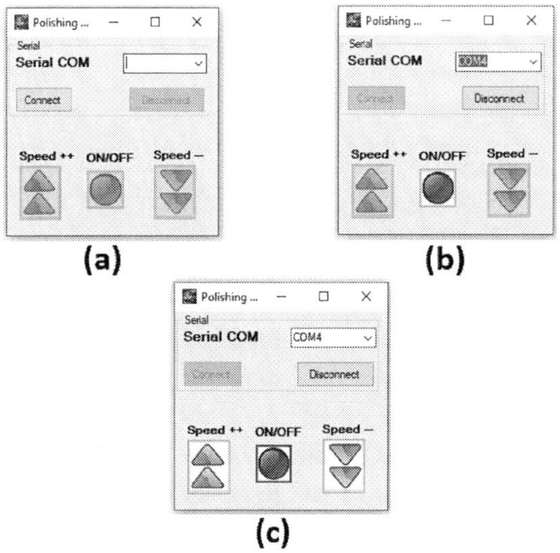

Figure 3 The control GUI software: (a) Software disconnected from the Arduino board, (b) software is connected to the Arduino board, (c) software is connected, and the motor is switched on

In addition to controlling the stage via push buttons, a C# graphical user interface Windows-based program can be used to control the stage; source code and the application setup files can be downloaded from GitHub [17]. **Figure 3** shows the GUI Windows-based software that controls the grinding-polishing stage through Arduino's USB cable.

Figure 4 shows the actual implementation of the grinding polishing stage according to the design in **Figure 1**.

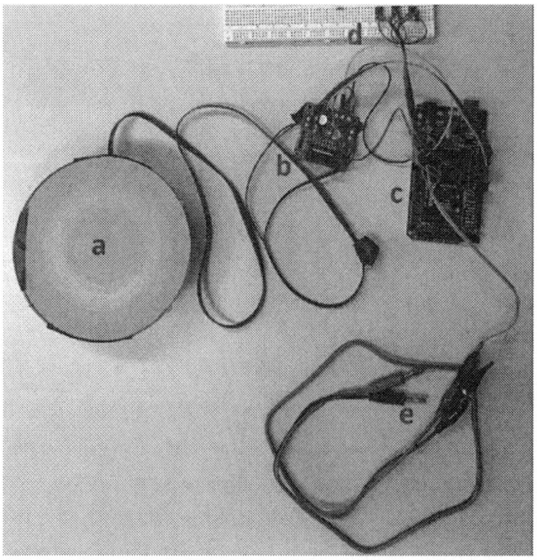

Figure 4 Implementation of the grinding-polishing stage: (a) Grinding polishing desk and motor, (b) motor's driver, (c) Arduino Mega, (d) Control push buttons, (e) Power source connections/battery connections

B- Plastic optical fibers (POF) and signal measurements

Plastic optical fibers have several applications in sensing and communications engineering. They have several advantages over glass optical fibers as plastic optical fibers have a lightweight, require simple and cheap components, are more flexible and resilient to bending, and don't require a laser source to operate. However, POF has a higher transmission loss than glass optical fiber [18].

Two POF (PMMA 0.2 mm POF, refractive index 1.49, numerical aperture 0.5) samples of a 65 mm length with and without grinding/polishing were compared in terms of the received signal to prove the grinding-polishing stage functionality. A simple setup was constructed to send, receive, convert and amplify the photodiode's received optical signal. **Figure 5** shows the used design and its components to measure the received signal from the POF in volts; the retrieved signal is measured by a voltmeter, as shown in the figure.

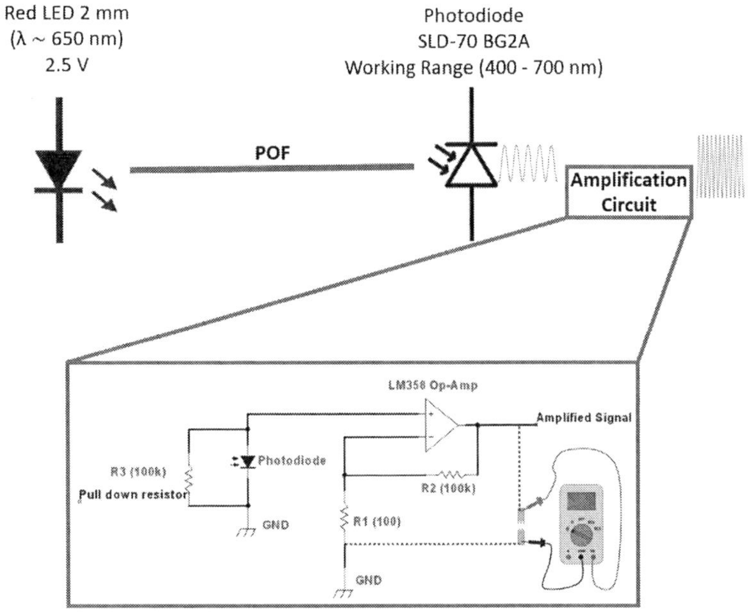

Figure 5 Signal conversion and amplification circuit

3. RESULTS AND DISCUSSION

The designed grinding and polishing stage successfully hosted commercial fiber optic films. POF sample of a 65 mm length was grinded and polished from each side for about 30 seconds at an average speed of 100 RPM; **Table 1** shows the obtained results comparing the polished and unpolished POF; the results were obtained using the setup in **Figure 4**.

Table 1 Received signal comparison of POF with and without using the stage

POF	Received Signal
Using the stage	0.78 V
Without polishing	0.24 V

The obtained results show the effectiveness of using the grinding-polishing stage as it saves processing time compared to manually polishing the fibers on the optical films.

4. CONCLUSIONS

The paper introduced an open-source grinding-polishing system for glass and plastic optical fibers; all designs and codes are available for download online, so other researchers can replicate, develop and enhance the proposed system. The system saves time and provides more accuracy and control on polishing speeds and consumed time. The results showed the system's effectiveness in improving the signal transmission over POF and decreasing the power loss due to imperfect non-polished fiber cross sections.

A promising future perspective would be incorporating more automation into the system in loading the fibers and approaching similar commercial systems but at a much lower budget to accomplish a broader concept of open-source devices.

ACKNOWLEDGEMENTS

This work was supported by the Ministry of Education, Youth and Sports in the Czech Republic under the "Inter Excellence-Action programme" within the framework of project "Micro-struCtural imaging as a Tool for modelinG fibrOus materiALS (μ-CT GOALS)" (registration number LTAUSA18135) and by the Ministry of Industry and Trade under the "OP PIK programme" within the framework of project "Hybrid optical fiber sensor for sensing the level and quality of vehicle operating fluids" (registration number CZ.01.1.02/0.0/0.0/20_321/0024437).

REFERENCES

[1] AMATYA, A. Review: Understanding Open Source Software Development. *The Computer Bulletin*. 2003, vol. 45, pp. 31-31.

[2] RUSEEL, L., STEELE, AL, GOUBRAN, R. Low-cost, rapid prototyping of IMU and pressure monitoring system using an open source hardware design. *IEEE I2MTC - International Instrumentation and Measurement Technology Conference, Proceedings*. 2012.

[3] ANZALONE, GC., ZHANG, C., WIJNEN, B., et al. A low-cost open-source metal 3-D printer. *IEEE Access*. 2013, vol. 1., pp. 803-810

[4] RENDON, MA., ANDRADE, L., TELICH, JE., et al. Dimensional error in rapid prototyping with open source software and low-cost 3D-printer. *Plast Reconstr Surg Glob Open*. 2018, vol. 6.

[5] PEARCE, JM., ANZALONE, NC., HELDT, CL. Open-Source Wax RepRap 3-D Printer for Rapid Prototyping Paper-Based Microfluidics. *SLAS technology*. 2016, vol. 21, pp. 510-516.

[6] GANG, B. *Handbook of Optical Fibers*. Springer, 2019.

[7] FANG, Z., CHINN, KK., QU, R., et al. *Fundamentals of Optical Fiber Sensors*. Wiley, 2012.

[8] PETRES, K. Polymer optical fiber sensors - A review. *Smart Materials and Structures*. 2011, vol. 20.

[9] MEMON, SF., WANG, R., STRUNZ, B., et al. A Review of Optical Fibre Ethanol Sensors: Current State and Future Prospects. *Sensors*. 2022, vol. 22, pp. 950-985.

[10] Fiber Optic Polishing Machine. *FS-20A*. [online]. [viewed: 2022-08-04]. Available from: https://www.fs.com/de-en/products/13325.html?currency=EUR

[11] TechOptics. *Domaille APM-HDC-5300 Polishing Machine*. [online]. [viewed: 2022-08-04]. Available from: https://www.techoptics.com/products/equipment/polishing-machines/polishing-machine-apm-hdc-5300/

[12] Fritzing, *Fritzing Software*, [online]. [viewed: 2022-08-09]. Available from: https://fritzing.org/

[13] Autodesk, *Inventor Autodesk*. [online]. [viewed: 2022-08-15]. Available from: https://www.autodesk.com/education/edu-software/overview?sorting=featured&filters=individual

[14] *Desk's STL design*. [online]. [viewed: 2022-09-02]. Available from: https://www.thingiverse.com/thing:5490618

[15] OPTOKON. [online]. [viewed: 2022-08-30]. Available from: https://www.optokon.com/

[16] Arduino code. *Grinding-polishing stage*. [online]. [viewed: 2022-09-01]. Available from: https://github.com/MohamedAbdelkader94/Grinding-Polishing-stage/blob/main/Stage_Code.ino

[17] *GUI Windows Application*. [online]. [viewed: 2022-09-01]. Available from: https://github.com/MohamedAbdelkader94/Grinding-Polishing-stage/blob/main/Control Windows Application.rar

[18] MARECELO, M., REGINA, A. *Plastic Optical Fiber Sensors*. CRC Press, 2019.

HYBRID FIBER-OPTIC/NANOFIBER SENSOR FOR BRAKE FLUID CONDITION MONITORING

Mayza IBRAHIM, Stanislav PETRÍK

Technical University of Liberec, Department of Advanced Materials, Institute for Nanomaterials, Advanced Technology and Innovation, Liberec, Czech Republic, EU, mayza.ibrahim@tul.cz, stanislav.petrik@tul.cz

https://doi.org/10.37904/nanocon.2022.4584

Abstract

Focus of this paper is to build and investigate an optical fiber sensor based on silica nanofibers prepared by a reliable and low cost electrospinning technique to detect water content in DOT-4 brake fluid. For the best of our knowledge, this is a novelty study of optical fiber sensor to detect water content in an aqueous substance using electrospinning nanofibers. The proposed sensor has a sensitivity of 5.9 nW/1% of water change.

Keywords: Optic fiber sensor, electrospinning, nanofibers, intensity modulation

1. INTRODUCTION

Brake fluid is very important to the brake system and safety of automobiles. Brake fluid is composed of polyethylene glycol which is hydroscopic. Hence it absorbs water from air. Since the brake fluid is not designed in a vacuum system within cars, the moisture easily finds its way to brake fluid. Moisture deteriorates the main function of any brake fluid. Moisture corrodes the main parts of the brake system. Moreover, it decreases the boiling point of brake fluid. Hence the brake fluid will vaporize faster before even reaching its original boiling point. Eventually, it leads to decrease or even failure of the brake capacity.

In automobile industry, oil engine sensors have attracted many researchers and engineers to develop many sensors. Concerning brake fluid condition monitoring, many researchers have focused on developing brake fluid testers. Nowadays at least three commercial brake fluid testers are available in which they mainly depend on physical methods. One of these methods depends on measuring directly the boiling point of the brake fluid, another depends on electrical methods (conductance or capacitance), and the third depends on optical method. These methods mainly depend on electric methods. Hence the brake fluid is usually checked manually with a certain time interval e.g., one or two years [1-3], whereas this checkup could be missed or not even a priority of a driver's vehicle maintenance routine. Hence it is utmost importance for on-line and safe method to monitor brake fluid condition [4-6].

A few studies have been reported for on-line monitoring [7-9]. However, they mainly depend on electric methods. It is not totally safe to depend on electricity when dealing with a flammable substance. Spark could be generated and cause fire which threaten people lives. Data from United states U.S. that more people die in car fires than in apartment fire each year where 1 out of 5 fires involve motor vehicles. U.S. fire department responded to an estimated 278,000 vehicles fire in U.S. during 2006. These fires caused an estimated 490 civilians death and 1,200 civilian injuries. 75 % of those fires were caused by bad maintenance, mechanical or electrical failures or malfunctions. Collision or over turns caused only 3 % of these fires [10]. Therefore, on-line safer method to monitoring brake fluid is utmost importance. Optic methods is safer, however many of optic sensors are still bulky, expensive, and complicated. Much research is going on to involve optic fibers in sensing applications.

Fiber optic sensor technology has been rapidly developed in the past 30 years due to the innovations in telecommunication, semiconductor and electronics sectors which have significantly reduced the prices of optical components and stimulated the development of optical fiber sensor [11]. Optical fiber sensors are capable of measuring a wide variety of physical properties, such as chemical changes, strain, electric and magnetic fields, pressure, temperature, displacement (position), radiation, flow, liquid level, vibrations, and light intensity. Optical fiber sensors exhibit a number of advantages over the conventional electrical and electronic sensors e.g., small sizes and weights, allow access to inaccessible places, permit remote sensing, and immune to radio frequency and electromagnetic interference. In order to improve the sensitivity of optical fiber sensors a layer of sensitive materials is coated on it.

Coating optical fiber surface with hygroscopic-sensitive material enhances the sensitivity. The refractive index of hygroscopic-sensitive material changes in accordance to the humidity or moisture level which change the power intensity of optical fiber. There are many methods for producing thin films for example vacuum evaporation, ion sputtering coating, sol-gel technology, layer-by-layer self-assembly (LBL), and the dip-coating method. Compared with these methods, electrospinning is a simple and effective method to form nanofibers layer.

Electrospinning manifests the capability of any material at the nanoscale especially concerning sensing materials. At nanoscale many features are easily accessible and even improved e.g., excellent mechanical properties, mainly flexibility, high porosity, and large surface area which increases the number of accessible sites for surface functionalization. Another important aspect from sensor point of view that electrospinning produces continuous nanofibers. This feature is important as the sensors are usually assembled in a certain measuring system which involves analog to digital conversion, therefore it should provide a continuous stable signal. Concerning about humidity sensor, the nanofibers layer processed by electrospinning has a larger specific surface area compared to conventional coating film, which can absorb a large number of water molecules [12,13]. Batool et al., 2013 [14] studied the effect of RH on dielectric response of SiO_2 nanofibers. However, it is rarely investigated the effect of RH on refractive index of SiO_2 nanofibers (NF).

In our study we developed a sensor based on hybrid nanofiber- optic fiber for detecting water content on brake fluid that could be later integrated into the vehicle's electronic system for on-line brake fluid monitoring

2. EXPERIMENTATION

Different concentration of DOT-4 (US Department of Transportation) [4-6] solutions were prepared by mixing pure DOT-4 with distilled water. Range of concentration is from 0 to 7 % V/V. The developed sensor

was used to determine the water content presented in brake fluid. The developed fiber optic sensor consists of fiber probe, PVP/SiO_2 composite NFs, Laser diode as a source of light, and Photodetector as a measuring device as pictorial on **Figure 1.**

The developed sensor based on reflective intensity-modulation [15-18]. PVP/SiO_2 composite NFs were produced according to the method descried in patent WO 2017/186201 [19]. The produced composite PVP/ SiO_2 composite NFs were thermal annealed at 300 °C for 6 h [20].

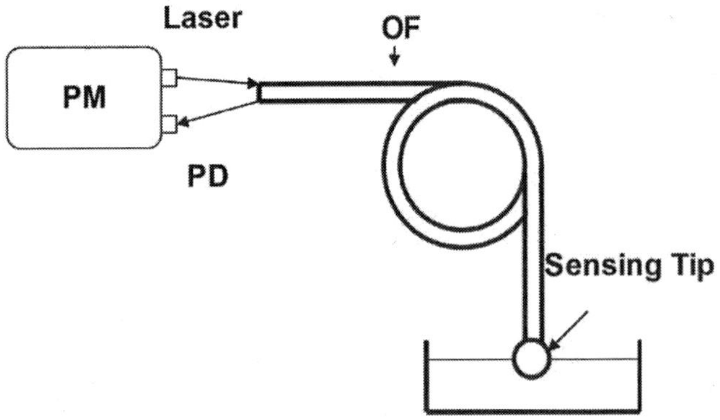

Figure 1 Experimental setup

3. CHARACTERIZATION OF THE COMPOSITE PVP/SIO_2 NFs

The obtained nanofibers composite were characterized by scanning electron microscopy (SEM). Fourier transform infrared microscopy (FTIR) was employed to study the composition of NFs before and after thermal treatment. Energy-dispersive X-ray spectroscopy (EDX) was also employed to investigate the percentage composition of fibers.

4. RESULTS AND DISSCUSSION

4.1. FTIR spectrum

Figure 2 shows the FTIR spectrum taken for the composite before and after thermal annealing. The peaks at 451 cm^{-1} and were assigned to Si-O vibration in TEOS [21]. A number of peaks between 3600 and 2400 long side with peaks at 1633 and 945 cm^{-1} were assigned to O-H vibrations. The peak at 945 cm^{-1} was assigned to Si-OH bond which is disappeared after thermal annealing along with OH bonds. The beaks at 1067 and 797 cm^{-1} were assigned to Si-O-Si bonds which became more intense after thermal annealing. The fibers after thermal annealing composed mainly of Si-O bonds which proved the formation of silica nanofibers.

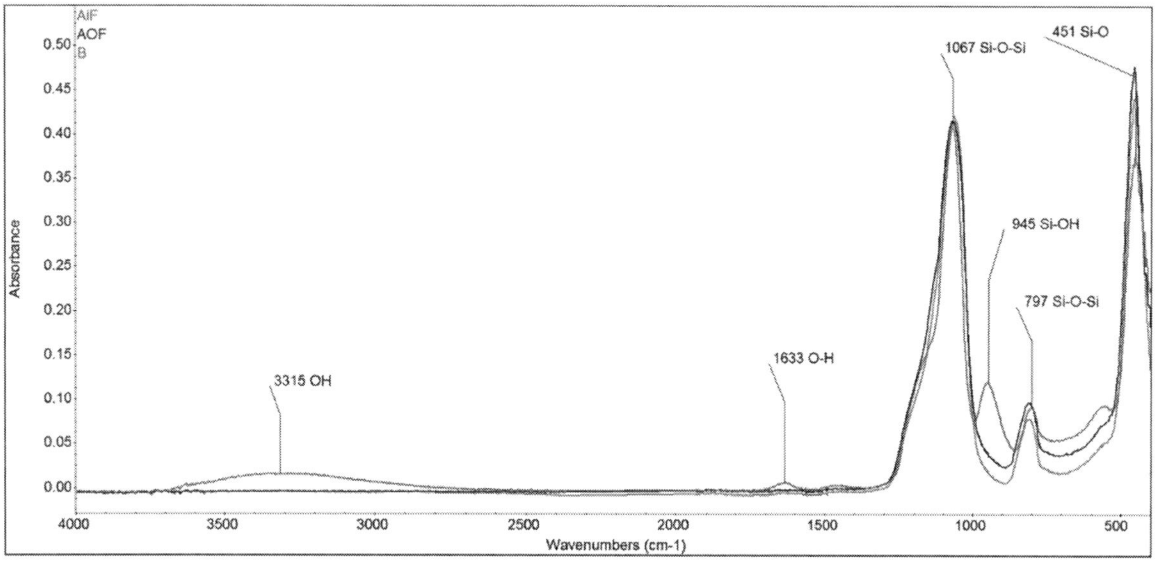

Figure 2 Fourier transform infrared microscopy spectrum taken for nanofibers before and after thermal annealing

4.2. Energy-dispersive X-ray spectroscopy characterization of fibers

Chemical composition of the composite before thermal annealing consisted of O with a percent of 63.7 %, Si of 33.1 % and C of 3.2 % as indicated in **Figure 3** and **Table 1**

Table 1 Chemical composition of the composite before thermal annealing

Element	Line Type	Wt%	Wt% Sigma	Atomic %
C	K series	1.96	0.16	3.25
O	K series	51.26	0.16	63.66
Si	K series	46.78	0.15	33.09
Total:		100.00		100.00

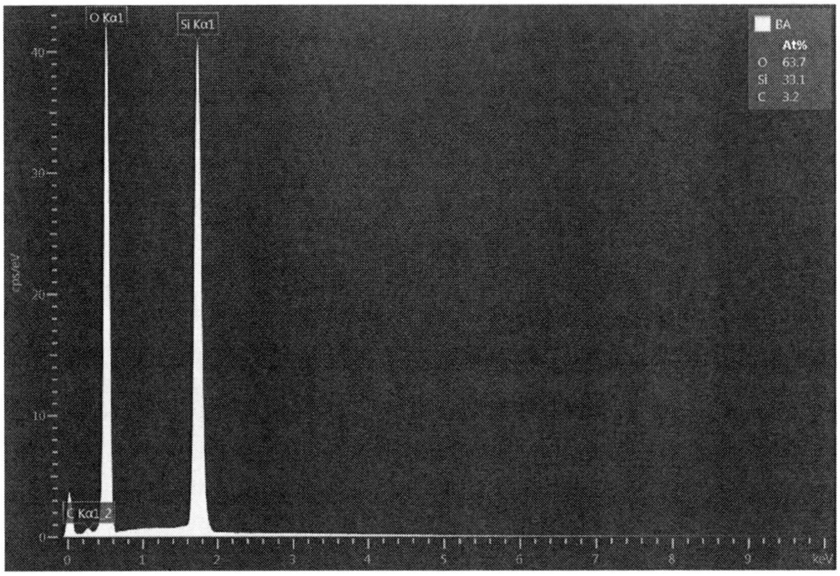

Figure 3 EDX spectrum of SiO_2/PVP Composite before thermal annealing

Chemical composition of the composite after thermal annealing consisted of O with a percent of 60.2 %, Si of 36.3 % and C of 3.4 % as indicated in **Figure 4** and **Table 2**

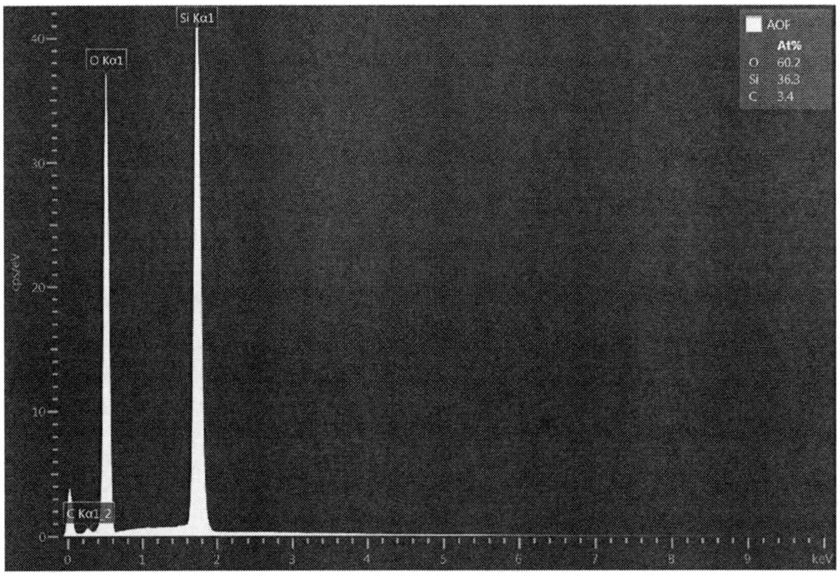

Figure 4 EDX spectrum of the composite NFs after thermal annealing

Table 2 Chemical composition of the composite NFs before thermal annealing

Element	Line Type	Wt%	Wt% Sigma	Atomic %
C	K series	2.02	0.16	2.41
O	K series	47.58	0.16	61.24
Si	K series	50.39	0.16	36.34
Total:		100.00		100.00

Since the thermal annealing was at 300 °C which didn't exceeded the thermal decomposition temperature of PVP, the percent of carbons didn't change. However, the percent of Si increased from 33 % to 36 % on the expense of O. This may be considered as a consequence of vaporization of moisture presented on NFs as FTIR confirms the disappear of hydroxyl group after thermal annealing. In general the chemical composition of SiO_2 NFs was optimally adjusted from the synthesizing process to keep the carbon percent as low as possible 3 % even before the thermal annealing. Hence silica structure was formed which mainly comprised of silicon and oxygen.

4.3. Results of proposed the hybrid-optic fiber sensor

As water content increased on DOT-4, the intensity of the reflected power decreased linearly with sensitivity of 5.9 nW/1% of water change as shown in **Figure 5**

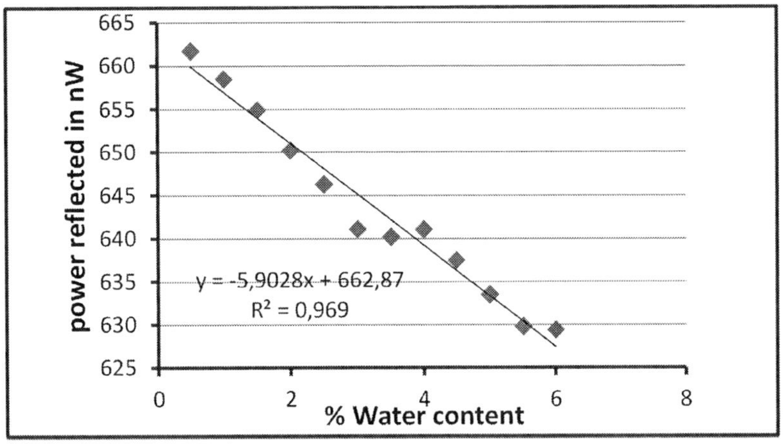

Figure 5 Intensity of reflected light as a function of water content of DOT-4

The power intensity of the reflected power changed linearly as a function of water content in DOT-4 as shown in **Figure 5**. The proposed sensor is based on reflected light intensity modulation. Based on Fresnel reflection, a proportion of lights are leaked out when the sensor is in the liquid rather than air. This amount of light depends on the refractive index of the liquid. For normal incidence, the reflectance simplifies to the following equation Eq. [1].

$$R = \left|\frac{n_1 - n_2}{n_1 + n_2}\right|^2 \quad [1]$$

- n_1, n_2 are the RIs of optical fiber and exterior medium; n_1 are known by the fabrication of the optical fiber; n_2 is variable according to the exterior medium

Since the exterior medium is Dot-4 with different water contents, those water molecules may be absorbed and concentrated in the pores of the silica nanofibers. This effect will alter the refractive index (RI) of the composite NFs; hence change the reflectance. As a result, the accumulation of the water molecules changed the refractive index of composite NFs. Hence, leads to the leakage of the light through evanescent field [22,23]

5. CONCLUSION

The proposed sensor depends on hybrid nanofiber-optic fiber which is simple, straightforward, small size and cost effective has a good repeatability and reasonable sensitivity. Moreover, it is safer to incorporate into car's electronic system for on-line monitoring of DOT condition. The sensitivity could be further enhanced by a special designed electronic circuit in order to enlarge the output signals and at the same time compatible with car's system.

ACKNOWLEDGEMENTS

This work was supported by the Ministry of Industry and Trade under the "OP PIK programme" within the framework of project "Hybrid optical fiber sensor for sensing the level and quality of vehicle operating fluids" (registration number CZ.01.1.02/0.0/0.0/20_321/0024437

REFERENCES

[1] J., KUNTNER, R., CHABICOVSKY, and B., JAKOBY. "Oil condition monitoring using a thermal conductivity sensor." In *Proc. GMe Forum (Vienna, USA)*, 2005, pp. 203-209.

[2] S. S., WANG. "Road tests of oil condition sensor and sensing technique." *Sensors and Actuators B: Chemical.* 2001, vol. 73, no. 2-3, pp. 106-111.

[3] B. JAKOBY, M. BUSKIES, M. SCHERER, S. HENZLER, H. EISENSCHMID, and O. SCHATZ. "A novel multifunctional oil condition sensor." in *Advanced Microsystems for Automotive Applications 2001. Springer.* 2001, pp. 157-165.

[4] C. WANG and K. SHIDA. "A new method for on-line monitoring of brake fluid condition using an enclosed reference probe." *Measurement Science and Technology.* 2007, vol. 18, no. 11, p. 3625.

[5] R. W. RADLINSKI, R. J. FORTHOFER, and J. L. HARVEY. "Operating performance of motor vehicle braking systems as affected by fluid water content." *SAE Transactions.* 1971, pp. 1037-1045.

[6] K.-H. DIETSCHE and M. KLINGEBIEL. "Bosch Automotive Handbook." *7th Editio, Robert Bosch GmbH.* 2007.

[7] E. IRION, K. LAND, T. GÜRTLER, and M. KLEIN. "Oil-Quality Prediction and Oil-Level Detection with the TEMIC QLT-Sensor Leads to Variable Maintenance Intervals." *SAE International, Warrendale, PA.* SAE Technical Paper 970847, Feb. 1997. Available from: https://doi.org/10.4271/970847.

[8] M. J. KAO, D. C. TIEN, C. S. JWO, and T. T. TSUNG. "The study of hydrophilic characteristics of ethylene glycol," *J. Phys.: Conf. Ser.* 2005, vol. 13, pp. 442-445. Available from: https://doi.org/10.1088/1742-6596/13/1/102.

[9] T. E. RYAN and T. HINZ. "Analysis of Water Content in Brake Fluid. Part I. Method Comparison: Karl Fisher Titration Versus Refractive Index." *SAE Transactions.* 1997, vol. 106, pp. 3399-3404.

[10] "Escape and Safety from Vehicle Fire." *Arrive Alive.* https://www.arrivealive.mobi/Escape-and-Safety-from-Vehicle-Fire (accessed Sep. 06, 2022).

[11] M. YASIN, H. AROF, and S. W. HARUN. *Fiber optic sensors.* BoD-Books on Demand, 2012.

[12] M. D. MORALES-ACOSTA, M. A. QUEVEDO-LOPEZ, H. N. ALSHAREEF, B. E. GNADE, and R. RAMÍREZ-BON. "Dielectric properties of PMMA-SiO2 hybrid films." In *Materials Science Forum*, 2010, vol. 644, pp. 25-28.

[13] T. WAGNER *et al.* "A high temperature capacitive humidity sensor based on mesoporous silica." *Sensors.* 2011, vol. 11, no. 3, pp. 3135-3144.

[14] S. S. BATOOL *et al.* "Silica nanofibers based impedance type humidity detector prepared on glass substrate." *Vacuum.* 2013, vol. 87, pp. 1-6.

[15] J. ZHENG and S. ALBIN. "Self-referenced reflective intensity modulated fiber optic displacement sensor." *Optical engineering.* 1999, vol. 38, no. 2, pp. 227-232.

[16] W. YAO *et al.* "Intensity Modulated Fiber Optic Sensor: A Novel Grid Measurement Unit." *IEEE Transactions on Industrial Informatics.* 2022.

[17] P. RORIZ, A. RAMOS, J. L. SANTOS, and J. A. SIMÕES. "Fiber optic intensity-modulated sensors: A review in biomechanics." *Photonic Sensors.* 2012, vol. 2, no. 4, pp. 315-330.

[18] D. A. KROHN. "Intensity modulated fiber optic sensors overview." In *Fiber Optic and Laser Sensors IV.* 1987, vol. 718, pp. 2-11.

[19] J. RUZICKOVA, M. TEJKL, A. KUDROVA, Z. VOSATKA, and J. BUK, "Precursor Fibers Intended for Preparation of Silica Fibers, Method of Manufacture Thereof, Method of Modification Thereof, Use of Silica Fibers." 2017.

[20] S. S. BATOOL *et al.* "Silica nanofibers based impedance type humidity detector prepared on glass substrate." *Vacuum.* 2013, vol. 87, pp. 1-6.

[21] Y. ZHAO, H. WANG, X. LU, X. LI, Y. YANG, and C. WANG. "Fabrication of refining mesoporous silica nanofibers via electrospinning." *Materials Letters.* 2008, vol. 62, no. 1, pp. 143-146.

[22] T. YUAN *et al.* "Humidity sensor based on micro optical fiber array fabricated by electrospinning." *Optics Communications.* 2018, vol. 427, pp. 517-521.

[23] G. LIU *et al.* "Low-loss prism-waveguide optical coupling for ultrahigh-Q low-index monolithic resonators." *Optica.* 2018, vol. 5, no. 2, pp. 219-226.

ELECTRICAL CHARACTERIZATION OF HYDROTHERMALLY SYNTHESIZED METAL OXIDE NANOWIRES WITH REGARD TO OXYGEN ADSORPTION/DESORPTION THERMODYNAMICS

[1]Petr SMISITEL, [2]Helena SIMUNKOVA, [1,2]Ondrej CHMELA, [3]Martha CLAROS, [1,4]Stella VALLEJOS, [1,2]Jaromir HUBALEK

[1]Central European Institute of Technology (CEITEC), Brno University of Technology, Czech Republic, EU, smisitel@vutbr.cz

[2]Department of Microelectronics, Faculty of Electrical Engineering and Communication (FEEC), Brno University of Technology, Czech Republic, EU, simunkova@vutbr.cz

[3]Escuela de Ingeniería Química, Pontificia Universidad Católica de Valparaíso, Valparaíso, Chile

[4]Instituto de Microelectrónica de Barcelona, IMB-CNM (CSIC), Spain, EU

https://doi.org/10.37904/nanocon.2022.4583

Abstract

Self-assembled metal oxide MnO_2 nanowires (NWs) were hydrothermally synthesized and electrically characterized. The nanowires were from 3 to 10 µm long and from 20 to 100 nm in diameter. The nanowires were suspended in water and deposited on gold interdigitated electrode (IDE) chip using dielectrophoresis (DEP) to align them perpendicularly across the electrodes. Screening of the MnO_2 material properties, such as semiconductor type, vacancies concentration and relative permittivity was performed by means of impedance and Mott-Schottky analyses. The conductivity was measured both in synthetic air and in nitrogen ambient. The tests consisted in measuring resistivity of the NWs in relation to temperature of the bottom-placed heater under the IDE chip. The temperature went from room temperature up to 300 °C. The resistivity changes were observed accounting for oxygen reduction on the NWs surface as the electrons were moving from the NWs to the oxygen. The resistivity was explored at a constant current arrangement test. Overall, we observed changes in the electrical properties of the wires upon oxygen adsorption, such as activation energy. Therefore, in future experiments, we might also explore the effect of other gases on the NW's electrical properties, e.g., ethanol, H_2, NO_2.

Keywords: MnO_2 nanowires, dielectrophoresis, gas sensing, resistivity, Mott-Schottky analysis

1. INTRODUCTION

Single NW devices seems to be a promising way to improve sensitivity and selectivity of metal-oxide (MOX) gas sensors thanks to high surface to volume ratio. MOX gas sensors rely on reactions of detected gas with oxygen adsorption layer formed on the NW surface in the air. As the reactions are often thermally activated the MOX sensors are typically heated by external heater causing relatively high power consumption. A future motivation of our study is a precise temperature control including a self-heating effect. Moreover, good understanding of oxygen adsorption layer formation and its influence on NW electrical properties is also highly desirable. In the air, oxygen species (O_2^-, O^-, O^{2-}) get adsorbed on the nanowire surface influencing the NW electrical properties. Oxygen acts as an oxidizing agent. By trapping electrons from the NW, the oxygen negative surface charge leads to upwards band bending and formation of electron lateral depletion layer (around the NW surface) in n-type MOX and hole accumulation layer in p-type MOX. The result is the formation of core-shell configuration: e.g., in p-type NW the core has the highest resistance while the main conducting channel is the shell formed by hole accumulation layer with increased free charge carrier concentration [1]. The oxygen adsorption might be disturbed by air pollutants. Therefore, synthetic air (SA) with constant composition is applied to eliminate a possible interference with air pollutants, as well as to maintain a stable humidity and create the same conditions as for measurements in nitrogen (namely the same constant flow of gas in measurement chamber). The second set of measurements was done using nitrogen as inert atmosphere. The purpose is to eliminate the oxygen adsorption providing so a reference measurement with no electron lateral depletion/hole accumulation layer allowing the evaluation of oxygen adsorption influence on NWs electrical properties. Additionally, impedance and Mott-Schottky analyses were performed to corroborate a type and vacancies concentration of the semiconductor. By polarizing the IDE / MnO_2 (metal/semiconductor) interface, a potential barrier either emerges or an existing one might narrow or broaden because of a direct potential V applied for a certain time. Considering a p-type semiconductor, a negative space charge layer arises on the side of positive potential due to a depletion (repulsion) of the positively charged carriers. The depletion layer (frontal contact depletion = on the interface with IDEs, not the same as lateral depletion) or let's say the space charge layer might be determined by the linear relationship between the inverse second power of the space charge capacitance C_{SC}^{-2} and the applied bias V, which is expressed by the Mott-Schottky equation [2]:

$$C_{SC}^{-2} = \frac{2}{\varepsilon_r \varepsilon_0 e N_A}\left(V_{FB} - V - \frac{kT}{e}\right) \quad (1)$$

where e is the electron charge, N_A is the acceptor density in the semiconductor (cm^{-3}), V is the applied voltage, and V_{FB} is the flat-band voltage. The Mott-Schottky equation is a valid tool to determine the type of a semiconductor and the acceptor density, as well as the flat-band voltage. The acceptor or donor density can be calculated from the slope of the C_{SC}^{-2} vs. V curve, and the flat-band voltage V_{FB} can be determined by extrapolation to $C_{SC} = 0$. For a p-type semiconductor response C_{SC}^{-2} vs. V should be linear with a negative slope that is inversely proportional to the acceptor concentration N_A in the film. For an n-type semiconductor the slope should be positive. Relative permittivity ε_r of the hydrothermally prepared nanowires might be estimated by knowing a frontal contact depletion layer thickness, as well as the space charge capacitance at a certain direct potential applied. Mott-Schottky analysis can be applied to a system of fixed vacancies (dopants) concentration [2]. Therefore, the C_{SC} versus V was measured at a high rate scanning of about 600 mV/sec, in order to avoid a change in the dimension or vacancy profile of the film (i.e. N_A cannot be a function of V). Moreover, no dependence on frequency is expected from Mott-Schottky theory [2], therefore all C/V analyses were carried out at high frequencies 10 kHz - 1 MHz.

2. EXPERIMENTAL

2.1. Dielectrophoretic preparation of aligned MnO_2 nanowires

A drop of MnO_2 NWs diluted water solution was placed on the ID electrodes connected to alternating voltage (sinusoidal with 5 V_{PP}, 9 MHz) for approx. 3 minutes. Afterwards, the drop was removed by compressed air. The nanowires long enough to connect the electrodes were captured while the majority of the shorter NW were removed with the rest of the drop leaving clean electrodes with 24 single-nanowire connections. Further pre-treatment was done directly in a flowing gas chamber. The sample was connected to constant 0.1 µA current flow and slowly heated up to 275 °C, where it was held for 15 minutes (until the resistance was stabilized). The result was approx. one order drop in resistance thanks to a strengthening of the NWs to the electrode contact.

2.2. Resistivity vs. temperature analysis

Resistivity in dependence on temperature was measured by using a custom build setup. Sample was placed on ceramic heater located in an enclosed chamber providing a slow constant flow of gas - either synthetic air (SA) or nitrogen was used. Direct measurements of NWs temperature were not possible, the readout from the bottom-placed heater was used. Sample was connected to the source of constant current (KEITHLEY 2401 Source Measure Unit) and temperature was driven from room temperature (25 °C) up to 300 °C and back to 25 °C again. The temperature was recorded in 10 degrees increments every 5 minutes. A stabilization time for reaching an adsorption/desorption equilibrium per temperature was hence 5 minutes. Both heating and cooling proceeded at the same rate. Before experiments in nitrogen, the setup equipment had to be pure of oxygen. Even after flushing the apparatus with nitrogen, oxygen keeps desorbing from inner walls of measuring chamber, tubing, sample itself, etc. Therefore at least 24 hours flushing is necessary to reach good quality oxygen free atmosphere. As a direct control of NWs temperature was not available, we needed to eliminate the self-heating effect that would cause difference between real NW temperature and value measured on the heater. Therefore, low probing current 0.1 μA was chosen, meaning the Joule heating power was lower than 1 nW per nanowire.

2.3. C/V, Mott-Schottky analysis

The impedance spectrum analyses were carried out dry (without electrolyte), performed using μAutolabIII potentiostate/FR analyzer (Metrohm). In all cases, a 10 mV sinusoidal excitation signal was employed to interrogate the capacitance. *C/V* behavior was determined in a direct bias range +/- 5V. Diverse formation times were tested: 60, 120 and 360 s, whereas the formation potential was always +5 V. Consequently, Mott-Schottky plots $1/C_{SC}^2$ vs. V were performed. The analyses were carried out in a frequency range of 1 MHz-10 kHz. Moreover, a broad range (1 MHz - 0.1 Hz) impedance analysis was performed as well. According to an impedance data simulation, the frontal contact depletion layer capacitance shows a CPE-like distribution behavior rather than that of a pure capacitor. The frequency dispersion is generally attributed to a capacitance dispersion (frequency-dependent capacitance) expressed in terms of a constant-phase element (CPE) [3]. The impedance of CPE is defined as follows [4]:

$$Z_{CPE} = \frac{1}{Q(j\omega)^{\alpha}} \qquad (2)$$

where, Q (in F cm^{-2} sα^{-1}) and α are the CPE parameters. α equals the slope of the log-log imaginary impedance plot vs frequency. Both parameters, as well as R from the equation below, were obtained by regression of a simple equivalent (R, Q) circuit implemented by an impedance analysis software (a graphical method). The so called effective equivalent capacitance C_{eff} value was then extracted from the parameters α, Q and R using the Hsu and Mansfeld formula [5]:

$$C_{eff} = Q^{1/\alpha} R^{(1-\alpha)/\alpha} \qquad (3)$$

For better clarity: $C = C_{eff} = C_{SC}$ in our study.

3. RESULTS

3.1. Synthesis and characterization of the MnO_2 nanowires

The nanowires were obtained by the hydrothermal synthesis method presented elsewhere [6]. The product of the hydrothermal synthesis is a brown powder with high density of nanowires as shown in SEM image (**Figure 1**(a)), in which several well-defined nanowires with smooth and uniform surface are observed. The synthesis procedure was repeatable usually obtaining the same yield (~ 55 %) and nanostructure morphology. The structures are crystalline, and their diameter and length vary from ~20 nm to ~100 nm and from ~3 to ~10 μm, respectively. The crystallinity was corroborated elsewhere [6] by a high-resolution TEM image and a XRD diffraction pattern recorded for the synthesized powder. The presence of intense diffraction peaks in the XRD pattern indicates that the obtained product is well crystallized.

3.2. Dielectrophoretic preparation of aligned MnO_2 nanowires

Separated and alone-standing MnO_2 nanowires were aligned perpendicularly to the ID gold electrodes by means of dielectrophoretic technique, see **Figure 1**(b) (described in the experimental part). After the preparation the nanowires were observed by scanning electron microscopy (SEM), see **Figure 1**(c). In total, there were 24 nanowires standing alone and connecting the two parallel ID electrodes all over the sample.

The nanowires surroundings were clean from impurities, such as nanowires flocks, or short nanowires crossing the long ones. The measurement area was therefore well defined, given by the length and diameter of the 24 single nanowires.

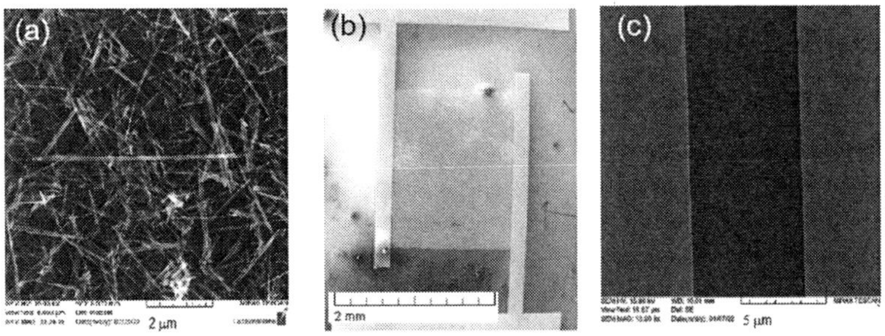

Figure 1 Scanning electron micrographs of: (a) MnO_2 nanowires, (b) gold interdigitated electrodes IDEs, (c) single nanowire aligned between two ID electrodes.

3.3. Resistivity thermal analysis

The resistance was measured in temperature range from room temperature to 300 °C and the influence of oxygen adsorption was evaluated.

Basic equation derived from concentration of thermally activated free charge carriers

$$\rho = \rho_0 \, e^{E_a/kT} \tag{4}$$

describes the thermal resistivity dependence in semiconductors, where ρ_0 is a material constant, E_a activation energy and k is Boltzman constant. This model is too simple to fully describe our measurement as there are other efect involved (oxygen adsorption, self-heating effect, Shottky contacts), however this simple equation well fits the data above aprox. 120 °C (the operation range of metal oxide gas sensors), see **Figure 2**(a). This well-predictabel behaviour is benefitial for potential sensing applications. Keeping in mind that MnO_2 is a p-type semiconductor, the adsorbed oxygen creates surface acceptor states resulting in lowering the MnO_2 activation energy (see **Table** 1), as well as overall increase of hole concentration under the surface (formation of hole acumulation layer in the outer shell-part of the MnO_2 NWs). Comparing the two experimental setups - concerning the synthetic air and the nitrogen ambients - we might conclude that the resistivity decreases in synthetic air compared to nitrogen, **Figure 2**(a). The overall result is decrease of the resistivity caused by the oxygen adsorption, which is typical for p-type MOX.

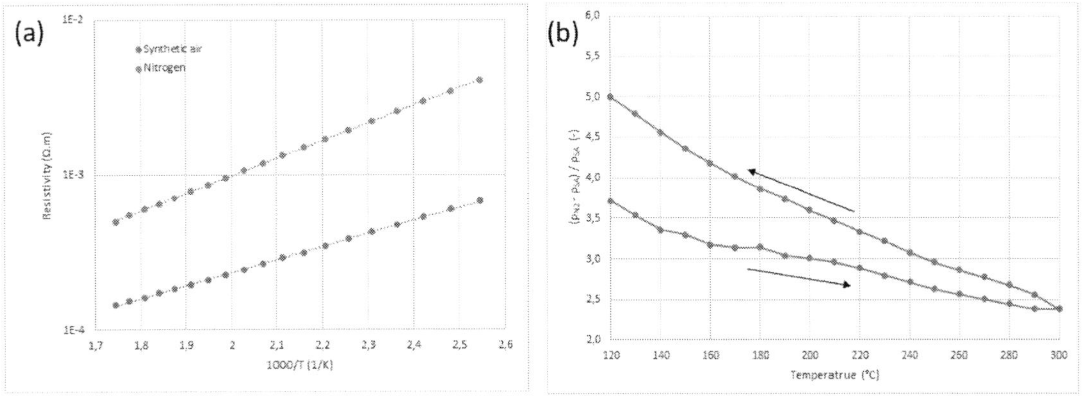

Figure 2 (a) The Arrhenius plot of resistivity vs. temperature range 120 - 300 °C fitted to equation (4), (b) relative resistivity change caused by removal of adsorbed oxygen both during heating and cooling.

Table 1 Parameters obtained by fitting the equation (4) to resistivity data in range 120 - 300 °C

	0.1 μA, SA	0.1 μA, N_2
ρ_0 (10^{-6} Ω.m)	4.9	11.1
E_a (eV)	0.17	0.23

The effect of oxygen adsorption on NWs resistance and its temperature dependence can be better estimated form relative resistance change between ρ_{N2} and ρ_{SA}, see **Figure 2**(b). The relative resistivity difference is decreasing with temperature, which can be explained by lower equilibrium concentration of adsorbents at higher temperatures [7]. Measurements were done both during heating and cooling. A significant hysteresis can be seen, with higher resistivity difference (corresponding to higher concentration of adsorbed oxygen) during the cooling compared to heating meaning that 5 minutes stabilization time was not sufficient to reach the equilibrium adsorbents concentration.

3.4. C/V, Mott-Schottky analysis

The relationship of the reciprocal capacitance square vs bias (Mott-Schottky analysis) of the MnO_2 NWs was employed to estimate a frontal contact depletion layer capacitance which had arisen during a formation by a direct potential V. Moreover, by this method we had approved that the MnO_2 NWs are a p-type semiconductor. Additionally, acceptors (probably Mn metal vacancies) concentration, depletion layer thickness and permittivity were estimated as well. To be reproducible with the C/V characterization, we had searched for a formation conditions (potential, time) of our nanowires. At a formation potential a rise of a quasi-stable depletion layer thickness assuming a rectangular distribution with a quasi-constant charge would be expected. At first, we applied a constant bias of +5 V for 60 seconds, which was further determined as insufficient. The positive charge carrier's depletion on the interface with the positively charged contact was not completed yet. Therefore, we prolonged the formation time up to 6 minutes. During the formation, the current passing through the structure did not change significantly, being about $7 \cdot 10^{-7}$ A, which is about 30 nA passing through one nanowire (24 NWs in total). After 360 seconds, we had obtained a depletion layer stable enough, which was approved by the C/V analysis, see **Figure 3**. The last formation step resulted in an almost constant shape of the Mott-Schottky plot in the bias range of +5 V to -1 V. The space charge acts as a pure capacitor at that moment and conditions. Increasing the potential towards negative values, a depletion layer thickness would narrow, and the space charge capacitance would increase. The linear part of the Mott-Schottky plot accounts for a semiconductive behaviors. A descend of the $1/C^2$ vs. V plot was observed in the negative bias range from -1 V to -5 V. The negative slope accounts for a p-type semiconductor and it is inversely proportional to the acceptor concentration N_A in the film.

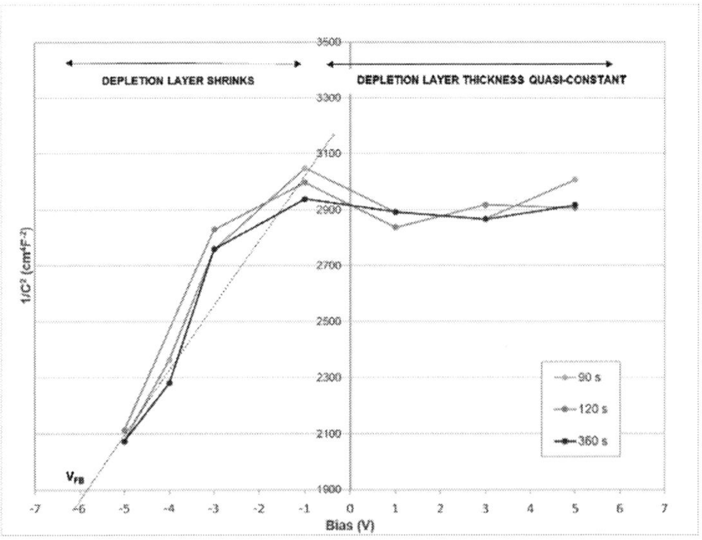

Figure 3 Mott-Schottky plots at various stabilization times (90, 120 and 360 s). Measured at high frequency range 10 kHz - 1 MHz. Capacitance obtained *via* a graphical simulation of the impedance data.

The acceptor concentration N_A (metal vacancies) might be calculated from the Mott-Schottky plot slope according to the equation (1). For that we need to know the MnO_2 relative permittivity. The relative permittivity was determined from the depletion layer capacitance, as well as from depletion layer thickness according to following equation:

$$\varepsilon_r = \frac{d}{A} \cdot \frac{C}{\varepsilon_0} \quad (5)$$

where, d is a nanowire diameter, A is a nanowire area, C is depletion layer capacitance and $\mathcal{E}_0$ is vacuum permittivity. The depletion layer thickness was estimated by SEM analysis after finishing the *C/V* experiments. Most probably, due to the potential formation trials, the nanowires got burned at the end of the *C/V* experiments. The current after burning the NWs off descended down the 10^{-12} A and the potential across them increased. Moreover, the SEM images showed that the nanowires burned near to the IDE electrode surface, see **Figure 4**. The nanowire might burn just at the end of the depletion layer. Considering the entire nanowire length, the place where the negatively charged depletion layer ends has a highest resistivity towards minority charge carriers (electrons) compared to the rest nanowire volume. The burning spot accounts for a place with the highest resistivity. The distance between the spot and IDE was about 730 nm. This length might be a thickness of the depletion layer. Consequently, the relative permittivity was calculated from equation (5) leading to about $1.6 \cdot 10^7$. The acceptor concentration N_A (metal vacancies) was evaluated in the range of $1.8 \cdot 10^{22}$ - $3.9 \cdot 10^{22}$ cm^{-3}. The higher the slope, the lower the concentration. The highest acceptor concentration was evaluated after the longest time of stabilization (360 s). To get a preciser acceptor concentration a more detailed analysis would be needed. Capacitances were determined from the high frequency range analyses 1 MHz-10 kHz (140 pF, i.e., 117 F/g). The capacitances were in good conjunction with other studies [8,9].

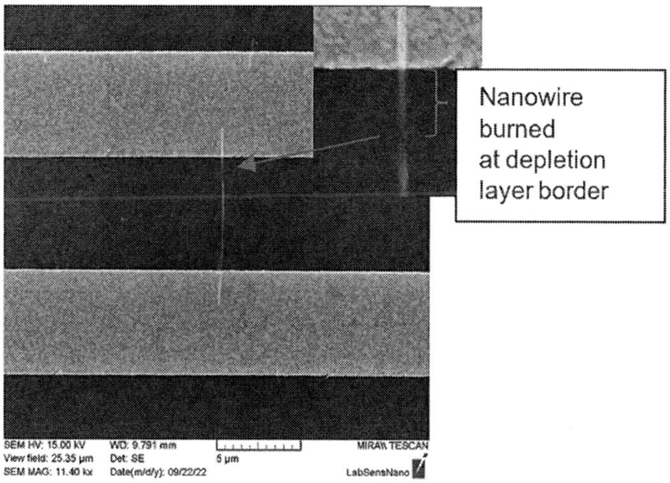

Figure 4 SEM micrograph and a detail of single MnO_2 nanowire after burning off.

4. CONCLUSION

In general, the study provides an overall electrical characterization of the hydrothermally synthetized MnO_2 nanowires. We approved that the MnO_2 nanowires are a p-type semiconductor, whereas permittivity and vacancies concentration were estimated. Resistivity of the NWs was tested in a reference synthetic air system in comparison to nitrogen ambient. The resistivity difference of the NWs in the two ambiences caused by oxygen adsorption was evaluated leading to its changes in relation to temperature variations. The knowledge gained in this study will serve as a reference platform for future gas sensing experiments.

ACKNOWLEDGEMENTS

The work was done under the Czech Grant Agency project no. 22-14886S and project no. CZ.02.2.69 / 0.0 / 0.0 / 19_073 / 0016948 under Quality Internal Grants of BUT.

REFERENCES

[1] KIM, H.-J., LEE, J.-H. Highly sensitive and selective gas sensors using p-type oxide semiconductors: Overview. *Sensors and Actuators B: Chemical.* 2014, vol. 192, pp. 607-627.

[2] MORRISON, S. *Electrochemistry at Semiconductor and Oxidized Metal Electrodes. 2. print.* New York: Plenum Press, 1984.

[3] JORCIN, J.-B., ORAZEM, M.E., PEBERE, N., TRIBOLLET, B. CPE analysis CPE analysis by local electrochemical impedance spectroscopy. *Electrochimica Acta.* 2006, vol, 51, pp. 1473-1479.

[4] LVOVICH, V.F. *Impedance Spectroscopy.* New Jersey: John Wiley & Sons, Inc., 2012.

[5] HIRSCHORN, M.E., ORAZEM, M.E., TRIBOLLET, B., VIVIER, V., FRATEUR, I., MUSIANI, M. Determination of effective capacitance and film thickness from constant-phase-element parameters. *Electrochimica Acta.* 2010, vol. 55, pp. 6218-6227.

[6] CLAROS, M., KUTA, J., EL-DAHSHAN, O. MICHALIČKA, J., JIMENEZ, Y.P., VALLEJOS, S. Hydrothermally synthesized MnO2 nanowires and their application in Lead (II) and Copper (II) batch adsorption. *Journal of Molecular Liquids.* 2021, vol. 325, pp. 115203.

[7] IWAMOTO, M., YODA, Y., YAMAZOE, N., SEIYAMA, T. Study of metal oxide catalysts by temperature programmed desorption. 4. Oxygen adsorption on various metal oxides. *The Journal of Physical Chemistry.* 1978, vol. 82, pp. 2564-2570.

[8] ZHU, J, HE, J. Facile Synthesis of Graphene-Wrapped Honeycomb MnO2 Nanospheres and Their Application in Supercapacitors. *ACS Applied Materials & Interfaces.* 2012, vol. 4, pp. 1770-1776.

[9] KIM, M., HWANG, Y., KIM, J. Super-capacitive performance depending on different crystal structures of MnO_2 in graphene/MnO_2 composites for supercapacitors. *Journal of Material Science.* 2013, vol. 48, pp. 7652-7663.

EFFECT OF TEMPERATURE ON SINUSOIDAL PULSE ANODIZATION OF ALUMINUM AND THE OPTICAL PROPERTIES OF THE RESULTED POROUS GRADIENT-INDEX FILTERS

[1]Blanka JANUSZEWSKA, [2]Maksymilian WŁODARSKI, [1]Małgorzata NOREK

[1]*Military University of Technology, Institute of Materials Science and Engineering, Faculty of Advanced Technologies and Chemistry, Warsaw, Poland, EU,*
blanka.januszewska@student.wat.edu.pl, malgorzata.norek@wat.edu.pl

[2]*Military University of Technology, Institute of Optoelectronics, Warsaw, Poland, EU,*
maksymilian.wlodarski@wat.edu.pl

https://doi.org/10.37904/nanocon.2022.4589

Abstract

Pulse anodization of aluminum is a promising method to produce porous photonic crystals (PCs) with tunable photonic stopbands (PSBs) by a fully programmable change of electrochemical conditions. The spectral position of the PSBs is mainly determined by thickness and porosity of alternate layers forming the periodic structure. Temperature is one of the most important factors influencing the growth of porous anodic alumina (PAA). It alters the rate of PAA formation and the porosity of PAA layers. In this work, the effect of the anodizing temperature on the characteristic PSBs of graded-index PCs produced by sinusoidal pulse anodization (SPA) is analyzed. Two different anodization periods (t_p) of the sinusoidal function were used whereas the amplitude and the number of cycles were kept constant. It was revealed that upon increasing the temperature from 5 to 30 °C the ion transport from electrolyte reservoir to the pore base becomes progressively hindered by increasing total PAA thickness which was manifested in the delayed current recovery to the initial values. As an effect, the periodic structures with inhomogeneous layers' thickness were formed providing broad and irregular PSBs in optical spectra. The quality of PSBs can be, however, significantly improved by post-processing pore broadening. Well-resolved and intensive peaks, which shift towards red part of the spectrum with increasing temperature and t_p, could be then obtained. After a proper modification of the pulse sequence (e.g. application of the current density-controlled mode) the PCs can be used in optical sensing platforms operating from UV to NIR spectral range.

Keywords: Anodization, porous anodic alumina, anodizing temperature, gradient-index filters, optical properties

1. INTRODUCTION

Porous anodic alumina (PAA) is a well-known template used in various fields of nanotechnology owing to fully controllable geometrical parameters, such as pore diameter or interpore distance, achieved by a simple change of anodization conditions (e.g. applied voltage, type of electrolyte, anodizing temperature) [1,2]. Upon application of pulse anodization the porosity of the material can be periodically modulated forming thus 1D photonic structure that prohibits light of a certain wavelength to propagate through the material. As a consequence of constructive interference of the light reflected from each boundary between low and high porosity layers, photonic stop band (PSB) is created which is manifested as a peak in a reflectance spectrum or a dip in a transmittance spectrum [3,4]. Various pulse anodization modalities were used to precisely modulate pore geometry in PAA in order to obtain high quality optical filters with PSBs in a given part of the spectrum [5-8]. A sinusoidal pulse anodization (SPA) was performed previously in sulfuric acid solution [9,10]. The effect of anodization amplitude, offset, period, time, and the acid concentration of a sinusoidal function on the quality of characteristic PSBs was systematically studied. It was shown that all parameters are statistically important in PSBs engineering, however, the anodization time appeared to be the most significant in defining the quality of photonic resonances. The influence of anodizing temperature was also analyzed in these works, however, the range of temperature variation was very narrow: from -2 to 6 °C. Here, we study the system behavior subjected to the SPA in oxalic acid solution in a broad temperature range: 5 - 30 °C. We demonstrate that both quality and the spectral position of PSBs are strongly affected by anodization temperature. By using this approach graded-index filters with photonic properties extended from UV up to NIR can be engineered.

2. MATERIALS AND METHODS

High-purity aluminium foil (99.9995% Al, Puratronic, Alfa-Aesar, Haverhill, MA, USA) was used to produce the PAA-based PCs. The aluminium substrate was electropolished in a mixture of ethanol and perchloric acid 4:1 at 25 V for 2.5 min at 0 °C. After the electropolishing, the samples were rinsed several times in distilled water and then in ethanol. Subsequently, the dried Al specimens were protected with acid-resistant paint at its the back and the edges. Next, the first anodization step was performed in a 0.3 M oxalic acid at 40 V for 20 h. After the first anodization alumina was selectively dissolved in a mixture of phosphoric acid and chromic acid at 65 °C for three hours. Then, PAA-based PCs were fabricated by sinusoidal pulse anodization with 150 cycles, amplitude (A) of 6 V (U_{max} = 52 V, U_{min} = 40 V), and two different anodization periods (t_p): 100 and 200 s (**Figure 1a**). The current density response was measured during the potentiostatic anodization performed under the time-controlled mode. A platinum grid served as a cathode. The distance between electrodes was kept constant (ca. 5 cm). A 1 L electrochemical cell was used with a powerful constant-temperature bath (with temperature stability ±0.01 °C) and vigorous stirring (250 rpm). The anodization temperature was varied between 5 °C, and 30 °C. A programmable DC power supply, model 62012P-600-8 Chroma, was employed to control the electrochemical parameters. Morphology of the samples was studied by a field-emission scanning electron microscope FE-SEM FEI Quanta 3D FEG. The transmission measurement were obtained using the Agilent CARY 7000 UV-Vis-NIR.

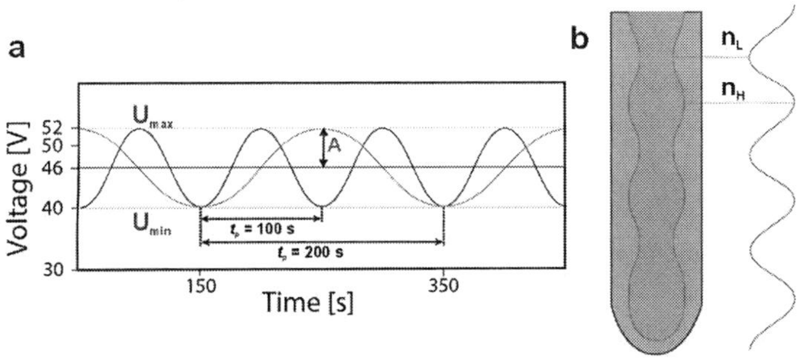

Figure 1 The parameters of the voltage sinusoidal function used in the pulse anodization (**a**) and a scheme of pore profiling resulted from the sinusoidal pulse anodization (SPA) (**b**)

3. RESULTS AND DISCUSSION

The application of the sinusoidal pulse anodization allows for pore shape profiling and thus obtaining periodically changing porosity with a gradual variation of refractive index from maximal (n_L) to minimal (n_H) values (**Figure 1b**). In **Figure 2**, voltage and current density vs. time transients recorded during sinusoidal pulse anodization of aluminum at temperatures of 5 -30 °C for the two different anodization periods (t_p = 100 and 200 s) are shown. Current density (red lines) tends to increase with increasing anodizing temperature for both t_p. Moreover, current recovery process becomes increasingly difficult when the samples are anodized above 25 °C for t_p = 100 s and above 15 °C for t_p = 200 s. This observation indicates the appearance of diffusional-related problems with ion transport from the electrolyte bulk to the pore base [11,12]. In this situation, the growth rate of porous anodic alumina occurs faster than the delivery of fresh ions to the reaction center. As an effect, the overall current does not reach the same value for the succeeding U_{max} in the following anodization cycles as the process proceeds. This, in turn, results in a progressive reduction of thickness of the subsequent low and high porosity segments [13]. Furthermore, the current oscillation during the sinusoidal pulse anodization with t_p = 200 s occurs within a much smaller j values than that during the anodization with t_p = 100 s. This is due to a milder voltage change when increasing to U_{max} and when decreasing to U_{min} (**Figure 1a**). At 30 °C, the amplitude of current oscillation decreases of more than two times at the end of the anodization as compared to the beginning of the process for both t_p. For t_p = 200 s, the irregular current behavior is observable already at temperature > 15 °C.

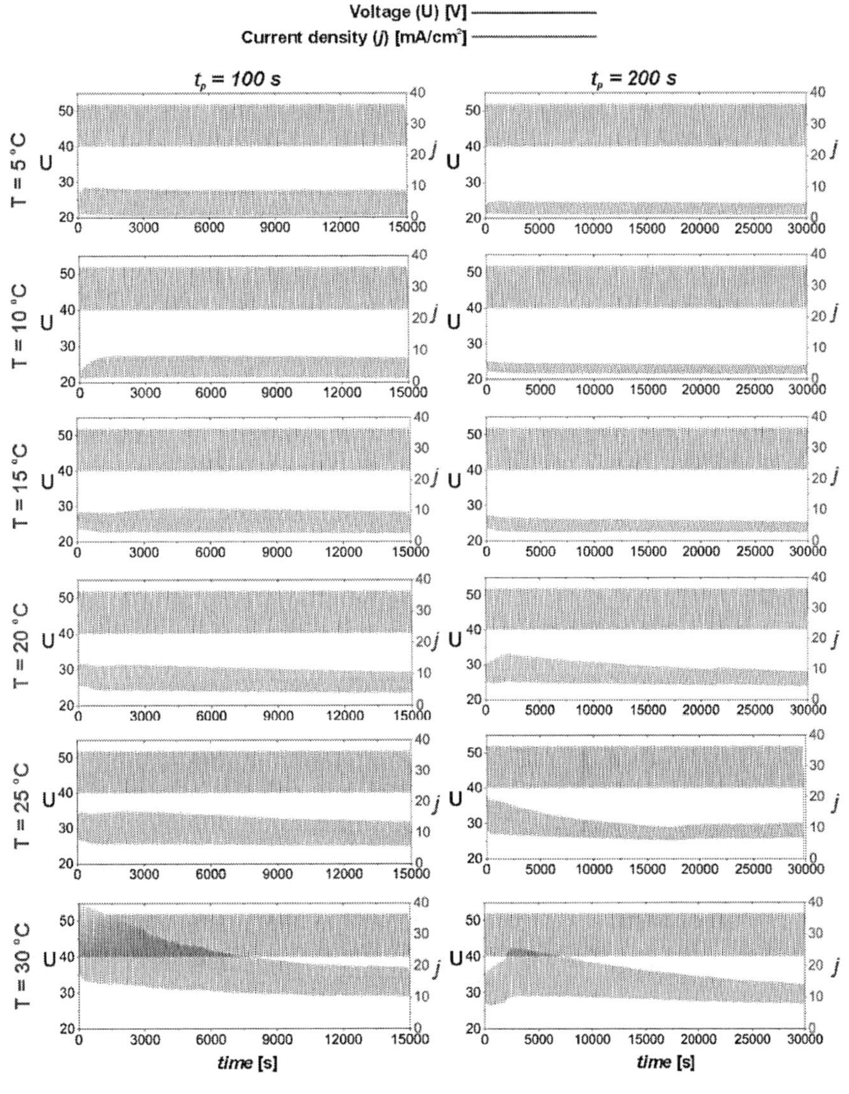

Figure 2 Voltage and current density vs. time transients recorded during sinusoidal pulse of aluminum at temperatures 5 -30 °C for two different anodization periods (t_p = 100 and 200 s)

In **Figure 3**, exemplary SEM images of porous gradient-index filters synthesized at 20 °C, are demonstrated. Periodic variation of applied voltage from U_{max} to U_{min} allows to distinguish separate segments of the thickness that depends on t_p. The thickness of separate segments as well as the total thickness of PAA prepared with t_p = 200 s (**Figure 3b**) is double the relevant thicknesses of PAA produced with t_p = 100s (**Figure 3a**).

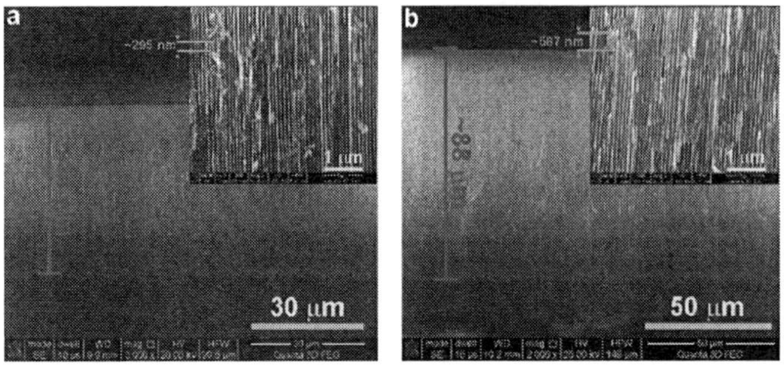

Figure 3 Exemplary SEM images of the porous gradient-index filters prepared at 20 °C for t_p = 100 s (**a**) and t_p = 200 s (**b**) (the insets: larger magnifications of the relevant images)

Optical spectra of the samples directly after the synthesis are shown in **Figure 4**. It can be seen that the quality of PSBs is rather low. Moreover, with temperature the peaks become progressively broadened and shift towards red part of the spectrum. This trend is opposite to that observed by Law et al. [10]. It should be, however, noticed that in the work [10] temperature was changing by 1 °C in the range -1 - 6 °C, which caused only a slight increase in the porosity (and thus the decrease of effective refractive index). This, in turn, according to the Bragg-Snell law [14], made the PSB to shift towards shorter wavelengths. In this experiment, owing to the much larger temperature variation, the increase of segment thickness was more significant than the increase in porosity and, therefore, the PSBs moved towards longer wavelengths.

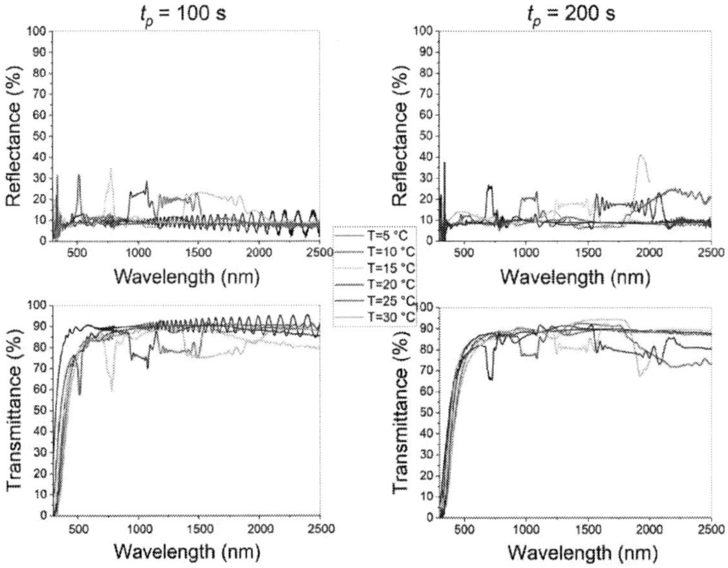

Figure 4 Reflectance and transmittance spectra of the gradient index optical filters produced at various temperatures (5-30 °C) directly after anodization (before pore broadening)

It was shown before that pore widening has a substantial effect on position and shape PSBs [15]. In **Figure 5**, the optical spectra of the same samples but after pore broadening (the samples were immersed in 5 wt% H_3PO_4 acid solution for 30 min at 30 °C) process are shown. As can be seen, the quality of PSBs was considerably improved. Well-resolved and symmetric first order PSB (λ_{PSB1}) can be distinguished that widen with increasing temperature. The spectral position of the λ_{PSB1} shifts to the red part of the spectrum with increasing both anodizing temperature and t_p (**Figure 6**). The shift of λ_{PSB1} is about 65 nm/°C and 52 nm/°C for t_p = 200 s and 100 s, respectively.

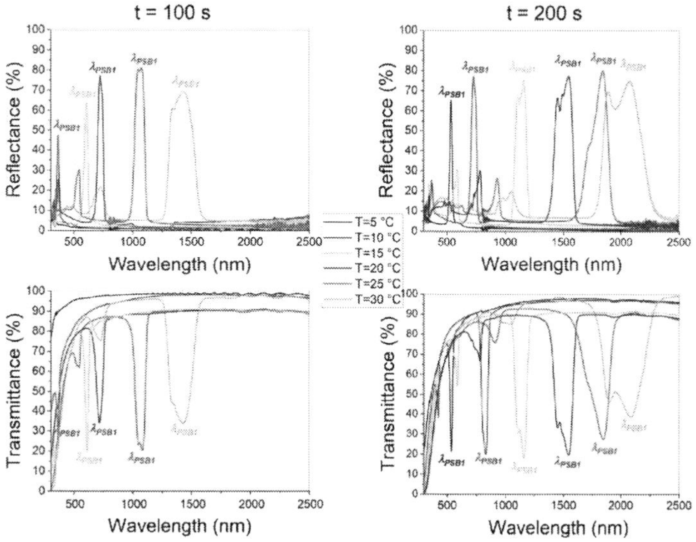

Figure 5 Reflectance and transmittance spectra of the gradient index optical filters produced at various temperatures (5-30 °C) and after pore broadening in 5 wt% H_3PO_4 acid solution

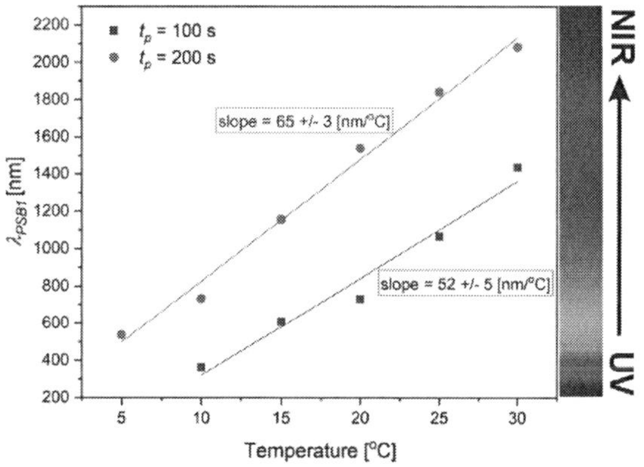

Figure 6 Spectral position of the PSBs as a function of anodizing temperature for the samples anodized with t_p = 100 s and 200 s

4. CONCLUSION

Gradient-index filters were synthesized by sinusoidal pulse anodization of aluminum. The effect of anodizing temperature in the range between 5 and 30 °C on electrochemical performance and optical properties of the optical filters was studied for two anodization periods (t_p): 100 and 200 s. It was show that with increasing the temperature the current recovery to the initial values recorded during the first anodization cycles becomes hindered by diffusion-related problems. The irregular current flow starts to be significant at temperature > 25 °C for anodization with t_p = 100 s and at temperature > 15 °C for the anodization with t_p = 200 s. This behavior is reflected in the optical characteristics of the samples. Directly after the synthesis the filters demonstrate very wide, irregular photonic stopbands (PSBs) of low intensity. Post-process pore widening improves considerably the quality of PSBs. Well-resolved and symmetric first order PSBs (λ_{PSB1}) can be distinguished that shift towards longer wavelengths with both increasing anodizing temperature and t_p with proportionality constant of about 65 nm/°C and 52 nm/°C for t_p = 200 s and 100 s, respectively. The λ_{PSB1} become, however, broadened as the anodization temperature increases. The approach shown in this work can be used for engineering porous optical filters for a broad spectral range spreading from VIS to NIR.

ACKNOWLEDGEMENTS

This research was funded by National Science Centre in Poland (NCN) [UMO-2019/35/B/ST5/01025].

REFERENCES

[1] DOMAGALSKI, J.T., XIFRE-PEREZ, E.; MARSAL, L.F. Recent advances in nanoporous anodic alumina: principles, engineering, and applications. *Nanomaterials.* 2021, vol. 11, pp. 430.

[2] RUIZ-CLAVIJO, A., CABALLERO-COLERO, O., MARTÍN-GONZÁLEZ, M. Revisiting anodic alumina templates: from fabrication to applications. *Nanoscale.* 2021, vol. 13, pp. 2227-2265.

[3] PAVESI, L. Porous silicon dielectric multilayers and microcavities. *Riv. Nuovo Cimento.* 1997, vol. 20, pp. 1-76.

[4] STARKEY, T., VUKUSIC, P. Light manipulation principles in biological photonic systems. *Nanophotonics.* 2013, vol. 2, pp. 289-307.

[5] BIAŁEK, E., SZWACHTA, G., KALISZEWSKI, M., NOREK, M. Charge density-versus time-controlled pulse anodization in the production of PAA-based DBRs for MIR spectral region. *Energies (Basel).* 2021, vol. 14, pp. 1-14.

[6] BIAŁEK, E., WŁODARSKI, M., NOREK, M. Fabrication of porous anodic alumina (PAA) by high-temperature pulse-anodization: Tuning the optical characteristics of PAA-based DBR in the NIR-MIR region. *Materials.* Vol. 13, pp. 1-18.

[7] ACOSTA, L.K., BERTO-ROSELLO, F., XIFRE-PEREZ, E., LAW, C.S., SANTOS, A., FERRÉ-BORRULL, J., MARSAL, L.F. Tunable nanoporous anodic alumina photonic crystals by Gaussian pulse anodization, *ACS Applied Materials and Interfaces.* 2020, vol. 12, pp. 19778-19787.

[8] LI, S.-Y.; WANG, J.; WANG, G.; WANG, J.-Z.; WANG, C.-W. Fabrication of one-dimensional alumina photonic crystals by anodization using a modified pulse-voltage method. *Mater. Res. Bull.* 2015, vol. 68, pp. 42-48.

[9] SANTOS, A., YOO, J.H., ROHATGI, C.V., KUMERIA, T., WANG, Y., LOSIC, D. Realisation and advanced engineering of true optical rugate filters based on nanoporous anodic alumina by sinusoidal pulse anodisation, *Nanoscale.* 2016, vol. 8, pp. 1360-1373.

[10] LAW, C.S., LIM, S.Y., LIU, L., ABELL, A.D., MARSAL, L.F., SANTOS, A. Realization of high-quality optical nanoporous gradient-index filters by optimal combination of anodization conditions. *Nanoscale.* 2020, vol.12, pp. 9404-9415.

[11] ROSLYAKOV, I. V., GORDEEVA, E.O., NAPOLSKII, K.S. Role of Electrode Reaction Kinetics in Self-Ordering of Porous Anodic Alumina. *Electrochimica Acta.* 2017, vol. 241, pp. 362-369.

[12] LEE, W., SCHWIRN, K., STEINHART, M., PIPPEL, E., SCHOLZ, R., GÖSELE, U. Structural engineering of nanoporous anodic aluminium oxide by pulse anodization of aluminium. *Nature Nanotechnology.* 2008, vol. 3, pp. 234-239.

[13] SZWACHTA, G., JANUSZEWSKA, B., WŁODARSKI, M., NOREK, M. From gradient-index to step-index filters: a switch between the two types of photonic crystals induced by the amplitude and period of sinusoidal function applied during high-temperature anodisation of aluminum. *Appl. Surf. Sci.* 2023, vol. 607, pp. 155031.

[14] OZIN, G.A., ARSENAULT, A. *Nanochemistry: A chemical Approach to Nanomaterials.* 2nd ed.; Royal Society of Chemistry: London, UK, 2015.

[15] RAHMANN, M.M., MARSAL, L.F., PALLARÈS, J., FERRÉ-BORRULL, J. Tuning the photonic stop bands of nanoporous anodic alumina-based distributed Bragg reflectors by pore widening, *ACS Applied Materials and Interfaces.* 2013, vol. 5, pp. 13375-13381.

ENHANCED GROWTH RATE OF DIAMOND FILMS AT LOW TEMPERATURE IN FOCUSED MICROWAVE PLASMA SYSTEM

Oleg BABČENKO, Irena BYDŽOVSKÁ, Jan FAIT, Ekaterina SHAGIEVA, Lukáš ONDIČ, Alexander KROMKA

Institute of Physics of the Czech Academy of Sciences, Prague, Czech Republic, EU,
babcenko@fzu.cz, fait@fzu.cz, shagieva@fzu.cz, bydzovska@fzu.cz, ondic@fzu.cz, kromka@fzu.cz

https://doi.org/10.37904/nanocon.2022.4587

Abstract

Diamond films are multifunctional materials with a wide range of extraordinary properties and a high application potential in various fields. In this work, we investigate the low temperature (< 500 °C) diamond films chemical vapor deposition on fused silica substrates realized in two different focused microwave plasma systems, i.e. a multimode clamshell cavity reactor (MCC) and a rotational ellipsoid cavity (REC) reactor. During the experiments, the methane to hydrogen ratio, in the hydrogen-rich process gas mixture, varied from 1 % to 15 % for MCC and from 1 % to 9 % for the REC reactor. For both deposition systems, the methane increase led to an increase in the diamond growth rate, e.g. it raised at least by a factor of 2.6 for the MCC reactor. Furthermore, the MCC reactor also allowed the growth of well-shaped diamond crystals at a methane to hydrogen ratio as high as 15 %. This finding is in good agreement with the Raman shift measurements, which also revealed a low content of non-diamond phases in the films making the MCC reactor more preferential for industrial uses. Moreover, the enhanced diamond growth at low temperatures is further advantageous for overcoating fused silica as well as thermally sensitive substrates which altogether opens new vistas for diamond application in optically related fields (optical elements, photonic crystals, sensors).

Keywords: Well-faceted diamonds, low temperature deposition, high growth rate, low non-diamond content, focused plasma

1. INTRODUCTION

Diamond is a material with a wide range of outstanding properties, such as high hardness and abrasion resistance, wide band gap, broad optical transparency, high thermal conductivity, high chemical stability, etc., potentially attractive for numerous applications [1,2]. Unfortunately, technological limitations, so far, hinder large monocrystalline diamond fabrication. In contrast, diamond films are less limited in terms of the large area synthesis [2,3]. Therefore, diamond films are investigated in various research areas as protective coatings, heat spreaders or chemically stable electrodes, or even opto-electronically active parts of sensors or complex optoelectronic devices including photonic structures, quantum devices, etc. [1,2,4].

Although diamond films are generally considered more flexible concerning the integration with other materials, the studies related to an adjustment of diamond deposition processes to the specific needs are still essential. For instance, diamond film deposition on thermally or chemically sensitive substrates requires the deposition conditions tailoring and/or application of protective layers [2,4,5]. In the case of diamond growth on glass or quartz substrates, the issue of thick diamond layer delamination due to thermal expansion coefficients mismatch and the appearance of residual stress needs to be taken into account [2,5]. Modification of substrate surface morphology or additional stress-relaxation layers as well as decrease of deposition temperature are the typical solutions for the mentioned issue [5-7]. Here it needs to be noted that for electronic applications the electrical properties of the resulting heterostructure are important while optical applications required optical properties preserving [1,2,6-9]. Moreover, the low deposition temperature means a decreased diamond growth rate that prolongs deposition time and/or changes the growth kinetics [5,10,11]. A higher amount of carbon-based species in the growth mixture commonly enhances the growth rate. However, it increases not only the amount of non-diamond content in the grown layers but carbon deposits on the chamber walls which often lead to plasma instability [2,9-11].

In this work, we realize the series of experiments on the low temperature (< 500 °C) diamond film chemical vapor deposition (CVD) at different methane concentrations. We investigate the diamond growth on fused silica substrates from H_2/CH_4 process gas mixture in two different focused microwave plasma systems, i.e. a multimode clamshell cavity reactor and a rotational ellipsoid cavity reactor. To facilitate the diamond growth in its early stage, i.e. beginning of the deposition process, thin diamond layers were pre-deposited on fused silica wafers in the large area linear antenna pulsed microwave plasma system [5,12], which were then used as the starting substrates for this study.

2. EXPERIMENT

The experiments with a focus on the low temperature (< 500 °C) diamond film CVD at different methane concentrations were realized in the multimode clamshell cavity (MCC) reactor (SDS6K, Seki Diamond Systems) and rotational ellipsoid cavity (REC) reactor (P6, Diamond Materials GmbH), i.e. two types of focused microwave (MW) plasma CVD system [1,3]. The fused silica (type JGS1) wafers cut to 1x1 cm^2 pieces were used as substrates. Before the diamond CVD process, substrates were seeded in an ultrasonic bath for 40 min using the water-based ultra-dispersed diamond powder suspension (NanoAmando, nominal particle size 5 nm), to receive high diamond nucleation densities [12,13]. Next, the set of samples was coated by the thin (95 nm) diamond layer in the large area linear antenna pulsed MW plasma system (AK 400, Roth & Rau) [5,12]. The 12 h CVD process was realized in the $H_2/CO_2/CH_4$ gas mixture (gas flows 200/20/5 sccm) at a pressure of 15 Pa (0.15 mbar) and microwave power 2x1800 W delivered from two sides of antennas. The average substrate temperature during the process was kept at 420 °C by resistive heating of the substrate holder. These samples, labeled as SD, were then used for diamond CVD growth in focused plasma systems while only seeded fused silica substrates, labeled as SN, were used in the experiments as reference.

The series of experiments in the MCC reactor was realized at the following deposition conditions: H_2/CH_4 process gas mixture with an amount of methane varied from 1 to 15 % of hydrogen flow (300 sccm), pressure 4 kPa (30 Torr), microwave power 2000 W and temperature maintained around 450-460 °C. The time of each CVD process was adjusted according to the results of preliminary experiments and methane amount in the mixture in order to have diamond layers of the total thickness 340±14 nm.

The diamond films deposition conditions in the REC reactor were as followed: H_2/CH_4 process gas mixture with the amount of methane varied from 1 to 9 % of hydrogen flow (300 sccm), pressure 3 kPa (30 mbar), microwave power 1500 W and temperature around 450 °C. As for MCC reactor, the time of each CVD process was adjusted according to the earlier experiments and methane amount in a mixture in order to have diamond layers of the total thickness 230±13 nm.

The diamond films thickness for experiments was evaluated by a non-destructive method using the home-made reflectance interferometry setup [12]. The grown films surface morphology was characterized by the field-emission scanning electron microscopy (FE-SEM, Tescan MAIA3). The grown films composition was studied by the Raman spectroscopy (Renishaw InVia Reflex Raman microscope) using the laser with an excitation wavelength of 442 nm.

3. RESULTS AND DISCUSSION

The typical morphology (investigated by SEM) and Raman spectrum of pre-grown diamond layer on SD samples are shown in **Figure 1**. The pre-grown layer is uniform and consisted of features with size less than

100 nm and no clear facets (**Figure 1a**) reported earlier [5]. The peaks and bands observed in the Raman spectrum of SD sample (**Figure 1b**) were attributed to: characteristic diamond peak centered at 1332 cm^{-1}, D-band centered at 1355 cm^{-1}, the G-band around 1590 cm^{-1} and a weak band around 1490 cm^{-1} the most probably related to transpolyacetylene (TPA) residues at grain boundaries [13-15].

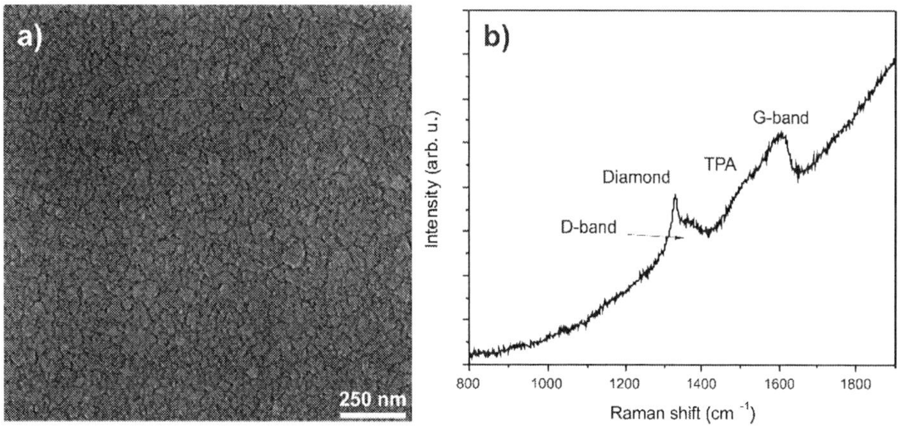

Figure 1 The SEM images of pre-grown diamond layer morphology (a) and measured Raman spectrum (b).

The morphology of the diamond films grown on SD samples (fused silica with thin pre-grown diamond layer) at different methane to hydrogen ratios in the MCC reactor are shown in **Figure 2**. It can be seen that the diamond layer grown from mixture with CH_4/H_2 ratio 1 % (low methane concentration) is not uniform, although consisted of fine-grained crystallites. In the grown layer the large (up to 500 nm) diamond crystallites are protrude from the layer of smaller (150-250 nm) crystallites with randomly distributed voids in between. Increasing CH_4/H_2 ratio up to 6 % resulted in a slight decrease of crystal sizes while voids in layer disappeared. Further increase of CH_4/H_2 ratio to 15 % decreased the crystal sizes down to approximately 50 and 400 nm for small and large crystals, respectively. Nevertheless, the crystallites remain well-faceted and distinguishable. The decreasing of diamond crystal sizes is attributed to the shift of diamond growth kinetic toward the higher diamond re-nucleation yields [2,5,13]. Whilst, high plasma density and still enough high amounts of atomic hydrogen promote the lateral diamond growth [3].

Figure 2 The SEM images of diamond films grown on SD samples at different methane to hydrogen ratios in the MCC reactor.

Raman spectra from layers grown on SD samples in the MCC reactor (**Figure 3a**) reveal characteristic features of thin diamond films. In particular, the spectrum is dominated by a diamond peak centered at 1332 cm^{-1}, and reveals also the D-band at 1355 cm^{-1}, the G-band around 1590 cm^{-1} and two bands centered at 1490 cm^{-1} and a weaker band at 1150 cm^{-1} attributed to TPA located at grain boundaries [13-15]. The best signal-to-noise ratio and the narrower diamond peak were observed for the diamond film grown at 1 % of CH_4/H_2. However, all spectra from diamond layers grown in the MCC reactor normalized to diamond peak (not shown) were very similar and demonstrated a very low amount of non-diamond carbon phases (with slight variation of TPA and G-band contribution). It agrees with the SEM observation of diamond films with well-faceted crystals in CH_4/H_2 ratio range from 1 to 15 %.

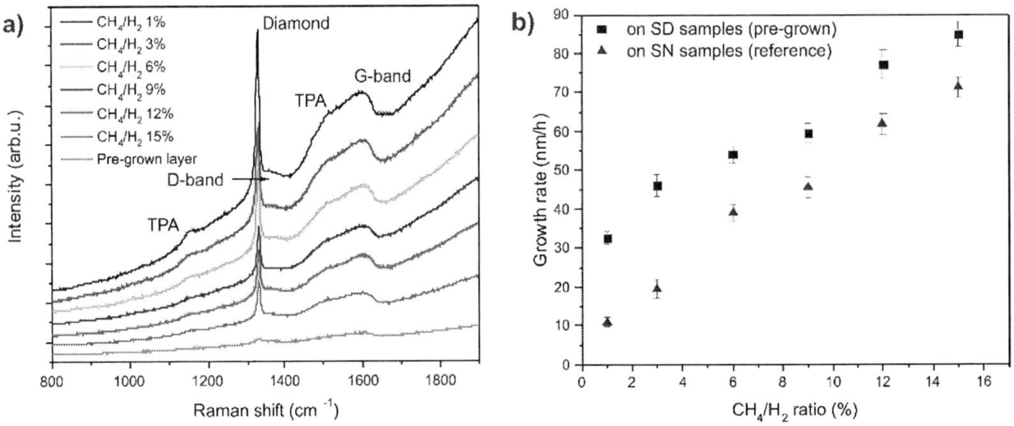

Figure 3 As-measured Raman spectra of layers grown on SD samples (a) and the growth rates of diamond films grown on SD and SN samples (b) at different methane to hydrogen ratios in the MCC reactor.

Concerning the growth rate (**Figure 3b**), it monotonically increases for both sample types, i.e. SD samples with the pre-grown diamond layer and SN - the seeded reference samples, respectively. The increase in the growth rate is attributed to a higher amount of carbon growth species due to used higher amount of methane [3,11]. The growth rate for SD samples raised from 32.6 to 85 nm/h (i.e. the enhancement factor of 2.6) for the increase of CH_4/H_2 ratio from 1 % to 15 %. The lower growth rate values observed for the reference SN samples (from 11 nm/h for 1 % CH_4/H_2 to 71.3 nm/h for 15 % CH_4/H_2) are affected by the prolonged incubation time (also known as delay time) at the early stage of the diamond CVD growth [6]. It is known that after passing that delay time the growth rate values on such samples as SN will increase [6] while using of SD samples with the pre-grown diamond layers allow to significantly shorten the incubation period. Finally, it needs to be noted that at higher CH_4/H_2 ratio the delay time is shorter which is indicated by the decreasing of the difference between growth rate for SN and SD samples with increase of CH_4/H_2 ratio. Keeping in mind the indication on the re-nucleation yields increasing and growth kinetic change the growth rate trend should not be linearized.

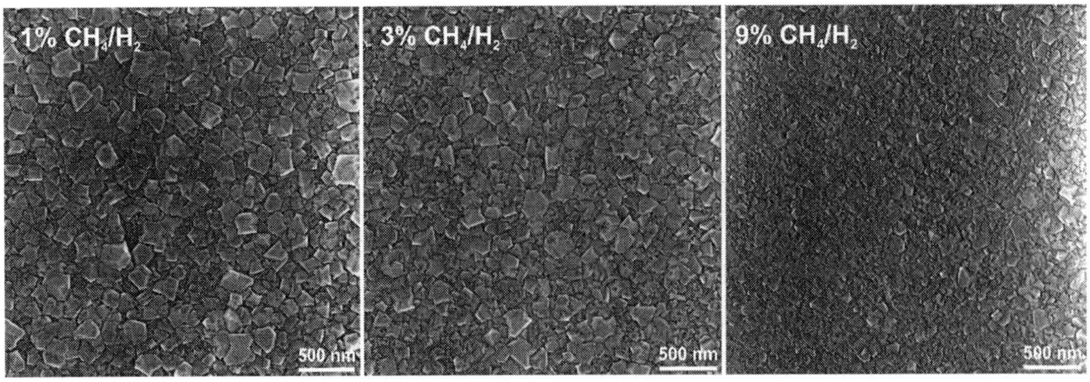

Figure 4 The SEM images of diamond films grown on SD samples at different methane to hydrogen ratios in the REC reactor.

Surface morphologies of diamond films grown on SD samples at different methane to hydrogen ratios in the REC reactor are shown in **Figure 4**. In the case of low temperature diamond deposition with a CH_4/H_2 ratio from 1 to 3 % in the REC reactor, the diamond films consisted of randomly oriented, densely packed, well-faceted diamond crystallites with sizes from 50 nm to 300 nm. With the increase of methane to hydrogen ratio to 9 % the amount of the small-sized (50 nm and smaller) diamond crystallites also increases, while crystallites with sizes above 150 nm were not observable, which should be attributed to the increase in the diamond re-nucleation rate [2,5,13]. Contrary to the MCC reactor (**Figure 2**), the shift of the growth kinetic towards the high re-nucleation yield in the REC reactor begins at much lower methane concentrations, which is more or less related to differences in the reactor designs [3].

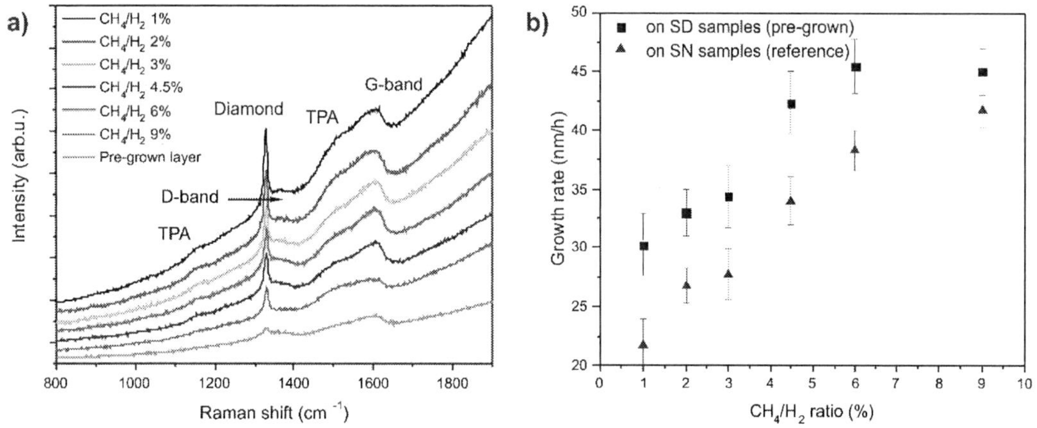

Figure 5 As-measured Raman spectra of layers grown on SD samples (a) and growth rates of diamond films grown on SD and SN samples (b) at different methane to hydrogen ratios in the REC reactor.

Raman spectra of samples grown in the REC reactor are depicted in **Figure 5a.** Similarly to the MCC reactor, all spectra reveal thin diamond films features consisting of the diamond peak (1332 cm^{-1}), D- and G-bands (1355 cm^{-1} and 1590 cm^{-1} respectively), and TPA bands (1490 cm^{-1} and 1150 cm^{-1}) [13-15]. The spectra normalized to diamond peak (not shown) reveal increase of non-diamond phases with the increase of the methane amount in the mixture. Nevertheless, the diamond peak in spectra remains clearly detectable for all used CH_4/H_2 ratios which agrees with observation of Catledge et al. for temperature 425 °C [10]. Therefore, we believe that our findings indicate on reasonable non-diamond carbon species etching during the CVD process in REC reactor at temperature around 450 °C up to CH_4/H_2 ratio at least 9 %. The growth rate (**Figure 5b**) for SD samples increased from 30.2 to 45.4 nm/h with methane to hydrogen ratio increase from 1 % up to 6 %, and then stays close to the saturated value of 45 nm/h for 9 % CH_4/H_2. The maximum enhancement of the growth rate in the REC reactor is 1.5 found for 6 % of methane to hydrogen ratio. For SN samples the growth rate increases from 21.7 to 41.7 nm/h with CH_4/H_2 ratio increase from 1 % up to 9 %. Similar to previous, the slightly lower growth rate values for SN samples are attributed to delay time at the early stage of the diamond CVD [6]. However, in contrast with the MCC reactor, it seems that the growth rate trend for REC reactor approaches the saturation at CH_4/H_2 ratio above 6 %. It corresponds to the SEM observations (**Figure 4)** indicated the diamond growth kinetic changing and the re-nucleation yield increasing for higher CH_4/H_2 ratio.

Finally, it should be noted, that for both reactors types the high methane concentration in the gas mixtures (CH_4/H_2 ratio for MCC reactor from 15 % and for REC reactor from 6 %) resulted in increased carbon-based film deposition on the chamber walls and quartz MW window which negatively influenced the plasma stability [2,3]. This is especially more critical for the REC reactor, where carbon deposit on the quartz jell bar suppresses the microwave propagation into the chamber and often results in the plasma ball movement ("*jump*") to the quartz bell jar, its later overheating and destroying. Therefore, although the further growth rate increase with increase of CH_4/H_2 ratio in MCC reactor is possible, the experiments on the diamond growth with higher methane amount in the mixture were estimated as unsafe because endanger the systems functionality.

4. CONCLUSION

We investigated the low temperature (< 500 °C) diamond film deposition on fused silica in the hydrogen-rich process gas mixtures in two different focused microwave plasma systems. For both deposition systems we observed an increase in the diamond growth rate with the increase of methane amount in the growth mixture. Moreover, during CVD with high methane to hydrogen ratio we also observed increase of carbon films deposit on chamber walls which is a limiting factor for the long-term deposition process. The diamond growth rate on the substrates with pre-deposited (growth enhancing) diamond layers for CVD in MCC reactor increases by factor of 2.6 with the methane to hydrogen ratio increase from 1 % to 15 %. For REC reactor the growth rate on the same type of substrates increases by factor 1.5 with the methane to hydrogen ratio increase from 1 % to 6 % and then saturates. The Raman shift measurements on diamond films grown in the MCC reactor with CH_4/H_2 ratio in range from 1 % to 15 % revealed a low content of non-diamond phases in all grown layers. It corresponds with the SEM observation of diamond films with well-shaped crystals in all cases. On the other hand, the CH_4/H_2 ratio increasing from 1 % to 9 % for REC reactor resulted to evident increase in number of diamond crystals with small sizes and also increase in amount of non-diamond carbon in the films. It was attributed to the change of diamond film growth kinetic and increase of the diamond re-nucleation yield that for REC reactor begins at lower methane concentrations than for MCC reactor. Considering all mentioned observation, we believe that the design of the MCC reactor is more favorable for well-facetted diamond films deposition at high CH_4/H_2 ratio than REC reactor. We proposed, that the enhanced diamond growth rate at low temperatures in MCC system will be advantageous for overcoating of fused silica, thermally sensitive substrates, e.g. optical elements, photonic crystals, sensors, etc.

ACKNOWLEDGEMENTS

This work was supported by the research bilateral project from the Slovak Research and Development Agency and Ministry of Education, Youth and Sports of Czech Republic, grant No. LUASK22147, the Operational Programme Research, Development and Education financed by European Structural and Investment Funds and the Czech Ministry of Education, Youth and Sports project No. CZ.02.1.01/0.0/0.0/16_019/0000760 - Solid21, and by the project Lumina Quaeruntur of the Czech Academy of Sciences. This work used the large research infrastructure CzechNanoLab supported by the LM2018110 project.

REFERENCES

[1] WERNER, M and LOCHER, R. Growth and application of undoped and doped diamond films. *Reports on Progress in Physics.* 1998, vol. 61, no. 12, p. 1665-1710. Available from: https://doi.org/ 10.1088/0034-4885/61/12/002.

[2] GRACIO, J J, FAN, Q H and MADALENO, J C. Diamond growth by chemical vapour deposition. *Journal of Physics D: Applied Physics.* 2010, vol. 43, no. 37, p. 374017. Available from: https://doi.org/ 10.1088/0022-3727/43/37/374017.

[3] SILVA, F, HASSOUNI, K, BONNIN, X and GICQUEL, A. Microwave engineering of plasma-assisted CVD reactors for diamond deposition. *Journal of Physics: Condensed Matter.* 2009, vol. 21, no. 36, p. 364202. Available from: https://doi.org/10.1088/0953-8984/21/36/364202.

[4] COSTELLO, M.C., TOSSELL, D.A., REECE, D.M., BRIERLEY, C.J. and SAVAGE, J.A. Diamond protective coatings for optical components. *Diamond and Related Materials.* 1994, vol. 3, no. 8, p. 1137-1141. Available from: https://doi.org/10.1016/0925-9635(94)90108-2.

[5] IZAK, T., BABCHENKO, O., POTOCKY, S., REMES, Z., KOZAK, H., VERVENIOTIS, E., REZEK, B. and KROMKA, A. CHAPTER 13. Low Temperature Diamond Growth. *Nanoscience & Nanotechnology Series.* Cambridge: Royal Society of Chemistry, 2014.

[6] BOGDANOWICZ, R., ŚMIETANA, M., GNYBA, M., FICEK, M., STRAŇÁK, V., GOLUŃSKI, Ł., SOBASZEK, M. and RYL, J. Nucleation and growth of CVD diamond on fused silica optical fibres with titanium dioxide interlayer. *physica status solidi (a).* 2013, vol. 210, no. 10, p. 1991-1997. Available from: https://doi.org/10.1002/pssa.201300096.

[7] T.-AIROLDI, V., BORGES, C. F. M., MOISAN, M. and GUAY, D. High optical transparency and good adhesion of diamond films deposited on fused silica windows with a surface-wave sustained plasma. *Applied Optics.* 1997, vol. 36, no. 19, p. 4400. Available from: https://doi.org/10.1364/AO.36.004400.

[8] VERHOEVEN, H., BOETTGER, E., FLÖTER, A., REISS, H. and ZACHAI, R. Thermal resistance and electrical insulation of thin low-temperature-deposited diamond films. *Diamond and Related Materials.* 1997, vol. 6, no. 2-4, p. 298-302. Available from: https://doi.org/10.1016/S0925-9635(96)00681-4.

[9] KOIDL, P. and KLAGES, C.-P. Optical applications of polycrystalline diamond. *Diamond and Related Materials.* 1992, vol. 1, no. 10-11, p. 1065-1074. Available from: https://doi.org/10.1016/0925-9635(92)90076-Z.

[10] CATLEDGE, Shane A, VOHRA, Yogesh K and MIRKARIMI, Paul B. Low temperature growth of nanostructured diamond on quartz spheres. *Journal of Physics D: Applied Physics.* 2005, vol. 38, no. 9, p. 1410-1414. Available from: https://doi.org/10.1088/0022-3727/38/9/013.

[11] BUSHUEV, E.V., YUROV, V.Yu., BOLSHAKOV, A.P., RALCHENKO, V.G., KHOMICH, A.A., ANTONOVA, I.A., ASHKINAZI, E.E., SHERSHULIN, V.A., PASHININ, V.P. and KONOV, V.I. Express in situ measurement of epitaxial CVD diamond film growth kinetics. *Diamond and Related Materials.* 2017, vol. 72, p. 61-70. Available from: https://doi.org/10.1016/j.diamond.2016.12.021.

[12] VARGA, M., REMES, Z., BABCHENKO, O. and KROMKA, A. Optical study of defects in nano-diamond films grown in linear antenna microwave plasma CVD from $H_2/CH_4/CO_2$ gas mixture. *Physica status solidi (b).* 2012, vol. 249, no. 12, p. 2635-2639. Available from: https://doi.org/10.1002/pssb.201200154.

[13] WILLIAMS, O.A., NESLADEK, M., DAENEN, M., MICHAELSON, S., HOFFMAN, A., OSAWA, E., HAENEN, K. and JACKMAN, R.B. Growth, electronic properties and applications of nanodiamond. *Diamond and Related Materials.* 2008. Vol. 17, no. 7-10, p. 1080-1088. Available from: https://doi.org/10.1016/j.diamond.2008.01.103.

[14] FERRARI, A. C. and ROBERTSON, J. Origin of the 1 1 5 0 − cm − 1 Raman mode in nanocrystalline diamond. *Physical Review B.* 2001, vol. 63, no. 12, p. 121405. Available from: https://doi.org/10.1103/PhysRevB.63.121405.

[15] KLAUSER, F., STEINMÜLLER-NETHL, D., KAINDL, R., BERTEL, E. and MEMMEL, N. Raman Studies of Nano- and Ultra-nanocrystalline Diamond Films Grown by Hot-Filament CVD. *Chemical Vapor Deposition.* 2010, vol. 16, no. 4-6, p. 127-135. Available from: https://doi.org/10.1002/cvde.200906827.

EFFECT OF SPIN COATING ON ELECTRICAL PROPERTIES OF MXENE FILMS DEPOSITED FROM NON-AQUEOUS SOLVENTS

[1,2]Oksana GUTSUL, [3]Ondrej SZABO, [1]Rene PFEIFER, [1]Kezia SASITHARAN, [3]Rajisa JACKIVOVA, [4]Vsevolod SLOBODYAN, [3]Alexander KROMKA, [1]Bohuslav REZEK

[1]Czech Technical University in Prague, Faculty of Electrical Engineering, Prague, Czech Republic, EU

[2]Bukovinian State Medical University, Chernivtsi, Ukraine, gutsul@bsmu.edu.ua

[3]Institute of Physics of the Czech Academy of Sciences, Prague, Czech Republic, EU

[4]Yuriy Fedkovych Chernivtsi National University, Chernivtsi, Ukraine

https://doi.org/10.37904/nanocon.2022.4590

Abstract

We investigated the effect of spin coating parameters on the electrical properties of Ti_3C_2 MXene thin films deposited from non-aqueous suspensions in N,N-dimethyl formamide (DMF) and N-methyl-2-pyrrolidone (NMP) on gold interdigitated electrodes (IDE). The electrical properties of DMF-MXenes and NMP-MXenes films are characterized by impedance spectroscopy (4 Hz - 8 MHz at 1 V) using gold IDE with 25 μm gap. The electrical conductivity of MXene films decreases with increasing spin coating speed from 300 to 900 rpm. The series resistance (R_s) and double layer capacitance remain similar (C_{dl}). In all cases, MXenes deposited from DMF have five orders of magnitude higher electrical conductivity (lower R_{ct}) than MXene films deposited from NMP. It is correlated with the thin film morphology obtained by scanning electron microscopy (SEM). These findings can be useful for possible application of MXenes as charge transport layers in hybrid photovoltaic devices.

Keywords: MXene, thin films, impedance spectroscopy, electrical conductivity, spin coating

1. INTRODUCTION

MXenes are a new class of two-dimensional materials that can be denoted by the general formula $M_{n+1}X_nT_x$, where M is a transition metal, X is carbon or nitrogen, and T is a surface functional group (-OH, -O, and/or -F) [1]. Modified 2D MXenes are promising materials in energy conversion and storage, such as lithium-ion batteries and supercapacitors, with the possibility of using them to develop new electrode designs. Due to the microporous structure of MXenes, the process of transferring ions to the oxidative regenerative centers is much faster, which allows faster recharging of batteries. So, MXenes are also attractive in terms of the development of sustainable energy technologies due to a wide range of extraordinary properties, such as high electrical conductivity, hydrophilicity, excellent thermal stability, large surface area, etc. In the form of highly concentrated suspensions of MXenes in water, due to the functional groups on the surface, its hydrophilicity and stability during storage is ensured [2]. MXenes have a significant advantage over other nanomaterials, including reduced graphene oxide, for obtaining conductive nanocomposites, precisely due to their high conductivity [3-5]. MXenes are characterized by dependence between the electrical conductivity and their morphology, for example, ultrathin, compact MXenes films with highly aligned scales exhibit very high electrical conductivity. Grain boundaries can also affect electron hopping transport behavior, smaller flakes correspond to more particle boundaries and therefore more interparticle contacts, resulting in lower electrical conductivity of the films. This is best confirmed by the significantly higher conductivity of the flakes obtained by the spin coating method, which allows for more compact films (easier electron hopping).

An urgent task given by the promising use of MXenes in organic or perovskite solar cells is to study the conductive properties and structural features of MXenes thin films obtained from polar non-aqueous solvents.

This work aims to determine the effects on the electrical conductivity of MXenes deposited from suspensions of NMP and DMF on a sensor platform with gold IDE depending on the spin coater rotation speed by impedance spectroscopy. Impedance spectroscopy is used as an effective method for studying the electrical characteristics of interfaces in a wide frequency range [6-10]. The surface morphology of MXene thin films in terms of IDE sensor coating is evaluated from secondary electron images obtained by SEM.

2. EXPERIMENTAL METHODS

2.1. Preparation of MXenes in polar aprotic solvents

A solvent replacement method was used to prepare the suspensions. A known volume (0.5 mL) of the 0.9 mg/mL dispersion of Ti_3C_2based MXenes in deionized water was centrifuged at a rotation speed of 13000 rpm (15115 g) for 30 minutes, after which the supernatant above the sediment was pipetted off. The precipitated MXene was re-dispersed in polar aprotic solvents - NMP and DMF, respectively. Then the second stage of centrifugation was carried out at 13000 rpm (15115 g) for 5 minutes, the supernatant was removed by pipette and NMP or DMF solvent was added to 1 mL. This is a simplified and faster procedure than the previously reported [11]. To prevent particle aggregation, the resulting suspension was sonicated in an ultrasonic bath at 37 kHz for 30 minutes and stored for further analysis.

2.2. Sample Preparation

The interdigitated microelectrode sensor platform was used for the preparation and characterization of the MXenes thin films. Multilayer NiCr/Ni/Au interdigitated electrodes were deposited on a ceramic substrate of 5.5 × 8.8 × 0.6 mm^3. The IDE electrode structure consists of repeated lines/gaps with the width of 25 μm for both. The IDE was first cleaned with ethanol, then 1 μL of MXenes from NMP and DMF suspensions was applied by drop casting, followed by spin coating (Ossila Ltd). Samples with deposited thin films of MXenes from NMP and DMF suspensions were prepared at three different spin coating rotation speeds (300, 600 and 900 rpm) for 60 seconds, followed by evaporation of the liquid at T=100 °C to form a dry thin film of MXenes.

2.3. Morphology Characterizations

The scanning electron microscope (SEM, MAIA3 Tescan) was used for the characterization of the surface morphology of the prepared samples. All presented SEM images of samples were acquired at 10 kV in the regime of secondary electrons.

2.4. Electrical Impedance Measurements

The IM 3536 LCR METER Hioki device was used to measure impedance spectra. All measurements were carried out in the frequency range from 4 Hz to 8 MHz at a constant voltage of 1 V. The sample was placed in a Faraday cage, the electrodes we fixed by a clamp, and connected to the LCR meter. LCR Meter Software was used to measure the impedance spectroscopy. Experimental Nyquist plots of the impedance $-Z_{im}=f(Z_{rel})$ were constructed to analyze the electron transport processes occurring at the interface of the gold IDE sensor and DMF-MXenes or NMP-MXenes.

3. RESULTS AND DISCUSSION

3.1. Methodology of MXene thin films preparation

For this study, we prepared samples of MXenes thin films deposited from NMP and DMF suspensions by spin coating on the surface of the IDE sensor. A drop of the suspension with a volume of 1 μl was deposited on a gold IDE sensor platform followed by spin coating at different spin speeds: 300, 600, 900 rpm for 60 seconds and evaporation at T=100 °C. A schematic representation of all stages of the preparation process of the studied samples is shown in **Figure 1a**. The spin coating method was used to obtain a thin coating of MXenes deposited on the surface of the IDE sensor. The general four stages of the spin coating process are shown: 1 - application of a drop of MXenes suspension on the substrate; 2 - the process of rotation of the sample with the suspension applied to its surface in the form of a drop, during which there is a uniform distribution of the layer of liquid applied to the surface; 3 - in the process of spin coater rotation there is a uniform distribution of liquid and MXenes on the surface of the IDE. A detailed schematic representation of the spin coating process is shown in **Figure 1b**.

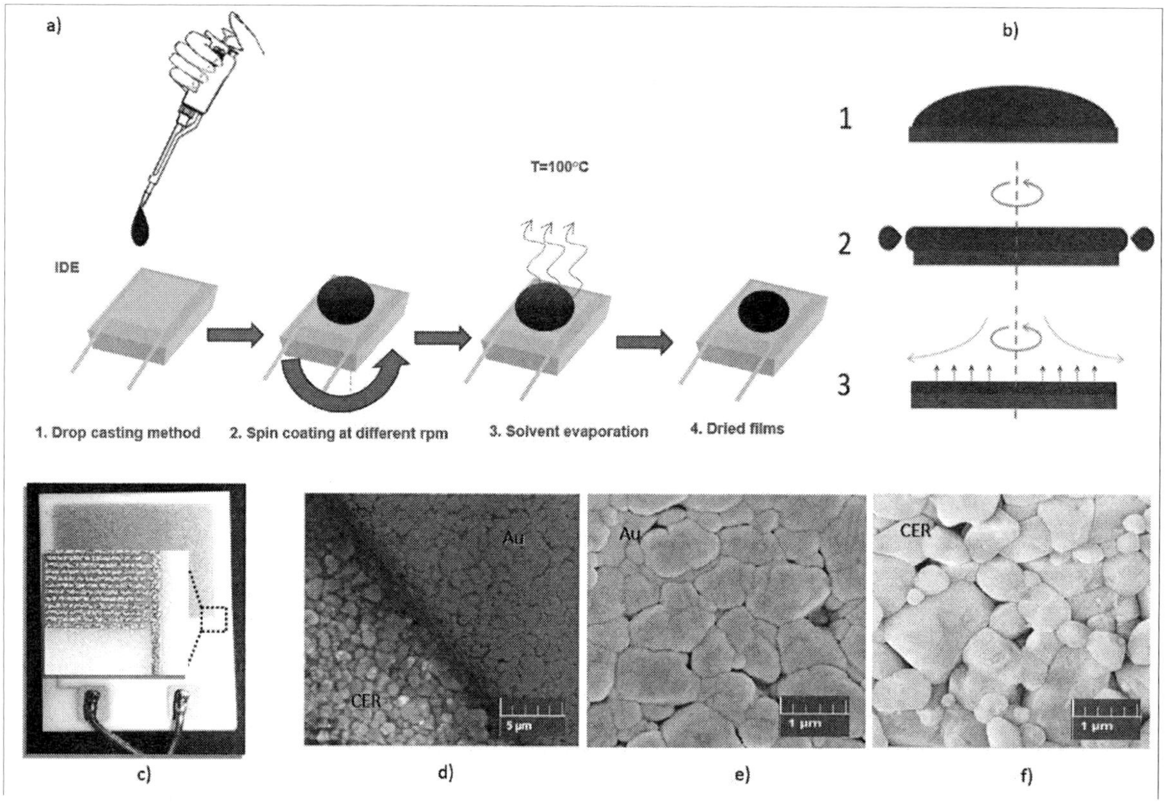

Figure 1 Scheme of MXene layer deposition process stages (a) and detail of the spin coating process (b). Photographs of bare gold IDE impedance sensor (c), its SEM surface morphology the border between Au electrode and ceramic substrate (d), and detail of gold electrode (e), ceramic substrate surface (f)

A photo of a bare IDE with an enlarged detail for better visualization of the structure of the gold electrodes on the ceramic surface is shown in **Figure 1c**. In order to compare the surface morphology of the deposited MXenes on the surface in terms of the uniformity of the sensor surface coating, the secondary electron images of the bare sensor were first obtained in three positions: at the interface between the gold electrode (Au) and the ceramic (CER), a scale of 5 μm, and the surface of the gold electrode and the between electrode space, the surface of the ceramic substrate **Figures 1e-f**.

3.2. Surface Morphology of MXene films

To analyze the morphology of MXenes thin films deposited from NMP and DMF suspensions on the surface of gold IDE by spin coating at different rotational speeds: 300, 600 and 900 rpm, SEM images of the samples were obtained, which are shown in **Figures 2 a,b** (the scales bar are 1 mm for the overview image on the left and 1 μm for the detailed image on the right). For the investigated samples, thin homogeneous coatings are observed at lower rotational speeds of 300 and 600 rpm for both types of MXenes suspensions (NMP and DMF). For a rotation speed of 900 rpm, inhomogeneities and breaks are observed on the surface of the obtained films. By analyzing the acquired SEM images, it can be concluded that the spin coating method for allows obtaining more homogeneous MXenes coatings at lower rotational speeds (300 and 600 rpm) than at higher speeds (900 rpm).

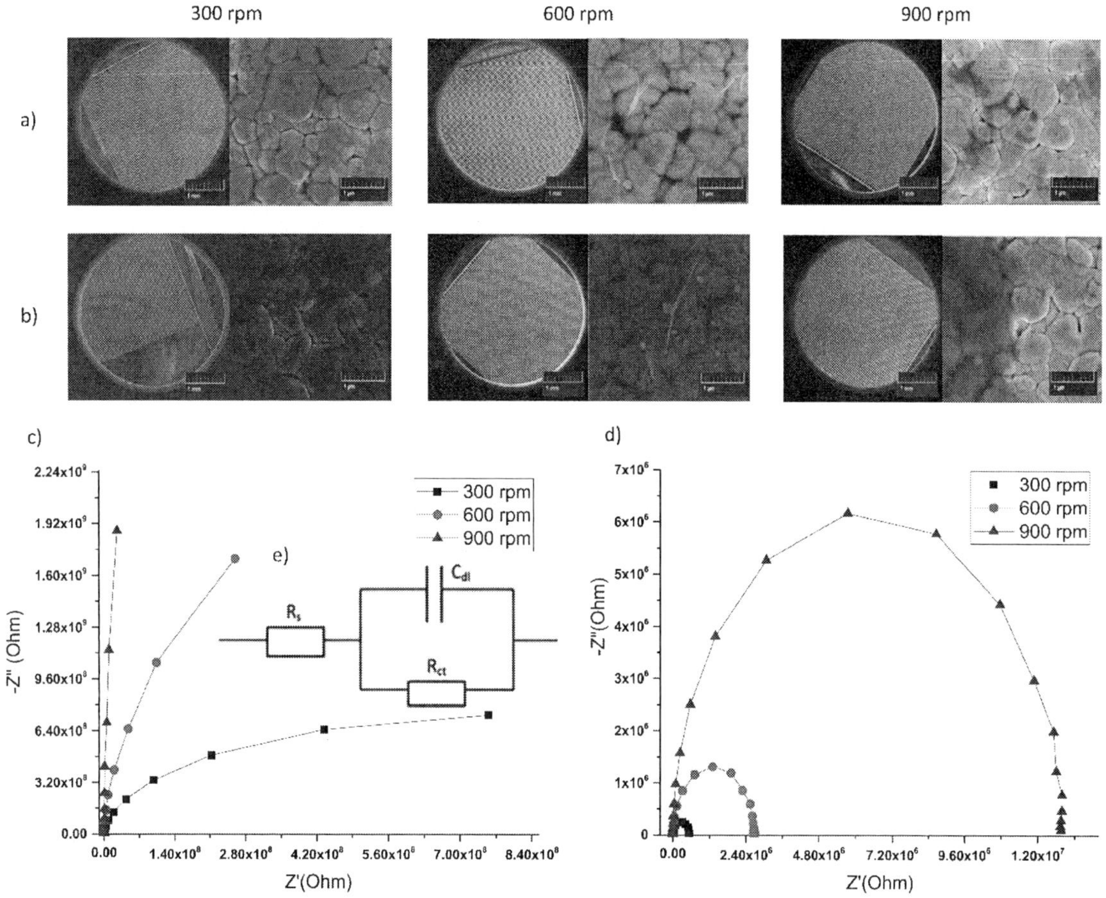

Figure 2 SEM morphology of the MXenes thin films deposited from NMP (a) and DMF (b) at different spin coating speeds (300, 600, 900 rpm). Nyquist plots of NMP-MXene (c) and DMF-MXene (d) obtained by Electrical Impedance Spectroscopy. Scheme of the equivalent model of the circuit (e)

3.3. Electrical properties of MXenes films

The results of electrical impedance spectroscopy measurements of MXenes thin films deposited from NMP and DMF suspensions on a gold IDE sensor platform at different spin coater rotation speeds (300, 600 and 900 rpm) are shown in **Figures 2c** and **2d**. Analyzing the Nyquist plots, we can see a tendency of decreasing of electrical conductivity of MXenes thin films with increasing rotation speed for MXenes deposited from both types of suspensions. For the NMP-MXenes films, the Nyquist plot shows only a small part of the semicircle in the given frequency range, indicating that the role of the charge transfer resistivity component decreases. The radius of the semicircle arcs approaches the real part of the impedance. The decrease in the radius of curvature with decreasing rotation speed indicates an increase in electrical conductivity. The impedance spectra were fitted using series resistance R_s, charge transfer resistance R_{ct} and double layer capacitance C_{dl} using the equivalent circuit model shown in **Figure 2e**. The resulting values of R_s, R_{ct} and C_{dl} parameters for all NMP-MXenes and DMF-MXenes samples are summarized in **Table 1**.

Table 1 Evaluated parameters R_s, R_{ct}, C_{dl} of MXenes deposited from NMP and DMF suspensions on the gold IDE platform

Speed of rotation (rpm)	NMP-MXenes			DMF-MXenes		
	R_s (Ω)	R_{ct} (Ω)	C_{dl} (pF)	R_s (Ω)	R_{ct} (Ω)	C_{dl} (pF)
300	58.00	0.15×10^{10}	26.6	62.92	5.25×10^{5}	21.3
600	63.42	6.63×10^{10}	21.3	61.29	26.6×10^{5}	22.7
900	63.57	6.63×10^{10}	20.9	53.73	1280×10^{5}	23.4

In general, the MXenes thin films formed at different spin coating speeds are highly resistive and there is a tendency to increase the electrical conductivity of the films with decreasing spin coating speed. At the spin coating speeds of 300 rpm, MXenes thin films deposited from both types of NMP and DMF suspensions had the lowest resistance. Comparing the total resistivity of MXenes thin films deposited from DMF and NMP solutions at different rpm, it can be seen that DMF-MXenes had better conductive properties (by 5 orders of magnitude) than NMP-MXenes. The series resistance R_s and double layer capacitance C_{dl} are about similar and remain similar independent of spin coating speed. The series resistance R_s is also quite low, around 60 Ω for both types of MXenes. Thus, the determining factor for R_s and C_{dl} are not the differences in surface functional groups that were indicated previously by Raman spectroscopy [11]. It is known that the grain boundaries can also influence the behavior of electron hopping, smaller flakes correspond to a larger number of particle boundaries and hence more inter-particle contacts, which leads to a decrease in the electrical conductivity of the films. However, the average size of DMF-MXenes and NMP-MXenes particles is similar (36 nm and 42 nm) and thus the effect of internal structure can also be excluded. It is thus the MXene/Au-electrode interface that limits the charge transfer and leads to the very high resistance R_{ct} despite low series resistance R_s. During the spin coating and drying process the MXene flakes are most likely not sufficiently well connected to the IDE electrode below them. The lower R_{ct} of DMF-MXenes compared to NMP-MXenes may be due to better electrode coverage, at the lowest 300 rpm speed, as indicated by darker SEM images.

4. CONCLUSIONS

The study of spin coating parameters' effect on MXenes thin films deposited from suspensions of DMF and NMP on a gold/ceramic IDE sensor platform. For the spin coating speeds 300-900 rpm, series resistance (related to actual electrical conductivity) and double layer capacitance (related to actual microstructure) of MXenes remained similar for both types of suspensions independent of rotation speeds. The highest charge transfer conductivity, i.e. the lower charge transfer resistance, was systematically achieved at the lowest rotation speed of 300 rpm, irrespective of NMP or DMF solvent. We assume that it is mainly due to the highest density and uniformity of MXenes distribution by spin coating on the IDE surface as shown by the SEM images. Interestingly, MXenes spin coated from DMF solution had better charge transfer compared to NMP. This difference can be explained by better coverage and contact with the electrodes (again noticeable in SEM images) due to differences in the surface functional groups of NMP-MXenes and DMF-MXenes. The conclusions made in this work regarding the dependence of electrical properties on the spin coating process parameters may serve as a guide for the possible use of MXenes in non-aqueous solvents as charge transporting layers in organic-based photovoltaic devices.

ACKNOWLEDGEMENTS

The work was financially supported by the MSMT/EU project CZ.02.1.01/0.0/0.0/15_003/0000464 (CAP). This work used the large research infrastructure CzechNanoLab supported by the LM2018110 project. The authors thank Polymer Institute, Slovak Academy of Sciences, and Drexel University for providing MXene materials.

REFERENCES

[1] M., ALHABEB, K., MALESKI, B., ANASORI, P., LELYUKH, L., CLARK, S., SIN, Y. ,GOGOTSI. Guidelines for Synthesis and Processing of Two-Dimensional Titanium Carbide (Ti3C2Tx MXene). Chem. Mater. 2017, vol. 29 pp. 7633-7644. Available from: https://doi.org/10.1021/acs.chemmater.7b02847.

[2] B., AKUZUM, K., MALESKI, B., ANASORI, P., LELYUKH, N.J., ALVAREZ, E.C., KUMBUR, Y., GOGOTSI. Rheological Characteristics of 2D Titanium Carbide (MXene) Dispersions: A Guide for Processing Mxenes. *ACS Nano*. 2018, vol. 12, pp. 2685-2694. Available from: https://doi.org/10.1021/acsnano.7b08889.

[3] C., ZHANG, L., MCKEON, M.P., KREMER, S.-H., PARK, O., RONAN, A., SERAL-ASCASO, S., BARWICH, C.Ó., COILEÁIN, N., MCEVOY, H.C., NERL, B., ANASORI, J.N., COLEMAN, Y., GOGOTSI, V., NICOLOSI. Additive-free MXene inks and direct printing of micro-supercapacitors. *Nat. Commun*. 2019, vol. 10, p. 1795. Available from: https://doi.org/10.1038/s41467-019-09398-1.

[4] H., RIAZI, S.K., NEMANI, M.C., GRADY, B., ANASORI, M., SOROUSH. Ti 3 C 2 MXene-polymer nanocomposites and their applications. *J. Mater. Chem. A*. 2021, vol. 9, pp. 8051-8098. Available from: https://doi.org/10.1039/D0TA08023C.

[5] A. IQBAL, J. HONG, T.Y. KO, C.M. KOO. Improving oxidation stability of 2D MXenes: synthesis, storage media, and conditions. *Nano Converg*. 2021, vol. 8, no. 9. Available from: https://doi.org/10.1186/s40580-021-00259-6.

[6] H.S., MAGAR, R.Y.A., HASSAN, A., MULCHANDANI. Electrochemical Impedance Spectroscopy (EIS): Principles, Construction, and Biosensing Applications. *Sensors*. 2021, vol. 21, p. 6578. Available from: https://doi.org/10.3390/s21196578.

[7] Z., LI, S., WANG, L., LIU, P., LIU, F., CHEN, X., ZHENG. Electrochemical Impedance Biosensor for the Determination of Lipopolysaccharide Using Peptide as the Recognition Molecule. *J. Braz. Chem. Soc*. 2019, vol. 30, pp. 1762-1768. Available from: https://doi.org/10.21577/0103-5053.20190081.

[8] I., MORA-SERÓ, G., GARCIA-BELMONTE, P.P., BOIX, M.A., VÁZQUEZ, J., BISQUERT. Impedance spectroscopy characterisation of highly efficient silicon solar cells under different light illumination intensities. *Energy Environ. Sci*. 2009, vol. 2, pp. 678-686. Available from: https://doi.org/10.1039/B812468J.

[9] R., RAMADAN, R.J., MARTÍN-PALMA. The Infiltration of Silver Nanoparticles into Porous Silicon for Improving the Performance of Photonic Devices. *Nanomaterials*. 2022, vol. 12, p. 271. Available from: https://doi.org/10.3390/nano12020271.

[10] CHANGJIE SHEN, LIBO WANG, AIGUO ZHOU, BO WANG, XIAOLONG WANG, WEIWEI LIAN, QIANKU HU, GANG QIN and XUQING LIU. Synthesis and Electrochemical Properties of Two-Dimensional RGO/Ti3C2Tx Nanocomposites. *Nanomaterials*. 2018, vol. 8, no. 2, p. 8. Available from: https://www.ncbi.nlm.nih.gov/pmc/articles/PMC5853712/.

[11] K., SASITHARAN, J., KULIČEK, Y., SOYKA, M., PROCHAZKA, M., OMASTOVÁ, B., REZEK. Microstructure and opto-electronic effects in MXenes spincoated from polar aprotic solvents on ITO. In: *NANOCON 2021, Conf. Proceedings*. 2021. Available from: https://doi.org/10.37904/nanocon.2021.4307.

SESSION B
INDUSTRIAL AND ENVIRONMENTAL APPLICATIONS OF NANOMATERIALS

Chairmen

Prof. Dr. Radek ZBORIL	Palacky University in Olomouc, CATRIN-RCPTM, Czech Republic, EU
Prof. Dr. Miroslav CERNIK	Technical University of Liberec, Czech Republic, EU
Prof. Dr. Miloslav PEKAR	Brno University of Technology, Czech Republic, EU

SUPERHYDROPHILIC SURFACE MODIFICATION OF PTFE HOLLOW-FIBER MEMBRANE WITH ADVANCED BIOFOULING PROPERTIES FOR WATER PURIFICATION

[1,2]Hadi TAGHAVIAN, [1,2]Miroslav CERNIK, [1,2]Lukáš DVORAK

[1]*Technical University of Liberec, Institute for Nanomaterials Advanced Technologies and Innovation, Liberec, Czech Republic, EU, hadi.taghavian@tul.cz*

[2]*Technical University of Liberec, Faculty of Mechatronics, Informatics and Interdisciplinary Studies, Liberec, Czech Republic, EU*

https://doi.org/10.37904/nanocon.2022.4575

Abstract

Irrespective of all affirmative physiochemical properties of polytetrafluoroethylene (PTFE) hollow fiber (HF) membrane, its promotion to be served for water purification function is significantly threatened by biofouling which is known as the primary problem in the membrane technology. The hydrophobic nature of PTFE attracts biofoulants leading to the pore blockage and lowering of the flux and functional life. Thus, this investigation aimed to devise an innovative silan-mediated polydopamine (PDA) coating incorporated with ZnO nanoparticles (NP) to trim the surface of the PTFE HF membrane and subsequently increase its functional property and biofouling resistivity. SEM images and EDX spectroscopies showed the time-dependent growth of PDA coating and elemental analysis on the surface, respectively. The water contact angle results illustrated the successful transformation of the surface characteristics of the PTFE HF membrane from hydrophobic (app. 130°) to superhydrophilic (0°) after surface modification. An exclusive dead-end cell was devised to comparatively evaluate the permeation and protein separation of surface-modified and pure PTFE HF membranes under constant pressure. Filtration tests' results, alongside the microbial assessments, revealed that the modified membranes showed a lower tendency towards biofouling. The outcomes of this study shed light on addressing the key challenges of today's demands for harnessing membrane biofouling and remarkably expand the PTFE HF membrane's potential to be utilized in advanced water treatment technology.

Keywords: Superhydrophilicity, polytetrafluoroethylene, zinc oxide nanoparticle, hollow fiber membrane, surface treatment, biofouling resistivity

1. INTRODUCTION

Organic hollow fiber (HF) membranes are unanimously considered the promising separation approach in water treatment due to its cost effective, cost effective, higher efficiency during the water filtration process, and space-saving which then ensure cleaner water purification [1-2]. However, the hydrophobic nature of these HF membranes are prone to the adsorption of biofoulants which consist of proteins, humic acids, and microorganisms which will then lead to membrane blockage [3-4]. Although polytetrafluoroethylene (PTFE) hollow fiber (HF) membrane contains comparatively higher chemical, thermal and mechanical properties than all other HF membranes [5], its widespread promotion in water application is affected by the feasible, permanent and efficient surface modification [6-7]. However, there is still a knowledge gap in the long-lasting and biofouling resistant modified coating for PTFE HF membrane to meet the requisites of the water treatment technologies. Accordingly this study tries to introduce the promising surface modification on the basis of a nature-inspired polydopamine (PDA) coating due to its remarkable properties [8-9] and incorporation with zinc oxide nanoparticles (ZnO NPs) as an advanced biofouling resistant approach for PTFE HF membranes.

2. MATERIALS AND METHODS

2.1. Surface modification method

The surface modification method of the PTFE HF membrane was based on the hierarchical steps of surface polymerization of PDA and immobilization of ZnO NPs on the surface, respectively. PDA coating initiated by immersion of PTFE HF (Dongyang Hanchen Membrane Technology) membranes in the Tris-base (Tris(hydroxymethyl)-aminomethane, Penta) buffer solution and In-Situ polymerization of dopamine hydrochloride (99%, Alfa Aesar) at a certain adjusted pH = 8.5. Then, the membranes were washed intensively three times with deionized water (DW) to remove non-reacted monomers. ZnO NPs immobilized on the membranes' surface followed by dispersion and ultrasonication in Ethanol, and addition of (3-Aminopropyl)triethoxysilane (APTES, TCI) 2 % v/v to the solution. The specimens were gently washed three times again and kept in DW for further analysis.

2.2. Surface characterization

Scanning electron microscopy (SEM) images were used to analyze the surface morphology of the treated PTFE HF membranes. Simultaneously, energy Dispersive X-ray Spectroscopy (EDX) analysis was carried out to quantitatively measure the elemental composition on the surface area of specimens after modifications. Water contact angle measurement was performed to evaluate the surface hydrophilicity of PTFE HF membranes before and after surface treatment. ATR-FTIR spectrophotometer was utilized to assess the spectra of the chemical bonds for pristine and surface-treated membranes.

2.3. Filtration performance

The particular homemade dead-end cell specially devised for HF membranes was conducted to determine membrane hydraulic performance. The pressure-driven cell coupled with PTFE HF membrane modules (34 cm^2) was fed by DW. The measurements were performed at 1 bar, and room temperature. Meanwhile, the flux (J) and flux recovery ratio (FRR) before and after modification against filtration of deionized water (DW) and bovine serum albumin (BSA) were determined by using the below equations:

$$J\left(\frac{L}{M^2.H}\right) = \frac{V}{A\,.\,t} \qquad (1),$$

$$FRR = \left(\frac{J_{w2}}{J_{w1}}\right) * 100 \qquad (2)$$

Initial filtration of DW was performed to determine the flux, J_{w1}, at 1 bar for 1 h. Then, the cell was emptied, refilled with BSA, and filtrated for 1 h at 1 bar. Subsequently, the flux recovery ratio (FRR) was determined after gently three times washing the membranes and refilter the DW for another 1 h to measure J_{W2}.

2.4. Antimicrobial assessments

Surface modified PTFE HF membranes incorporated with different concentrations of ZnO NPs (0.5 mg/mL, 1 mg/mL, and 1.5 mg/mL) were qualitatively tested under static contact condition according to the AATCC method 147 [10] and quantitatively under dynamic contact condition according to the standard ASTM E2149 by inoculation with gram-positive *Staphylococcus spp.* CCM 2446 and gram-negative *Escherichia coli* CCM 7395 (obtained from the Czech Collection of Microorganisms, Masaryk University in Brno, Czech Republic).

2.4.1. Antimicrobial activity under static contact conditions

To evaluate the inhibition zone of surface-modified PTFE HF membranes an initial concentration of 10^5 cells inoculum/1ml of bacteria suspensions were prepared by using a Densi-La-Meter® II densitometer (Erba Lachema). Then, four lines of the given bacterial suspensions ranging from non-diluted, 10 times, 10^2, and 10^3 times dilution were prepared in parallel on the agar plates (PCA). Tested HF membranes were placed perpendicularly with the bacteria lines on the agar plates.

Sample plates were cultivated beneath 37 °C for 48 h. After incubation, the contact area between the tested membranes and bacteria lines were qualitatively assessed based on three factors: growth of bacteria under the HF membrane, leaking from the modified HF membrane into the agar plates, and reduction zones between the edge of the bacteria lines and samples. Consequently, the reducing inhibition zones sign a qualitative degree of antimicrobial activity.

2.4.2. Antimicrobial activity under dynamic contact conditions

The quantitative measurements of each specimen was carried out at the certain contact time based on the following steps. The initial suspension solutions of the test organisms were prepared in sterile saline solution (8.5 g NaCl / 1l H2O), and the concentration of bacteria equal to the 10^5 colony-forming-unit (CFU) / 1 ml was obtained. CFU were determined after 0, 1, 3, 6, and 24 hours contact times.

Each sample piece (0.5 g) was put in the 25 ml bacterial inoculum solution and thoroughly agitated on a shaker at 120 rpm. Subsequently, for the specified time intervals of 0, 1, 3, 6, and 24 hours, samples were pipetted into Petri dishes and poured with tempered PCA agar (BioRad). The same procedures were repeated for 10 times and 100 times dilutions of each sample. The samples-contained-agar Petri dishes were incubated at 37 °C for 48 hours. Finally, individual colonies of microorganisms were counted and recorded in.

3. RESULTS AND DISCUSSION

3.1. Surface characterization

The SEM images of the PTFE HF membranes before and after treatment illustrated that thickness of the PDA layer is time-dependent in a way that by increasing the polymerization time, PDA coating significantly increase **(Figure 1)** util it uniformly cover whole of the membrane's surface after 24 h **(Figure 1, d)**. It was also deduced that PDA coating is not a practicable technique to immobilize NP on the surface, and as it does, APTES was successfully exploited to immobilize ZnO NPs on the PTFE@PDA 24h HF membranes **(Figure 1, e)**.

EDX analysis was also performed to distinguish various amounts of immobilized ZnO NPs on the membrane surface. Accordingly, samples containing ZnO = (0.5 mg/mL, 1 mg/mL and 1.5 mg/mL) showed homogenous elemental distribution on the surface with the values equal to 1.72 %wt, 8.44 %wt and 18.95 %wt, respectively.

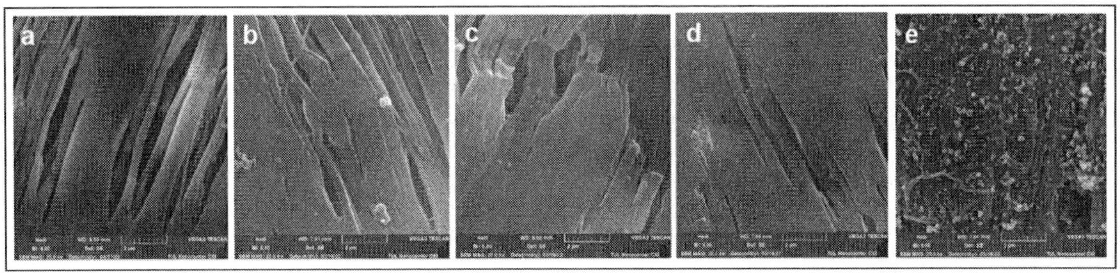

Figure 1 SEM surface images of: a) Pure PTFE HF membrane, b) PTFE@PDA 4h, c) PTFE@PDA 14h, d) PTFE@PDA 24h and e) PTFE@PDA 24h incorporated ZnO = 1 mg/mL

3.2. Water contact angle

Water contact angle results of the pristine and surface-treated PTFE HF membranes are shown in the **Figure 2**. It was deduced that the transformation of the surface characteristics of the PTFE HF membrane from hydrophobic to superhydrophilic can be accessible by decreasing the water contact angle from (app. 130°) to (0°), followed by increasing the PDA polymerization time up to 24 h and uniform growth of PDA NP on the surface. Time-dependency of the surface hydrophilicity of the membrane to PDA polymerization time was also proved by considering the reduction of the water contact angle degrees from (app. 89°), (app. 68°), and (0°) after increment in PDA polymerization time from 4 h, 14 h, and 24 h.

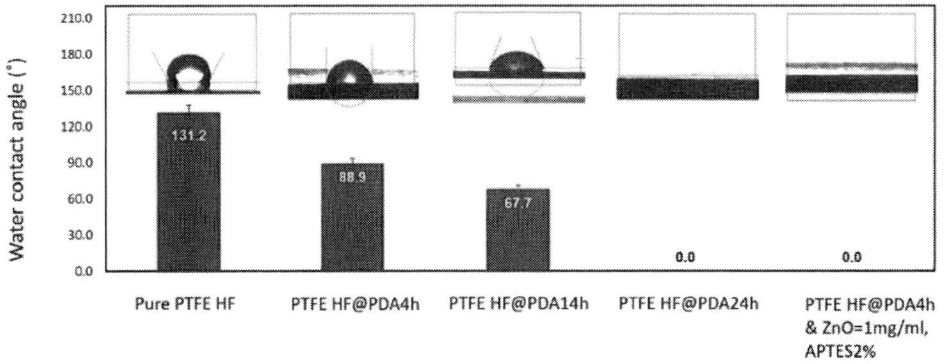

Figure 2 Effects of polymerization time and surface treatment on water contact angle degrees

3.3. ATR-FTIR spectroscopy

Successful polymerization of PDA and also APTES-aided immobilization of ZnO NPs on the surface of PTFE HF membranes were proved by comparing the obtained spectra of FT-IR. As it is shown in the **Figure 3 a)**, two peaks at 1140 cm^{-1} and 1200 cm^{-1} attributed to the C-F bond of the PTFE. New spectra appeared on the modified membranes between 1300 cm^{-1} to 1700 cm^{-1}, and 2400 cm^{-1} to 3600 cm^{-1} assigned to the O-H, N-H, and C-H bonds and hydrogen bonds of PDA, respectively. The broad peak at 1010 cm^{-1} is related to the vibration of Si-O bonds of the APTES crosslinking agent. These results indicated the successful modification of the membranes through PDA coating and ZnO NPs incorporation on the surface of PTFE HF.

3.4. Flux recovery measurement

Fluxes of pristine and modified PTFE HF membranes during DW and BSA filtration operation are shown in the **Figure 3 b)**. The initial DW fluxes at the first 1 h (J_{w1}) for pure PTFE HF and PTFE@PDA24 incorporated 1mg/mL ZnO NPs are app. 1854 (L/m2.h) and app. 226 (L/m2.h), respectively. The secondary DW fluxes after filtration of BSA are always lower than the initial one owing to the adsorption and deposition of some protein molecules on the membrane surface, contributing to the increment of the fouling rate. However, as it is shown in the **Figure 3 b)**, the FRR for the treated membrane is significantly improved by recovering the app. 46 % of the initial flux compared to the non-treated membrane, with an 8 % recovery ratio.

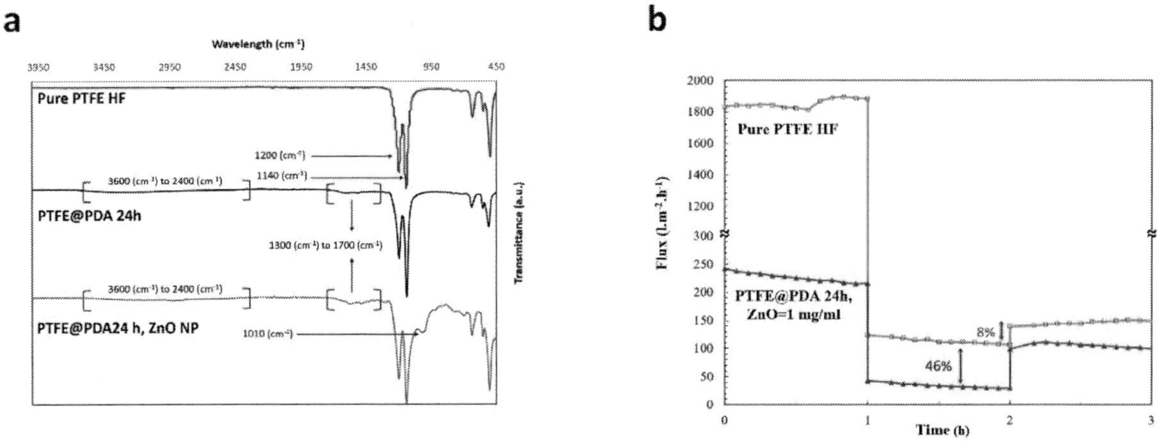

Figure 3 a) FT-IR spectra of pure and treated membranes. b) Flux Vs process time diagrams

3.5. Antimicrobial assessments

Qualitative analyses of the PTFE HF membranes after 48 h at 37 °C incubation sign the non-activity of both pure PTFE and PTFE@PDA 24h, and also low antimicrobial activity of the surface treated PTFE HF membranes. Samples containing ZnO = 0.5 mg/mL did not show inhibition properties; nonetheless, those treated samples which incorporated with ZnO = 1mg/mL and ZnO=1.5 mg/mL showed low inhibition zone for both tested bacteria of *E. coli* and *Staphylococcus spp.* (**Figure 4**).

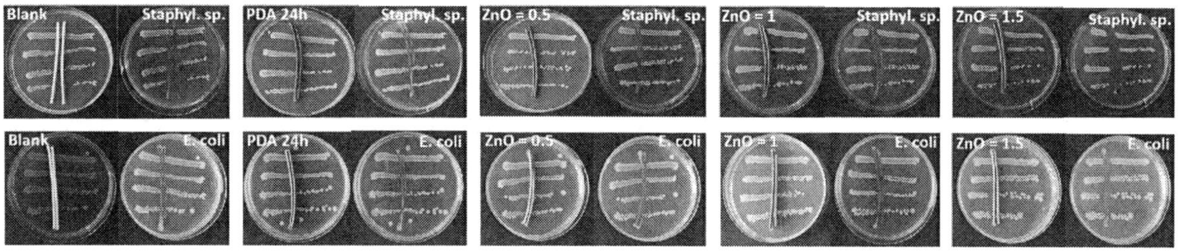

Figure 4 Inhibition zones of specimens against different concentrations of bacterial suspensions

According to the dynamic contact procedure, it can be deduced that the surface-treated PTFE HF membranes by ZnO NPss (containing 0.5 mg/mL, 1 mg/mL, and 1.5 mg/mL) showed favorable antimicrobial activities compared to the blank sample (**Figure 5**). Besides, samples containing 1 mg/mL and 1.5 mg/mL ZnO NPs showed the highest antimicrobial activity. Our observations, even after three times repetitions of experiments, proved the non-antibacterial properties of PDA coating.

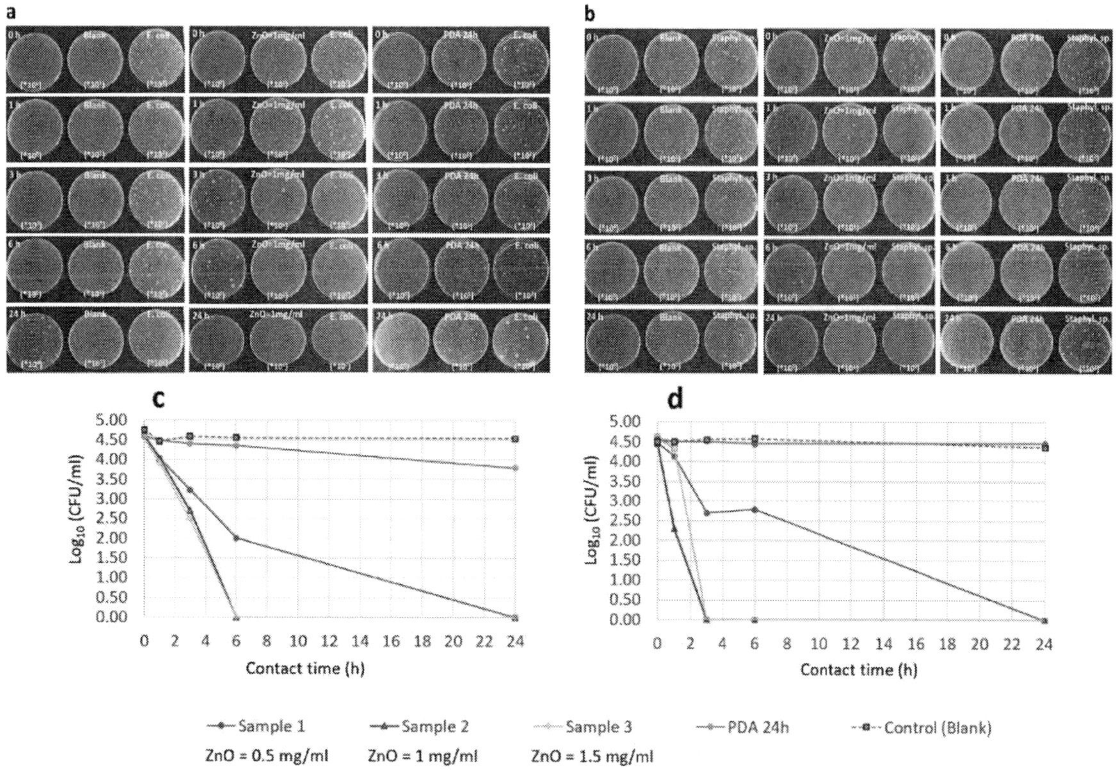

Figure 5 Colony-forming unit (CFU) assays. a) images & c) diagrams of *E. coli* and b) images & d) diagrams of *Staphylococcus spp.* cells grown by inoculation of 10^5, 10^4 and 10^3 cells with blank PTFE HF, PTFE@PDA 24 h, and PTFE@PDA 24 incorporated ZnO=1mg/mL, as the optimum specimen, at various contact time intervals from 0 h to 24 h.

4. CONCLUSION

The surface analysis experiments affirmed satisfactory transformation of the surface characteristics of PTFE HF membrane as one of the most hydrophobic polymers to become super-hydrophilic. In addition, the anti-biofouling capability of the treated membranes was proved by the antimicrobial experiments. Finally, flux and flux recovery filtration tests were performed as a supplement of the experiments to demonstrate the potential capability of the surface-treated PTFE HF membrane to serve in the submerged filtration process.

ACKNOWLEDGEMENTS

This work was partly supported by the Student Grant Scheme at the Technical University of Liberec through project nr. SGS-2022-3037.

REFERENCES

[1] M. R., MOHD RAMLI, A. L., AHMAD, and C. P., LEO. 'Surface Modification of Polytetrafluoroethylene Hollow Fiber Membrane for Direct Contact Membrane Distillation through Low-Density Polyethylene Solution Coating'. *ACS Omega*. 2021, vol. 6, no. 7, pp. 4609-4618. Available from: https://doi.org/10.1021/acsomega.0c05107.

[2] M., CZIEBOROWSKI, A. J. B., KEMPERMAN, E., ROLEVINK, J., BLOM, T. VISSER, and B. PHILIPP. 'A two-step bioluminescence assay for optimizing antibacterial coating of hollow-fiber membranes with polydopamine in an integrative approach'. *Journal of Microbiological Methods*. 2022, vol. 196, p. 106452. Available from: https://doi.org/10.1016/j.mimet.2022.106452.

[3] A. K., WARDANI, D., ARIONO, SUBAGJO, and I. G., WENTEN. 'Hydrophilic modification of polypropylene ultrafiltration membrane by air-assisted polydopamine coating'. *Polymers for Advanced Technologies*. 2019, vol. 30, no. 4, pp. 1148-1155. Available from: https://doi.org/10.1002/pat.4549.

[4] H., JANG, S.-J., PARK, and J., KIM. 'Response surface methodology to investigate the effects of operational parameters on membrane fouling and organic matter rejection in hard-shell encased hollow-fiber membrane', *Chemosphere*. 2022, vol. 287, p. 132132. Available from: https://doi.org/10.1016/j.chemosphere.2021.132132.

[5] H., XU et al. 'Development of Antibacterial PTFE Hollow Fiber Membranes Containing Silver-carried Zirconium Phosphate as Air Filtration Units for the Removal of Ultrafine Particles'. *Fibers and Polymers*. 2022, vol. 23, no. 2, pp. 423-435. Available from: https://doi.org/10.1007/s12221-022-3510-z.

[6] W., LIU, H., LIN, J., WANG, Q., HAN and F., LIU. 'Polytetrafluoroethylene (PTFE) hollow fibers modified by hydrophilic crosslinking network (HCN) for robust resistance to fouling and harsh chemical cleaning'. *Journal of Membrane Science*. 2021, vol. 630, p. 119301. Available from: https://doi.org/10.1016/j.memsci.2021.119301.

[7] H., SONG, H., YU, L., ZHU, L., XUE, D., WU, and H., CHEN. 'Durable hydrophilic surface modification for PTFE hollow fiber membranes'. *Reactive and Functional Polymers*. 2017, vol. 114, pp. 110-117. Available from: https://doi.org/10.1016/j.reactfunctpolym.2017.03.010.

[8] F. SUN, J. LU, Y. WANG, J. XIONG, C. GAO, and J. XU. 'Reductant-assisted polydopamine-modified membranes for efficient water purification'. *Frontiers of Chemical Science and Engineering*. 2021, vol. 15, no. 1, pp. 109-117. Available from: https://doi.org/10.1007/s11705-020-1987-9.

[9] Z. AN, F. DAI, C. WEI, Y. ZHAO, and L. CHEN. 'Polydopamine/cysteine surface modified hemocompatible poly(vinylidene fluoride) hollow fiber membranes for hemodialysis'. *Journal of Biomedical Materials Research Part B: Applied Biomaterials*. 2018, vol. 106, no. 8, pp. 2869-2877. Available from: https://doi.org/10.1002/jbm.b.34106.

[10] E. PINHO, L. MAGALHÃES, M. HENRIQUES, and R. OLIVEIRA. 'Antimicrobial activity assessment of textiles: standard methods comparison'. *Annals of Microbiology*. 2011, vol. 61, no. 3, pp. 493-498. Available from: https://doi.org/10.1007/s13213-010-0163-8.

ANTIMICROBIAL ACTIVITY OF GEOPOLYMERS WITH METAL MICROPARTICLE ADDITIVE

Vojtěch RŮŽEK, Lucie SVOBODOVÁ, Totka BAKALOVA, Martina RYVOLOVÁ

Department of Materials, Technical University of Liberec, Liberec, Czech Republic, EU, vojtech.ruzek@tul.cz

https://doi.org/10.37904/nanocon.2022.4576

Abstract

This paper investigates the geopolymers with metal microparticles (metal powder) as additives and assesses the antimicrobial effect on these composites. Samples of geopolymers were prepared with 4 % silver, copper, or nickel microparticle additive. The antimicrobial properties were based on a leaching test from composites, and the inhibition zone around *Escherichia coli* and *Micrococcus luteus* bacteria cells was investigated. Microparticles, especially silver and copper, were confirmed effective as antimicrobial reagents in geopolymer composites, with copper microparticles also being favourable over silver due to its significantly lower price.

Keywords: Geopolymer, antimicrobial property, bacteria cell, microparticle, metal

1. INTRODUCTION

Geopolymers are materials formed by the polycondensation of various precursors (such as fly ash [1] or metakaolin [2]) in a strongly alkaline environment [3]. They are a potential alternative to materials based on ordinary portland cement (OPC), which is the most widely used construction material and a base for concrete [4]. When compared to OPC based materials, geopolymers have higher compressive strength [5], resistance against high temperatures [6], lower thermal conductivity [7] etc. However, like OPC based materials [8], geopolymers are also susceptible to microbially induced degradation (MIB) [9], especially in wet or humid environment. Geopolymers exhibit significant antimicrobial properties in dry state due to their high alkalinity. In humid environment, the geopolymer or concrete surfaces may be colonized by alkali-resistant bacteria species (usually sulfur-oxidizing bacteria [9]). These bacteria produce acidic compounds (like hydrogen sulfide or sulphuric acid), causing degradation of the material and allowing colonization by other microorganisms (including other bacteria, fungi, lichen [10], and algae [11]). Various methods of protecting the surface of geopolymers or concrete were investigated (usually epoxide coatings or metal nanoparticle additives), with the aim to increase the resistance of geopolymers or concrete against biogenic acids or to improve their antimicrobial or biocidal properties [12].

This study used silver, copper, or nickel microparticles (in the form of a powder) instead of metal nanoparticles. Their antimicrobial effect in geopolymer was investigated as a potential alternative to nanoparticles and other antimicrobial agents (while being effective, nanoparticles may pose a serious risk to the environment and are more expensive). The antimicrobial effect was tested on both gramnegative (*Escherichia coli*) and grampositive (*Micrococcus luteus*) bacteria.

2. MATERIALS AND METHODS

2.1. Geopolymer Samples Preparation

The geopolymer samples were prepared using locally sourced Baucis L_K metakaolin-base and potassium-based activator, both manufactured by ČLUZ, a.s Czech company [13]. In addition, three types of microparticles were used to prepare geopolymer samples, namely silver, copper and nickel. The microparticle size was: silver (made by PkChemie) and copper (made by Fischema) were both sized below 45 µm in diameter, and nickel microparticles (made by Selkat ireneusz Katarzynski) in the range of 3-7 µm. Each set of samples contains one type of microparticle, except for the control sample. The composition of geopolymer samples is shown in **Table 1**.

Table 1 Geopolymer samples composition (materials are listed in weight proportion to the geopolymer base).

Material	Weight proportion
Geopolymer base	100
Activator	90
Microparticles	4

After hardening, the samples were cut to roughly 3x3x1 cm samples and analyzed on confocal microscope S neox, manufactured by Sensofar, with 20x magnification and image mode (without 3D surface mapping). It was confirmed that microparticles are dispersed in geopolymer matrix, as they were visible on the surface. The confocal images are shown in **Figure 1**.

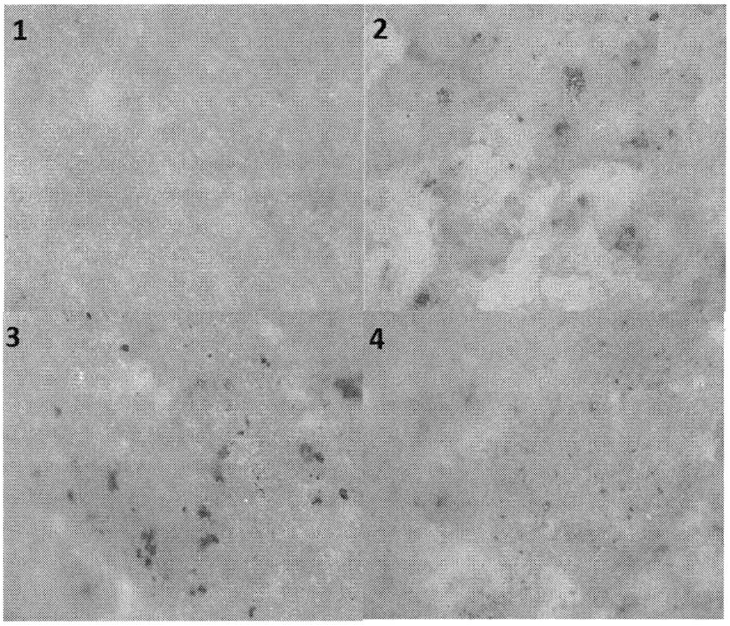

Figure 1 Images of geopolymer sample surface with 20x magnification, 1 = Pure geopolymer, 2 = Geopolymer with silver microparticles, 3 = geopolymer with copper microparticles, 4 = geopolymer with nickel microparticles

2.2. Antimicrobial Activity Assessment Tests

Gram-negative *Escherichia coli* and gram-positive *Micrococcus luteus* bacteria cells were bought from the Czech microorganism's collection [14] and were incubated for 24 hours at 37 °C before assays. Bacterial inoculum was prepared with a concentration of 10^5-10^6 CFU/mL (Colony forming units per milliliter), applied on an agar (Mueller-Hinton) surface in the Petri dish, and spread using a glass spreader. The bacteria cells were incubated in an agar medium for 15 minutes at 25 °C. The antimicrobial activity was investigated using a quantitative disk diffusion test in an agar medium (modified Kirby-Bauer test). The tested samples were applied in the middle of the Petri dishes and incubated for 24-48 hours (by microorganism type); each measurement was repeated four times.

Two types of antibiotics (Cefazolin 30 µg and Gentamicin 10 µg) were used as a positive control (with expected inhibition) in the form of antibiotic-infused paper discs; pure bacteria cells in saline solution were used as a negative control sample (with no inhibition). The antimicrobial activity of geopolymers with added microparticles was tested in the form of leach in saline solution (0.9 % NaCl). The leachate was prepared five days in advance. The leachate properties are specified in **Table 2**.

Table 2 Geopolymer leach (a geopolymer with no microparticles was used as a control sample).

Sample	Geopolymer weight (g)	Saline solution volume (ml)	Weight/volume ratio (%)	pH (t=0)	pH (t=5 min)
Control (1)	17.90	179	10	6.99	9.93
Ag (2)	15.02	150	10	6.99	9.65
Cu (3)	20.88	208	10	6.99	9.23
Ni (4)	22.65	227	10	6.99	9.09

The microparticle leachates were tested in a saline solution prepared five days before testing. In this medium, silver microparticles quickly sedimented (under 5 minutes) and significantly agglomerated even after thorough stirring. The agglomeration and sedimentation rate of copper microparticles was lower than silver microparticles. Nickel microparticles did not agglomerate and sedimented slowly (presumably due to their smaller size). The prepared mediums are specified in **Table 3**.

Table 3 Microparticle medium.

Microparticle	Weight (g)	Saline solution volume (ml)	Weight/volume ratio (%)	pH (t=0)
Ag (a)	0.317	13.17	2.41	6.99
Cu (b)	0.305	13.05	2.34	6.99
Ni (c)	0.620	16.20	3.83	6.99

The antimicrobial activity is proved when a visible inhibition zone around the tested area exists; the area without cells (colony forming units) may be easily measured. Image analysis software ImageJ (the National Institute of Health) was used to measure the diameters and areas of the inhibition zones. In each image, the background and agar color was changed to black, while bacterial colonies were colored to the scale of brown (to allow easy comparison and measurement of inhibition zones). Due to inhibition zones not always being perfectly circular, the area was measured instead of a diameter, and the average diameter was calculated from the measured area (according to the equation for the circle's volume). Basic statistical analysis (average values, standard deviation, exclusion of outliers) was calculated. Example images of inhibition zones are shown in **Figure 2**.

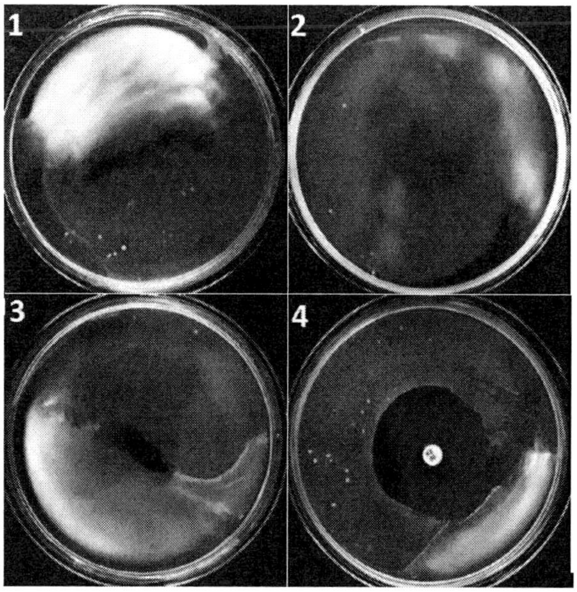

Figure 2 Inhibition zones for *Escherichia coli*; 1 = Silver microparticles geopolymer leach, 2 = Copper microparticles geopolymer leach, 3 = Nickel microparticles geopolymer leach, 4 = Cefazolin 30 µg

3. RESULTS AND DISCUSSION

Figure 3 shows the diameters of inhibition zones for each sample, while **Figure 4** shows antimicrobial activity of samples as a percentage value relative to the activity of the antibiotic (Cefazolin 30 µg). The positive control samples (Cefazolin or Gentamicin) proved the highest average antimicrobial activity (diameter of inhibition zones). In contrast, the negative control samples (pure bacteria cells in saline solution) showed no effect, per expectations.

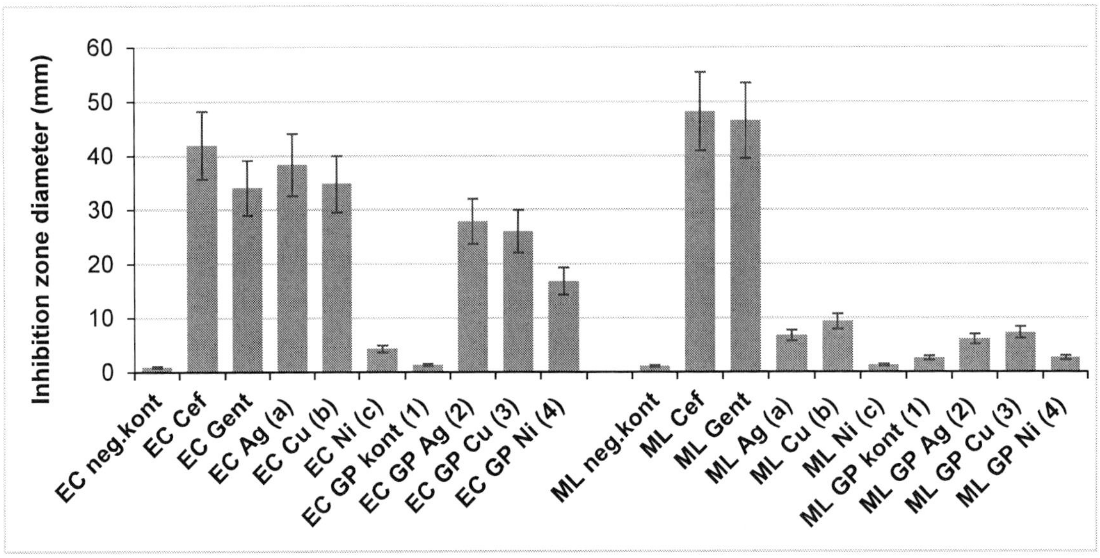

Figure 3 Inhibition zone diameters. EC = *Escherichia coli*, ML = *Micrococcus luteus*; neg.cont = Negative control sample; Cef = positive control sample Cefazolin 30 µg; Gent = Positive control sample Gentamicin 10 µg. Marked identically to Tables 2 and 3.

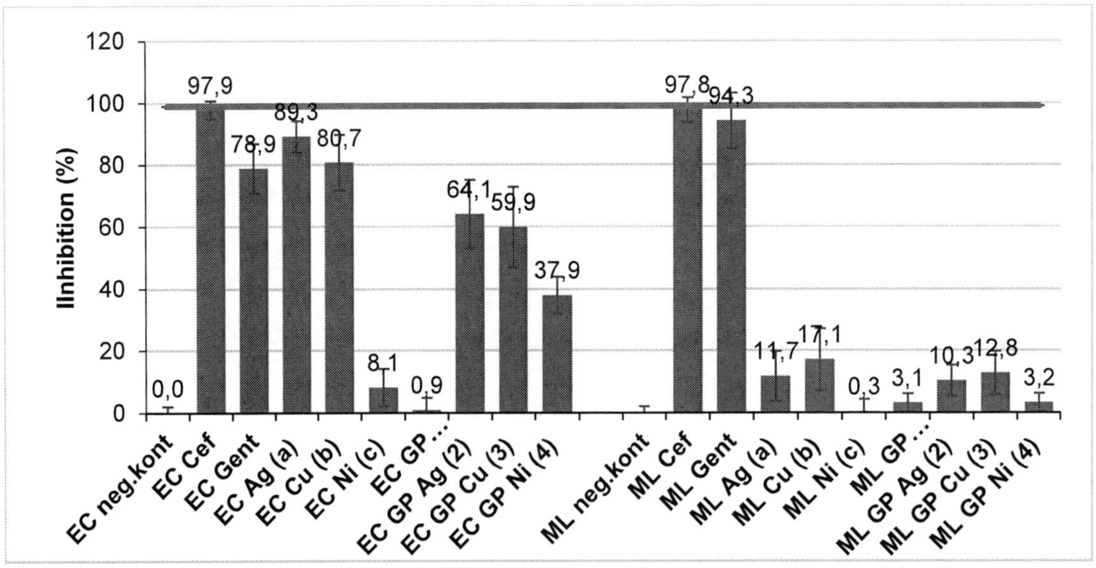

Figure 4 Graph of inhibition effect of antimicrobials compared to antibiotics; EC = *Escherichia coli*, ML = *Micrococcus luteus*; neg.cont = Negative control sample; Cef = positive control sample Cefazolin 30 µg; Gent = Positive control sample Gentamicin 10 µg. The red line in the graph shows 100% antimicrobial activity. Marked identically to Tables 2 and 3.

For *Escherichia Coli* (gram-negative bacteria cells), silver and copper microparticles in the geopolymer matrix (in **Figure 1** marked as EC GP Ag (2) or EC GP Cu (3)) have shown a high antimicrobial activity of up to 64 % and 60 %. Pure silver microparticles in saline solution (in **Figure 1** marked as EC Ag (a)) had a slightly higher antimicrobial activity than copper microparticles (EC Cu (b)) by about 9 %, although within a margin of error. The antimicrobial activity of pure nickel microparticles (EC Ni (c), despite their smaller size) is much lower, reaching only 8 %. In the geopolymer matrix (EC GP Ni (4)), these particles' effect is almost 38 % (it may be a synergistic effect of particles and high pH). Despite its high pH, pure geopolymer leach (EC GP kont (1)) does not affect *Escherichia Coli* growth.

For *Micrococcus Luteus* (gram-positive bacteria cells), the antimicrobial activity of microparticles was much weaker compared to *Escherichia Coli,* both in pure form and as incorporated in the geopolymer matrix (in **Figure 1** marked as ML). Copper microparticles have achieved the highest antimicrobial activity, about 17 % in saline solution (ML Cu (b)) and almost 13 % in geopolymer matrix ((ML GP Cu (3), relative to the Cefazolin). Nickel microparticles (pure ML Ni (c) or in geopolymer matrix ML GP Ni (4)) and pure geopolymer leach (ML GP kont (1)) did not affect *Micrococcus Luteus* growth (values are lower than 5 %).

Results indicate that microparticles, especially silver and copper, may be an appropriate alternative to nanoparticles to reach a high antibacterial activity (even as additives in geopolymer). While the inhibition effect on *Micrococcus luteus* was much lower than the effect on *Escherichia coli*, microparticles inhibited bacterial cell growth all the time. Microparticles may prevent the colonization of the geopolymer matrix surface even against gram-positive bacteria, which is a key factor in preventing microbially-induced degradation. The disadvantage at present is that the microparticles may leach out of the geopolymer matrix.

The nickel microparticles reached higher antimicrobial activity as part of geopolymer composite than by themselves. It is possible that, while weak on their own, nickel microparticles and alkaline geopolymer environment have a synergistic effect, giving nickel microparticle geopolymer composite higher antimicrobial activity than its components. However, it may also indicate higher rate of leaching out of microparticles into the environment. As such, leaching rate of microparticles should be tested in further studies. However, despite the improved antimicrobial effect of nickel microparticle geopolymer composite, the antimicrobial activity of silver and copper microparticles geopolymer composites was still significantly higher when compared to nickel microparticle geopolymer composite.

Despite their antimicrobial effectiveness, silver microparticles are significantly more expensive than other metal microparticles, including similarly effective copper. Silver microparticles were bought for around 1850 EUR/kg, while copper microparticles were bought for around 75 EUR/kg. With current data, the significant difference in price makes copper microparticles the best choice from the three tested types, due to their high antimicrobial activity and relatively low price. However, it is currently unknown whether copper or silver microparticles antimicrobial activity would persist with lower microparticle content in geopolymer matrix, as high antimicrobial activity of low silver microparticle content may justify their use as antimicrobial agent in geopolymers, despite higher price.

4. CONCLUSION

This paper investigated silver, copper and nickel microparticles as a potential antimicrobial additive to geopolymers. The high antimicrobial activity of silver and copper microparticles was confirmed by testing (a) pure microparticles in saline solution and (b) leaching test of geopolymer with microparticles as additives. Silver microparticles are, however, very expensive, making copper microparticles more suitable as antimicrobial geopolymer additive. Further studies will be focused on finding assessment options for the antibacterial effectiveness of these highly alkaline materials (testing the geopolymer sample surface itself instead of leach). The effect of microparticle content on mechanical properties and their release rate from the geopolymer matrix for environmental acceptability assessment will be investigated, as well as antimicrobial activity of geopolymers with lower microparticle content.

ACKNOWLEDGEMENTS

This work was supported by the Student Grant Competition of the Technical University of Liberec under the project No. SGS-2022-5066.

The project „Development of geopolymer composites as a material for protection of hazardous wrecks and other critical underwater structures against corrosion" registration number TH80020007 were obtained through the financial support Technology Agency of the Czech Republic within the Epsilon Program, in the Call 2021 M-ERA.Net2

REFERENCES

[1] KRISTÁLY, F., SZABÓ, R., MÁDAI, F., DEBRECZENI, Á., and MUCSI, G. Lightweight composite from fly ash geo-polymer and glass foam. *J. Sustain. Cement-Based Mater.* 2020, vol. 10, pp. 1-22. Available from: https://doi.org/10.1080/21650373.2020.1742246.

[2] RŮŽEK, V., LOUDA, P., BUCZKOWSKA, K., JUST, P., PRAŁAT, K., CIEMNICKA, J. and PRZEMYSŁAW, P. Modifying geopolymer wettability by plasma treatment and high-carbon fly ash. *Front. Built Environ.* 2022, vol. 8, 991496. Available from: https://doi.org/10.3389/fbuil.2022.991496.

[3] RŮŽEK, V. *Effect of combustion product addition and plasma treatment on surface properties of geopolymers.* Diploma thesis. Technical university of Liberec, 2020.

[4] NGUYEN, V. V., LE, V. S., LOUDA, P., SZCZYPIŃSKI, M. M., ERCOLI, R., VOJTĚCH, R., et al. Low-density geopolymer composites for the construction industry. *Polymers.* 2022, vol. 14, 304. Available from: https://doi.org/10.3390/polym14020304.

[5] GAILITIS, R., KORNIEJENKO, K., SPRINCE, A., and PAKRASTINS, L. Comparison of the long-term properties of foamed concrete and geopolymer concrete in compression. *AIP Conf. Proc.* 2020, vol. 2239, 020012. Available from: https://doi.org/10.1063/5.0007787.

[6] HEAH, C. Y., LIEW, Y. M., MOHD, M. A. B. A., and KAMARUDIN, H. Thermal resistance variations of fly ash geopolymers: Foaming responses. *Sci. Rep.* 2017, vol. 7, 45355. Available from: https://doi.org/10.1038/srep45355.

[7] PRAŁAT, K., CIEMNICKA, J., KOPER, A., BUCZKOWSKA, K. E., and ŁOŚ, P. Comparison of the thermal properties of geopolymer and modified gypsum. *Polymers.* 2021, vol. 13, 1220. Available from: https://doi.org/10.3390/polym13081220.

[8] WEI, S., JIANG, Z., LIU, H., ZHOU, D., SANCHEZ-SILVA, M. Microbiologically induced deterioration of concrete--a review. *Braz J Microbiol.* 2014, vol. 44, no. 4, pp. 1001-1007. Published 2014 Mar 10. Available from: https://doi.org/10.1590/S1517-83822014005000006.

[9] ALLAHVERDI, A., ŠKVÁRA, F. Sulfuric acid attack on hardened paste of geopolymer cements Part 1. Mechanism of corrosion at relatively high concentrations. *Ceramics - Silikaty.* 2005, vol. 49, pp. 225-229.

[10] SALVADORI, O., CASANOVA, A. The Role of Fungi and Lichens in the Biodeterioration of Stone Monuments. *The Open Conference Proceedings Journal.* 2016, vol. 7, pp. 39-54. Available from: https://doi.org/10.2174/2210289201607020039.

[11] JAYAKUMAR, S., SARAVANANE, R. Biodeterioration of coastal concrete structures by Macro algae - Chaetomorpha antennina. *Materials Research.* 2009, vol. 12, no. 4, pp. 465-472.

[12] BUCZKOWSKA, K., RUZEK, V., LOUDA, P., BOUSA, M., YALCINKAYA, B. Biological Activities on Geopolymeric and Ordinary Concretes. *J Biomed Res Environ Sci.* 2022 July 14, vol. 3, no. 7, pp. 748-757. Available from: https://doi.org/10.37871/jbres1509, Article ID: JBRES1509. Available at: https://www.jelsciences.com/articles/jbres1509.pdf

[13] Baucis LNk (2022). České lupkové závody a.s. Available at: https://www.cluz.cz/cz/baucis-lk (accessed 10. 9, 2022).

[14] Katalog kultur: bakterie a archaea. Czech collection of microorganisms. Available at: https://ccm.sci.muni.cz/katalog-kultur/bakterie-a-archaea (accessed 10.9.2022)

FERROELECTRIC JANUS MONOLAYER APPLICATION

[1]BEZZERGA Djamel, [2]SAHNOUN Mohammed

[1]*University of Relizane, Algeria, bezzergadjamel@yahoo.fr*
[2]*Department of physics, Faculty of Science, University of Mascara, Algeria, msahnoun@univ-mascara.dz*

https://doi.org/10.37904/nanocon.2022.4595

Abstract

We have performed an *ab initio* study of janus monolayers, a SiX state with as a direct wide bandgap semiconducting ferroelectric material, two-dimensional, 2D, silicon carbide has the potential to bring revolutionary advances into optoelectronic and electronic devices. It can overcome current limitations with silicon, bulk SiC, and gapless graphene. In addition to SiC, which is the most stable form of monolayer silicon carbide, other compositions, i.e., XC, are also predicted to be energetically favorable. Depending on the stoichiometry and bonding, monolayer SiC may behave as a semiconductor, semimetal or topological insulator. With different Si/C ratios, the emerging 2D silicon carbide materials could attain novel electronic, optical, magnetic, mechanical, and ferroelectric properties that go beyond those of graphene, silicene, and already discovered 2D semiconducting materials. This paper summarizes key findings in 2D SiX and provides insight into how changing the arrangement of silicon and carbon atoms in SiC will unlock incredible electronic, ferroelectric properties. It also highlights the significance of these properties for electronics, optoelectronics, and energy devices.

Keywords: Ferroelectric, Janus monolayer, quantum-mechanical, DFT

1. INTRODUCTION

One of the most intensively studied materials is Janus materials [1]. The Janus monolayer include two combinations of different chemical elements and some of them possess a suitable ferroelectric -structural transitions that are often accompanied by significant changes of their properties [2]. In particular, the off-stoichiometric SiX (X = C, Ge, Sn) compounds exhibit unusual ferroelectric and piezoelectric [3] The Stoichiometric SiX possess a long range in-plane ferroelectric arrangement of moments below the Curie temperature, then become a paraelectric compound above the Curie temperature [4].

Exceptionally high carrier mobility and zero band gap semi metallic [5] conductivity are only two of the peculiar optical and electrical characteristics of this 2D silicon material, which also contains a single Dirac point where the valence and conduction bands cross. It is one of the strongest materials known to man and incredibly mechanically flexible. Because it may be used in nanoscale energy harvesting, sensors, and actuators, piezoelectricity in 2D van der Waals materials has drawn a lot of attention. However, strain and electric polarization are restricted to the basal plane in every system examined so far, which restricts the functionality of piezoelectric devices. The freshly synthesized Janus MXY (M=Mo or W, X/Y=S, Se or Te) monolayer [6] and multilayer structures with large out-of-plane piezoelectric polarization materials are described in this study based on ab initio calculations.

2. COMPUTATIONAL METHODOLOGY

We used DFT implemented within the Quantum-ESPRESSO 6.8[5] ab initio software package. Depending on the material simulated, ion cores were treated using projector augmented-wave (PAW) pseudopotentials. Electron exchange and correlation effects were described using the generalized gradient- corrected Perdew-

Burke-Ernzerhof (PBE) approximation [6]. Periodic boundary conditions were employed in all three dimensions using an orthorhombic unit cell containing either two silicon atoms and two carbon atoms for SiC. Note the hexagonal primitive cell was not used in order to align the direction of polarization along one of the lattice vectors (i.e., the x-axis). A 20 Å cell height was used in the z-direction to prevent periodic images from interacting with each other.

The electronic wave function was expanded in a plane wave basis set with an energy cutoff of 60 Ry. The charge density was expanded in a basis set with a 240 Ry plane wave cutoff. Brillouin zone sampling was done using a Monkhorst- Pack mesh [7] of 12× 12× 1 k-points. All ionic relaxations and cell optimizations were performed using a threshold of of 10^{-4}Ry/Bohr for the force and 10^{-6} Ry for the unit cell energy.

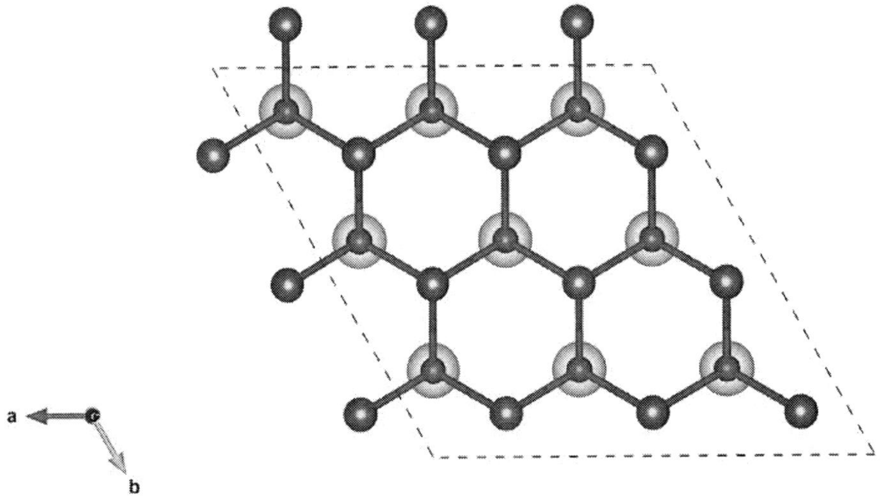

Figure 1 Schematic visualizations of 12-atom conventional supercell of Janus-structure

3. RESULTS

Figure 1 shows the 2D Janus SiX optimized crystal structures. In order to create the Janus SiX structures (X = C, Ge, Sn), the kind of atoms at one side must be swapped out for a different element. 2d materials that exhibit spontaneous electric polarization at $Pmn2_1$ phase (fig 01), we found the calculated strong in-plane polarization (x direction [1 0 0]) In **Figure 2**, we show that there is a linear relationship between uniaxial strain and polarization change in the x. The slope of the corresponding polarization shift with applied uniform strain along the x-direction of the rectangular cell may be used to calculate the piezoelectric coefficients e_{11}. Using the computed e_{11}, value, the corresponding clamped-ion coefficient d_{11} is produced. The polarizations along the axe is computed using the Berry phase method included in quantum espresso code. **Table 2** shows the clamped-ion findings for the e_{11}, d_{11}, coefficients.

Table1 Calculated and band gap Eg (eV), lattice constant (Å), and the spontaneous electric polarization (Ps) in pico-coulombs per metre (pC/m) of SiC, GeC, and SnC monolayers with Local-density approximation (LDA) and generalized gradient approximation (GGA).

material	Ps(pC/m) GGA	Ps(pC/m) LDA	Eg (eV)	Lattice constant (Å)
SiC	71.43	71.81	2.21	3.11
SiGe	54.24	53.87	1.85	3.46
SiSn	44.55	42.34	1.76	3.62

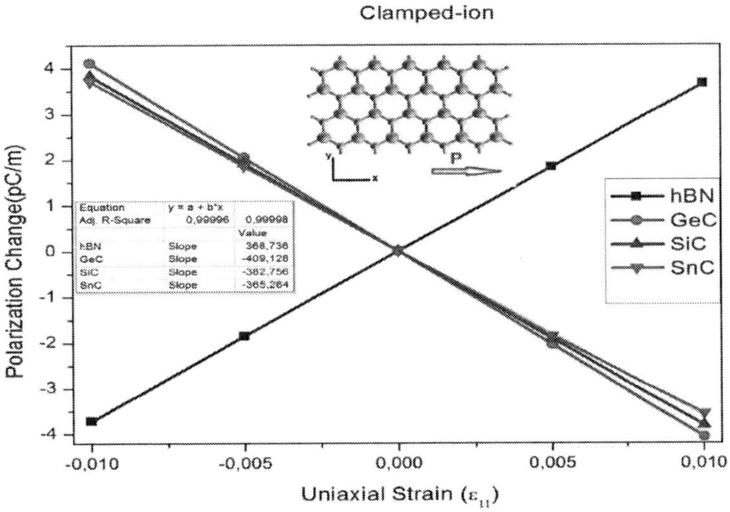

Figure 2 Applying an uniaxial in-plane strain to SiC induces a change in the polarization perpendicular to the sheet. At low strains, this relationship is linear where the slope gives the e_{11} piezoelectric coefficient.

Table 2 Calculated clamped-ion piezoelectric coefficients, e_{11} and d_{11}, of SiC, GeC, and SnC monolayers. The piezoelectric coefficient of h-BN and MoS_2 monolayers calculated by Duerloo et al. [8], are listed for comparison

Material	e_{11} (10^{-10} C·m^{-1})	d_{11} (pm·V^{-1})
SiC	-3.83	-2.46
GeC	-4.09	-2.90
SnC	-3.65	-3.2
h-BN	3.68	1.48
h-BN cal. [9]	3.71	1.50
MoS_2 cal. [9]	3.06	2.91

4. CONCLUSION

We have performed a density functional theory study of new ferroelectric Janus monolayer structures, our research focuses on SiX type monolayers with (X = C, Ge, Sn). We use First Principles method based on DFT, and our findings of optimization of the free parameters and lattice parameters of the three compounds are placed in the ranges of confusing results for the others in the same family, thus we have determined that the three SiX compounds are piezoelectric materials. The electronic characteristics were investigated, and the results revealed that they are semi-conductors. to ensure the state's mechanical stability. Fundamentally, we have a problem with these compounds being stable in the hexagonal phase at the price of the elastic constant. The obtained results encourage us to extend our work on this type of material and to study the optical and thermodynamic properties of these compounds.

ACKNOWLEDGEMENTS

Computational resources were provided by the Ministry of Education, of the Algeria Republic under Development and Innovations Infrastructures. Sub-figures 1 were visualized using the VESTA software [9].

REFERENCES

[1] GUO, Y., ZHOU, S., BAI, Y. and ZHAO, J. Enhanced piezoelectric effect in Janus group-III chalcogenide monolayers. *Applied Physics Letters.* 2017, vol. 110, 163102.

[2] RUIZ-VARGAS, C S, ZHUANG, H. L., HUANG, P. Y., VAN DER ZANDE, A. M., GARG, S., MCEUEN, P. L., MULLER, D. A., HENNIG, R. G. and PARK, J. Softened elastic response and unzipping in chemical vapor deposition graphene membranes. *Nano letters.* 2011, vol. 11, pp. 2259-63.

[3] LI, W. and LI, J. Piezoelectricity in two-dimensional group-III monochalcogenides. *Nano Research.* 2015.

[4] Vi *et al.* Direct evidence of the magnetoelastic interaction in Ni_2MnGa magnetic shape memory system. *Applied Physics Letters.* 2013, vol. 102, Art. No. 201906.

[5] GIANNOZZI, P., BARONI, S., BONINI, N., CALANDRA, M., CAR, R., CAVAZZONI, C., & WENTZCOVITCH, R. M. (). QUANTUM ESPRESSO: a modular and open-source software project for quantum simulations of materials. *Journal of physics: Condensed matter.* 2009, vol. 21, no. 39, 395502.

[6] BROWN, P. J.; *et al.* The crystal structure and phase transitions of the magnetic shape memory compound Ni_2MnGa. *Journal of Physics: Condensed Matter.* 2002, vol. 14, pp. 10159-10171.

[7] J. P. PERDEW, K. BURKE, M. ERNZERHOF. Generalized Gradient Approximation Made Simple. *Phys. Rev. Lett.* 1996, vol. 77, 3865.

[8] DUERLOO, K. A. N., ONG, M. T., & REED, E. J. Intrinsic piezoelectricity in two-dimensional materials. *The Journal of Physical Chemistry Letters.* 2012, vol. 3, no. 19, pp. 2871-2876.

[9] MOMMA, K., IZUMI, F. VESTA 3 for three-dimensional visualization of crystal, volumetric and morphology data. *Journal of Applied Crystalography.* 2011, vol. 44, pp. 1272-1276.

ZnO/Cu_2O HETEROJUNCTIONS TREATED GLASS SURFACE FOR PHOTOCATALYTIC AND SELF-CLEANING APPLICATIONS

[1]Milan MASAR, [1]Hassan ALI, [1]Muhammad YASIR, [1]Jakub SEVCIK, [1]Michal URBANEK, [1]Tomas SOPIK, [1]Michal MACHOVSKY, [1]Ivo KURITKA

[1]*Tomas Bata University in Zlin Centre of Polymer Systems, Zlin, Czech Republic, EU,* masar@utb.cz

https://doi.org/10.37904/nanocon.2022.4598

Abstract

Here we report the synthesis route for glass surface functionalization by ZnO nanorods and nanowires-based Cu_2O heterojunction. In the first step, ZnO elongated structures were grown on a seeded substrate following the classic hydrothermal method, and their thickness can be controlled by polyethyleneimine (PEI) concentration. In the second step, ZnO/Cu_2O heterojunctions were fabricated by immersion of ZnO-grown substrate upside-down into the copper sulfate solution with glucose as a reducing agent. After characterization of prepared ZnO/Cu_2O heterojunction by SEM, TEM, and XRD, its application capability for waste-water treatment was successfully demonstrated on Estriol (E_3) hormone degradation under UV light by using a continuous Drip Flow Biofilm Reactor. Furthermore, functionalized glasses were shown to be effective for designing self-cleaning surfaces. The photocatalytic-induced self-cleaning ability was demonstrated by resazurin-based smart ink and *tert*-butyl alcohol-based methylene blue ink.

Keywords: ZnO, self-cleaning, photocatalytic, Estriol

1. INTRODUCTION

In recent years, the persistent release of toxicants has been of grave concern to humans and the environment. Many bioactive toxicants are dangerous for human life, even in small concentrations. Adverse physiological effects have been reported among the commonly discharged steroidal hormones such as Testosterone, Estrogen, 17ß-Estradiol (E_2), and Estriol (E_3). These contaminants are in the category called endocrine-disrupting chemicals (EDCs). Interfering these toxicants with the physiological functions of hormones potentially causes numerous illnesses [1]. Recent technologies currently in use at the wastewater treatment plant (WWTP) are ineffective in eliminating EDCs and usually cause secondary pollution.

On the other hand, photocatalysis is considered the most cost-energy-effective approach for the remediation of environmental pollutants [2]. The essential strategy is to prepare a surface containing the multifunctional semiconductor-based photocatalyst. Surface functionalization is a widely adopted technique for surface modification that allows researchers to customize surfaces to further applications. The functionalized glass was promising in applications that may benefit from photocatalytic-induced self-cleaning, day-night switchable hydrophilicity/hydrophobicity, and antibacterial [3]. According to these observations, many different synthesis routes for ZnO yield varied morphological configurations, crystallinity, and surface-to-volume ratio, such as nanorods, nanoforests, and nanospheres, were described [4 - 6]. One of the most promising synthesis routes for glass surface functionalization by ZnO nanorods and nanowires is presented. ZnO wire structures were grown on a seeded substrate following the classic hydrothermal method, and their thickness can be controlled by polyethyleneimine (PEI) concentration [7]. Next, ZnO/Cu_2O heterojunctions were fabricated by immersion of ZnO-grown substrate upside-down into the copper sulfate solution with glucose as a reducing agent [8].

This work presented two different morphologies of ZnO, i.e., nanorods and nanowires, which are synthesized and coupled with Cu_2O as a heterojunction. Furthermore, the photodegradation of the E_3 hormone in a closed cross-flow reactor showed its successful degradation. In addition, the photocatalytic activity of prepared samples under UV irradiation was evaluated by a photocatalytic test kit called "Explorer" (Ink Intelligent, Belfast, United Kingdom). This kit is suitable for assessing moderate active surfaces such as photocatalytic glass and similar products. The experiment was held according to the ISO 21066:2018 (Fine ceramics (advanced ceramics, advanced technical ceramics) - Qualitative and semiquantitative assessment of the photocatalytic activities of surfaces by the reduction of resazurin in a deposited ink film). [9] Finally, a new approach to blue-dye (Methylene blue, MB) degradation on the sample surface was demonstrated.

2. MATERIALS PREPARATION AND MEASUREMENTS METHODS

All chemicals were used as bought without any further purification. First, a solution for the seed layer was prepared using ZnO dissolved in a mixture of peroxide and ammonia (ZnO_2 seed solution). Microscope slides (Cat. No. 7105 - one side frosted) were covered by this solution via cotton wool. ZnO_2 evaporates rapidly from the surface. Slides covered in this way were annealed at 500 °C for 30 minutes in a Muffle furnace. Next, a standard hydrothermal method synthesized ZnO one-dimension elongated structures - nanorods (ZnOnr) and nanowires (ZnOnw). For the ZnO nanostructure growth (growth solution) were used zinc nitrate hexahydrate (0.05M; Sigma Aldrich), hexamethylenetetramine (0.05M; Penta), and polyethyleneimine (0.016M; Sigma Aldrich) - for ZnOnw preparation. Seeded slides were placed upside-down for growth at 94.1 °C for 12 h. Seeded substrates were immersed in the solution after the initial precipitation of the growth solution to avoid the first massive accumulation of unwanted ZnO structures (for two hours). By adding the amount of PEI, nanowires were obtained. The samples were rinsed with de-ionized water, dried, and annealed at 500 °C for 30 minutes in a Muffle furnace to remove PEI and improve ZnO crystallinity. For the fabrication of ZnO/Cu_2O heterojunctions, a method reported in the literature using glucose as a reducing agent was employed [8].

Morphology of prepared samples was performed via NovaNanoSEM 450 microscope (The Netherland, FEI company). Microscopic images were taken using an ETD (topographic contrast) detector at an accelerating 5 kV voltage. Distance up to sample (WD - Working Distance) 5 mm.

The crystalline phase composition and diffraction lines position of all as-prepared samples was determined by the X-ray diffractometer MiniFlex 600 (Japan, Rigaku) with a Co-Kα X-ray source working at 40 kV and 15 mA. The detailed morphology and composition of samples were observed by JEM-2100 TEM (Jeol Ltd., Japan) at an operational voltage of 250 kV.

All UV-light-dependent experiments used UVA light (UVP-XX-15BLB, 15W, PN 95-0042-11, Analytik Jena). The light intensity at the top of the microscopic slides with samples was optimized by radiometer RM-22 (Opsytec Dr. Gröbel GmbH) to be 2 mW/cm2.

Each photodegraded E_3 sample of $ZnOnw/Cu_2O$, and $ZnOnr/Cu_2O$ heterojunctions on the functionalized glass surface were evaluated using HPLC in triplicates, and mean values were reported.

The photocatalytic activity with a hole-dependent mechanism was evaluated by the photocatalytic test kit Explorer (Ink Intelligent, Belfast, United Kingdom). The photocatalytic activity based on the electron pathway was measured by MB discoloration.

3. RESULTS AND DISCUSSION

The morphology, surface structure composition, and detail of ZnO/Cu_2O heterojunction, microscopy slide, and Cu_2O microparticles are shown in **Figure 1**. It shows that ZnO wires and rods are arranged in a uniform array and oriented orthogonally to the plane of the glass slides. The length was estimated to be 3 microns and the thickness to be 100 nm, according to the TEM detailed image. The morphology of $ZnOnr/Cu_2O$ heterojunction shows many Cu_2O particles with dimensions ranging from 500 nm to 1 micron entangled in ZnOnr. Observed Cu_2O microparticles are not growing at the top of nanorods but instead seem to be pierced through the nanorods. The addition of PEI resulted in significant lengthening and tapering of rods resembling nanowires, ZnOnw.

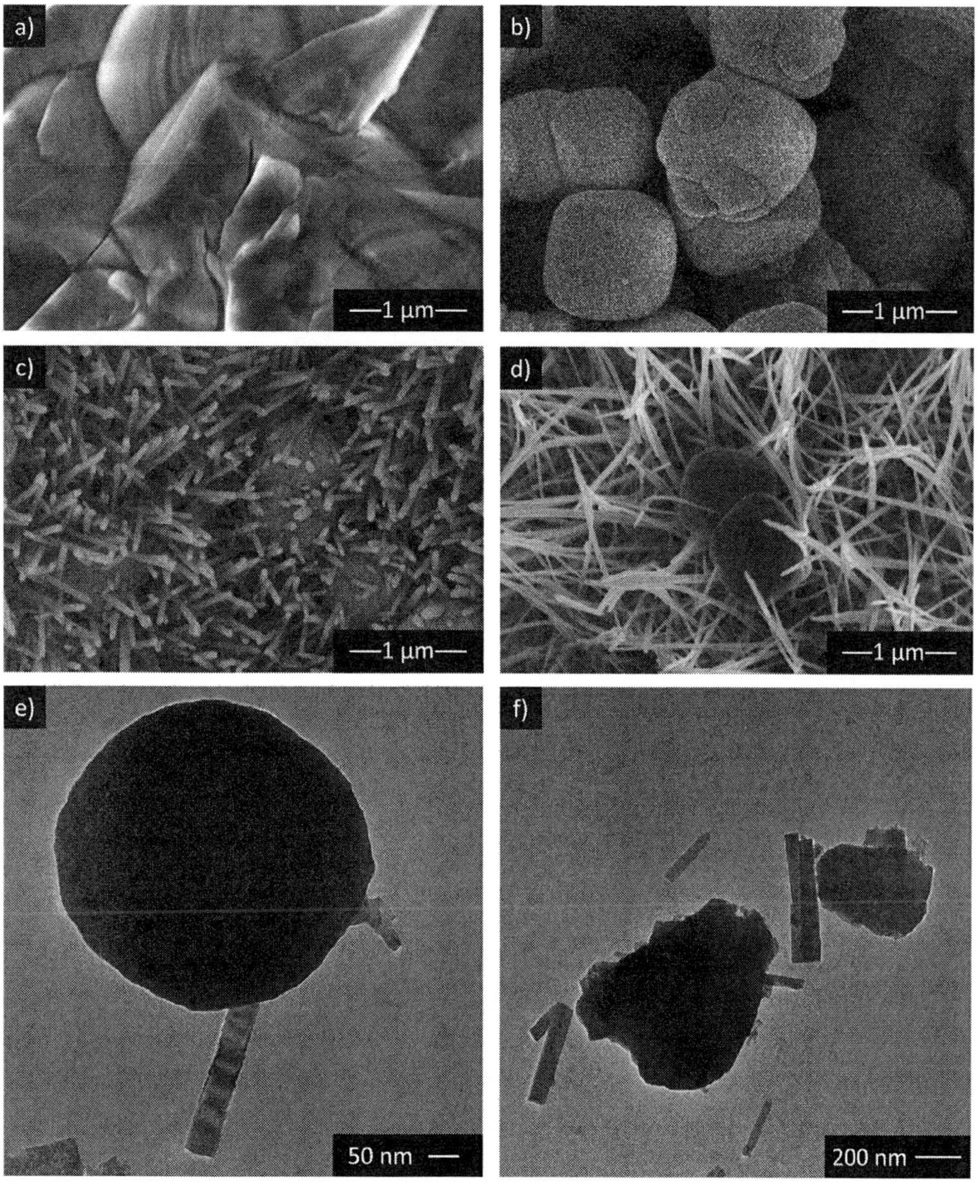

Figure 1 SEM images of a Microscopy slide (a), Cu_2O particles (b), ZnOnr/Cu_2O (c), ZnOnw/Cu_2O (d), and detailed TEM images of separated rods with Cu_2O particles (e, f).

Figure 2 shows the E_3 hormone degradation rate and total efficiency. The UV-photogenerated E_3 degradation test was conducted using a continuous Drip Flow Biofilm Reactor DFR 110-4PET (Biosurface Technologies Corporation). It consists of 4 parallel channels made from polyethylene terephthalate (PET). Nevertheless, the system was slightly modified by replacing the PET cover with borosilicate glass. The scheme of this apparatus as used can be seen in **Figure 3**. Each channel capable of holding one standard glass microscope was circulated with 40 mL of E_3 solution (0.2 mg/L concentration) using a peristaltic pump PFP 5408 (VWR) equipped with silicon tubing at a flow rate of 10 mL.min^{-1}. [7] All heterojunctions were able to degrade the E_3 hormone. However, the difference in their performance is evident. Although, ZnOnw/Cu_2O heterojunction is more efficient than ZnOnr/Cu_2O due to their higher aspect ratio and surface area. In the case of hormone E3, there was no detectable degradation on the clear microscopy slide under UV irradiation.

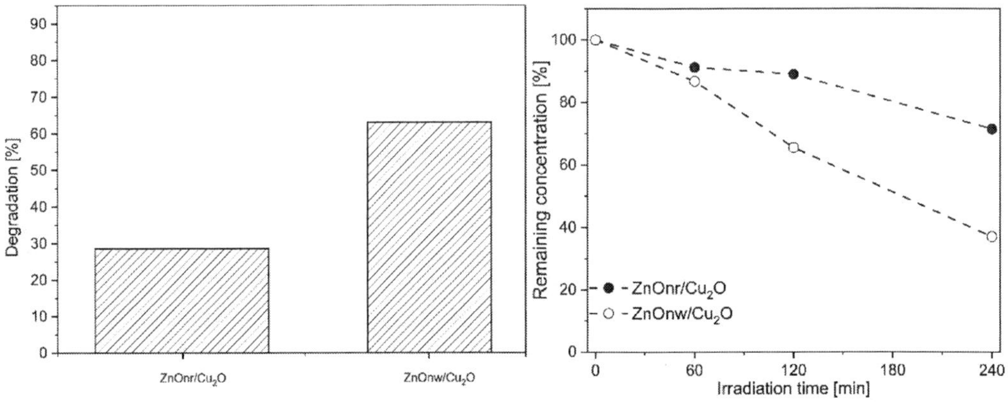

Figure 2 Total degradation efficiency in 240 min and E_3 degradation by ZnOnr/Cu_2O and ZnOnw/Cu_2O heterojunction under UV irradiation in given time intervals.

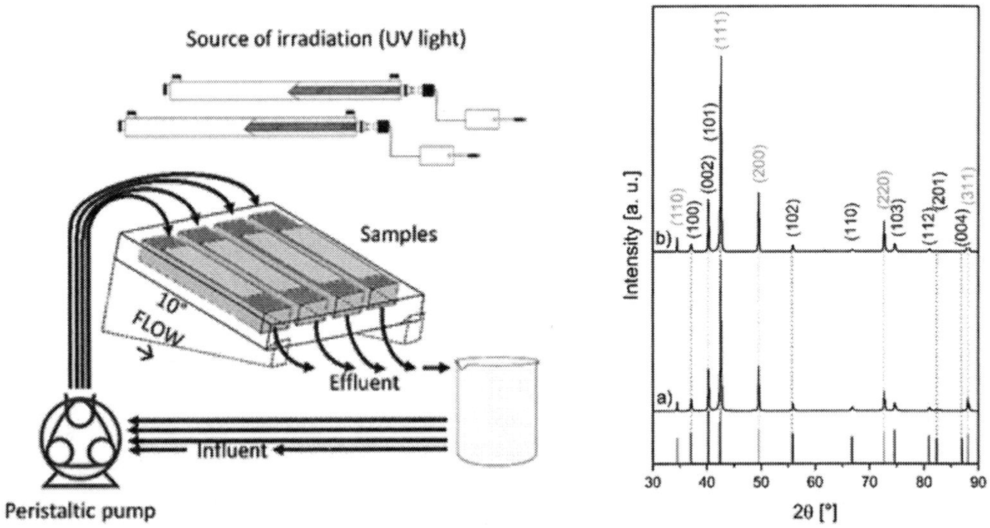

Figure 3 Left part: scheme of the used system for hormone degradation; right part: X-ray diffractogram patterns of ZnOnr/Cu_2O (a) and ZnOnw/Cu_2O heterojunction.

All as-prepared samples exhibit the diffraction peaks of (1 0 0), (0 0 2), (1 0 1), (1 0 2), (1 1 0), (1 0 3), (1 1 2), (2 0 1), and (0 0 4) of ZnOnr/Cu_2O and ZnOnw/Cu_2O, which are consistent with wurtzite ZnO hexagonal phase structure (PDF Card No.: 01-073-8765). In the case of nanowires, the dominant diffraction peak for (0 0 2) indicates a high degree of orientation in the c* axis, which means vertical growth to the substrate surface. [10,11] Besides diffraction peaks, (1 1 0), (1 1 1), (2 0 0), (2 2 0), and (3 1 1) were identified for Cu_2O cubic phase structure (PDF Card No.: 01-071-3645). No other phase was detected. Diffractograms of as-grown ZnOnr/Cu_2O and ZnOnw/Cu_2O heterojunctions are shown in **Figure 3** - right part.

The explorer ink color changes from blue to violet for both heterojunctions are visible in **Table 1** after 15 min of UV-light irradiation. These changes indicate significant photocatalytic activities. This color change of the explorer ink resulted from the irreversible reduction of resazurin (Rz) to resofurin (Rf). The explorer ink consists of glycerol and hydroxyethyl cellulose. Glycerol generates electrons to kill photogenerated holes effectively, and hydroxyethyl cellulose is a thickening agent for more effortless electron transfer. [12,13] The color change occurs almost immediately after five minutes of UV irradiation for both samples. The second prepared ink containing *tert*-butyl alcohol and MB was prepared to determine the pollutant's absorption rate onto the surface of the prepared substrates. *Tert*-butyl alcohol is used as a scavenger of free OH radicals. In the case of an untreated microscopy slide, there was no activity to the applied ink on the microscopy slide surface. In contrast, the ink was degraded entirely on surfaces covered with ZnOnr/Cu_2O and ZnOnw/Cu_2O heterojunctions. This phenomenon may indicate the absorption of the prepared heterojunctions, especially Cu_2O. [14]

Table 1 Test of prepared layers under UV irradiation for two different ink

Irradiation time	Tested Ink	ZnOnr/Cu_2O	ZnOnw/Cu_20	Microscopic slide
0 min	*Tert*-butyl alcohol + Methylene Blue	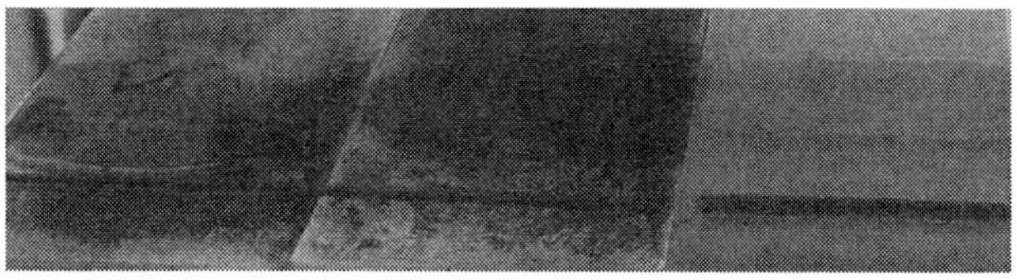		
	Explorer (Resazurin)			
5 min	*Tert*-butyl alcohol + Methylene Blue	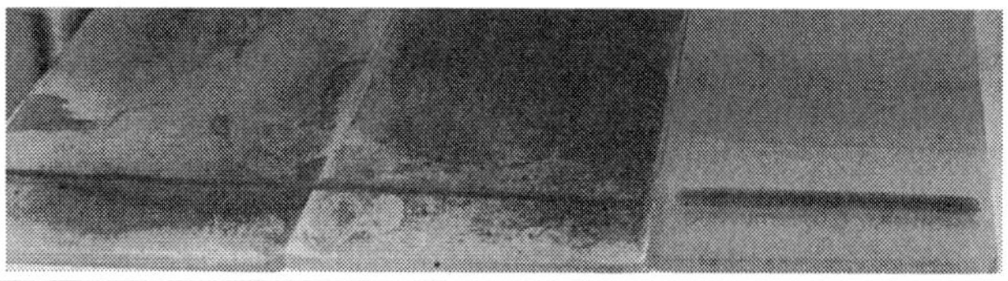		
	Explorer (Resazurin)			
10 min	*Tert*-butyl alcohol + Methylene Blue	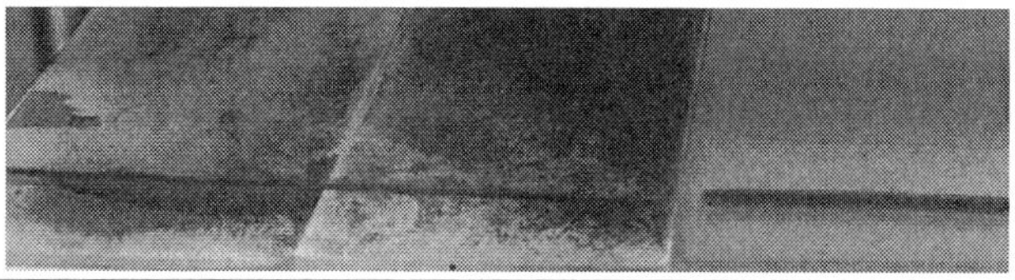		
	Explorer (Resazurin)			
15 min	*Tert*-butyl alcohol + Methylene Blue	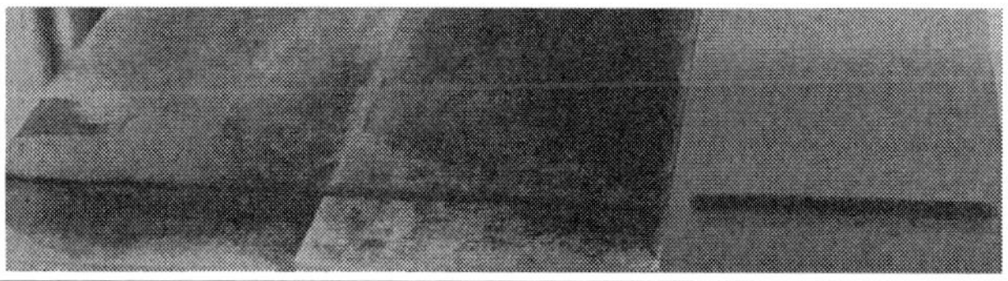		
	Explorer (Resazurin)			

4. CONCLUSION

The ZnO nanorods- and nanowires-based Cu_2O heterojunction were successfully synthesized, characterized, and were compelling enough for the photocatalytic degradation of E_3 hormone with the overall range of efficiencies for nanorods were ~ 30.0 %, and for nanowires 65.0 %. Hence, the as-prepared heterojunctions in this study can be used in a continuous cross-flow closed system to eliminate micro-pollutants such as E_3 from wastewater. In addition, the photocatalytic activity of prepared heterojunctions was assessed by monitoring the rate of resazurin reduction in a motif layer, which may indicate the photocatalytic-induced self-cleaning ability of functionalized glass. Next, using non-commercial prepared ink, an OH radical scavenger, prove the surface structures' absorption.

ACKNOWLEDGEMENTS

Research obtained funding from the Ministry of Education, Youth and Sports of the Czech Republic for project LTT20010 and grant no. RP/CPS/2022/007.

REFERENCES

[1] KAHN, Linda G. et al. Endocrine-disrupting chemicals: implications for human health. *The Lancet Diabetes & Endocrinology*. 2020, vol. 8, no. 8, pp. 703-718. ISSN 2213-8587. Available from: https://www.sciencedirect.com/science/article/pii/S2213858720301297.

[2] DJURIŠIĆ, Aleksandra B., Yanling HE a Alan M. C. NG. Visible-light photocatalysts: Prospects and challenges. *APL Materials*. 2020, vol. 8, no. 3, pp. 030903. Available from: https://doi.org/10.1063/1.5140497.

[3] ANSARI, Ali a P. I. IMOUKHUEDE. Plenty more room on the glass bottom: Surface functionalization and nanobiotechnology for cell isolation. *Nano Research*. 2018, vol. 11, no. 10, pp. 5107-5129. ISSN 1998-0000. Available from: https://doi.org/10.1007/s12274-018-2177-7.

[4] ZHANG, Y. Y. et al. Synthesis, Characterization, and Applications of ZnO Nanowires. *Journal of Nanomaterials*. 2012, vol. 2012. ISSN 1687-4110.

[5] WITKOWSKI, B. S. Applications of ZnO Nanorods and Nanowires - A Review. *Acta Physica Polonica A*. 2018, vol. 134, no. 6, pp. 1226-1246. ISSN 0587-4246.

[6] VAYSSIERES, L. Growth of Arrayed Nanorods and Nanowires of ZnO from Aqueous Solutions. *Advanced Materials*. 2003, vol. 15, no. 5, pp. 464-466. ISSN 0935-9648. Available from: https://doi.org/10.1002/adma.200390108.

[7] YASIR, M. et al. ZnO nanowires, and nanorods based ZnO/WO3/Pt heterojunction for efficient photocatalytic degradation of estriol (E3) hormone. *Materials Letters*. 2022, vol. 319. ISSN 0167-577X.

[8] HAN, Xingrong et al. Rapid and template-free synthesis of Cu2O truncated octahedra using glucose as green reducing agent. *Materials Letters*. 2018, vol. 210, pp. 31-34. ISSN 0167-577X. Available from: https://www.sciencedirect.com/science/article/pii/S0167577X17313095.

[9] ISO 21066:2018 Fine ceramics (advanced ceramics, advanced technical ceramics) - Qualitative and semiquantitative assessment of the photocatalytic activities of surfaces by the reduction of resazurin in a deposited ink film.

[10] GUO, Z. Q. et al. A simple XRD method for determining crystal orientation and its distribution. *Journal of Inorganic Materials*. 2002, vol. 17, no. 3, pp. 460-464. ISSN 1000-324X.

[11] HUANG, Qiuliu et al. Effect of polyethyleneimine on the growth of ZnO nanorod arrays and their application in dye-sensitized solar cells. *Journal of Alloys and Compounds* [online]. 2011, vol. 509, no. 39, pp. 9456-9459. ISSN 0925-8388. Available from: https://www.sciencedirect.com/science/article/pii/S0925838811015131.

[12] MILLS, Andrew a Nathan WELLS. Reductive photocatalysis and smart inks. *Chemical Society Reviews*. 2015, vol. 44, no. 10, pp. 2849-2864. ISSN 0306-0012. Available from: http://dx.doi.org/10.1039/C4CS00279B.

[13] ROCHKIND, M., S. PASTERNAK a Y. PAZ. Using dyes for evaluating photocatalytic properties: a critical review. *Molecules*. 2014, vol. 20, no. 1, pp. 88-110. ISSN 1420-3049.

[14] WARDMAN, Peter. Reduction Potentials of One-Electron Couples Involving Free Radicals in Aqueous Solution. *Journal of Physical and Chemical Reference Data*. 1989, vol. 18, no. 4, pp. 1637-1755. ISSN 0047-2689. Available from: https://doi.org/10.1063/1.555843.

PLASMA TREATMENT OF TEXTILES FOR PROMOTED ADHESION WITH NANOFIBROUS LAYER IN INDUSTRIAL PRODUCTION OF COMPOSITE STRUCTURES

[1]Michal MÉSZÁROS, [1]Jan LUŠTINEC, [1]Jakub MELICHAR, [2]Iva HAMANOVÁ, [2]Miroslav TICHÝ, [2]Marie PILÍKOVÁ, [3]Jozef RÁHEĽ, [3]David PAVLIŇÁK, [3]Oleksandr GALMIZ, [4]Marcela MUNZAROVÁ, [4]Michal BITTNER

[1]SVÚM a.s., Polymers and Fluoroplastics Technology Department, Čelákovice, Czech Republic, EU, meszaros@svum.cz

[2]VÚB a.s., Department of Special Textiles, Ústí nad Orlicí, Czech Republic, EU, inform@vubas.cz

[3]Masaryk University, Faculty of Science, Department of Physical Electronics, Brno, Czech Republic, EU

[4]Nano Medical s.r.o., Liberec, Czech Republic, EU

https://doi.org/10.37904/nanocon.2022.4596

Abstract

Elementary nanofibrous layers show insufficient mechanical properties, especially tensile strength and abrasion resistance. In order to overcome these disadvantages, they are combined with carrier substrates and cover materials. A critical factor of the usability of nanofibrous layers in composite structures is to ensure sufficient adhesion between the carrier medium, the nanofibrous layer and the cover material. The adhesion between the layers is fundamental for the processability of the composite structures into the final products. To ensure sufficient adhesion of the individual components, additional adhesives are commonly used, which, in addition to the ecological burden, bring additional technological steps and costs to the production. One of the possibilities for achieving sufficient adhesion without the use of adhesives is the utilization of the physical component of adhesion by means of plasma activation of surfaces prior to electrospinning.

Keywords: Electrospinning, adhesion, nanofibrous layer, plasma treatment

1. INTRODUCTION

The goal of the research project was focused on the development of innovative nanohybrid structures, whereby the nanofibers were produced by electrospinning and deposited onto the carrier substrate.

The project had two main goals:

1) Ensuring the adhesion between the carrier substrate and the nanofibrous layer by the means of plasma surface modification

2) Preparation of hybrid structures based on textile components from various polymers with surface modification by the means of plasma treatment

The mechanical connection of textile laminates or multi-layer media is usually carried out by applying an adhesive solution onto the substrate before covering it with nanofibers, or by one-sided or two-sided coating with a powder adhesive. To ensure the adhesion of individual layers without the use of an adhesive component, the possibility of modifying the surface layers with plasma treatment was investigated. Research was carried out on the adhesion of contact surfaces of textiles made of different fibers, the surface of which was subjected to the effect of plasma, and the development of special hybrid textile structures. Carrier textiles in the form of woven fabric, knitted fabric and non-woven fabric were used as the carrier substrate.

The effect of plasma treatment dwells in the creation of active particles (ions, excited atoms, radicals, etc.). In our case, the active particles of the working gas created in the discharge caused specific reactions on the polymer surface, where new functional groups were induced to polymer chain (especially hydroxyl groups). This can lead to change in the behavior of surface (i.e. from non-polar to polar, from hydrophobic material to hydrophilic), or can generally improve the amount of chemical and physico-chemical interactions between interacting surfaces leading to better adhesive strength and increase in the wettability.

2. EXPERIMENTAL

2.1. Selection and production of carrier and cover layers for combined nanohybride textile structures

For the carrier substrates, two types of knitted fabrics (weft-knit and warp-knit), four types of woven fabrics (from monofilaments and multifilaments) and five types of nonwoven textiles (spunbond and melt-blown) were produced. In terms of material composition, polyamide, polyester, polypropylene and polyethylene fibers were selected. **Table 1** provides an overview of selected textiles and their material composition.

A spunbond nonwoven textile of mass per unit area weight of 20-30 $g \cdot m^{-2}$ made of 100% polypropylene was used as the covering layer for the production of final layered textile structures with nanofibers.

Table 1 List of selected textile substrates

Sample	Description	Material (wt%)
1	Knitted fabric, weft-knit, 50 $g \cdot m^{-2}$	100 % polypropylene
2	Fabric, plain weave, 65 $g \cdot m^{-2}$	100 % polypropylene
3	Nonwoven, melt-blown, 25 $g \cdot m^{-2}$	100 % polypropylene
4	Nonwoven, bicomponent, spunbond, 25 $g \cdot m^{-2}$	70 % polypropylene (core) 30 % polyethylene (shell)
5	Nonwoven, melt-blown, 35 $g \cdot m^{-2}$	100 % polypropylene
6	Fabric, plain weave, 65 $g \cdot m^{-2}$	100 % polyamide
7	Nonwoven, bicomponent, spunbond, 25 $g \cdot m^{-2}$	50 % polypropylene (core) 50 % polyethylene (shell)
8	Nonwoven, bicomponent, spunbond, 25 $g \cdot m^{-2}$	80 % polyethylene terephthalate (core) 30 % polyethylene (shell)
9	Knitted fabric, warp-knit, 70 $g \cdot m^{-2}$	100 % polyamide

2.2. Selection and production of nanofibrous layers

Selected nanofibrous layers based on polyamide 6 and polyvinylidene fluoride were deposited onto carrier substrates. The deposition of the nanofibrous layer by electrospinning was performed on a Nanospider™ NSLAB 500 machine (Elmarco, Czech Republic) in an arrangement with a short wire electrode.

Electrospinning conditions for polyamide 6: 55 kV applied high voltage, 140 mm electrode-to-collector distance, 6 rpm spinning electrode rotation speed. The nanofibers were applied from a 14 % solution of polyamide 6 prepared from the polymer ULTRAMID B24 (BASF) dissolved in a 2:1 mixture of concentrated acetic acid and formic acid in one layer at a textile feed speed of 25 $mm \cdot min^{-1}$.

Electrospinning conditions for polyvinylidene fluoride: 55 kV applied high voltage, 140 mm electrode-to-collector distance, 6 rpm spinning electrode rotation speed. The nanofibers were applied in one layer at a textile feed speed of 60 $mm \cdot min^{-1}$. Solution composition was 6.75 g of polymer in 50 ml dimethylacetamide with addition of 1 ml tetraethylammonium bromide.

2.3. Research and development of combined textiles structures (nanofibrous layers on carrier substrates)

The aim of the work was preparation of textile composites consisting of carrier substrates made of nonwoven textiles, woven fabrics and knitted fabrics, nanofibrous layers and covering nonwoven textiles. Research into the effect of plasma pretreatment of carrier substrates on adhesion of nanofibrous layer was crucial. Adhesion between nanofibers and carrier depends on both materials and is generally low. For applications where high adhesion is required, glue is added to the product, which practically always leads to a deterioration in breathability. The goal was to achieve sufficient adhesion between the supporting substrate and the nanolayer without the use of any auxiliary adhesive (glue) while barrier functions and improved breathability are preserved.

For the intended purpose, the effect of plasma treatment of carrier substrates with different compositions was tested. The effect of Diffuse Coplanar Surface Barrier Discharge (DCSBD) and Rolling Surface Barrier Discharge (R-SBD) generated plasma in comparison with materials without treatment was investigated.

DCSBD treatment [1]

The plasma was generated under the discharge conditions of 400 W, 15 kHz, air and at atmospheric pressure. The duration of the plasma treatment ranged from 10 to 60 sec, depending on the type of material. The requirement of complete wetting was used as a measure of sufficient plasma treatment - a drop of distilled water with a volume of 5 μl placed on the surface of the treated fabric had to be absorbed into the fabric immediately. The time between the DCSBD treatment and the application of the nanofibrous layer was never longer than 60 min. **Figure 1** shows the DCSBD scheme and the discharge itself during processing. The purpose of this treatment is to create non-specific chemically active or polar groups on the surface of the textile, increasing the chemical activity and free surface energy of the material (wettability).

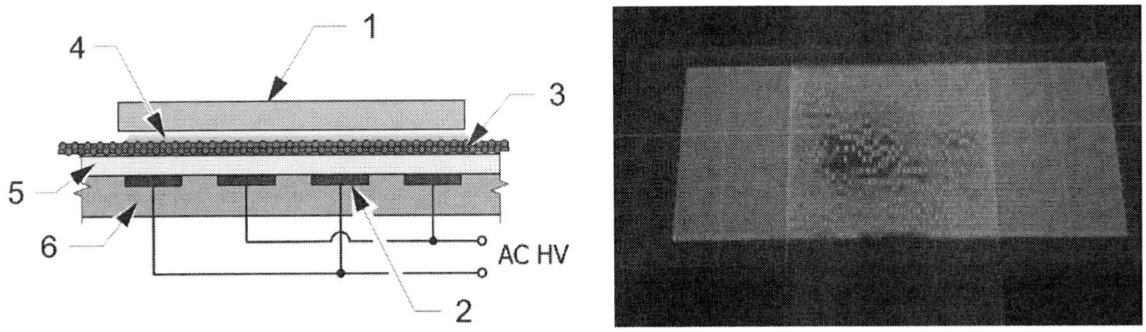

Figure 1 Plasma treatment of textiles using DCSBD. 1 - glass pressure block, 2 - coplanar electrodes, 3 - textile treated with plasma, 4 - discharge plasma, 5 - ceramic electrical barrier, 6 - cooling liquid

R-SBD treatment [2]

The plasma was generated under conditions of discharge of 130 W, 8 kHz, in a mixture of nitrogen (6 $l \cdot min^{-1}$) and propane-butane (0.6 $l \cdot min^{-1}$), at atmospheric pressure (see **Figure 2**). The textile was guided through the discharge space by a simple winding device. The duration of plasma treatment was 30 seconds. The purpose of the treatment was to create a thin layer of a non-specific plasma polymer on the surface of textile, with a significant representation of amino groups. A well-known property of this polymer is the formation of a strongly hydrophilic (water-wettable) surface, which is stable over time. A 5 μl drop of distilled water placed on the surface of the treated fabric immediately soaked into the treated fabric. The time between this plasma treatment and the application of the nanofibrous layer was 3-4 days.

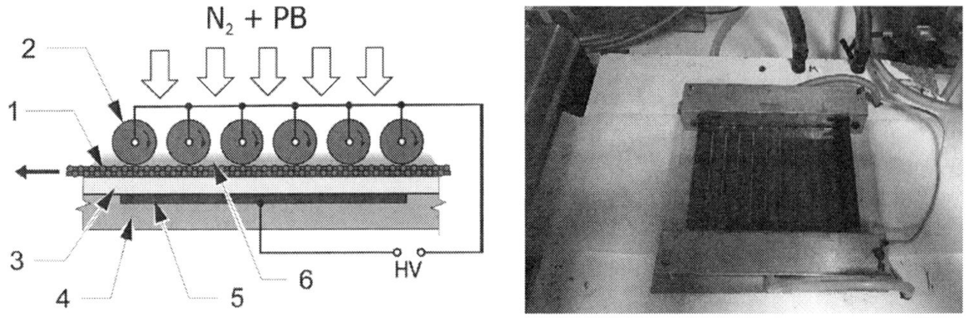

Figure 2 Plasma treatment of textiles using R-SBD. 1 - textile treated with plasma, 2 - roller electrodes, 3 - mica electrical barrier, 4 - cooling liquid, 5 - collection electrode, 6 - discharge plasma

2.4. Testing

Important part of the work was research and development of testing methodology and evaluation of nanofibrous structures. The samples were subjected to microscopic analyses (light microscopy, scanning electron microscopy - SEM) in order to determine the structure of the substrates and nanofibrous layers, to evaluate the condition of the surfaces after the separation of the nanofibrous layer. With the help of elemental analysis (using EDS - electron dispersion spectrometry) the composition of nanofibers was verified. Nanofiber mass per unit area was determined by gravimetry. The most important area of testing was adhesion tests of the nanofibrous layer before and after plasma treatment of the substrate. The evaluation of the tear-off behavior was carried out using modified method based on the adhesive tape measurements, the testing and determination of the peel force was done using a constant rate of extension (CRE) machine with 10 N force sensor.

Loop test

Tests were performed by measuring the force required to tear the deposited nanofiber layer from an area of 25 mm × 25 mm. A TA.XTplusC Texture Analyser (Stable Micro Systems, UK) with a 5 N load cell was used for the measurement. A method used for these measurements called "loop test" is a modified test based on the adhesion measurement for the adhesive tapes. The sample with cut 25 mm × 25 mm sections was glued to a plastic holder with double-sided adhesive tape on the side of the substrate, which was then fixed in the lower jaws of the measuring device. A 10 cm long test adhesive tape with sufficiently high adhesion (in accordance with ISO 2409:2003) was placed in a controlled manner into the upper movable jaws so as to form a loop (see **Figure 3**). By keeping the same length (10 cm) and the shape of the loop, a reproducible pressure force of the test adhesive tape can be achieved. The tape was not pressed to the sample in any other manner. Subsequently, the jaws were moved away, and the value of the applied force vs. jaw position was recorded. The speed of movement of the jaw head was 5 mm·s^{-1}. The trigger force was 1 g (i.e., 10 mN). As the parameter for comparison, the area under the obtained curve corresponding to the mechanical work necessary to tear off the layer W (N·mm) was calculated. A minimum of 10 measurements were carried out for each sample. Other evaluated parameters were average force (F_{avg}) and maximal force (F_{max}).

Figure 3 Illustrative photo of the adhesive force measurements by the loop test method

2.5. Verification of plasma treatment application in production conditions

The pilot test was carried out at NanoMedical s.r.o. using spinning machine Nanospider™NS 8S1600U, see **Figure 4**. Polyamide 6 based nanofibers were deposited onto a 100 % polypropylene spunbond nonwoven fabric substrate that had been plasma treated using DCSBD method at four winding speeds. **Table 2** lists the production parameters.

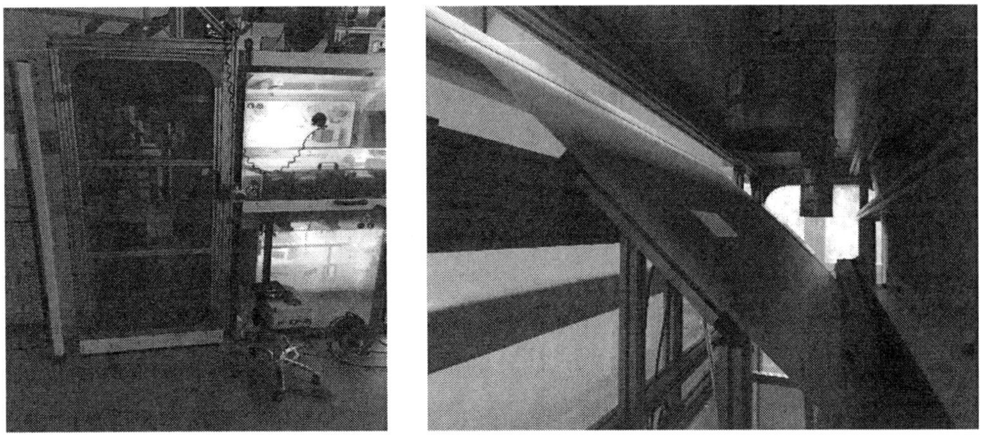

Figure 4 Integration of DCSBD plasma system into the production line

Table 2 Technical parameters of plasma treatment in production conditions

Sample	Voltage on the spinning electrode	Winding speed	Plasma exposure time
L1S7711	75 kV	2.42 $m \cdot min^{-1}$	1.98 s
L1S7712	65 kV	1.80 $m \cdot min^{-1}$	2.67 s
L1S7713	55 kV	1.23 $m \cdot min^{-1}$	3.90 s
L1S7714	45 kV	0.82 $m \cdot min^{-1}$	5.85 s

3. RESULTS

Graph below (see **Figure 5**) shows the absolute difference in the mechanical work (ΔW) necessary to tear off the layer between samples without treatment and plasma treated samples. Results of loop tests for production trials are displayed as comparison of parameter W between untreated and treated samples (see **Figure 6**).

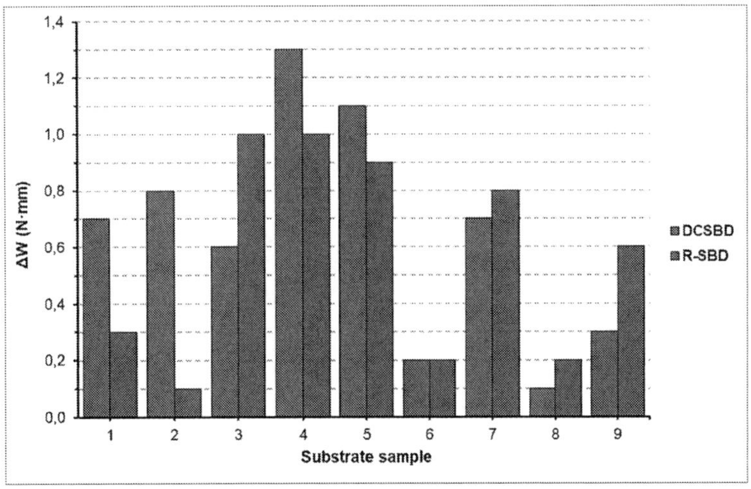

Figure 5 Absolute difference in the mechanical work necessary to tear off the polyamide 6 based nanofibrous layer between substrates without treatment and plasma treated samples

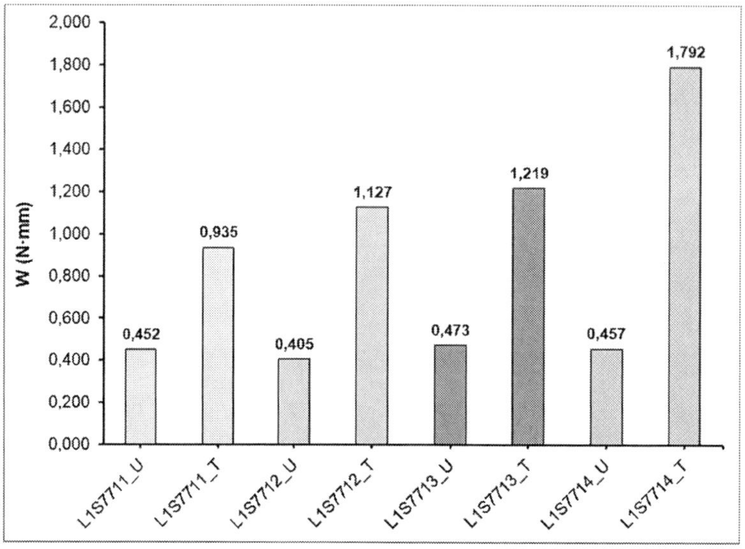

Figure 6 Comparison of work necessary to tear off the nanofibrous layer from untreated (left column) and plasma treated samples (right column of the same color) obtained in the production line trials

4. CONCLUSION

Based on the results of adhesion testing it can be said that both types of tested plasma treatments lead to an improved adhesion between tested textile substrates and nanofibrous layers. The choice of the most suitable type of plasma treatment must therefore be decided by technological aspects, such as the speed of plasma treatment, time stability of the effect, operational requirements, or energy and price requirements. The R-SBD treatment offers a slightly better energy balance (130 W vs 400 W) and temporal stability of the effect achieved (weeks vs days), but requires nitrogen gas management, which increases the operating costs of the treatment. With DCSBD, this problem disappears, because the DCSBD discharge works in a normal atmosphere (air). For both plasma generators, it is necessary to take into account the removal of gaseous products. System for DCSBD plasma treatment of carrier substrates was successfully integrated into industrial line for nanofibrous layer production. Achieved improved adhesion allowed transport and further processing of textiles with nanofibrous layer without any need for additional adhesives.

ACKNOWLEDGEMENTS

This research was solved within the framework of Collective Research networking CORNET - 28th Cornet Call, project "Hybrid Textile Structures based on Nanofibres with Barrier Effects against Bacteria and Viruses (NanoHyb)", project number EG19_263/0018773. This R&D project from programme OP PIK was realized in years 2019 - 2023 with state support, project number CZ.01.1.02/0.0/0.0 /19_263/0018773; support program Cooperation - Clusters - Call VI.

REFERENCES

[1] PAVLIŇÁK, D., GALMIZ, O., PAVLIŇÁKOVÁ, V., POLÁČEK, P., KELAR, J., STUPAVSKÁ, M., ČERNÁK, M. Application of dielectric barrier plasma treatment in the nanofiber processing. *Materials Today Communications.* 2018, vol. 16, pp. 330-338. Available from: https://doi.org/10.1016/j.mtcomm.2018.07.010.

[2] DVOŘÁKOVÁ, H., ČECH, J., STUPAVSKÁ, M., PROKEŠ, L., JURMANOVÁ, J., BURŠÍKOVÁ, V., RÁHEĽ, J., SŤAHEL, P. Fast Surface Hydrophilization via Atmospheric Pressure Plasma Polymerization for Biological and Technical Applications. *Polymers.* 2019, vol 11, no. 10, pp. 1-15. Available from: https://doi.org/10.3390/polym11101613.

PRODUCTION TECHNOLOGY OF A FLAT NANOFIBROUS STRUCTURE USING THE BELT ELECTRODE FOR AC ELECTROSPINNING

[1]Ondrej BATKA, [1]Jan VALTERA, [1]Josef SKRIVANEK, [1]Ondrej FRIEDRICH

[1]*TUL - Technical University of Liberec, Department of Textile Machine Design, Liberec, Czech Republic, EU, ondrej.batka@tul.cz*

https://doi.org/10.37904/nanocon.2022.4602

Abstract

The article deals with new technology for the production of a flat nanofibrous structure using AC electrospinning and the development of a new type of spinning electrode. Simulations of the distribution of the electric field around the electrode were performed. The dependencies of the electric field intensity on the design parameters of the electrode were determined. Based on the performed simulations, a suitable technical and technological solution of the electrode for the production of a flat nanofibrous structure was designed. The functionality of the designed technological system was verified experimentally. The results of the experiments demonstrated the high potential of the designed system for the production of flat nanofibrous structures.

Keywords: AC electrospinning, electric field, nanofibers, spinning electrode, belt electrode

1. INTRODUCTION

Electrospinning is currently one of the established methods used for the industrial production of nanofibers. Nanofiber products have the potential for application in many fields of human activity [1-4]. One type is AC-electrospinning [5,6], which uses an alternating electric field. This principle is currently used mainly for producing narrow strips of nanofibers or for producing nanofiber yarns [7-10]. Currently, there is an effort to use AC-electrospinning to produce flat nanofiber products with large widths. This contribution deals with researching and developing an electrode suitable for creating flat nanofibrous products.

Based on successful experiments with discontinuously rewinding a textile with a polymer solution over an electrically charged edge, the aim of this task was to design a functional model of a new type of AC electrode with a continuously rewound belt. The assumption here is that the rewound belt will enable the transport of the polymer solution from the reservoir to the upper part, where nanofibers will form due to the supercritical value of the electrical intensity. The subsequent rewinding of the belt and its wading in the polymer bath ensure continuous cleaning and dissolution of any fibers on the surface of the belt. The width of the belt determines the length over which the fibers are formed, which enables the production and application of nanofibers to the underlying textile in an adequate amount and uniformity for subsequent use, e.g. in filtration, technical textiles, etc. As part of this task, based on the detailed results of the electric field simulation, a strip AC electrode including drive and gearbox was designed. The functionality of the manufactured laboratory equipment was experimentally verified, including the measurement of the specific gravity of the nanofibrous layer.

2. SIMULATION

The analysis of the electric field was used as support for the design of the optimal geometry of the electrode during the AC-electrospinning. The electric field distribution was analyzed using the finite element method in the Autodesk Simulation Mechanical software 2015. The FEM model was created as a linear 2D problem. **Figure 1 a)** shows a detailed FEM model. Its dimensions and shape correspond as closely as possible to the real model. A detailed model of the device consists of a belt electrode (steel plate, wading roller and a wound belt), a device frame, a reservoir with a polymer solution and an air environment. **Figure 1 c)** shows a simplified model that was used in the variational analysis of the strip electrode geometry. Spinning by AC electrospinning does not require the use of an oppositely charged collector, as a virtual collector is created around the electrode due to air ionization, and the charge travels to the far end with a different potential. For this reason, in the model, this virtual collector is created as an equidistant boundary around the electrode at a distance of 50 mm with a boundary condition of 0 V. A boundary condition of 25 kV was applied to the electrode and the polymer, which corresponds to the supplied electric voltage. The relative permittivity of 109 was entered for the electrode. The surrounding environment consists of air with a relative permittivity of 1. The model also contains a polymer solution with a relative permittivity of 20 and HDPE with a value of 2.3.

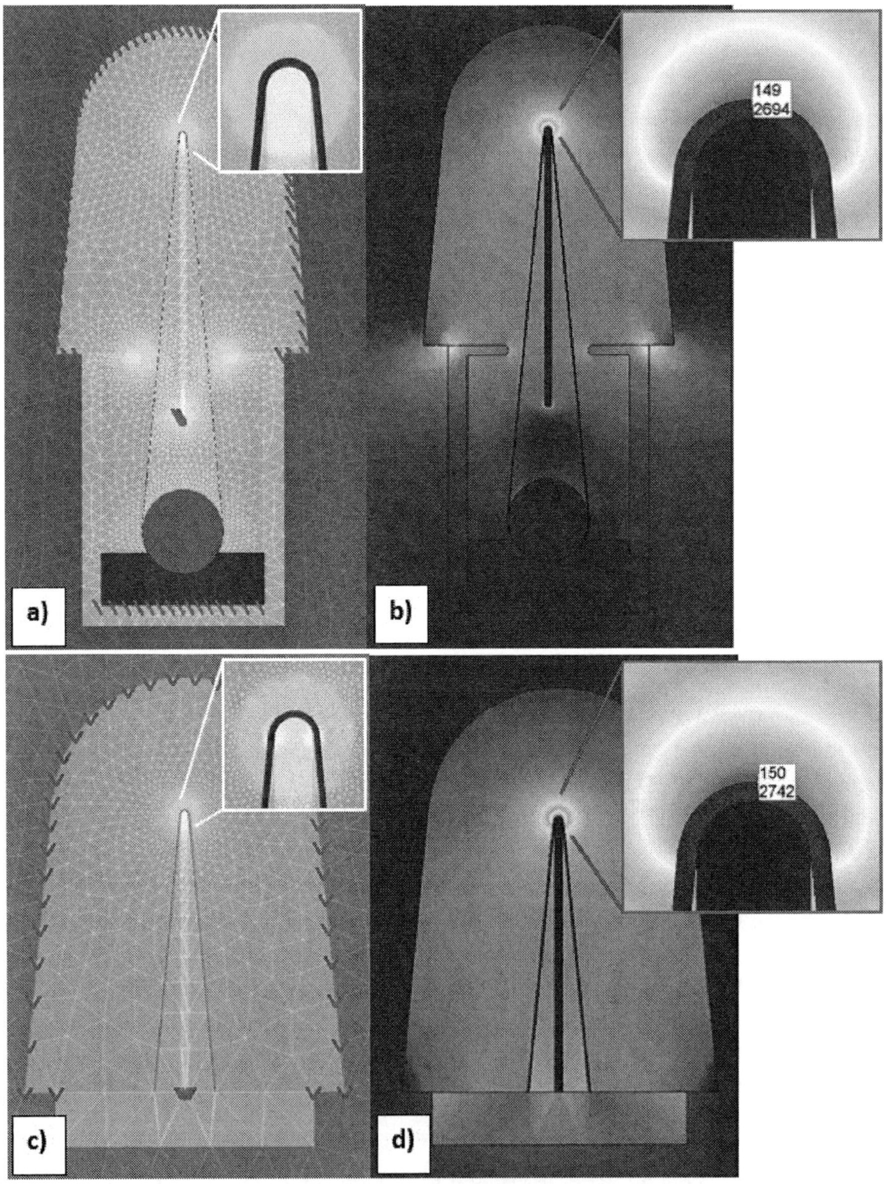

Figure 1 detailed FEM model a), electric field distribution for detailed model b), simplified FEM model c), electric field distribution for simplified model b)

The first simulation was was performed with a detailed model **(Figure 1 a)).** The goal was to determine the electric field's overall distribution in the proposed device's vicinity. A voltage of 25 kV was applied to the electrode and the polymer. **Figure 1 b)** shows this model's electric field intensity distribution. At the top of the electrode where electrospinning is desired, the magnitude of the intensity is 2694 V/mm.

The second task was performed with a simplified model whose geometry is shown in **Figure 1 c)**. This model was created due to the complexity of adjustments when changing the parameters on the first model and the speed of calculation. In the case of a simplified model with an applied voltage of 25 kV, the magnitude of the intensity on the spinning area of the electrode is 2742 V/mm, see **Figure 1 d)**. The difference between the detailed and the simplified model is slight; therefore, the simplified model was used for further analyses.

Three analyzes were performed with the simplified model. A sketch with the dimensions of the simplified model is shown in **Figure 2 a)**. Material properties and boundary conditions were used from the detailed model. During the first analysis, the electrode height parameter H was varied from 25 to 150 mm at a constant thickness t = 3 mm and radius R = 1.5 mm. The distance H is measured from the container's top to the electrode's highest part. **Figure 3 a)** shows the intensity distribution at the top of the electrode when the height H changes. At H = 25 mm, the intensity value is 2863 V/mm. As the value of height H increases, the intensity is almost constant. For further analyses, the electrode height H = 100 mm was chosen. In the second analysis, the radius R at the top of the electrode was changed from 0.25 to 2.5 mm at a constant height of the electrode H. As the radius increased, the thickness of the electrode t also increased, depending on t = 2R. Model of the top of the tip is shown in **Figure 2 b)**. The graph in **Figure 3 b)** shows the dependence of the intensity at the top of the electrode as the radius changes. At the size of the radius R = 0.25 mm, the intensity at the top of the electrode reaches a value of 4407.5 V/mm. As the radius value increases, the intensity value decreases exponentially. In the third analysis, the electrode thickness t was varied at a constant electrode height H and radius R = 0.5 mm, according to the sketch in **Figure 2 c)**. **Figure 3 c)** shows the dependence of the intensity at the top of the electrode when the electrode thickness t changes from 1 to 5 mm. At a thickness of t = 1 mm, the intensity reaches a value of 3889 V/mm. As the thickness increases, the intensity value decreases exponentially.

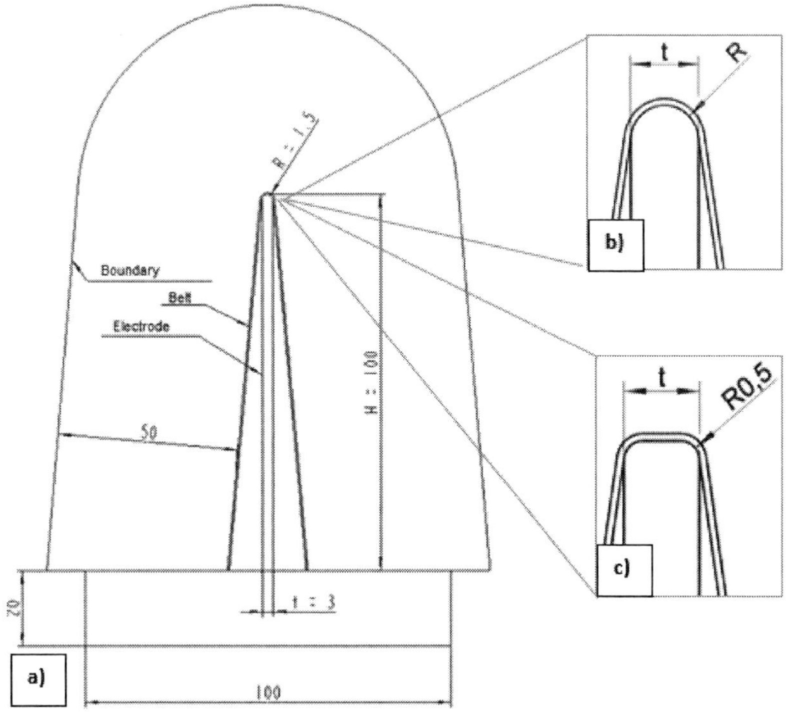

Figure 2 sketch of model

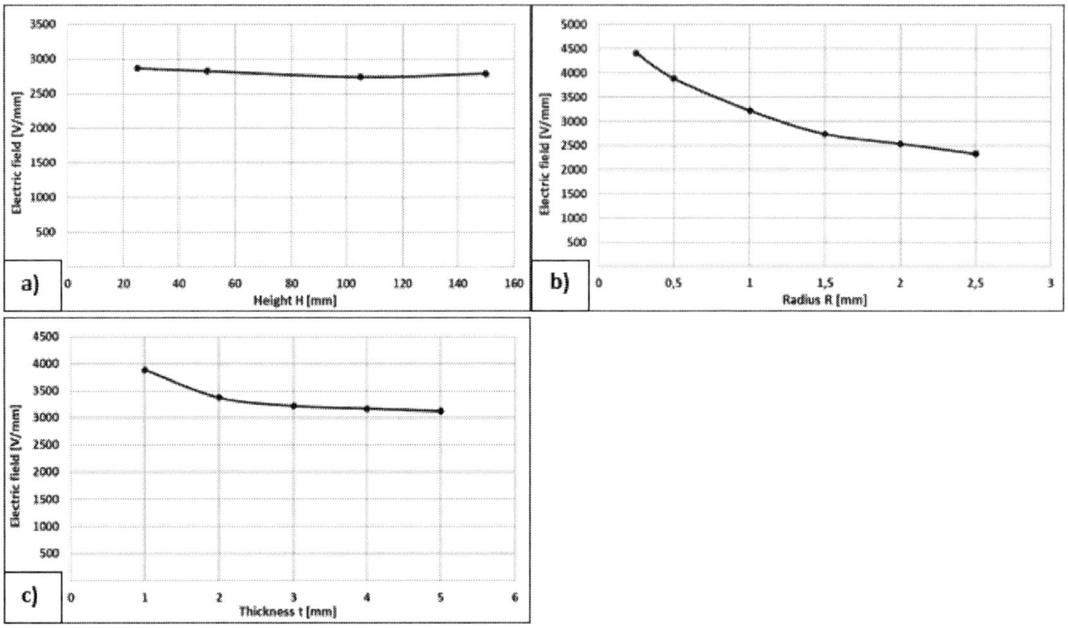

Figure 3 a) dependence of the electric field on the height of the electrode, **b)** dependence of the electric field on the radius of the electrodes tip, **c)** dependence of the electric field on the thickness of the electrode

3. DESIGN

Figure 4 shows a 3D model of a device for producing nanofibers constructed in the Creo Parametric program. An asynchronous motor with a belt transmission previously used in tests of other methods of alternating spinning was used to drive the device. The engine speed could be regulated using a frequency converter. The output speed from the belt mechanism was transmitted using a worm gear. This achieved optimal speeds and torque for driving the electrode belt. The output shaft of the worm gearbox drives the belt drive cylinder. At the same time, this shaft works as isolation of the high voltage between the AC electrode and the grounded worm gear. The belt drive ensures the shaft rotation on which the roller is mounted. The shaft is fixed using sliding bushings in the container. In the container, there is a polymer solution, without which it would not be possible to spin, as well as two drain holes equipped with screws. The movement is given to the belt through the notched roller, and the polymer is also applied over it.

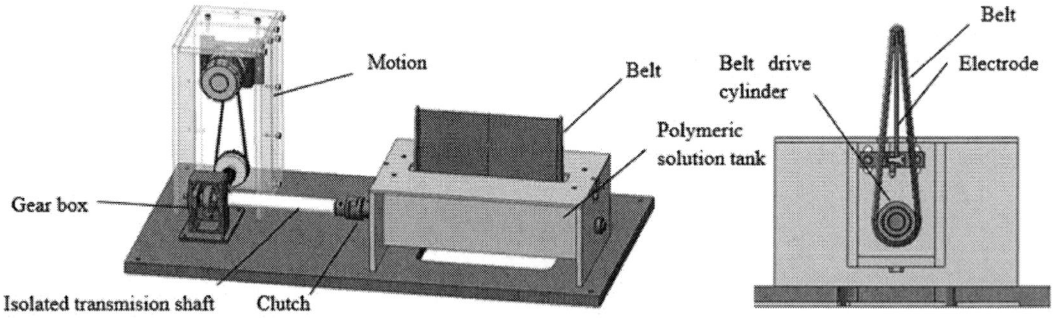

Figure 4 design of electrode

An electrode is placed above the roller, which is made of sheet metal and has a rectangular shape with rounded edges with dimensions derived from the results of the electric field simulation. During the process, a belt is moved over the upper edge of the electrode. The construction of the electrode is designed so that it can be moved vertically due to belt tension. A lid covers the electrode's frame with the polymer solution's reservoir with a hole for the belt.

4. EXPERIMENT

The designed experimental device was built and used for the experiment was carried out in order to verify the functionality of the model and whether it is possible to produce nanofibers in the required width. A polymer solution of polyvinyl butyral dissolved in ethanol was used for the experiment (10%wt PVB Movital B60H in 98%vol ethanol Technisolv). During initial tests, different belt materials were verified. First, a strip of plastic foil was tested. The problem was that the lower knurled shaft did not have enough traction on the foil and the foil also cut into the edge of the electrode. In the end, a non-woven fabric was chosen as the belt. The non-woven fabric belt easily moved over the electrode's edge and absorbed the polymer well. The produced nanofibers were applied to a rotating drum placed above the electrode. The experiment is shown in **Figure 5**.

Figure 5 experiment

Various electrode widths and thicknesses were tested. For the first experiment, an electrode with a thickness of 3 mm and a width of 240 mm was produced. At these parameters, self-spinning did not occur. For these parameters, the intensity value is 2739 V/mm (**Figure 5**). This intensity is close to the critical spinning value of 2150 V/mm, at which spinning no longer occurs. In the second experiment, an electrode with a thickness of 1 mm and a width of 240 mm was used. At these parameters, a stable spinning process was already taking place. The effort was to produce nanofibers as wide as possible. Therefore, an electrode with a thickness of 1 mm and a width of 320 mm was used for the next experiment. At these parameters, the productivity of nanofiber production was found to be 2 g*min-1*m-1.

5. CONCLUSION

This research aimed to create a model, perform analyses, design the structure and then verify it experimentally. An electrode with a height of 100 mm, a thickness of 1 mm and a radius of 0.5 mm was selected from the analysis. With these electrode parameters, the value of the intensity on the spinning area was 3889 V/mm. Subsequently, the device's design for producing nanofibers was designed. An electrode with a width of 320 mm was used, and the productivity of nanofiber production was measured to be 2 g*min-1*m-1. This productivity can be further increased by using several electrodes in a row.

The advantage of this design is self-cleaning, when the polymer on the electrode does not dry out due to the constant wetting of the strip in the reservoir with the solution. With this design, the polymer is in the container, but it is possible to design such a design where the polymer is wiped off, and the fresh polymer is applied. The construction of the functional model is designed with an electrode width of 320 mm. Structurally, however, it is possible to design a strip electrode with a longer usable length, enabling the continuous production of flat nanofibrous material with a larger width. The disadvantage is spinning from the edges of the electrode, where as a result of higher values of the electric field intensity, there is a more intensive production of fibres and the edges of the textile are more massive. This problem, which is common in textiles, is usually solved by cutting off the edges of the textile. The fineness of the produced layer was measured from the obtained samples, which showed very good uniformity of the layer in the center and edges of the produced nanofibrous layer. A more detailed measurement of fineness and the design of a new device will be the subject of further work in this area of research

ACKNOWLEDGEMENTS

This publication was written at the Technical University of Liberec as part of the project with the support of the Specific University Research Grant 2021-5029, as provided by the Ministry of Education, Youth and Sports of the Czech Republic in the year 2021.

REFERENCES

[1] LANNUTTI, J., D. RENEKER, T. MA, D. TOMASKO and D. FARSON. Electrospinning for tissue engineering scaffolds. *Materials Science and Engineering: C* [online]. 2007, vol. 27, no. 3, Next Generation Biomaterials, pp. 504-509. ISSN 0928-4931. Available from: https://doi.org/10.1016/j.msec.2006.05.019.

[2] *Electrospun Nanofibers for Energy and Environmental Applications* [online]. [vid. 2022-09-01]. Available from: https://link.springer.com/book/10.1007/978-3-642-54160-5.

[3] BENDKOWSKA, W. Use of nanotechnology in the textile industry. *Przeglad Wlokienniczy*. 2003, vol. 5, pp. 17-21. ISSN 1230-0381.

[4] KALAYCI, V., M. OUYANG a K. GRAHAM. Polymeric nanofibres in high efficiency filtration applications. *Filtration*. 2006, vol. 6, no. 4, pp. 286-293. ISSN 1479-0602.

[5] KOCIS, Lubomir, Pavel POKORNY, David LUKAS, Petr MIKES, Jiri CHVOJKA, Eva KOSTAKOVA, Jaroslav BERAN, Martin BILEK, Jan VALTERA, Evzen AMLER, Matej BUZGO and Andrea MICKOVA. *Method for production of polymeric nanofibers by spinning of solution or melt of polymer in electric field, and a linear formation from polymeric nanofibers prepared by this method* [online]. EP2931951B1. Available from: https://patents.google.com/patent/EP2931951B1/en?inventor=valtera&oq=inventor:valtera.

[6] POKORNY, P., L. KOCIS, J. CHVOJKA, D. LUKAS, J. BERAN, M. BILEK and K. PEJCHAR. Alternating current electrospinning method for preparation of nanofibrous materials. In: *NANOCON 2013 - Conference Proceedings, 5th International Conference*. 2013, pp. 302-304. ISBN 978-80-87294-47-5.

[7] BERAN, Jaroslav, Jan VALTERA, Martin BÍLEK, Ondřej BAŤKA, Josef SKŘIVÁNEK, Petr ŽABKA, Jiří KOMÁREK, David LUKÁŠ, Pavel POKORNÝ, Eva KUŽELOVÁ-KOŠŤÁKOVÁ, Petr MIKEŠ, Jiří CHVOJKA, Tomáš KALOUS and Filip SANETRNÍK. A linear fibre formation with a case of polymeric nanofibres enveloping the supporting linear formation constituting the core, the method and equipment for its production [online]. CZ306428B6. Available from: https://patents.google.com/patent/CZ306428B6/en

[8] VALTERA, J., T. KALOUS, P. POKORNY, O. BATKA, M. BILEK, J. CHVOJKA, P. MIKES, E.K. KOSTAKOVA, P. ZABKA, J. ORNSTOVA, J. BERAN, A. STANISHEVSKY and D. LUKAS. Fabrication of dual-functional composite yarns with a nanofibrous envelope using high throughput AC needleless and collectorless electrospinning. *Scientific Reports* [online]. 2019, vol. 9, no. 1. Available from: https://doi.org/10.1038/s41598-019-38557-z.

[9] MADHESWARAN, Divyabharathi, Šárka HAUZEROVÁ, Jan VALTERA, Maxim LISNENKO, Manikandan SIVAN, Ondřej BAŤKA, Věra JENČOVÁ, David LUKÁŠ, František JANDA and Eva KUŽELOVÁ KOŠŤÁKOVÁ. Braided threads with AC electrospun nanofibers for hygienic and medical applications - production and properties. In: *NANOCON 2021* [online]. 2021, pp. 252-257. Available from: https://doi.org/10.37904/nanocon.2021.4355.

[10] SHYNKARENKO, Andrii, Dekel AZULAY, Sarka HAUZEROVA, Andrea KLAPSTOVA, Michal MOUCKA, Vera JENCOVA and David LUKAS. *Fabrication of Micro/Nanofibrous Scaffolds using a Robotic Manipulator and Their Application for Tissue Engineering* [online]. 2021. [vid. 2021-10-28]. ISSN 2693-5015. Available from: https://doi.org/10.21203/rs.3.rs-112972/v1

INFLUENCE OF THE SPINNING ELECTRODE MATERIAL ON THE FIBER STRUCTURE

Josef SKRIVANEK, Ondrej BATKA, Pavel HOLEC, Jan VALTERA

TUL - Technical University of Liberec, Department of Textile Machine Design, Liberec, Czech Republic, EU, josef.skrivanek@tul.cz

https://doi.org/10.37904/nanocon.2022.4601

Abstract

The article focuses on the issue of the spinning process in AC electrospinning. The conducted study aimed to identify the possible influence of the material type of the spinning electrode on the fibres formed. As part of the work, studies were carried out on our own test spinning equipment intended for the production of nanofibers. The electrodes were made of three different materials: stainless steel, aluminium alloy and plastic. Productivity experiments and individual partial nanofibrous structures related to individual electrodes were performed on these electrodes. The mentioned structures were examined using an SEM microscope and also statistically evaluated. Other parts of the article are simulations of the electrostatic field for the used electrodes.

Keywords: Electrospinning, electrode, nanofibers, electrostatic field

1. INTRODUCTION

The issue of nanofibers and their use is currently among the sought-after directions of primary as well as applied research. They have high added value and broad application use. The most common applications of nanofibers are in the fields of filtering water [1], air [2] and other fluids [3], medical applications [4], etc. [5]. The production of nanofibers can be divided into two basic directions. The essential representative of the first group is production using melt-blown technology, which currently reaches fibre diameters of around 500 nm [6]. The second direction of production is processes that are directly focused on the production of nanofibers. A wide variety of processes are used to create nanofibers. The first and by far the most used process is the production of nanowires using direct current (DC). Another lesser-known method of creation is using alternating current (AC). A minor production method includes drawing [7] and force spinning [8]. Some of the methods can be combined with each other. For example, a combination of force spinning and electrospinning or electrospinning and melt-blowing.

The most widely used and well-known method of producing nanofibers is electrospinning. This process of creating nanofibers is a breeding ground for the scientific community, especially in the field of electrostatic field research [9]. Of the spinning methods listed above, DC-electrospinning is the most well-known and widespread method of producing nanofibrous structures. Another type of electrospinning is AC-electrospinning; compared to DC-electrospinning, it is a very productive method of creating nanofiber structures. To implement the method, it is necessary to use a high-voltage source that changes polarity during the formation of fibres. This phenomenon of polarity change creates a nanofibrous structure in which the positive/negative polarity alternates, and the effect of the mutual attraction of the fibres is achieved. It is necessary to point out that the use of high voltage is necessary for electrospinning. This fact entails several technological disadvantages and pests, especially in the area of safety and a limited range of materials for the preparation of the spinning solution [10].

In this study, AC electrospinning was used to perform experiments and create nanolayers using a conical electrode. Three basic materials from which the spinning heads of the electrodes were completed and tested during the experiment. Microscopic images and histograms related to the individual electrodes used are evaluated in the study. The obtained nanofibrous layers are comprehensively described in terms of different structures.

2. METHODS AND EXPERIMENTS

Equipment design

The main task in this part was the creation of a suitable technical background for implementing experiments. **Figure 1** shows schematically the equipment used during the experiment. The main part of the device is the spinning unit 1, which consists of the spinning electrode 2 placed on the dosing tube 3. This tube is used to place the spinning head in the required position, and secondly, with its help, the polymer solution is transported. The spinning tube is part of pumping device 4, which contains dosing device 5 and solution stock 6. The goal of the solution supply was to create a continuous layer of solution on the spinning head and tube, which flows around their outer shell. In this way, the liquid state of the solution is ensured, and there are no dry areas that cause product defects. An AC high-voltage source - Trek 50/12 high-voltage amplifier 7 - was connected to the dosing tube. The control of the dosing device and the drive of the collection rotary collector 8 was ensured by means of the connected PLC module (Programmable Logic Controller) 9. The resulting nanofibrous formation 10 was applied to some textile 11. During the experiment, the identical value of the effective voltage of 30 kV sinusoidal current with a frequency of 50 Hz was set on the high voltage source.

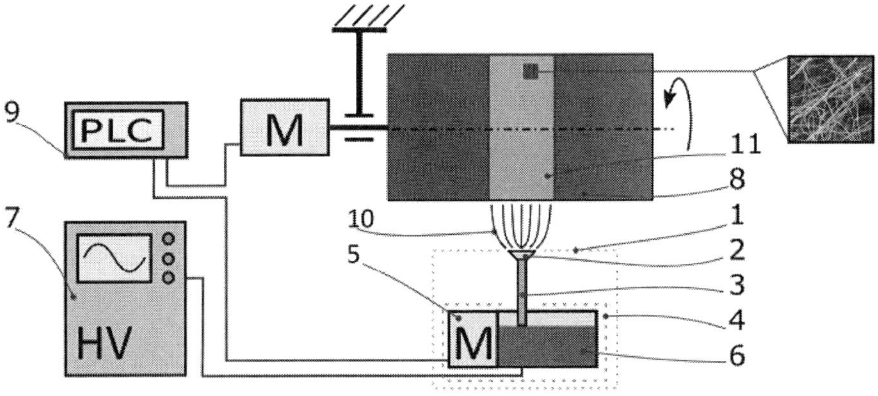

Figure 1 Scheme of the technical equipment used during the experiment

The next step was the design of a suitable shape of the electrode head or the spinning electrode head. A basic CAD geometry of the head was created in Creo 7.0.1.0 software. Furthermore, a mathematical model was created using geometry, and a series of analyses was performed with the aim of simulating a charge distribution in an electrostatic field. Based on a series of analyses of the distribution of the electrostatic field, the optimal geometry of the electrode head was determined. The analysis was performed for metal and plastic materials. The results of the analysis are shown in **Figure 2**. The Autodesk Simulation Multiphysics software was the primary tool for designing a suitable shape for the spinning head.

Simulation

Simulations were performed in FEM software Autodesk Simulation Mechanical 2015. The task was simulated as 2D axisymmetric, where the model's axis corresponds to the electrode's axis. **Figure 2 a)** shows the FEM model used for the simulations. The model consists of the following parts. A metal rod (1) with a relative permittivity of 109, a conical electrode (2) with a relative permittivity of 109 for duralumin and steel and 3.4 for PA6 material, a polymer solution (3) covering the electrode with a relative permittivity of 20, a disk (4) that simulates a lid and a container with a polymeric solution made of polyethene with a relative permittivity of 2.4 and the surrounding air (5) bounded by a spherical surface with a diameter of 2500 millimetres with a relative permittivity of 1. A boundary condition of a voltage value of 30 kV and an electrical stiffness of 109 A/V was applied to the bottom of the rod (6) and at the air boundary (7) with a voltage value of 0 V and an electrical stiffness of 109A/V.

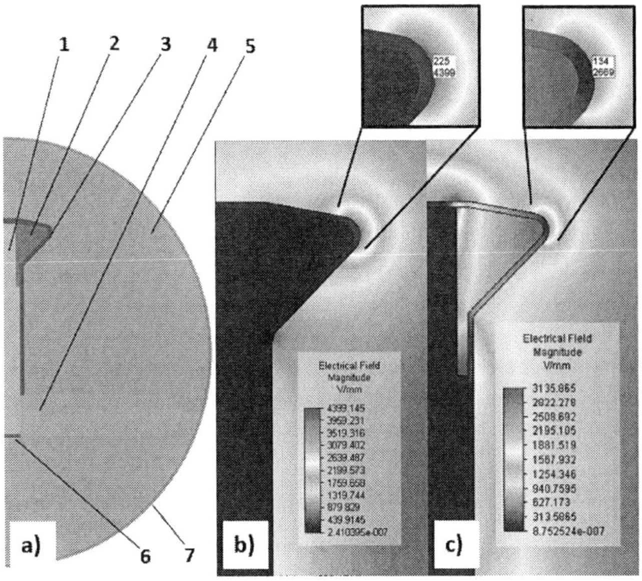

Figure 2 Distribution of electrical field

Experiment

Based on the analyses performed, real electrode models were produced using CAD models. The experiment was carried out using the methodology described in the previous paragraph with a set of spinning electrodes, see **Figure 3**. The assembly modification during the experiment consisted of replacing the spinning heads, which were geometrically identical and differed in the materials used. The relative humidity was 55 %, and the ambient temperature was 24 °C during all experiments.

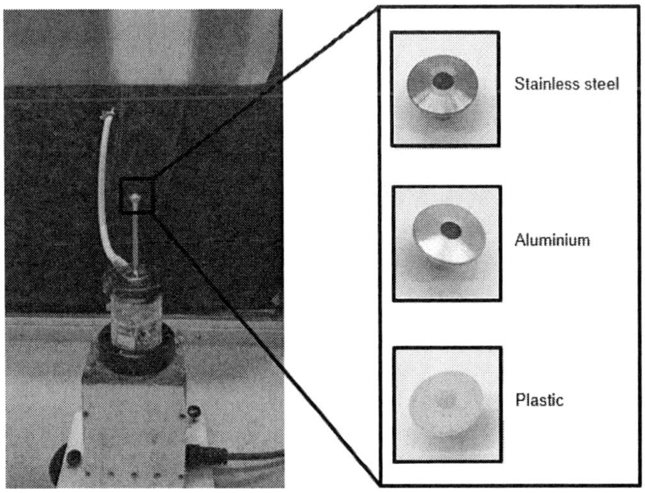

Figure 3 Nanofibrous layers

Preparation of solution

Fibrating solution concentration 10 wt. % was prepared by dissolving polyvinyl butyral (PVB - CAS 63148 65 2, Mowital B60H, Kuraray, Germany, Hattersheim) in 96% denatured ethanol (CAS 64 17 5; TechniSolv (1:1:1), VWR, Belgium, Leuven). Dissolution was carried out at room temperature for 24 hours using a magnetic stirrer.

Analysis of nanofibrous layers

Nanofibrous layers prepared by alternating electrospinning of a 10% PVB solution using three different types of electrodes were coated with a 10 nm layer of gold in a Quorum apparatus (Lewes, UK). Subsequently, images of individual samples were taken on a Vega 3 scanning electron microscope (SEM) (TESCAN, Brno, Czech Republic) at different magnifications. The applied voltage was 10 kV. Fibre diameters of the acquired layers were measured in ImageJ software (v. 3.3.1, Garching, Germany). In total, 250 diameters were measured for each analyzed sample.

3. RESULTS

In the first step, the spinning process was monitored, especially concerning the stability of the nanofibrous formation moving in the free environment between the place of creation and the place of collection of fibres. The tests showed that the compactness of the structure increases as the intensity of the electric field increases in the area where fibres are formed. Further, with the increase in intensity, there was a decrease in the misalignment from the desired trajectory of the movement of the fibres through the free environment.

The results of the SEM analysis are shown in **Figure 4**. By spinning the PVB solution, nanofibrous layers of good quality were prepared. Fibre branching and drop defects were noted, but their amount was minimal and did not affect the further usability of the product. The resulting fibre structure was fluffier and bulkier than that of fibres prepared by the direct spinning method. No differences were observed between samples prepared using electrodes of different materials. Neither the rate nor the representation of individual defects has changed. From the SEM images, it can be concluded that the material of the electrode has no direct effect on the electrospinning process.

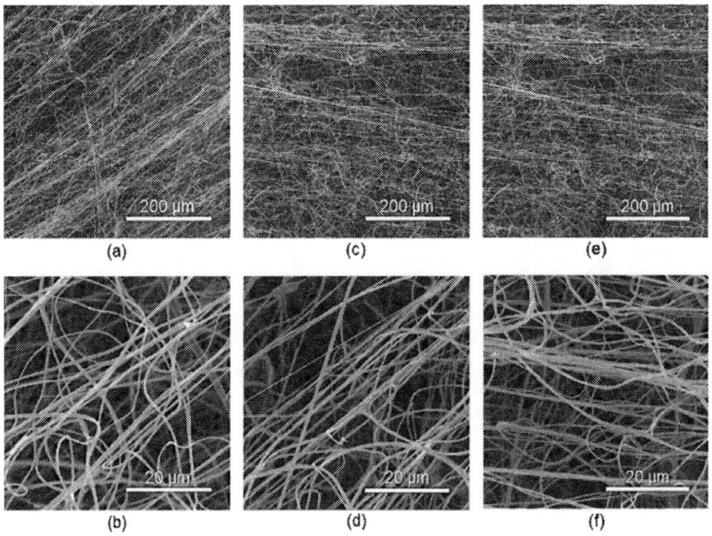

Figure 4 Images of nanofibrous layers (top) and their details (bottom) prepared using electrodes A (a, b), B (c, d) and C (e,f). It is clear from the images that the electrode material did not affect the resulting structure or quality of the nanofibers prepared by electrospinning.

Figure 5 shows the results of measuring the fibre diameters of samples 1, 2 and 3. Fibres with diameters in the range of 400 to 600 nm were most frequently represented. The smallest fibres reached approximately 200 nm, the largest then less than 1000 nm. The distribution of fibre diameters was almost the same for all three histograms. It can thus be judged that the material of the electrode did not particularly affect the diameters of the emerging fibres.

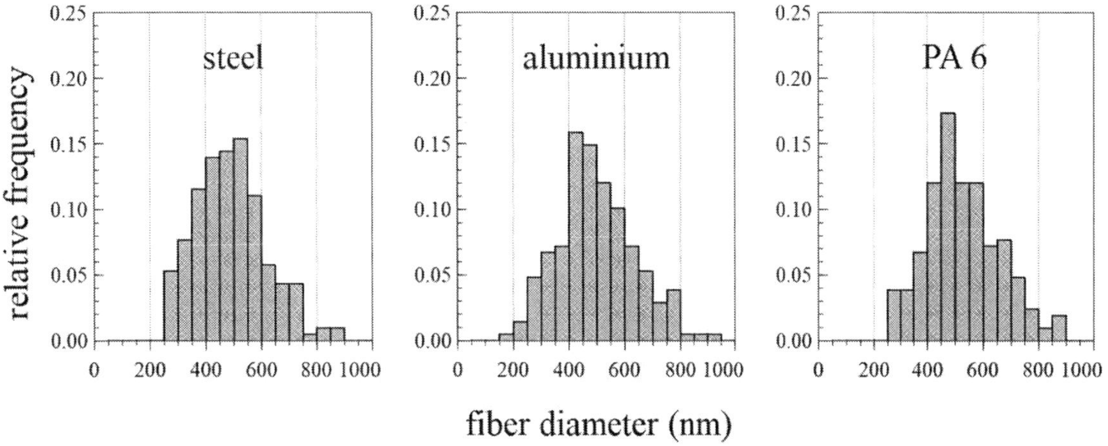

Figure 5 Histograms of fibre diameters of nanofibrous layers were prepared using electrodes of different materials. The graphs show that the material of the electrode did not fundamentally affect the diameters of the prepared nanofibers.

4. CONCLUSION

The work was focused on the possibilities of influencing the spinning process and the character of the nanofibrous structure by the type of spinning electrode material used. A device was assembled to allow the repeatability of the experiments during spinning. The only variable parameter was the type of electrode material. The geometry of the electrode was identical. Three basic materials resistant to the selected type of solution were selected, namely stainless steel (DIN 1.4401), aluminium (DIN 1725-1) and plastic (PA6). First, simulations of the distribution of the electric field intensity in the area of the electrodes were performed. For both metal materials (stainless steel and aluminium) the distribution of the electric field intensity was almost identical. The plastic electrode showed a lower level of intensity compared to the metal electrodes. The experiments proved the assumptions from the performed simulations and that with the increasing intensity of the electric field, higher stability of the moving nanofibrous formation is achieved from the place of its origin, i.e. from the electrode to the place of storage, i.e. the collector. From the results of the analysis of SEM images and created histograms, it was found that the selected base materials do not have a significant effect on the obtained structures. Research into the material type of spinning electrodes will continue with the aim of mapping a wider field of materials.

ACKNOWLEDGEMENTS

This publication was written at the Technical University of Liberec as part of the project with the support of the Specific University Research Grant 2021-5029, as provided by the Ministry of Education, Youth and Sports of the Czech Republic in the year 2021.

REFERENCES

[1] FAHIMIRAD, S., FAHIMIRAD, Z., SILLANPÄÄ, M. Efficient Removal of Water Bacteria and Viruses Using Electrospun Nanofibers. *Science of Total Environ.* 2021, vol. 751. Available from: https://doi.org//10.1016/j.scitotenv.2020.141673.

[2] FASDAS WEI, Z., SU, Q., YANG, J., ZHANG, G., LONG, S., WANG, X. High-Performance Filter Membrane Composed of Oxidized Poly (Arylene Sulfide Sulfone) Nanofibers for the High-Efficiency Air Filtration. *Journal of Hazardous Materials.* 2021, vol. 417. Available from: https://doi.org/10.1016/j.jhazmat.2021.126033.

[3] OZBEY UNAL, B., GEZMIŞ YAVUZ, E., ERYILDIZ, B., KOSEOGLU-IMER, D., KESKINLER, B., KOYUNCU, I. Boron Removal from Geothermal Water by Nanofiber-Based Membrane Distillation Membranes with Significantly Improved Surface Hydrophobicity. *Journal of Environmental Chemical Engineering.* 2020, vol. 8. Available from: https://doi.org/10.1016/j.jece.2020.104113

[4] JIANG, Y., MA, D., JI, T., SAMEEN, D.E., AHMED, S., LI, S., LIU, Y. Long-Term Antibacterial Effect of Electrospun Polyvinyl Al-cohol/Polyacrylate Sodium Nanofiber Containing Nisin-Loaded Nanoparticles. *Nanomaterials.* 2020, vol. 10. Available from: https://doi.org/10.3390/nano10091803.

[5] SRIVASTAVA, Y., LOSCERTALES, I., MÁRQUEZ, M., THORSEN, T. Electrospinning of Hollow and Core/Sheath Nanofibers Using a Microfluidic Manifold. Microfluid. *Nanofluidics.* 2008, vol. 4, pp. 245-250. Available from: https://doi.org/10.1007/S10404-007-0177-0.

[6] ELLISON, J.; PHATAK, A.; GILES, D.; MACOSKO, CH.; BATES, F. Melt blown nanofibers: Fiber diameter distribution and onset of fiber breakup. *Polymer.* 2007, vol. 48, pp. 3306-3316.

[7] WIDIYANDARI, H.; ADE PUTRA, O.; PURWANTO, A.; ABIDIN, Z. Synthesis of PVDF/SiO2 Nanofiber Membrane Using Electrospinning Method as a Li-Ion Battery Separator. *Materials Today Proceedings.* 2021, vol. 44, pp. 3245-3248. Available from: https://doi.org/10.1016/j.matpr.2020.11.448.

[8] WEI, Z.; SU, Q.; YANG, J.; ZHANG, G.; LONG, S.; WANG, X. High-Performance Filter Membrane Composed of Oxidized Poly (Arylene Sulfide Sulfone) Nanofibers for the High-Efficiency Air Filtration. *Jounal of Hazardous Materials.* 2021, vol. 417. Available from: https://doi.org/10.1016/j.jhazmat.2021.126033.

[9] SKRIVANEK J., KOMAREK J., BATKA O. Research of electrodes enabling the production of nanofibres with defined fineness. *NANOCON 2020 - Conference Proceedings, 12 th International Conference on Nanomaterials - Research and Application,* 2020, pp. 145-150. ISBN 978-80-87294-97-7. https://doi.org/10.37904/nanocon.2020.3730

[10] SKRIVANEK J., BATKA O., VALTERA, J. Analysis of electric field on the circular collector. *NANOCON 2019 - Conference Proceedings, 11 th International Conference on Nanomaterials - Research and Application,* 2019, pp. 204-209. ISBN 978-80-87294-95-6. Available from: https://doi.org/10.37904/nanocon.2019.8671.

PREPARATION AND TESTING OF HYDRO INSULATING SILICONE SEALING WITH ADDITION OF THERMOCHROMIC DYE DEA-CuCl4

[1]Hana KŘÍŽOVÁ, [1]Zuzana HRUBOŠOVÁ

[1]Technical University of Liberec, Institute for Nanomaterials, Advanced Technologies and Innovations, Liberec, Czech Republic, EU, hana.krizova@tul.cz

https://doi.org/10.37904/nanocon.2022.4591

Abstract

Nanoparticles of thermochromic complex pigment bis- (diethylammonium) tetrachloridocuprate (II) with chemical formula $[NH_2(C_2H_5)_2]_2CuCl_4$ (abbreviated DEA-$CuCl_4$) was prepared by crystallization and subsequent grinding in a cryo-mill. Nanoparticles were incorporated into a flexible hydro insulating transparent silicone matrix. The surface properties, especially surface hydrophobicity and elasticity of the resulting composite, as well as color indication of thermal changes and resistance to mold were tested.

Keywords: DEA-$CuCl_4$, bis-(diethylammonium) tetrachloridocuprate (II), thermochromic dye, antifungal, silicone

1. INTRODUCTION

Smart functional dyes are the new trend in many application fields such as textile industry, medicine, automotive, robotics, biotechnology etc. Reversible color changes characterizing a chromism can occur depending on physical or chemical stimulations, but thermochromism is the most studied of all chromisms.

In this study, we focused on the incorporation of nanoparticles of the inorganic thermochromic pigment bis-(diethylammonium) tetrachlorocuprate (II) into a silicone matrix to create a smart multifunctional composite material which combines the waterproofing and water-repellent properties of silicone [1] with the thermochromic effect of the inorganic dye DEA-$CuCl_4$. Exterior and interior applications of silicone sealants often expose this material to conditions ideal for biofilm growth on their surfaces: biofilm is created by bacteria and fungi which thrive in moisture and heat in the areas difficult to clean. For this reason, these materials (called "sanitary silicones") are often treated with a biocide with the purpose to block the growth of biofilm on their surface. Although the antibiotic properties of the DEA-$CuCl_4$ dye have not been described yet, it could give the composite the ability to resist bacterial biofilm and mold growth. The DEA-$CuCl_4$ dye contains copper (II) thermochrome and many studies have already demonstrated the antifungal and antibacterial effects of various copper compounds in both nanoparticle and solution form [2-6].

1.1. DEA-$CuCl_4$

Bis-(diethylammonium) tetrachloridocuprate (II) with chemical formula $[NH_2(C_2H_5)_2]_2CuCl_4$ (abbreviated DEA-$CuCl_4$) is an inorganic complex crystalline dye that exhibits discontinuous thermochromism. Its bright green colour changes rapidly to yellow at temperatures above approximately 55 °C. This colour change corresponds to a change in the coordination geometry of the tetrachlorocuprate anion (which is the chromophore of the compound) from a deformed square plane to a deformed tetrahedron [7,8]. Diethylammonium cations form hydrogen bridges via N-H···Cl interactions with tetrachlorohydrate anions, leading to the formation of a two-dimensional planar hydrogen bridge network [9]. Although the cause of the colour changes is the aforementioned change in the geometry of the tetrachlorocuprate anion, in fact, the weakening of the hydrogen bonds and the stretching of the ethyl chains of the organic component (diethylammonium cations) due to increasing temperature are the driving forces behind this change [10]. Although the process is reversible, it has recently been found that the dynamic phase transition process is not perfectly reversible - the green-to-yellow transition is driven primarily by the weakening of hydrogen bonds, whereas during the yellow-to-green transition, the reconstitution of hydrogen bonds falls behind the rearrangement of alkyl chains [11].

1.2. Silicone

Silicone rubber is an inorganic - organic elastomer. Chemically it is a polysiloxane that consist of an inorganic silicon-oxygen backbone chain with two organic groups (mostly methyls) attached to each silicon center. Properties include low chemical reactivity, low toxicity, high biocompatibility, heat resistance in the temperature range -100 to 300 °C and water repellency. It is used e.g., as insulation, sealing, coating and specialty products in construction, electronics, food, medical and cosmetic surgery, automotive, etc. One component condensation acetoxy curing silicones are used most often. These silicones cure with the presence of atmospheric moisture and generally have a very fast cure time (within 24 hours). Unfortunately, although these silicones adhere very well to most materials, the acetic acid could be corrosive to metals [12].

2. EXPERIMENTAL PART

2.1. Chemicals and materials

Diethylammonium hydrochloride (Sigma Aldrich), copper (II) chloride dihydrate (Sigma Aldrich), transparent acetoxy silicone (Den Braven), transparent sanitary acetoxy silicone (Monton), rice agar in Petri dishes (Viamar International).

2.2. Synthesis of DEA-CuCl4

Bis-(diethylammonium) tetrachloridocuprate (II) was prepared in the laboratory from aqueous solutions of diethylammonium hydrochloride and copper (II) chloride dihydrate, compounds were mixed in stoichiometric proportions according to equation (1). The resulting solution was left in a wide glass dish for several days at room temperature when crystallization occurred. After a few days, the entire bottom of the glass dish was covered with a continuous layer of light green crystalline dye. This crystalline pigment was subsequently cooled to -5 °C and then ground in a laboratory cryo-mill (Retsch) using a zirconium oxide grinding jar and balls. Two 30-second cycles of impact ball milling with a frequency of 20 Hz in dry grinding mode allowed the preparation of DEA-CuCl4 pigment nanoparticles with a size of 236 ± 44 nm (95% confidence interval).

$$2[(C_2H_5)_2NH_2]Cl + CuCl_2 \cdot 2H_2O \rightarrow 2[(C_2H_5)_2NH_2]_2CuCl_4 + 2H_2O \quad (1)$$

2.3. Preparation of silicone/DEA-$CuCl_4$ composite

Three samples with a diameter of 3.5 cm and a thickness of about 0.5 cm were prepared based on a transparent neutral silicone paste with 0.02 wt.% (sample A) and 0.2 wt.% (sample B) of crushed DEA-$CuCl_4$ nanoparticles. The reference sample (sample 0) contained only silicone without the addition of DEA-$CuCl_4$ nanoparticles. The samples containing DEA-$CuCl_4$ were thoroughly homogenized, and all samples were left at room temperature for 4 days to dry. After drying, these samples were measured and tested for: hardness measurement, color reversal of the composite, surface hydrophobicity (measurement of the contact angle of the water droplet), and testing for resistance to mold (*Penicillium* spp.) growth.

2.4. Shore hardness measurement

Silicone rubber is often applied as a sealing compound that retains a certain degree of elasticity even after curing (drying). It was investigated whether the addition of DEA-$CuCl_4$ nanoparticles to silicone rubber in amounts up to 0.2 wt.% would significantly affect its hardness. For this purpose, a Shore durometer measurement according to [ASTM D2240] is standardly performed. Especially for the measurement of flexible rubbers, thermoplastic elastomers and semi-rigid plastics, the Shore A method is suitable. In this measurement, a hardened steel rod of 1.4 mm diameter with 35° truncated cone of 0.79 mm diameter is pressed into the material, spring force approximately 8.05 N. The Shore A hardness scale 0 - 100 measures the hardness of flexible rubbers that range from very soft and flexible to hard with almost no flexibility at all. Each specimen was measured 10 times and the results averaged.

2.5. Thermochromic properties measurement

Measurement of the reversible thermochromic properties of the formed composite was performed. For this purpose, the composite was repeatedly heated and cooled, and the thermal changes were recorded using a Fluke thermal imaging camera. An RGB-2000 color analyzer (10 bits) with 45° geometry was used to record the color changes.

2.6. Surface hydrophobicity measurement

The hydrophobicity of silicone surfaces is well known, which is why silicone rubbers are often used as waterproofing seals. Investigation was focused on the addition of DEA-$CuCl_4$ nanoparticles significantly affects this hydrophobicity. To this end, the contact angle of the water droplet with the surface of the samples was measured using a USB 2.0 VGA Video Device and evaluated with the See System for surface energy measurement software (version 6.3) using the Li-Neumann model. The result was obtained from 10 measurements of each sample.

2.7. Fungal growth resistance testing

The antifungal activity test of the composite was arranged as a classic disc diffusion antimicrobial assay. The samples were tested in contact with spores of *Penicillium* spp. The surface of rice agar in petri dishes was inoculated with 1 ml of fungal spore suspension, which was obtained by simply rinsing the fungus growing on organic waste. A sample of the composite was loosely placed on the surface and the sample was left in the dark at room temperature for 48 hours. The growth of CFUs in the vicinity of the test samples was monitored and the average width of the following rings was measured: a clearing ring (denoted as "halo zone" in **Table 1)** where the number of CFUs grown was reduced, and a completely clear ring around the samples (denoted as "zone of clearing" in **Table 1**) where the growth of CFUs of bacteria or mold was completely suppressed. For comparison, an additional sample X was made of so-called sanitary silicone sealant, which is a commercial silicone rubber that contains a fungicide to inhibit mold growth and is intended for use e.g., in bathrooms.

3. RESULTS AND DISCUSSION

The results of the tests performed on samples 0, A and B are summarized in **Table 1**. As regards the hardness of the samples, the Shore A values were around 20, which corresponds to values of relatively soft and flexible rubber. The confidence intervals of pure silicone and sample A overlap, indicating that the low content of dye nanoparticles (here 0.02 wt. %) practically does not affect the hardness and flexibility of silicone. However, the confidence interval values of sample B are slightly higher, and it can be concluded that the hardness of the composite increases slightly with higher content of dye nanoparticles in the silicone matrix. As for surface hydrophobicity, the contact angle of the water droplet with the surface of the composites exceeded 90° for all samples, which is the limit at which the material can be described as hydrophobic. However, the confidence intervals of the water droplet contact angles overlap for all three samples, therefore it can be concluded that the addition of DEA-$CuCl_4$ dye nanoparticles to the silicone did not significantly affect the surface hydrophobicity even at a dye content of 0.2 wt.%.

During heating and cooling of the composite, the color changes of samples B were recorded using RGB coordinates (**Figure 1**). A significant change in the RGB-coordinates values in the region around 55 °C corresponds to the phase transition of DEA-$CuCl_4$ and the color change of the composite.

Table 1 An overview of the test results of samples 0, A and B

Sample	Shore A	Contact angle of water	Phase transition temperature	Halo zone (zone of clearing)
0	20.7 ± 0.6	93.0° ± 7.5°	-	0
A	19.0 ± 1.3	94.5° ± 5.5°	55 °C	8-12 mm (2 mm)
B	22.5 ± 0.5	93.5° ± 4.5°	55 °C	8-12 mm (2 mm)

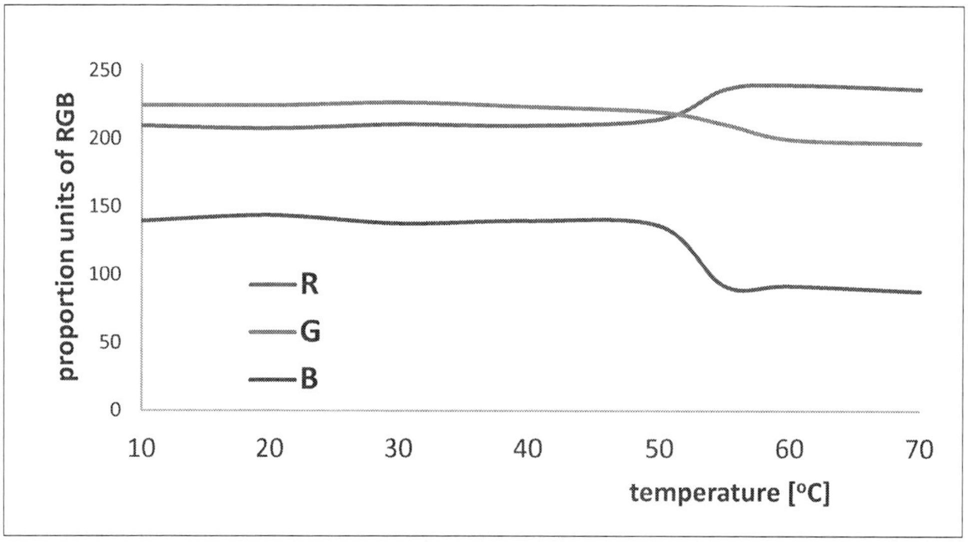

Figure 1 Changes in RGB values of the composite in the temperature range 10 to 70 °C

Figure 2 Sample 0 on rice agar inoculated by *Penicillium*

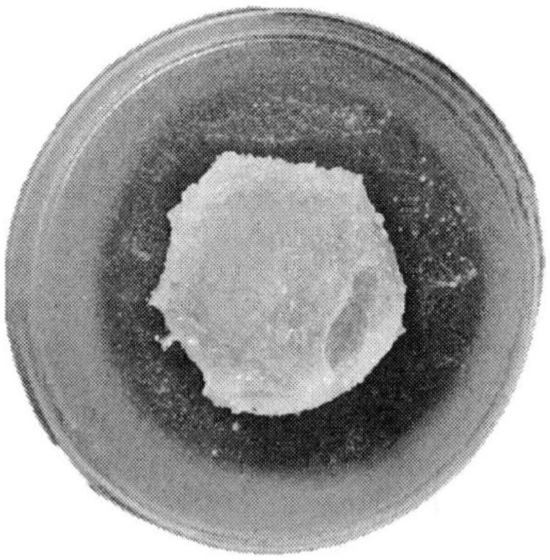

Figure 3 Sample B on rice agar inoculated by *Penicillium*

Figure 4 Sample X on rice agar inoculated by *Penicillium*

When the samples were contacted with rice agar inoculated with *Penicillium* spp. spores, a distinct clearing (halo zone) of 8-12 mm width occurred in samples A, B and X within 48 hours, with a completely clear ring of approximately 2 mm width (zone of clearing) in the immediate vicinity of the samples. No clearing was observed for sample 0. Interestingly, the inhibition zones were approximately the same in samples A and B regardless of the DEA-$CuCl_4$ concentration (**Figures 2 - 4**). A similar width of inhibition zones was observed in sample X. This suggests that the DEA-$CuCl_4$ content of 0.02 wt.% is already sufficient to inhibit fungal growth and that the diffusion of the dye, like the diffusion of the fungicide from the silicone matrix into the surroundings, is limited. The test results support our expectation of the DEA-$CuCl_4$ dye antifungal effects due to its cuprate (II) chromophore content and it can replace fungicides that are standardly added to so-called sanitary silicones. It is also necessary to draw attention to the issue of instability of this system. We observed that the samples changed color and decreased thermochromism over several days. DEA-$CuCl_4$ is sensitive to corrosive environments, acids and atmospheric oxygen or the presence of base metals due to its copper cation content. The use of so-called neutral silicones, which are one-component oxime or alkoxy silicones intended for contact with copper and zinc, did not solve this issue either, because the dye in contact with these systems immediately lost its thermochromic properties. This issue could be solved, for example, by protective encapsulation of dye particles, which is beyond the scope of this study.

4. CONCLUSION

We created a silicone composite with the addition of 0.02 and 0.2 wt.% of the thermochromic dye DEA-$CuCl_4$ in the form of nanoparticles with a size of approximately 236 ± 44 nm. While the dye content of 0.2 wt.% slightly increased the hardness of the silicone rubber, the surface hydrophobicity of the composite did not change. The thermochromic properties of the dye were maintained even after repeated heating-cooling cycles in the temperature range 10 to 80 °C with a temperature phase transition around 55 °C. It was demonstrated that the presence of the DEA-$CuCl_4$ pigment nanoparticles can replace the addition of a fungicide, which is added to commercial silicone hydro insulating sealing for wet applications and that the composite inhibited growth of *Penicillium*. Hydrophobic surface, waterproofing properties, color indication of thermal transition in 55 °C and resistance to mold growth could make this composite an ideal insulating material for indoor and outdoor applications in warm and humid areas with large temperature ranges (e.g., sauna, spa, tropical areas). However, the issue of dye degradation still needs to be improved as copper is reduced in contact with the external environment and the dye loses its thermochromic properties over time.

ACKNOWLEDGEMENTS

This work was supported by Iceland, Liechtenstein and Norway through the EEA Grants and the Technology Agency of the Czech Republic within the framework of project "inherently Flexible Aerogels for energy effiCient structurES (i-FACES)" (Grant number TO01000311).

REFERENCES

[1] CHANG, J.W., GORUR, R.S. Surface recovery of silicone rubber used for HV outdoor insulation. In: *IEEE Transactions on Dielectrics and Electrical Insulation*. 1994, vol. 1, no. 6, pp. 1039-1046.

[2] KŘÍŽOVÁ, H., ROTKOVÁ, J. Green Synthesis of Copper-Based Nanostructures Using Tannic Acid and Testing of Their Antibacterial Properties. In: *Nanocon* - 6th International Conference on Nanomaterials. Brno: Tanger, 2014, pp. 621-626.

[3] CIOFFI, N., TORSI, L., DITARANTO, N., TANTILLO, G., GHIBELLI, L., SABBATINI, L., TRAVERSA, E. Copper nanoparticle/polymer composites with antifungal and bacteriostatic properties. *Chemistry of Materials*. 2005, vol. 17, no. 21, pp. 5255-5262.

[4] DANG, J., ZHANG, Y., DU, Z., ZHANG, H., LIU, J. Antibacterial properties of PES/CuCl2 three-bore hollow fiber UF membrane. *Water Science and Technology*. 2012, vol. 66, no. 4, pp. 799-803.

[5] CHATTERJEE, A.K., CHAKRABORTY, R., BASU, T. Mechanism of antibacterial activity of copper nanoparticles. *Nanotechnology*. 2014, vol. 25, no. 13, 135101.

[6] CHALANDAR, H.E., GHORBANI, H.R., ATTAR, H., ALAVI, S.A. Antifungal effect of copper and copper oxide nanoparticles against Penicillium on orange fruit. *Biosciences Biotechnology Research Asia*. 2017, vol. 14, no. 1, pp. 279-284.

[7] CHOI, S., LARRABEE, J.A. Thermochromic tetrachlorocuprate (II): an advanced integrated laboratory experiment. *Journal of Chemical Education*. 1989, vol. 66, no. 9, p. 774.

[8] KAPUSTYANYK, V.B., KORCHAK, Y.M. Thermochromic phase transition in [NH2 (C2H5) 2] 2CuCl4 Crystals. *Journal of Applied Spectroscopy*. 2000, vol. 67, no. 6, pp. 1045-1049.

[9] ALDRICH, E.P., BUSSEY, K.A., CONNELL, J.R., REINHART, E.F., OSHIN, K.D., MERCADO, B.Q., OLIVER, A.G. Crystal structure of the thermochromic bis (diethylammonium) tetrachloridocuprate (II) complex. *Acta Crystallographica Section E: Crystallographic Communications*. 2016, vol. 72, no. 1, pp. 40-43.

[10] KELLEY, A., NALLA, S., BOND, M.R. The square-planar to flattened-tetrahedral CuX42−(X= Cl, Br) structural phase transition in 1, 2, 6-trimethylpyridinium salts. *Acta Crystallographica Section B: Structural Science, Crystal Engineering and Materials*. 2015, vol. 71, no. 1, pp. 48-60.

[11] XIE, D., XU, J., CHENG, H., WANG, N., ZHOU, Q. The role played by amine and ethyl group in the reversible thermochromic process of [(C2H5) 2NH2] 2CuCl4 probing by FTIR and 2D-COS analysis. *Journal of Molecular Structure*. 2018, vol. 1161, pp. 267-272.

[12] DE BUYL, F. Silicone sealants and structural adhesives. *International Journal of Adhesion and Adhesives*. 2001, vol. 21, no. 5, pp. 411-422.

MOBILITY OF SULFONAMIDE ANTIBIOTIC SULFAPYRIDINE IN DIFFERENT TYPES OF SOIL

[1]Petra ZÁVODSKÁ, [1]Denisa JANEBOVÁ, [1]Jan FUČÍK, [1]Martina KLUČÁKOVÁ, [1]Miloslav PEKAŘ, [2]Victory JAQUES, [2]Tomáš ZIKMUND, [2]Jozef KAISER

[1]*Faculty of Chemistry, Brno University of Technology, Brno, Czech Republic, EU, Petra.Zavodska@vut.cz*
[2]*CEITEC, Brno University of Technology, Brno, Czech Republic, EU*

https://doi.org/10.37904/nanocon.2022.4593

Abstract

This contribution focuses on monitoring the mobility of sulfapyridine in different soil types. The main difference in the used types of soil was the content of soil organic matter. Simple sorption and desorption experiments in buffer solutions with different pH values were made. Adsorbed and desorbed amount of sulfapyridine was measured using UV-VIS and liquid chromatography. The water solution of sulfapyridine was used for experiments in real conditions when a plastic tube was put into the soil and watered with an aqueous antibiotic solution. For the following days, tubes were taken out of soils, separated and adsorbed amount was measured with UV-VIS. In addition, computed tomography was used to make models for molecular diffusivity. From these measurements, the diffusion coefficient was determined along with adsorbed and desorbed amount of sulfapyridine in different types of soils and in different values of pH. It was confirmed that the pH value and type of soil has an influence on the sorption of used antibiotic.

Keywords: Soil, sorption, desorption, sulfapyridine, tomography

1. INTRODUCTION

Pharmaceuticals are used widely in the world to treat human and animal diseases. But they also have an impact on the Environment. Due to incomplete absorption and metabolism, an unutilized part of drugs is discarded into the Environment through urine or faeces [1]. Pharmaceuticals have been found in the Environment in low concentrations (from nanogram to microgram per litre). Not only can they get into the environment by urine but also by improper disposal or use. Sometimes, the drugs are discarded via the toilet, or the drain or are thrown into waste. If the drugs are not eliminated or adsorbed by soils, they may get into the water and all the aquatic environment [2,3,4].

Sulfapyridine is a sulfonamide antibiotic. This term is generic for derivates of paraaminobenzenesulfonamide structure is shown on the **Figure 1**). Sulfapyridine was firstly found in the thirties of 20th century. Sulfonamides in general act as competitive inhibitors to para-aminobenzoic acid (PABA). Not only do they act as inhibitors to PABA but also have a wide range of antimicrobial activity against Gram-positive and Gram-negative bacteria. In the drug therapy, they had only a bacteriostatic effect, which means they only prevented the growth of bacteria. Therefore, they had a very limited role as single agents [5,6].

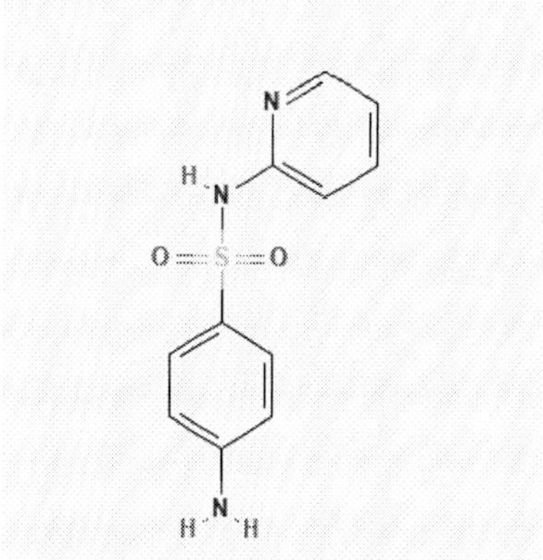

Figure 1 Chemical structure of Sulfapyridine [7]

The main motivation for this work was the fact that sulfapyridine and many other sulfonamides antibiotics can get through wastewater treatment in unchanged form. As was mentioned, many antibiotics are excreted in large quantities into the environment as active compounds. A significant amount of antibiotics gets into the water, that is used for watering or fertilizing agricultural soil. In soil, most of the antibiotics undergo a biological and physicochemical transformation, but they still have their antimicrobial activity and thus have a potentially toxic effect on the bacteria in soils. They can have an impact on soil enzyme activity as well. This enzyme activity is involved in the bioavailability of nutrients necessary for plants [8].

2. MATHERIAL AND METHODS

2.1. Sorption experiments

Firstly, simple sorption and desorption experiments were performed. Three types of soil, differed by the content of soil organic matter (SOM), were used (**Table 1**). Approximately 0.5 g of each soil was put into a centrifuge tube and filled with 25 ml of drug solution with a concentration 10 mg/l. As a drug solution Britton-Robinson buffer (BRB), with pH 3, 5, 7, and 10, was used.

Table 1 Amount of organic matter in soil samples

Sample	A	B	C
SOM (w.%)	5.75	1.17	6.86

For precisely 48 hours, samples were mixed on a magnetic stirrer. After the period, samples were centrifuged. After 20 minutes, samples were taken out. The supernatant was filtered using 0.45 µm syringe filters and measured with UV/VIS. Samples in BRB with pH 7 and 10 needed to be diluted three times. Part of the supernatant was filtered using 0.22 µm syringe filters and ten times diluted for liquid chromatography measurements (LC).

Sediment from centrifuge tube was transferred to Petri dish and dried. Dry sediment was put back into the clean centrifuge tube along with 25 ml of distilled water. For another 48 hours, samples were mixed on the magnetic stirrer. After the period, for another 20 minutes, they were centrifuged. The supernatant was filtered using 0.45 µm syringe filters for UV/VIS measurements and 0.22 µm for LC measurements. Once again, samples in pH 7 and 10 were three times diluted for UV/VIS. Samples for LC were ten times diluted.

2.2. Sorption experiments in real conditions

Further experiments were performed to simulate real conditions. Plastic syringes (volume of 150 ml) were put into the soil and watered with the drug solution (concentration 1 mmol/l). Syringes were covered with parafilm and for the next days left in the soil. 11 syringes in total were put into the soil. At different times the syringes with the soil were taken out and the soil was divided into 6 parts. Approximately 5 g from each part was put into 25 ml of tap and distilled water and into an $MgCl_2$ solution (concentration 1 mol/l). The leachates were filtered using the 0.45 um syringe filters and measured with UV/VIS. From these measurements, diffusion coefficients were obtained.

Figure 2 Transparent plastic tube used for sampling a CT measuring

2.3. CT measurements

In addition, diffusion coefficients were used in computed tomography (CT). Samples were taken from permanent grassland near the Brno University of Technology, Faculty of Chemistry in Brno, in an intact condition using a transparent plastic tube (**Figure 2**). Samples were transferred to the CT lab and measured by GE phoenix v|tome|x L240. CT measurements were used for creating a model of molecular diffusivity.

3. RESULTS AND DISCUSSION

3.1. Sorption experiments

Sorption experiments were made as described in section 2.1. Using four pH values (3, 5, 7, and 10±0,05) and three types of soil with different amounts of SOM (**Table 1**) showed differences in adsorbed and desorbed amounts of sulfapyridine between soil types and even between values of pH.

As shown in the **Figure 3 (A)** adsorbed amount reaches up to 260 µg/g. There is no evident trend while comparing pH values. All samples have a maximum at pH 3 (sample A 259 µg/g, sample B 198 µg/g, and sample C 194 µg/g). Except for sample C, they have their minimum at pH 7, which is close to the pKa of sulfapyridine (8.43). Sample A has its minimum equal to 154 µg/g, while sample B is equal to 132 µg/g. As mentioned before, sample C has its minimum at pH 10 equal to 57 µg/g. The **Figure 3 (B)** shows the adsorption efficiency of soil samples. As mentioned before, all samples reached their maximum at pH 3. The sample A even reaches 100 % of adsorption efficiency, while the efficiency of the sample B is 67.6 % and the efficiency of the sample C is only 63.3 % The minimum value of efficiency of the sample A is equal to 47.8 %, while the sample B equals 36 % at pH 7 and the minimum for the sample C equal to 15.2 % at pH 10. Therefore, it's apparent, that the best pH value for sulfapyridine adsorption equals 3.

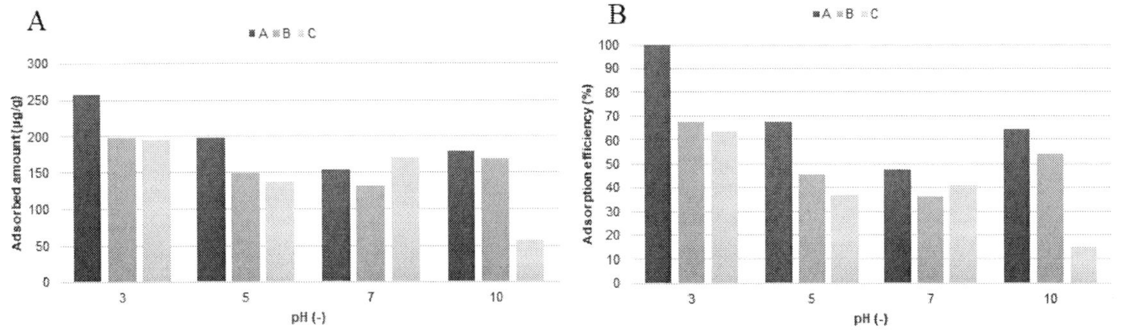

Figure 3 Adsorbed (A) amount of sulfapyridine in three types of soil in different values of pH and adsorption efficiency (B)

Comparing the efficiency of the soil within each sample it's evident that the sample A has the largest adsorbed amounts and the highest adsorption efficiency while the sample C, which has the highest content of SOM, has the lowest efficiency along with adsorbed amounts. Therefore, adsorption does not depend only on the content of SOM, but on the other soil properties (one of them may be the content of calcium).

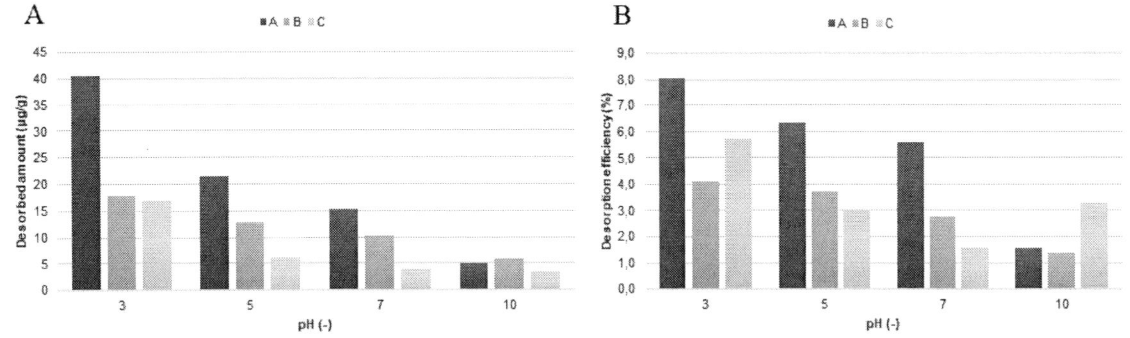

Figure 4 Desorbed amount (A) of sulfapyridine and desorption efficiency (B)

Desorbed amounts are shown in the **Figure 4** (A). There is an evident trend when all the samples have their maximum at pH 3 (the sample A 40.6 ug/g, sample B 17.8 ug/g, and the sample C 16.9 ug/g) and the amounts are decreasing to pH 10, where all samples have their minimum (the sample A 4.9 ug/g, the sample B 5.9 ug/g, and the sample C 3.4 ug/g). It's apparent that the sample A has the highest desorbed amounts, while the sample C has the lowest amounts. Therefore, increasing the pH reduces the desorption of sulfapyridine. The **Figure 4** (B) compares desorption efficiencies. As in the case of adsorption, the sample A has the highest desorbed amounts (at pH 3 up to 8.1 %). The sample A then is the best adsorbent within the samples but also the best desorbent. Contrary to the sample A, sample C is considered the worst adsorbent and also the worst desorbent, which means that the sample C was able to adsorb sulfapyridine molecules and retain them.

3.2. Sorption experiments in real conditions

Sorption experiments were made as described in the section 2.2. Using second Fick's law for nonstationary diffusion, three diffusion coefficients (distilled water, tap water, $MgCl_2$) were obtained (**Table 2**). $MgCl_2$ was used for its ability to adsorb instead of sulfapyridine. Comparing the coefficient obtained from our measurement with results from [9], where the measurements were carried out in diffusion cells using a gel as a membrane, it's apparent that the coefficients are quite similar. In the $MgCl_2$ solution, the diffusion coefficient has the largest value, which supports the assumption of adsorbing instead of sulfapyridine. In tap water the diffusion coefficient is higher than in distilled water, because in tap water there are some minerals (such as magnesium, calcium or chlorine), that can affect adsorption of sulfapyridine.

Table 2 Diffusion coefficients for distilled water, tap water and $MgCl_2$

	D (m^2/s)
Distilled water	$4.23 \cdot 10^{-10}$
Tap water	$4.39 \cdot 10^{-10}$
$MgCl_2$	$5.59 \cdot 10^{-10}$

3.3. CT measurements

For CT measurements the uppermost part of the soil was taken and measured with CT. Using VG studio and a scan of soil the model of molecular diffusivity was made. The **Figure 5** (B) shows the model of molecular diffusivity of soil sample of sulfapyridine in water. The **Figure 5** (A) shows the cut-out part of the model and it's apparent, that molar concentration decreases from top to bottom. The **Figure 5** (C) shows streamlines, as possible pathways of solution through the soil.

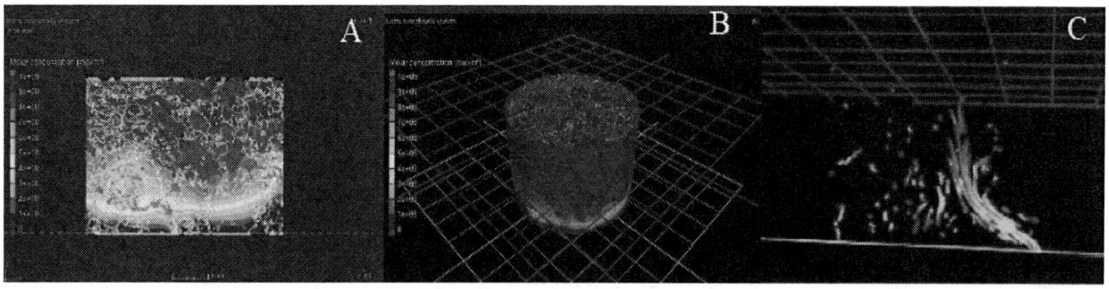

Figure 5 Model of molecular diffusivity (A, B) with streamlines (C)

4. CONCLUSION

Sorption experiments in the lab and real conditions were made along with CT measurements. The first experiments show the ability of soil to adsorb sulfapyridine with high efficiency and its ability to hold it within its structure. The soil A with 5.75 wt. % of organic matter turned out to be the best for the adsorption of sulfapyridine when at pH 3 adsorption reached 100 % of efficiency. Contrary to this, the soil C with 6.86 wt.% of organic matter turned out to be the worst adsorbent. The efficiency of the soil C reached only 63.4 %. Comparing soils shows that not only the content of organic matter is responsible for adsorption, but also other parameters which might be the content of calcium or magnesium. Desorption experiments showed that the best adsorbent (the soil A) is also the best desorbent, thus it releases much more molecules of sulfapyridine than the other two samples (desorption efficiency of the soil A reaches 8.1 %). On the contrary the sample B, that turned out to be worse adsorbent than the soil A, was the worst desorbent, thus it adsorbs sulfapyridine and holds it in its structure (desorption efficiency reaches only 4.1 %).

Experiments in real conditions were made to compare diffusion coefficients obtained from measurements in diffusion cells. Using diffusion cells in publication [9] writers got a diffusion coefficient equal to $4.75 \cdot 10^{-10}$ m^2/s. In our experiments in the soil column, diffusion coefficients were equal to $4.23 \cdot 10^{-10}$ m^2/s in distilled water, $4.39 \cdot 10^{-10}$ m^2/s in tap water and $5.59 \cdot 10^{-10}$ m^2/s. Results showed the possibility of using this method for measuring the diffusion coefficients.

In addition, the CT measurements were made. Using the VG studio to create a model of molecular diffusivity showed possible ways of diffusion of sulfapyridine solution. Both CT and real condition experiments are only preliminary results and will be develop in next years, but it's apparent that the usage is possible.

Experiments in real conditions in soil columns are new and for now, have some advantages (diffusion in real soil instead of gel) but also some disadvantages (the need for intact and large space in the field). But it's important to keep monitoring the drugs in soils and their behaviour.

ACKNOWLEDGEMENTS

The authors acknowledge partial support from VUT specific interfaculty junior grant 2022 n°7903.

REFERENCES

[1] DRILLIA, Panagiota, Katerina STAMATELATOU and Gerasimos LYBERATOS. Fate and mobility of pharmaceuticals in solid matrices. *Chemosphere.* 2005, vol. 60, no. 8, pp. 1034-1044.

[2] KÜMMERER, Klaus. 2010. Pharmaceuticals in the Environment. *Annual Review of Environment and Resources.* 2010, vol. 35, no. 1, pp. 57-75.

[3] AUS DER BEEK, Tim, Frank-Andreas WEBER, Axel BERGMANN, Silke HICKMANN, Ina EBERT, Arne HEIN and Anette KÜSTER. 2016. Pharmaceuticals in the environment-Global occurrences and perspectives. *Environmental Toxicology and Chemistry.* 2016, vol. 35, no. 4, pp. 823-835.

[4] CHRISTENSEN, F.M. Pharmaceuticals in the Environment-A Human Risk? *Regulatory Toxicology and Pharmacology.* 1998, vol. 28, no. 3, pp. 212-221.

[5] Sulfonamides. *Meyler's Side Effects of Drugs.* 16th. Elsevier, 2016, pp. 555-569.

[6] CHRISTENSEN, S.B. Drugs That Changed Society: History and Current Status of the Early Antibiotics: Salvarsan, Sulfonamides, and β-Lactams. *Molecules.* 2021, vol. 26, 6057.

[7] PubChem. *Sulfapyridine.* [online]. 2005. [viewed: 2022-09-15]. Available from: https://pubchem.ncbi.nlm.nih.gov/compound/Sulfapyridine

[8] MOLAEI, Ali, Amir LAKZIAN, Rahul DATTA, Gholamhosain HAGHNIA, Alireza ASTARAEI, MirHassan RASOULI-SADAGHIANI and Maria T. CECCHERINI. Impact of chlortetracycline and sulfapyridine antibiotics on soil enzyme activities. *International Agrophysics.* 2017, vol. 31, no. 4, pp. 499-505.

[9] CHEN, Chang-Er, Hao ZHANG, Guang-Guo YING and Kevin C. JONES. 2013. Evidence and Recommendations to Support the Use of a Novel Passive Water Sampler to Quantify Antibiotics in Wastewaters. *Environmental science & technology.* 2013, vol. 47, no. 23, pp. 13587-13593.

STUDY ON THE INFLUENCE OF BIOCHAR ON PHYSICO-CHEMICAL PROPERTIES OF SOIL

Nikola POLÁŠKOVÁ, Michal KALINA, Šárka SOVOVÁ, Jiří SMILEK

Brno University of Technology, Faculty of Chemistry, Materials Research Centre, Brno, Czech Republic, EU,
Nikola.Polaskova@vut.cz, smilek@fch.vut.cz

https://doi.org/10.37904/nanocon.2022.4599

Abstract

The paper is focused on assessment of the influence of biochar on the soil physico-chemical properties (specific gravity, bulk density, porosity, water retention capacity and maximum water holding capacity, oxidizable carbon content, total nitrogen content, phosphorus, potassium and magnesium content). The selected and examined soil types were: cambisol, fluvisol, chernozem and regosol. There were two types of biochar used for analysis – NovoTerra and Sonnenerde. Soil samples were dried to constant weight and adjusted to grain-size fraction under 2 mm, respectively 0.25 mm. The oxidizable carbon content was measured by oxidation using a mixture of potassium dichromate and sulphuric acid (Walkley-Black method), the total nitrogen content was measured by Kjeldahl method and P, K and Mg content was measured using Mehlich 3 extraction and ICP-OES spectrometer. The results show that biochar has a significant influence on soil characteristics depending on the type of biochar and soil we use for biochar application.

Keywords: Biochar, soil, cambisol, chernozem, fluvisol, regosol

1. INTRODUCTION

Biochar is a solid product of thermal decomposition of organic material with limited oxygen access at relatively low temperatures (300-700 °C) [1]. This process is called pyrolysis. Biochar is a material rich in carbon, nitrogen (in the form of nitrite, nitrate and ammonium ions), Mg, K, Na and P content, which are nutrients that are necessary for plant growth [2]. The main advantages of biochar are the ability to positively affect water retention [2], to retain substances due to its micropores and large surface area (therefore it can protect environment from contaminants) [3]. It can also play a role in fighting climate changes – it is quite stable in a long term and it can take part in sequestering carbon from the atmosphere [4]. Due to number of advantages biochar is applicable in agriculture as an auxiliary soil and plant substance to increase soil quality and fertility. The application of biochar into soil leads to decrease of its acidity, increase of cation exchange capacity and water holding capacity, therefore it is suitable environment for microbes [5].

Another fertilizer option is for example foliar application of fertilizers and biostimulants, which complements standard root fertilizer treatments [6]. In combination with the biochar application, it could be an effective method of improving the soil properties and increasing the uptake of nutrients by the plant at the same time.

As mentioned, biochar is a product of biomass pyrolysis. We can define biomass as an organic matter, especially plant-based biomass (spent peat, woody green waste, tree bark, applewood, flax shives, spent coffee grounds etc.) [7], but animal-based biomass (e.g. insect frass or chicken manure) can also be used [7,8]. The properties of biochar do not depend only on source biomass, but also on the production conditions – highest treatment temperature, pyrolysis time, heating rate, pressure or post treatment [9]. Generally, biochar as a material consists of aromatic compounds, which cause an increase of its stability in soil. By measuring molar ratio of oxygen and carbon we are able to estimate half-life of biochar. If the ratio is O: C < 0.2, the minimum half-life is about 1,000 years. This ratio can be influenced by production temperature [10].

Generally, there is number of properties and characteristics which are evaluated and measured, e.g. soil reactions, available P content, K, Mg and Ca content, but also micronutrients content (B, Cu, Fe, Mn, Zn and other heavy metals) and soil organic matter. Results show the decrease of high-quality agricultural soil in the Czech Republic [11], therefore it is important to deal with this issue. The main goal of this study is to compare soils before and after biochar application. The effect of biochar on modified and unmodified soil is studied in soil which is planted with corn plants. After 100 days the analysis of soil (and the biochar [12]) was performed.

2. MATERIALS AND METHODS

2.1. Samples

Four types of soil were selected for this study – cambisol, fluvisol, chernozem and regosol. These soils represent soils of the Czech Republic. Cambisol was collected from Náměšť nad Oslavou (28. 5. 2020), fluvisol was collected from Iváň (28. 5. 2020), chernozem was collected from Žabčice u Brna (28. 5. 2020) and regosol was collected from Pánov (28. 5. 2020). Samples were collected from topsoil (0-20 cm below surface).

Two types of biochar were selected for this study – NovoTerra (NovoCarbo GmbH, the source biomass was residual softwood wood chips) and Sonnenerde (Sonnenerde GmbH, the source biomass was grain husks, sunflower husks and fruit sludge).

One kg of soil and 20 g of biochar in the polypropylene bags was weighed into the pots. Corn was planted in the soil and the growing process took about 100 days. After the cultivation, the biochar and the plants were removed from the soil and the analysis of the soil was performed. All samples were dried at 105 °C for 3 days, homogenised and passed through a 2mm sieve before further analysis.

2.2. Determination of specific gravity ρ_s, bulk density ρ_d and porosity P

For specific gravity, labeled pycnometer with stopper was weighed. Then it was filled with distilled water and weighed again. The soil sample was added into the pycnometer to a quarter of its height and weighed. The pycnometer with soil was filled with distilled water to a half of its height and heated on the stove to boiling. After cooling down the pycnometer with soil sample and water was fully filled with adjusted distilled water (water in beaker was heated to boiling and then covered with watch glass beaker cover and cooled down) and weighed together with the stopper. Specific gravity was calculated according to the equation (1).

$$\rho_s = \frac{m_1}{m_1 + m_2 - m_3} \cdot (\rho_l - \rho_g) + \rho_g \tag{1}$$

where:

m_1 - the mass of the sample dried to a constant weight (g)
m_2 - the mass of the pycnometer filled with distilled water (g)
m_3 - the mass of the pycnometer with boiled sample and distilled water (g)
ρ_l - the density of water ($g \cdot cm^{-3}$)
ρ_g - the density of air ($g \cdot cm^{-3}$)
ρ_s - the specific gravity ($g \cdot cm^{-3}$)

For bulk density, weighed sample was added into a graduated cylinder. Bulk density was calculated according to the equation (2).

$$\rho_d = \frac{m_{sample}}{V} \tag{2}$$

where:

m_{sample} - the mass of the sample added into a graduated cylinder (g)
V - the volume of the sample added into a graduated cylinder (cm^3)
ρ_d - the bulk density ($g \cdot cm^{-3}$)

For porosity determination the following equation (3) was used:

$$P = \frac{\rho_s - \rho_d}{\rho_s} \cdot 100 \quad (3)$$

where:

ρ_s - the specific gravity ($g \cdot cm^{-3}$)
ρ_d - the bulk density ($g \cdot cm^{-3}$)
P - the porosity (%)

2.3. Determination of maximum water holding capacity MWHC and water retention capacity WRC

For maximum water holding capacity the soil sample was weighed and added into plastic bottles. Sample was excessively saturated with water. Opened bottle was reversed and put on 4x folded filter paper. After 1 hour of suction the filter paper was changed. After 1 hour the sample was weighed. Maximum water holding capacity was calculated according to the equation (4).

$$\mathrm{MWHC} = \frac{a - b}{a} \cdot 100 \quad (4)$$

where:

a - the mass of the sample after 2 hours of suction (g)
b - the mass of the sample dried to a constant weight (g)
MWHC - maximum water holding capacity (%)

For retention water capacity the suction of water was continued for another 22 hours, then the sample was weighed.

$$\mathrm{WRC} = \frac{c - b}{c} \cdot 100 \quad (5)$$

where:

b - the mass of the sample dried to a constant weight (g)
c - the mass of the sample after 24 hours of suction (g)
WRC - water retention capacity (%)

2.4. Determination of oxidizable carbon content C_{ox}

For oxidizable carbon content the sample was passed through a 0.25mm sieve and weighed (approximately 0.1 g). The sample, 5 ml of $K_2Cr_2O_7$ solution and 7.5 ml of H_2SO_4 was added into a beaker, covered with Petri dish and heated in dryer at 135 °C for 30 min. After cooling down, the solution with the sample was quantitatively added into the 100ml volumetric flask and fully filled with demineralized water. After 60 min of suspension, sedimentation the solution was centrifugated at 3,500 rpm for 10 min. For making of calibration solutions there was used a glucose. Glucose solutions was prepared by adding specific amount of glucose into the 100ml volumetric flask and the flask was fully filled with $K_2Cr_2O_7$ solution. These glucose solutions were oxidized the same way as sample solution (without another addition of $K_2Cr_2O_7$ solution). Then the absorbance of calibration solutions and sample solutions was measured at the wavelength 585 nm.

2.5. Determination of total nitrogen content N_{tot}

For total nitrogen content the sample was weighed (approximately 0.2-1.0 g). The sample and 4 ml of mixture of salicylic acid and sulfuric acid was mixed and let to react overnight. Then 0.5 g of copper(II) thiosulfate pentahydrate was added to a sample and cautiously heated until the foaming stopped. The mixture was cooled down. Then 1.1 g of catalyst (mixture of K_2SO_4, $CuSO_4 \cdot 5H_2O$ and TiO_2) was added to the mixture and the mixture was heated at 360-400 °C for 2 h until clarification. After cooling down, 20 ml of demineralized water was added to the mixture. The mixture was added into a distillation flask. 5 ml of H_3BO_3 was added into receiving flask and 20 ml of NaOH solution was added into distillation flask continuously during distillation. After distilling 40 ml of distillate, 2-3 drops od mixed indicator was added to distillate and titrated with HCl.

2.6. Determination of P, K and Mg content

For P, K and Mg content determination, 10 g of sample and 100 ml of extraction solution (Mehlich 3) was added into a PE bottle and the mixture was extracted on a rotary shaker for 10 min. After the extraction the mixture was centrifugated at 4,000 rpm for 5 min. Calibration standard solutions for P, K and Mg content determination was prepared at different concentrations. Mixed calibration standard solution was prepared by adding different amounts of standard solution and 20 ml of concentrated extraction solution (Mehlich 3) into a 100ml PE volumetric flask and the flask was fully filled with distilled water. An auxiliary calibration standard solution was prepared by adding different amounts of standard solution into a 100ml PE volumetric flask and the flask was fully filled with distilled water. Then 50, 25 and 10 ml of the auxiliary calibration standard solution was added into 3 100ml PE volumetric flasks. 20 ml of concentrated extraction solution (Mehlich 3) was added into each flask and the flasks were fully filled with distilled water. All solutions (the set of calibration solutions and the set of sample solutions) were measured on ICP-OES spectrometer.

3. RESULTS AND DISCUSSION

3.1. The influence of biochar on physical properties of soil

The resulting values of physical properties of soil are shown in **Table 1**. The specific gravity is increased in all types of soil after NovoTerra biochar application, on the contrary Sonnenerde biochar increased the specific gravity only in cambisol. Depending on the results of the carbon content after the biochar application (see **Figure 1**), it can be assumed that the Sonnenerde biochar compared to the NovoTerra biochar increased the carbon content, so much that it decreased the specific gravity, which is dependent on the carbon content.

Table 1 Physical properties of soil before and after biochar application - ρ_s is the specific gravity ($g\cdot cm^{-3}$), ρ_d is the bulk density ($g\cdot cm^{-3}$), *P* is the porosity (%), MWHC is the maximum water holding capacity (%) and WRC is the water retention capacity (%); C is cambisol, F is fluvisol, CH is chernozem, R is regosol, N is NovoTerra biochar and S is Sonnenerde biochar.

Sample	ρ_s ($g\cdot cm^{-3}$)	ρ_d ($g\cdot cm^{-3}$)	*P* (%)	MWHC (%)	WRC (%)
C	2.39±0.05	1.01±0.01	57.7±1.9	17.9±0.9	15.7±0.6
C+N	2.41±0.05	0.96±0.01	60.1±1.7	23.4±0.4	20.1±1.9
C+S	2.44±0.04	0.86±0.03	64.8±0.5	18.4±1.9	15.4±1.2
F	2.35±0.04	1.09±0.03	53.7±1.1	14.9±0.9	12.8±0.6
F+N	2.37±0.06	1.13±0.03	52.1±2.0	15.7±1.4	12.9±0.7
F+S	2.33±0.01	1.05±0.06	55.2±1.9	15.2±1.0	13.5±0.6
CH	2.53±0.04	1.21±0.05	52.2±1.3	19.3±0.3	17.7±1.3
CH+N	2.56±0.04	1.31±0.05	48.8±1.1	19.4±1.7	17.7±3.0
CH+S	2.52±0.01	1.24±0.07	51.0±3.0	16.1±4.0	17.1±1.4
R	2.41±0.05	1.44±0.03	40.0±2.0	13.2±0.8	10.0±0.7
R+N	2.42±0.07	1.42±0.04	41.1±0.7	18.8±1.3	18.0±1.3
R+S	2.28±0.04	1.42±0.07	37.9±4.0	15.2±2.0	11.2±1.2

In the case of bulk density, Sonnenerde repeatedly reduced the resulting value of this parameter, because it was able to increase the carbon content of the sample more than NovoTerra biochar. Even in this case, the bulk density is dependent on the carbon content (see **Figure 1**). The bulk density describing soil compaction has an influence on other physical properties such as porosity or water retention capacity, which also correlates.

A more significant increase in porosity occurred in the case of cambisol. This increase can be explained by the fact that the increased carbon content and other supplied minerals from the biochar caused that the plants had a larger root system, which aerated the soil more and thus created more pores. For both mentioned water capacities, we again encounter the fact that the carbon content correlates with the increased water retention capacity (according to the results).

3.2. The influence of biochar on chemical properties of soil

The effect of biochar on the oxidizable carbon content is shown in **Figure 1**. In general, there was an increase in the carbon content after the biochar application. However, it is worth noting that Sonnenerde biochar was more effective than NovoTerra biochar. This can be explained by the fact that NovoTerra biochar was made from biomass with a higher content of more stable organic matter, which was more difficult to release into the soil than in the case of Sonnenerde biochar.

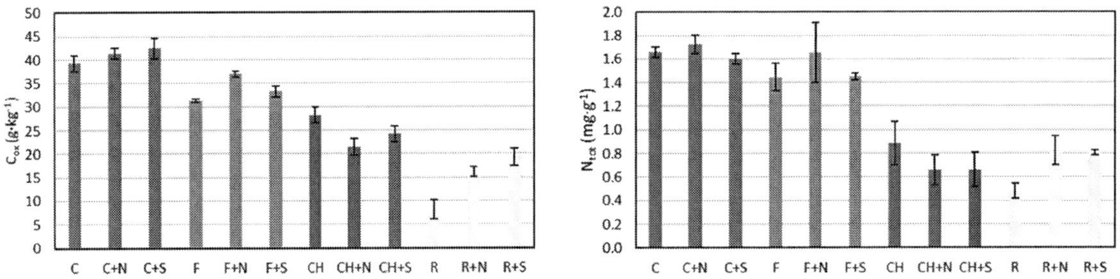

Figure 1 Oxidizable carbon content C_{ox} (left) and N_{ox} (right) in soil before and after biochar application

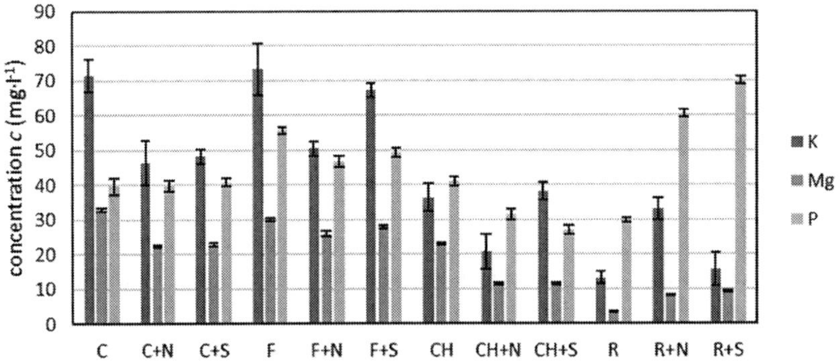

Figure 2 P, Mg and K content in soil before and after biochar application

In the case of cambisol and fluvisol, there was no significant influence on total nitrogen content after biochar application (**Figure 1**). NovoTerra biochar seems to be more efficient to increase the N_{tot} content, while Sonnenerde biochar did not significantly affect it. Chernozem showed a decrease in the total nitrogen content in the case of use of both biochars and regosol showed an increase in N_{tot} content in both cases.

In the **Figure 2**, there is shown the decrease in the P and Mg content after biochar application in the case of cambisol, fluvisol and chernozem. Regosol showed an increase in both mentioned parameters. Furthermore, it can be observed that there were no changes in the P content in the case of cambisol after biochar application. The increase in the case of fluvisol was not significant. Chernozem showed significant reduction in P content, while Sonnenerde biochar had a greater influence on this result. On the other hand, there was a very significant increase in the P content of the regosol and the Sonnenerde biochar had a greater influence on this increase again.

4. CONCLUSION

Generally, it is not easy to describe the effects of biochar on soil properties because of the complexity of the subject of this study. The influence of biochar very much depends on the type of biochar and also on the type of soil. There is an increase of C_{ox} and N_{tot} content after biochar application to the soil. The influence of biochar on P, Mg and K content in soil after biochar application varies depending on the type of soil.

Biochar is more effective in the case of low-quality soils (in this study, this type was represented by regosol). In the case of high-quality soils (cambisol and fluvisol), biochar is not so efficient and the influence is not so obvious. In the future, there is a need for long-term experiments and studies that include the effect of field conditions (larger scale, applications in agriculture, economic review – cost and crop yield, etc.).

ACKNOWLEDGEMENTS

This research was funded by Czech Science Foundation of Czech Republic by grant number GJ20-28208Y.

REFERENCES

[1] NAN, H., YIN, J., YANG, F., LUO, Y., ZHAO, L., CAO., X. Pyrolysis temperature-dependent carbon retention and stability of biochar with participation of calcium: Implications to carbon sequestration. *Environmental Pollution.* 2021, vol. 287. Available from: https://doi.org/10.1016/j.envpol.2021.117566.

[2] YANG, C. D., LU, S. G. Effects of five different biochars on aggregation, water retention and mechanical properties of paddy soil: A field experiment of three-season crops. *Soil and Tillage Research.* 2021, vol. 205. Available from: https://doi.org/10.1016/j.still.2020.104798.

[3] ZHANG, M., WANG, J., BAI, S. H., ZHANG, Y., TENG, Y., XU, Z. Assisted phytoremediation of a co-contaminated soil with biochar amendment: Contaminant removals and bacterial community properties. *Geoderma.* 2019, vol. 348, pp. 115-123. Available from: https://doi.org/10.1016/j.geoderma.2019.04.031.

[4] International Biochar Initiative. *About biochar.* [online]. 2018. [viewed: 2020-11-03]. Available from: https://biochar-international.org/biochar/.

[5] TAN, S., NARAYANAN, M., HUONG, D. T. T., ITO, N., UNPAPROM, Y., PUGAZHENDHI, A., CHI, N. T. L., LIU, J. A perspective on the interaction between biochar and soil microbes: A way to regain soil eminence. *Environmental Research.* 2022, vol. 214. Available from: https://doi.org/10.1016/j.envres.2022.113832.

[6] SMILKOVA, M., SMILEK, J., KALINA, M., KLUCAKOVA, M., PEKAR, M., SEDLACEK, P. A Simple technique for assessing the cuticular diffusion of humic acid biostimulants. *Plant Methods.* 2019, vol. 15. Available from: https://doi.org/10.1186/s13007-019-0469-x.

[7] LATAF, A., JOZEFCZAK, M., VANDECASTEELE, B., VIAENE, J., SCHREURS, S., CARLEER, R., YPERMAN, J., MARCHAL, W., CUYPERS, A., VANDAMME, D. The effect of pyrolysis temperature and feedstock on biochar agronomic properties. *Journal of Analytical and Applied Pyrolysis.* 2022. Available from: https://doi.org/10.1016/j.jaap.2022.105728.

[8] SUN, L., GONG, P., SUN, Y., QIN, Q., SONG, K., YE, J., ZHANG, H., ZHOU, B., XUE, Y. Modified chicken manure biochar enhanced the adsorption for Cd2+ in aqueous and immobilization of Cd in contaminated agricultural soil. *Science of The Total Environment.* 2022, vol. 851 (part 2). Available from: https://doi.org/10.1016/j.scitotenv.2022.158252.

[9] WEBER, K. Properties of biochar. *Fuel*. 2018, vol. 217, pp. 240-261. Available from: https://doi.org/10.1016/j.fuel.2017.12.054.

[10] SPOKAS, K. Review of the stability of biochar in soils: predictability of O:C molar ratios. *Carbon Management*. 2010, vol. 1, pp. 289-303. Available from: https://doi.org/10.4155/cmt.10.32.

[11] HRUŠKA, M. Situační a výhledová zpráva: Půda. *Ministerstvo zemědělství*. 2018. Available from: https://eagri.cz/public/web/file/611976/SVZ_Puda_11_2018.pdf

[12] KALINA, M., SOVOVA, S., HAJZLER, J., KUBIKOVA, L., TRUDICOVA, M., SMILEK, J., ENEV, V. Biochar Texture – A Parameter Influencing Physicochemical Properties, Morphology, and Agronomical Potential. *Agronomy*. 2022, vol. 12, no. 8. Available from: https://doi.org/10.3390/agronomy12081768.

HOW ORGANIC MATTER CAN AFFECT IMMOBILIZATION OF DICLOFENAC IN SOILS

[1]Martina KLUČÁKOVÁ

[1]*Brno University of Technology, Faculty of Chemistry, Brno, Czech Republic, EU,*
klucakova@fch.vutbr.cz

https://doi.org/10.37904/nanocon.2022.4592

Abstract

Natural organic matter is a complex pool of organic substances which can affect the transport of both beneficial and harmful species in soil systems. The occurrence of pharmaceuticals as soil pollutants has become one of the new environmental threats. The mobility of drug contaminants in soils and their availability for plants is largely controlled by their sorption. Soils have a "self-cleaning" ability which depends on the content of organic matter. Interactions between drugs and soil organic mattes can lead to the partial immobilization of these pollutants and suppression of their mobility in soil resulting in the pollution of drinking water resources, crop and consequently food contamination. In this contribution, the soils with different contents of organic matter were selected to assess their sorption abilities to widely used pharmaceutical diclofenac (non-steroidal anti-inflammatory drug that is frequently prescribed to reduce inflammation and pain). Stability of formed complexes and drug leachability in water were also studied to determine the immobilization degree of pharmaceutical in studied soils. The effect of organic matter on binding ability of soils as well as the leachability of diclofenac at different pH values are compared and discussed.

Keywords: Organic matter, diclofenac, adsorption, immobilization

1. INTRODUCTION

Diclofenac is widely used as nonsteroidal anti-inflammatory drug leaving residues in the environment. It has a low removal efficiency ranging from 5 to 81% in conventional sewage treatment and has been found in wastewater treatment plant effluents in many countries worldwide [1-4]. It is considered as a contaminant of emerging concern and has been included in the list of priority hazardous compounds [2,5] and Watch List of EU Decision 2015/495 [6-9]. Diclofenac seems to have the highest level of acute toxicity among other non-steroidal anti-inflammatory drugs tested [1,10]. Diclofenac is relatively persistent in soils under anaerobic conditions with a negligible degradation [11,12] while better degradation was observed under oxic conditions [12,13]. Recent works indicated that degradation was the main attenuation process of diclofenac during the wetting and drying cycles in soil, while adsorption was the only process controlling the fate of diclofenac under continuous infiltration conditions [12,14]. It means that adsorption would dominate the environmental fate of diclofenac in water saturated soils. Understanding the adsorption mechanism in the soil is significant for determining its ecotoxicity, migration, and conversion. Many studies have focused on the removal of diclofenac from waters by different low-cost and effective adsorbent materials [15-17]. Our contribution is focused on the adsorption of diclofenac directly by soil. The investigation of adsorption from aqueous solution onto soils is crucial for prediction of its possible immobilization, mobility and removing diclofenac contamination. In general, the interactions of pollutants in soils are affected by their physicochemical properties as well as soil properties (its structure and composition, content, and form of organic matter etc.). Other factors influencing interactions are, pH, ionic strength, temperature, contact time, presence of additional constituents, and (also) living organisms in soil systems [15-19].

Some studies dealing with the behaviour of diclofenac in soil were recently realized. Filep et al. [20] described the effect of chemical properties of pharmaceutics on sorption behaviour in soils and goethite. They stated that the main controlling factor for ionic compounds, such as diclofenac sodium, is their charge (and Coulomb forces). Drillia et al. [21] studied the adsorption of six drugs in two soil types. Their results showed that the adsorption on the soil of the low organic carbon was not dependent only on the organic content of the matrix but also on the other matrix properties and the dissociation degree of the compounds. Diclofenac had relatively high distribution coefficients obtained for both adsorption and desorption. It means that it is not strongly bonded and can be easily leached. Kohay et al. [22] studied efficiency of polycation-clay sorbents for diclofenac. They stated that the filtration of micro- and nano-concentrations of diclofenac by composite columns, in the presence of humic acids was more efficient than by granulated activated carbon. columns. In contrast, Yu and Bi [12] stated that the adsorption of diclofenac on soil is controlled by bonding with surface -OH groups of iron oxides. Their results indicated that adsorption coefficients of diclofenac onto soils can be well predicted by contents of extractable iron instead of total iron oxides contents. The influence of iron oxides in the adsorption of diclofenac should be important mainly in soils with relatively low content of organic matter. Zhang et al. [23] studied the sorption-desorption behaviour of several pharmaceuticals including diclofenac. Their experiments were realised with individual drugs as well as a mixture of them. The proportion of diclofenac adsorbed to the soil in the mixture system was 72 % and differed slightly from the individual adsorption. Diclofenac displayed strong sorption and low desorption in both the individual-compound and mixture-compound systems. De Mastro et al. [24] compared the adsorption of sulfamethoxazole and diclofenac on three agricultural soils with different mineralogical composition and organic matter content. Their results showed that diclofenac was adsorbed moderately while sulfamethoxazole negligibly, regardless the soils. While clay minerals had little influence on their adsorption, the organic matter supported interactions of soil with diclofenac. Similarly, *Graouer-Bacart et al. [25] confirmed that the* mobility of diclofenac in soils increases significantly with low organic matter content (and with high $CaCO_3$ content), which can lead to an enhanced risk for groundwater contamination.

As can be seen, the results and conclusions presented in studies can be different. The organic matter content as well as the presence of minerals can have different effect of the interactions of diclofenac with soils and its mobility in them. In general, the mobility of the pharmaceuticals in soils, and consequently their potential for contaminating groundwaters, is shown to depend on the amount of drug applied, the intensity of the "rain" events and the soil type and composition [21].

2. MATERIALS AND METHODS

Diclofenac (CAS 15307-79-6) was purchased from Sigma-Aldrich. Soil I (phaeozem) was sampled in the region Jablůnka, soil II (alluvial soil) in the region Černotín (both closed to river Bečva). Both samples are arable soils with periodically changed arable crops as wheat and corn. Main characteristics of soil samples are listed in **Table 1**.

Table 1 Characterization of soil samples

Sample	Total organic carbon (wt.%)	Humic /fulvic acids (-)	Cation exchange capacity (mmol/g)	Humification degree (%)	Humus content (wt. %)
Soil I	2.23	1.22	0.13	12.6	3.84
Soil II	1.73	0.88	0.21	12.1	2.98

Soil samples were mixed with the solution of diclofenac (0 - 2.5 mg/dm^3), stirred (48 h) and centrifuged (Hettich ROTINA 46 R; 5000 rpm, 10 minutes). The ratio between the solid powder sample and solution was 0.5 g: 25 cm^3. Supernatants were filtered (0.22 μm) to remove solid particles and analysed by means of UV/VIS spectrometry (HITACHI U-3900H) to determine the residual amount of diclofenac in solution after adsorption. Simultaneously pH values and conductivities of solutions were measured.

The stability of formed complexes was investigated by means of desorption experiments. Diclofenac-soil complexes were mixed with 10 cm^3 of deionized water, stirred, centrifuged, filtered, and analysed by the same methods as in the case of adsorption experiments.

To investigate the effect of pH on the adsorption, Britton-Robinson buffer solutions (at pH values equal to 5, 7, and 10 were used for the adsorption/desorption experiments. The experiments were realized only with the highest diclofenac concentration.

All experiments were realized at 25 °C (±1 °C). They were triplicated and average values are presented.

3. RESULTS AND DISCUSSION

In **Figure 1**, adsorption isotherms obtained for diclofenac adsorbed on two different soil samples are shown. It can be seen, that adsorbed amounts are much higher for soil I, which is richer in the total organic carbon (related to soil organic matter), humic acids and humus content (see **Table 1**). In contrast, the cation exchange capacity is higher for soil II. The efficiency of adsorption decreased with increasing initial concentration of diclofenac for both used soil samples. Their values ranged between 74 and 87 % for soil I; 24 and 36 % for soil II. Measured values of pH and conductivity were lower for soil II and similar as in the case of background experiments with soil and distilled water. Their dependence on the concentration of diclofenac was not significant. The pH values ranged between 7 and 7.8, conductivity between 0.2 and 0.5 mS cm^{-1}.

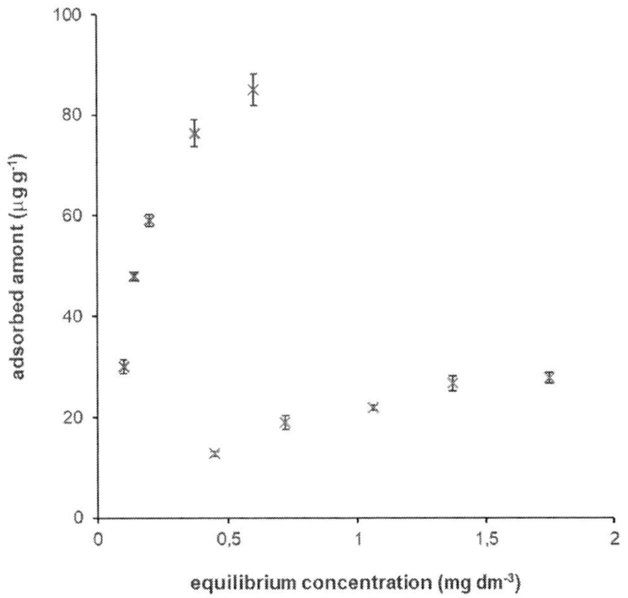

Figure 1 Adsorption isotherms obtained for soil I (red) and soil II (green).

Experimental data were fitted by Langmuir isotherm [26,27] frequently used for chemical adsorption

$$a = a_{max} \frac{bc}{1+bc}, \tag{1}$$

where a is adsorbed amount, a_{max} is adsorption capacity, c is equilibrium concentration and b is the ratio between adsorption and desorption rate constants ($b = k_{ads}/k_{des}$). Parameters of adsorption isotherm (a_{max} and b) can be determined on the basis of the linearized form of equation 1:

$$\frac{1}{a} = \frac{1}{a_{max}b}\frac{1}{c} + \frac{1}{a_{max}}. \tag{2}$$

The values of adsorption capacity a_{max} were determined as 117.6 $\mu g\ g^{-1}$ for soil I and 48.6 $\mu g\ g^{-1}$ for soil II. The ratio b was equal to 4.72 for soil I and 0.82 for soil II. The results indicated that soil I has not only much higher adsorption capacity but also that the adsorption is stronger in comparison with desorption one and the mobile fraction should predominate in complex of diclofenac with soil II.

Desorption experiments showed that the content of mobile fraction is relatively low for both used soil samples. A measurable desorption was indicated only for the highest amount of diclofenac adsorbed on soil I and two highest amounts of diclofenac adsorbed on soil II. However, it was confirmed that diclofenac can be desorbed more from its complex with soil II. The difference between both soil samples is less significant than can be deduced from the coefficient b in Langmuir isotherm (equation 2), which is much higher for soil I. The reason of this discrepancy can be explained by the character of Langmuir isotherm. The isotherm is based on the adsorption and desorption kinetics valid for the simplest reaction mechanism, where substance S is adsorbed on the adsorbent A (its active site) forming complex SA and can be desorbed in reverse reaction: $S + A \leftrightarrow SA$. In equilibrium, the rates of adsorption and desorption are equal and the adsorption amount as well as the concentration in solution do not change with time. Since soil is very complex system containing many different constituents, it is impossible to expect that all active sites in soil are equivalent. In contrast, soil can contain many different active sites with different bond strengths. Therefore, the above-mentioned mechanism of simple surface reaction can be considered as an idealized situation for the adsorbent which only one type of active sites (mutually independent). It means that adsorption parameters based on equation 1 and 2 can be considered as effective values including all types of active sites in given soil sample and all proceeding surface reactions.

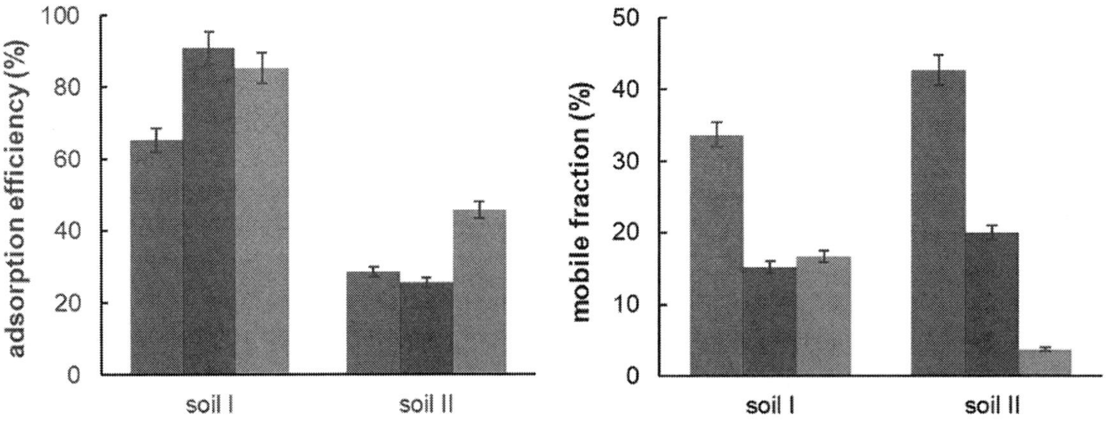

Figure 2 Adsorption efficiency and content of mobile diclofenac fraction in soils at pH 5 (red), 7 (blue) and 10 (green)

In **Figure 2**, The efficiencies of adsorption and desorption are compared. The adsorption efficiency was calculated as % adsorbed from the total amount of diclofenac in solution mixed with soil. The desorption efficiency as the amount leached from diclofenac-soil complex into water. The desorption efficiency thus can be considered as the mobile fraction of diclofenac which is leachable by water. As can be seen, the increase in pH resulted in the increase in adsorption efficiency and (also) in the increase in the bond strength. Our results are not with agreement with hypothesis of Filep et al. [20] that ionic compounds as diclofenac are bound to soil mainly by means of their charge and the importance of organic matter is negligible. In contrast, soil I richer in organic carbon was able to adsorb much higher amount of drug than soil II. It indicated that the organic matter can play important role in the immobilization of diclofenac in soil which agrees with conclusion s of other authors [22-25]. Our results also confirmed the significance of the properties of organic matter and soil matrix.

4. CONCLUSION

Adsorption and desorption experiments were carried out with two soil samples of different quality. Diclofenac as widely used anti-inflammatory pharmaceutical was chosen for this study. It was confirmed that soil organic matter plays key role in the mobility and bio-availability of pharmaceuticals in soils.

It was found that the immobilization ability of soils was strongly influenced by the content of organic matter. The stability of formed complexes determined by desorption experiments showed that formed complexes are relatively stable and the stability increases with the organic carbon content. The increase in pH of used solutions resulted in more effective adsorption and the suppression of drug leachability.

REFERENCES

[1] CHEN, G., DEN BRAVEN, M.W, COMELIS, A.M., VAN GESTEL, N.M., VAN STRAALEN, D.R., ROELOFS, D. Ecotoxicogenomic assessment of diclofenac toxicity in soil. *Environmental Pollution.* 2015, vol. 199, pp. 253-260. Available from: https://doi.org/10.1016/j.envpol.2015.01.032.

[2] STYLIANOU, K., HAPESHI, E., VASQUEZ, M.I., FATTA-KASSINOS, D., VYRIDES, I. Diclofenac biodegradation by newly isolated Klebsiella sp. KSC: Microbial intermediates and ecotoxicological assessment. *Journal of Environmental Chemical Engineering.* 2018, vol. 6, pp. 3242-3437. Available from: https://doi.org/10.1016/j.jece.2018.04.052.

[3] LUO, Z., GUO, W., NGO, H.H., NGHIEM, L.D., HAI, F.I., YHANG, J., LIANG, S., WAN, X.C. A review on the occurrence of micropollutants in the aquatic environment and their fate and removal during wastewater treatment. *Science of the Total Environment. 2014, vol. 473, pp. 619-641. Available from:* https://doi.org/10.1016/j.scitotenv.2013.12.065.

[4] ZHANG, Y., GEIßEN, S.U., GAL, C. Carbamazepine and diclofenac: removal in wastewater treatment plants and occurrence in water bodies. *Chemosphere.* 2008, vol. 73, pp. 1151-1161. Available from: https://doi.org/10.1016/j.chemosphere.2008.07.086.

[5] EUROPEAN COMISSION, 2012. Proposal for a Directive of the European Parliament and of the Council Amending Directives 2000/60/EC and 2008/105/EC as Regards Priority Substances in the Field of Water Policy. pp. 1-35.

[6] SATHISKUMAR, P., MEENA, R.A.A., PALANISAMI, T., ASHOKKUMAR, V., PALVANNAN, T., GU, F.L. Occurrence, interactive effects and ecological risk of diclofenac in environmental compartments and biota - a review. *Science of the Total Environment. 2020, vol. 698, pp. 134057. Available from:* https://doi.org/10.1016/j.scitotenv.2019.134057.

[7] LONAPPAN, L., BRAR, S.K., DAS, R.K., VERMA, M., SURAMPALLI, R.Y. Diclofenac and its transformation products: environmental occurrence and toxicity - a review. *Environment International. 2016, vol. 96, pp. 127-138. Available from:* https://doi.org/10.1016/j.envint.2016.09.014.

[8] SOUSA, J.C.G., RIBERIO, A.R., BARBOSA, M.O., PEREIRA, M.F.R., SILVA, A.M.T. A review on environmental monitoring of water organic pollutants identified by EU guidelines. *Journal of Hazardous Materials.* 2018, vol. 344, pp. 146-162. Available from: https://doi.org/10.1016/j.jhazmat.2017.09.058.

[9] LI, Y., ZHANG, L., LIU, X., DING, J. Ranking and prioritizing pharmaceuticals in the aquatic environment of China. *Science of the Total Environment. 2019, vol. 658, pp. 333-342. Available from:* https://doi.org/10.1016/j.scitotenv.2018.12.048.

[10] FENT, A., WESTON, A.A., CAMINADA, D. Ecotoxicology of human pharmaceuticals. *Aquatic Toxicology.* 2006, vol. 76, pp. 122-159. Available from: https://doi.org/10.1016/j.aquatox.2005.09.009.

[11] LIN, K., GAN, J. Sorption and degradation of wastewater-associated non-steroidal anti-inflammatory drugs and antibiotics in soils. *Chemosphere.* 2011, vol. 83, pp. 240-246. Available from: https://doi.org/10.1016/j.chemosphere.2010.12.083.

[12] YU, C., BI, E. Adsorption site-dependent transport of diclofenac in water saturated minerals and reference soils. *Chemosphere.* 2019, vol. 236, pp. 124256. Available from: https://doi.org/10.1016/j.chemosphere.2019.06.226.

[13] BERTELKAMP, C., VERLIEFDE, A.R.D., SCHOUTTETEN, K., VANHAECKE, L., VANDEN BUSSCHE, J., SINGHAL, N., VAN DER HOEK, J.P. The effect of redox conditions and adaptation time on organic micropollutant removal during river bank filtration: a laboratory-scale column study. *Science of the Total Environment. 2016, vol. 544, pp. 309-318. Available from:* https://doi.org/1 0.1016/j.scitotenv.2015.11.035.

[14] *SILVER, M., SELKE, S., BALSAA, P., WEFER-ROEHL, A., KUEBECK, C., SCHUETH, C.* Fate of five pharmaceuticals under different infiltration conditions for managed aquifer recharge. *Science of the Total Environment. 2018, vol. 642, pp. 914-924. Available from:* https://doi.org/10.1016/j.scitotenv.2018.06.120.

[15] *GULER, U.A., TUNCEL, E., ERSAN, M. Evaluation of factors affecting tetracycline and diclofenac adsorption by agricultural soils using response surface methodology. Environmental Progress & Sustainable Energy. 2022, vol. 2022, pp. e13939. Available from:* https://doi.org/10.1002/ep.13939.

[16] *YU, C., BAHASHI, J., BI, E.* Mechanisms and quantification of adsorption of three anti-inflammatory pharmaceuticals onto goethite with/without surface bound organic acids. *Chemosphere.* 2019, vol. 222, pp. 593-602. Available from: https://doi.org/10.1016/j.chemosphere.2019.01.155.

[17] MAGESH, N., ANNAM RENITA, A., SENTHIL, K.P. Practice on treating pharmaceutical compounds (antibiotics) present in wastewater using biosorption techniques with different biowaste compounds. A review. *Environmental Progress & Sustainable Energy.* 2020, vol. 39, pp. e13429. Available from: https://doi.org/10.1002/ep.13429.

[18] KLUČÁKOVÁ, M. Complexation of metal ions with solid humic acids, humic colloidal solutions, and humic hydrogel. *Environmental Engineering Science* 2014, vol. 31, pp. 612-620. Available from: https://doi.org/10.1089/ees.2013.0487.

[19] KLUČÁKOVÁ, M., PEKAŘ, M. Study of structure and properties of humic and fulvic acids. IV. Study of interactions of Cu^{2+} ions with humic gels and final comparison. *Journal of Polymer Materials.* 2003, vol. 20, pp. 155-162.

[20] FILEPA, T., SZABO, A., KONDOR, A.C., JAKAB, B., SZALAI, Z. Evaluation of the effect of the intrinsic chemical properties of pharmaceutically active compounds (PhACs) on sorption behaviour in soils and goethite. *Ecotoxicology and Environmental Safety.* 2021, vol. 215, pp. 112120. Available from: https://doi.org/10.1016/j.ecoenv.2021.112120.

[21] DRILLIA, P., STAMATELATOU, K., LYBERATOS, G. Fate and mobility of pharmaceuticals in solid matrices. *Chemosphere.* 2005, vol. 60, pp. 1034-1044. Available from: https://doi.org/10.1016/j.chemosphere.2005.01.032.

[22] KOHAY, H., IZBITSKI, A., MISHAEL, Y.G. Developing polycation-clay sorbents for efficient filtration of diclofenac: effect of dissolved organic matter and comparison to activated carbon. *Environmental Science & Technology. 2015, vol. 49, pp. 9281-9288. Available from:* https://doi.org/10.1021/acs.est.5b01530.

[23] ZHANG, Y., PRICE, G.W., JAMIESON, R., BURTON, D., KHOSRAVI, K. Sorption and desorption of selected non-steroidal anti-inflammatory drugs in an agricultural loam-textured soil. *Chemosphere.* 2017, vol. 174, pp. 628-637. Available from: https://doi.org/10.1016/j.chemosphere.2017.02.027.

[24] *DE MASTRO, F., CACACE, C., TRAVERSA, A., PALLARA, M., COCOZZA, C., MOTTOLA, F., BRUNETTI, G. Influence of chemical and mineralogical soil properties on the adsorption of sulfamethoxazole and diclofenac in Mediterranean soils. Chemical and Biological Technologies in Agriculture. 2022, vol. 9, pp. 34. Available from:* https://doi.org/10.1186/s40538-022-00300-8.

[25] *GRAOUER-BACART, M., SAYEN, S., GUILLON, E. Adsorption and co-adsorption of diclofenac and Cu(II) on calcareous soils. Ecotoxicology and Environmental Safety. 2016, vol. 124, pp. 386-392. Available from:* https://doi.org/10.1016/j.ecoenv.2015.11.010.

[26] *KLUČÁKOVÁ, M., KALÁB, M., PEKAŘ, M., LAPČÍK, L. Study of structure and properties of humic and fulvic acids. II. Complexation of* Cu^{2+} *ions with humic acid extracted from lignite. Journal of Polymer Materials. 2002, vol. 19, pp. 287-294.*

[27] *KLUČÁKOVÁ, M., OMELKA, L. Study of sorption of metal Ions on lignite and humic acids. Chemical Papers. 2004, vol. 58, pp.170-175.*

PLANT EXTRACT REDUCED AND STABILIZED GOLD NANOPARTICLES AS MODEL PARTICLES FOR FOLIAR APPLICATION OF FERTILIZERS

[1]Regina BURVE, [2]Augusta SZAMEITAT, [1]Jean-Claude GRIVEL

[1]*Department of Energy Conversion and Storage, Technical University of Denmark, Lyngby, Denmark, EU,* regbu@dtu.dk

[2]*Department of Plant and Environmental Sciences, University of Copenhagen, Frederiksberg, Denmark, EU*

https://doi.org/10.37904/nanocon.2022.4600

Abstract

Foliar application of nanofertilizers is currently regarded as a promising alternative to conventional soil fertilization. For successful development of nanofertilizers, firstly, a better understanding of interactions of nanoparticles with the leaf surface is needed. In the framework of this study a globally available cereal grain *Hordeum Vulgare* (barley) was chosen as model plant. Gold nanoparticles (Au NPs) were chosen as model particles for foliar application. Two synthesis methods were used: established photoreduction with citric acid and phytochemical synthesis using *Aegopodium Podagraria* (ground elder) extract. In both cases the reducing agent was acting also as surface stabilizer. Au NPs formation was confirmed by UV-VIS spectroscopy. To our knowledge it is the first time *A. Podagraria* extract was used for Au NPs synthesis. The morphology of the produced particles was characterized by transmission electron microscopy (TEM). The interaction of Au NPs with the adaxial side of barley leaves was studied by low-vacuum scanning electron microscopy (SEM). It was established that citrate stabilized Au NPs agglomerates evenly distribute in droplet-leaf contact areas avoiding stomates. Plant extract stabilized Au NPs form a film in the contact areas and accumulate on all leaf parts - stomates, trichomes and lamina. It was shown that Au NPs produced via green phytochemical synthesis have improved interaction with the leaf surface of barley.

Keywords: Au NPs, phytochemical synthesis, foliar fertilization

1. INTRODUCTION

One of the most important current challenges facing agriculture is to improve the efficiency and sustainability of fertilization procedures. Application of conventional fertilizers results in poor nutrient uptake due to low bioavailability and releasing rates overwhelming actual plant absorption rates. Furthermore, nutrients trapped in the soil are causing serious environmental constraints worldwide, including ground water pollution and eutrophication, soil quality degradation, and air pollution [1,2].

Foliar application of nanofertilizers is currently regarded as a promising alternative to conventional soil fertilization [3-5]. It was shown that NPs are capable of crossing plant mechanical barriers (e.g. cuticle and cell walls) with further translocation [6-8]. Moreover, NPs surfaces are easily modifiable with a variety of coatings and, similar to drug delivery, NP surface properties can theoretically be tuned to deliver them to specific leaf structures or tissues in plants [9,10]. This could also provide slow and controlled release of the nutrients in the desired locations of plants [8]. All this advantages combined with the application approach which bypasses soil have the potential to reduce unsustainable fertilization practices and increase yields through optimized nutrient management. However, despite the great potential of NPs as nanofertilizers, there remains limited understanding of how the physical and chemical properties (e.g. size, charge, solubility, coating, chemical composition) of NPs influence their interaction with the leaf surface. A better understanding of these NP-leaf interactions is needed for successful nanofertilizers development.

The leaf cuticle can be described as a composite membrane, which is mainly composed of cutin, an amorphous and non-soluble polyester formed by condensed polyhydroxylated fatty acids, which acts as a solid matrix for the deposition of the other components (e.g., polysaccharides and phenolic compounds). The leaf itself is covered with a layer of hydrophobic epicuticular waxes [11,12]. In general, there are three main natural pathways how NPs could penetrate cuticle layer - stomates, trichomes and pores [6,7,9].

In the framework of the study, a globally available cereal grain *Hordeum Vulgare* (barley) was chosen as model plant. Au NPs were chosen as model particles due their stability and absence of background signal within the plant. To evaluate the impact of surface modification on interaction with the leaf surface, two different agents were used - citric acid and *A. Podagraria* plant extract. Citrate reduced and stabilized Au NPs were synthesized by using an established photoreduction process [13]. Plant extract reduced and stabilized Au NPs were synthesized via phytochemical reduction also involving UV light. Phytochemical synthesis of NPs using plant extracts is a cost-effective and environmentally friendly approach [14,15]. The types of solvents used during extraction affect the type and amount of reducing agents extracted. According to [16] using water as solvent results in the extraction of the following compounds: tannins, terpenoids, saponins, starches, and polypeptides. Saponins are amphiphilic secondary plant compounds, which should improve interaction of metallic Au NPs with the hydrophobic surface of barley leaf.

The aim of the study is to investigate the interaction between these Au NPs as model particles for foliar application of fertilizers and the barley leaf surface. It is expected that the different nature of the reducing/stabilizing agents for Au NPs synthesis will result in different accumulation patterns on the barley leaf surface.

2. MATERIALS AND METHODS

2.1. *A. Podagraria* extract preparation

For plant extract preparation, the aerial parts of *A. Podagraria* plants were collected in May 2022 in Denmark, washed and air dried in the shadow for 2 weeks. The air-dried samples were ground until pieces with size around 0.2-0.5 mm were obtained. The extract was prepared via ultrasonification using DI water as extractant. For extraction, 1.0000 g of the ground material was weighted, and 30.00 ml of water was added. The sample was kept in an ultrasonic bath at room temperature for 20 min. The extraction process was repeated twice. The extract was filtered and stored at 2°C.

2.2. Synthesis and characterization of Au NPs

Photoreduction of the citrate reduced and stabilized Au NPs was performed as follows: a mixture of 0.039 mL of 0.052 M chloroauric acid trihydrate $HAuCl_4{\cdot}3H_2O$ (Sigma-Aldrich, ≥99.9% trace metal basis) and 1.0 mL of 0.10 M citric acid $C_6H_8O_7$ (anhydrous, Thermo Scientific, ACS, 99.5%) solutions was diluted with DI water until the total volume of the mixture was 10 mL. The final content of $HAuCl_4$ was 0.2 mM and the $HAuCl_4$:$C_6H_8O_7$ ratio was 1:50. Mixture was stirred at ambient conditions under 254 nm light irradiation (8 Watt) for 100 minutes.

For the plant extract reduced and stabilized Au NPs synthesis a mixture of 0.039 mL of 0.052 M $HAuCl_4$ solution and 1.0 mL of *A. Podagraria* plant extract (see 2.1. Section) was diluted with DI water until the total volume of the mixture was 10 mL. The final content of $HAuCl_4$ was 0.2 mM. The mixture was stirred at ambient conditions under 254 nm light irradiation for 100 minutes. Au NPs formation was confirmed by UV-VIS spectroscopy (VWR UV-3100PC). The produced NPs size and shape were characterized with TEM (Tecnai T20 G2 S-TEM, DTU National Centre for Nano Fabrication and Characterization) under 200 kV. For imaging, lacey carbon supported copper grids (grid size 300 mesh) were used. The particle size was calculated using the ImageJ software. The results are summarized in the Section 3.

2.3. Investigation of Au NPs and leaf surface interaction

Barley plants (*Hordeum vulgare L.* cv. KWS Irina) were cultivated in a greenhouse of the Department of Plant and Environmental Sciences, University of Copenhagen, Denmark with minimum day/night temperatures of 18 °C/15 °C and a day/night cycle of 16 h/8 h. Plants were provided with all required nutrients and 1.5 L water per week in the tray.

The Au NPs colloidal solutions were applied on two different undamaged, young, fully expanded leaves of barley crops. For that, few 10 µL droplets were directly applied on the adaxial side of the leaf and air-dried. The surface of the leaves was investigated with SEM (QFEG 200 Cryo ESEM, DTU National Centre for Nano Fabrication and Characterization) in low vacuum (60 Pa). Sample were prepared as follows: small pieces of leaves were secured on a sample holder with double-sided carbon tape. A leaf without applied Au NPs was used as control. For imaging secondary electron (SE) and back-scattered electron (BSE) detectors were used. The results are summarized in Section 3.

3. RESULTS AND DISCUSSION

Time-dependent UV-VIS spectra of the exposed solutions are shown in the **Figure 1** (A row - citrate reduction, B row - plant extract reduction). Au NPs formation was confirmed by the appearance of a gold corresponding LSPR absorption maximum with the intensity increasing with the irradiation time. For citrate reduced Au NPs after 100 min of irradiation, the LSPR absorption maximum was observed at 521 nm, whereas for plant reduced Au NPs at 534 nm. Both values correspond to in literature available data [13,14] and probably differ because of differences in surface properties.

A

B

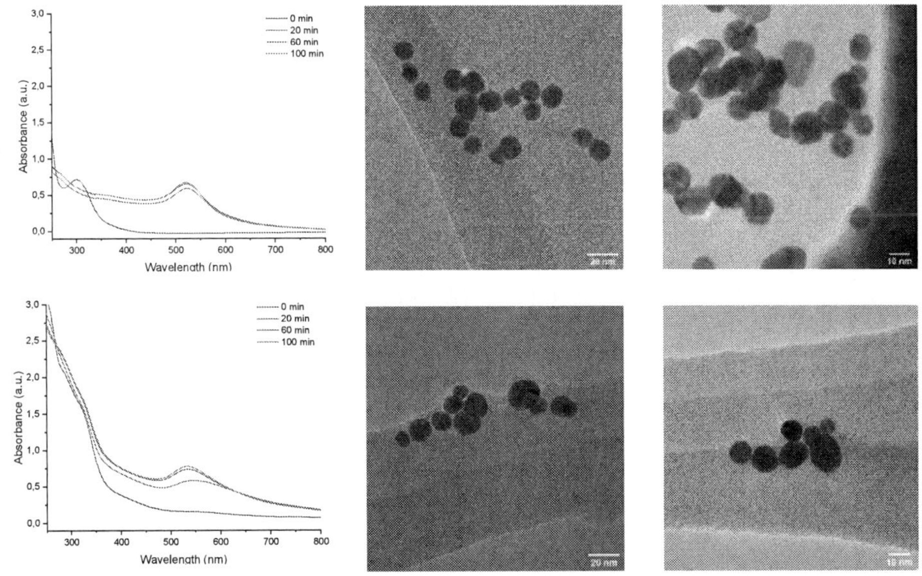

Figure 1 UV-VIS spectra of UV-exposed solutions and TEM images of resulting Au NPs particles: A - citrate reduced particles, B - plant extract reduced particles. Magnification: middle column - x285K, right column - x400K.

According to TEM images (see **Figure 1, row A**) the citrate reduced Au NPs are rounded-shaped and have diameters from 8 to 26 nm with 13.9 nm average size. Phytochemical reduction also resulted in the formation of rounded-shaped particles with size from 8 to 24 nm (see **Figure 1, row B**). The average size of the plant extract reduced Au NPs is 13.4 nm. For both samples, calculations were performed for 60 particles.

For investigating the interactions between Au NPs and the adaxial side of barley leaves, two different detectors were used: SE detector for surface characterization and BSE detector for getting NPs contrasting images. **Figure 2** shows SEM images of the control sample without Au NPs. The image in SE mode (**Figure 2A**) shows the adaxial side of the barley leaf surface with all leaf features - trichomes, stomates and lamina. Examples of leaf structural features are highlighted with red for stomata, yellow for trichome, and green for lamina. The identical image of the same area but in the BSE mode resulted in a darker image with low contrast as the BSE detector is more sensitive to elements with higher atomic number. This confirms that there are no heavy elements on the leaf surface.

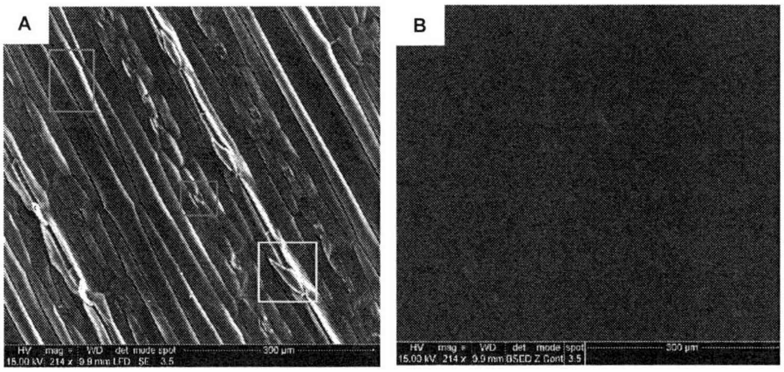

Figure 2 SEM images of control barley leaf adaxial side: A - in SE mode, B - in BSE mode. Examples of leaf structural features are highlighted with red for stomata, yellow for trichome, and green for lamina.

A SEM image of the adaxial side of barley leaf with applied citrate reduced NPs in SE mode is shown in the **Figure 3A**. It can be seen that citrate stabilized Au NPs agglomerates evenly distribute in the droplet-leaf contact area but avoiding stomates. The same area can be seen in BSE mode on the **Figure 3B**. There are no bright particles on and around stomates.

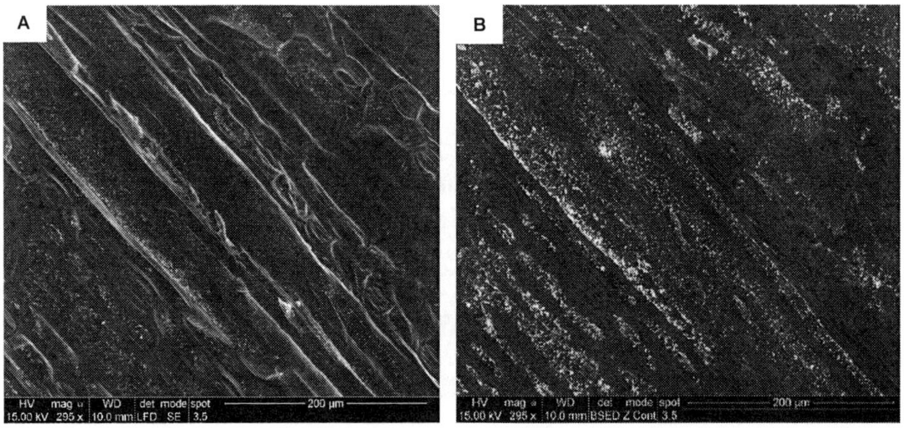

Figure 3 SEM images of citrate reduced Au NPs applied on the adaxial side of a barley leaf: A - SEM image in SE mode, B - SEM image in BSE mode.

SEM images of the adaxial side of a barley leaf with applied plant extract reduced NPs in SE mode are shown in **Figures 4A and 4C**. In the area where the applied droplet dried, the plant extract stabilized Au NPs formed a film. Contrasting SEM images in BSE mode confirm that Au NPs are distributed in this layer (see **Figures 4B and 4C**). It is shown that Au NPs are accumulating on all leaf parts: stomates, trichomes and lamina. An example of covered stomata is highlighted on the **Figure 4A**. It proves that Au NPs produced via phytochemical synthesis have improved interaction with the leaf surface of the model plant, probably due to amphiphilic compounds contained in the plant extract [16].

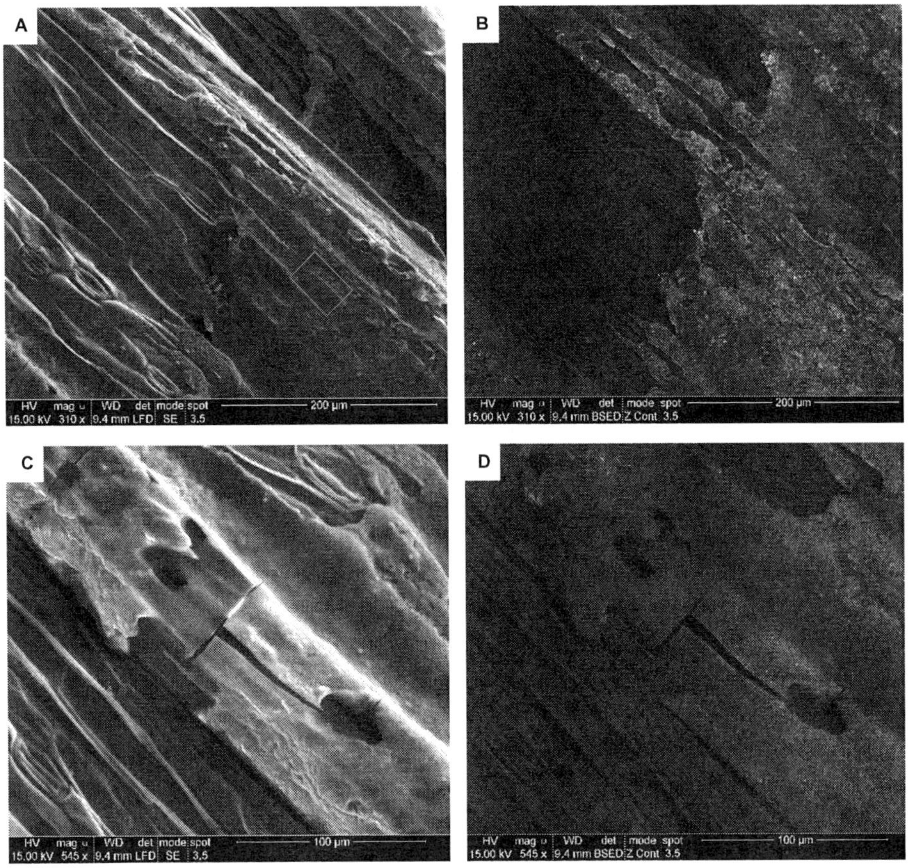

Figure 4 SEM images of plant extract reduced Au NPs applied on the adaxial side of a barley leaf: A, C - SEM image in SE mode; B, D - SEM image in BSE mode.

4. CONCLUSION

In this work, the interaction between Au NPs selected as model particles for foliar application of fertilizers and the barley leaf surface was studied. Au NPs were synthesized via reduction by UV light using two different reductants - citric acid and *A. Podagraria* leaf extract. The reductants performed also as NPs stabilizers. To our knowledge, it was the first time *A. Podagraria* extract was implemented for Au NPs synthesis. The formation of Au NPs was confirmed by UV-VIS spectroscopy. It was shown by TEM that in both cases rounded-shaped particles with around the same average size are formed: 13.9 and 13.4 nm for citrate and plant reduced Au NPs, respectively.

The different nature of the reducing/stabilizing agents resulted in different accumulation patterns on the barley leaf surface. It was established that citrate stabilized Au NPs agglomerates evenly distribute in the droplet-leaf contact area avoiding stomates. Plant extract stabilized Au NPs form a film in the contact area and accumulate on all leaf parts: stomates, trichomes and lamina. It was shown that Au NPs produced via green phytochemical synthesis using *A. Podagraria* extract as both reducing and stabilizing agent have an improved interaction with the leaf surface of the model plant.

ACKNOWLEDGEMENTS

The present work was supported by the project "Biocompatible nanofertilizers for targeted delivery and programmed release of essential mineral ions in crops" Nr. NNF21OC0066114, funded by Novo Nordisk Foundation Challenge Programme 2021 - Smart Nanomaterials for Application in Life-Science.

REFERENCES

[1] AZIZ. T., MAQSOOD, M.A., KANWAL, S., HUSSAIN, S., AHMAD., H.R., SABIR, M. Fertilizers and Environment: Issues and Challenges. In: *Crop Production and Global Environmental Issues*. Cham: Springer, 2015, pp. 575-598.

[2] ASHITHA, A., RAKHIMOL, K.R., JYOTHIS, M. Fate of the conventional fertilizers in environment. In: *Controlled Release Fertilizers for Sustainable Agriculture*. London: Elsevier, 2021, pp. 25-39.

[3] MAHAPATRA, D.M., SATAPATHY, K.C., PANDA, B. Biofertilizers and nanofertilizers for sustainable agriculture: Phycoprospects and challenges. *Science of The Total Environment*. 2022, vol. 803, p. 149990.

[4] DAS, S., BEEGUM, S. Nanofertilizers for sustainable agriculture. In: *Agricultural Nanobiotechnology. Woodhead Publishing Series in Food Science, Technology and Nutrition*. Cambridge: Elsevier, 2022, pp 355-370.

[5] KOPITTKE, P.M., LOMBI, E., WANG, P., SCHJOERRING, J.K., HUSTED, S. Nanomaterials as fertilizers for improving plant mineral nutrition and environmental outcomes. *Environmental Science: Nano*, 2019, vol. 6, pp. 3513-3524.

[6] BUENO, V., GAO, X., RAHIM, A.A., WANG, P., BAYEN, S., GHOSHAL, S. Uptake and Translocation of a Silica Nanocarrier and an Encapsulated Organic Pesticide Following Foliar Application in Tomato Plants. *Environmental Science & Technology*, 2022, vol. 56, pp. 6722-6732.

[7] ZHANG, Z., GUO, H., DENG, Y., XING, B., HE, L. Mapping gold nanoparticles on and in edible leaves in situ using surface enhanced Raman spectroscopy. RCS Advances, 2016, vol. 6, pp. 60152-60159.

[8] SZAMEITAT, A.E., SHARMA, A., MINUTELLO, F., PINNA, A., ER-RAFIK, M., HANSEN, T.H., PERSSON, D.P., ANDERSEN, B., HUSTED, S. Unravelling the interactions between nano-hydroxyapatite and the roots of phosphorus deficient barley plants. Environmental Science: Nano, 2021, vol. 8, pp. 444-459.

[9] SPIELMAN-SUN, E., AVELLAN, A., BLAND, G.D., CLEMENT, E.T., TAPPERO, R.V., ACERBO, A.S., LOWRY, G.V. Protein coating composition targets nanoparticles to leaf stomata and trichomes. Nanoscale, 2020, vol. 12, pp. 3630-3636.

[10] SPIELMAN-SUN, E., AVELLAN, A., BLAND, G.D., TAPPERO, R.V., ACERBO, A.S., UNRINE, J.M., GIRALDO, J.P., LOWRY, G.V. Nanoparticle surface charge influences translocation and leaf distribution in vascular plants with contrasting anatomy. *Environmental Science: Nano*, 2019, vol. 6, pp. 2508-2519.

[11] HEREDIA-GUERRERO, J.A., BENITEZ, J.J., DOMINGUEZ, E., BAYER, I.S., CINGOLANI, R., ATHANASSIOU, A., HEREDIA, A. Infrared and Raman spectroscopic features of plant cuticles: a review. *Frontiers in Plant Science*. 2014, vol. 5, no. 305.

[12] RACOVITA, R.C., JETTER, R. Composition of the epicuticular waxes coating the adaxial side of *Phyllostachys aurea* leaves: Identification of very-long-chain primary amides. *Phytochemistry*. 2016, vol. 130, pp. 252-261.

[13] SHIRAISHI, Y., TANAKA, H., SAKAMOTO, H., ICHIKAWA, S., HIRAI, T. Photoreductive synthesis of monodispersed Au nanoparticles with citric acid as reductant and surface stabilizing reagent. *RCS Advances*. 2017, vol. 7, pp. 6187-6192.

[14] UNAL, I.S., DEMIRBAS, A., ONAL, I., ILDIZ, N., OCSOY, I. One step preparation of stable gold nanoparticle using red cabbage extracts under UV light and its catalytic activity. *Journal of Photochemistry and Photobiology*. 2020, vol. 204, 111800.

[15] FLIEGER, J., FRANUS, W., PANEK, R., SZYMANSKA-CHARGOT, M., FLIEGER, W., FLIEGER, M., KOLODZIEJ, P. Green Synthesis of Silver Nanoparticles Using Natural Extracts with Proven Antioxidant Activity. *Molecules*. 2021, vol. 26, 4986.

[16] MOHAMAD, N.A.N., ARHAM, N.A., JAI, J., HADI, A. Plant Extract as Reducing Agent in Synthesis of Metallic Nanoparticles: A Review. *Advanced Materials Research*. 2014, vol. 832, pp. 350-355.

MODIFICATION OF GRAPHENE OXIDE SURFACE BY DOPAMINE HYDROCHLORIDE

[1,2] Karel KLOUDA, [2]Petra ROUPCOVÁ, [3]Hana KUBÁTOVÁ, [2]Bohdan FILIPI, [4]David KRSEK, [1]Kateřina BÁTRLOVÁ

[1]Occupational Safety Research Institute, Prague, Czech Republic, EU, klouda@vubp- praha.cz, batrlova@vubp-praha.cz;

[2]VSB - Technical University of Ostrava, Ostrava, Czech Republic, EU, karel.klouda@vsb.cz, petra.roupcova@vsb.cz, bohdan.filipi@vsb.cz;

[3]State Office for Nuclear Safety, Prague, Czech Republic, EU, hana.kubatova@sujb.cz

[4]National Institute of Public Health, Prague, Czech Republic, EU, krsek@szu.cz;

https://doi.org/10.37904/nanocon.2022.4594

Abstract

Polydopamine is an organic substance that is almost universally applicable to cover all kinds of materials, including nanomaterials. Considering the possible use of polydopamine in the water purification process, this article focuses on the change of selected properties of the chosen material (graphene oxide and the hybrid compound graphene oxide-C60) after being coated with polydopamine. Attention was paid to changes in thermal stability and surface morphology. In particular, changes in surface morphology could positively affect the sorption capabilities of the original materials in relation to organic pollutants.

Keywords: Metallurgy, steel, properties, applications, testing methods

1. INTRODUCTION

Dopamine (DA) is a biologically active substance from the catecholamine group. It is a neurotransmitter that is naturally produced in the nervous system of humans and animals. At the same time, it can be easily synthetized in large quantities. It is easy to prepare polydopamine (PDA) from dopamine and other catecholamines, [1,2]. It is currently being studied in detail for its unique properties, which are very similar to the adhesive proteins of marine shells. Adhesive proteins allow mussels to interact with the substrate in moist environments, and they are rich in 3,4-dihydroxyphenylalanine (DOPA) as well as cysteine and lysine [1,4]. The catechol side chain of DOPA is responsible for the strong bond between the surface of a bivalve leg and the substrate, as well as for the rapid solidification of the adhesive proteins. Catechol forms strong covalent and non-covalent interactions with the surface [5]. The synergistic effect of catechol and amine moiety further enhances biomolecular adhesion [1,3]. These strong interactions with surfaces are also generated by PDA. Therefore, PDA is seen as an organic substance that is almost universally applicable to cover all kinds of materials. The thickness of the PDA layer can be influenced by the reaction conditions (reaction time, temperature, pH) [1,5,6]. Following PDA application, new functional groups (mainly hydroxyl or other, depending on the pH) appear on the surface of a material increasing its hydrophilicity. These functional groups can be subsequently used for further surface modifications [3]. PDA demonstrates considerable chemical and environmental stability [7]. Pyrolysis at temperatures around 800 °C produces a material with a graphene-like structure, where a nitrogen atom is irregularly embedded in the carbon skeleton. The obtained material is electrically conductive [1,6]. PDA can be applied in many fields due to its potential to undergo further surface modifications and its ability to adhere to organic and inorganic substances. In biomedicine, for example, it is used to prepare nanocapsules for drug delivery or to reduce the immunogenicity of materials by improving their hydrophilicity and biocompatibility [8]. PDA can be widely used in wastewater treatment to remove heavy metals, organic pollutants, dyes and radioactive isotopes [9,10-12].

Our experiments focused specifically on the potential applications in water treatment. Graphene oxide (GO) [13,14] and its hybrid compounds [15,16] demonstrate sorption capabilities. It can be assumed that the reaction with DA will increase these abilities as a result of surface modification. Since we consider using both materials for sorption of organic pollutants present in water, it is necessary to have characteristics of both the input materials and of the materials with modified surfaces. For the pilot characterization we have chosen to use FTIR, TGA, DSC and microscopic methods (optical microscopy, TEM).

2. EXPERIMENTAL PART AND RESULTS

GO was prepared by oxidating finely ground graphite (0.025 mm) according to the classical Hummers method (H_2SO_4, $NaNO_3$, H_2O_2, $KMnO_4$, HCl) [17]. The oxidation product was repeatedly centrifuged until a negative reaction to sulfate ions was obtained. The prepared GO (0.389 g) was subsequently used for the reaction with dopamine hydrochloride (0.309 g). PDA was synthesized via DA self-polymerization. The reaction was carried out in water (30 ml) with an addition of hydrogen peroxide (30%; 0.4 ml). The suspension was agitated for 19 h and the pH was adjusted to 8-9 using $NaHCO_3$. The resulting suspension was washed with water and ethanol on a filter and then dried at 55 °C on a Petri dish. The hybrid compound GO-C_{60} (3:1) was prepared by the simultaneous oxidation of graphite with fullerene C_{60} in a weight ratio of 3:1. The procedure is described in Roupcová's dissertation [18]. The obtained GO-C_{60} (3:1) (0.350 g) was used for the reaction with dopamine hydrochloride (0.3 g). The reaction was carried out in water (30 ml) with an addition of hydrogen peroxide (30%; 0.4 ml) for 4 h. The pH was adjusted to 8-9 using $NaHCO_3$. The obtained suspension was washed with water and ethanol on a filter and then dried at 55 °C on a Petri dish.

2.1. Characterization of materials by FTIR

Fourier transform infrared spectroscopy (FTIR) of GO and GO-PDA in KBr pellets was conducted on Digilab Excalibur FTS 3000 MX model spectrometer (USA) with a scanning range of 4000-600 cm^{-1}. The obtained spectra are presented in **Figure 1**.

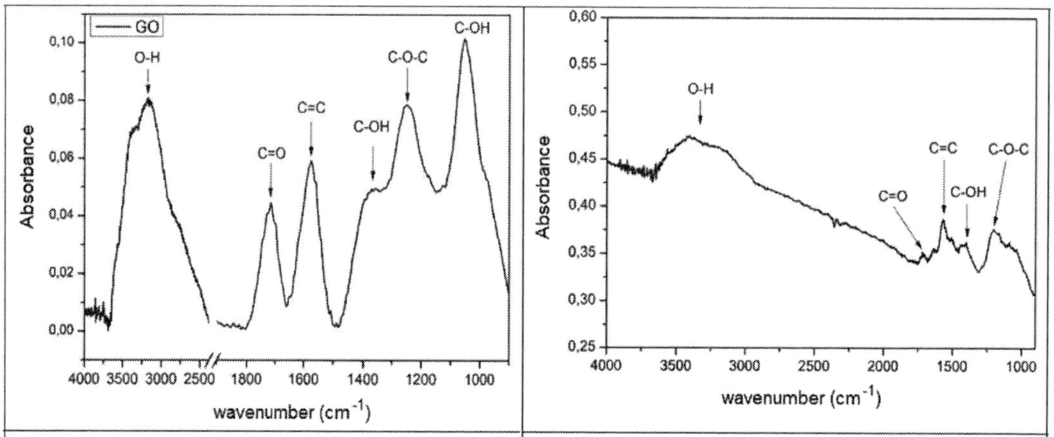

GO	GO-PDA

Figure 1 The FTIR spectra of GO and product formed on reaction of GO and PDA (GO-PDA)

The GO spectrum is characterized by intense bands indicating C=O valence vibrations (ketones and carboxylic acids, especially at the edges of graphitic structures), C=C (planar vibration of sp^2 hybridized bonds of the aromatic ring of unoxidized graphene), C-OH, C-O-C (epoxides) and C-OH (carboxylic acids) functional groups at 1720 cm^{-1}, 1575 cm^{-1}, 1380 cm^{-1}, 1250 cm^{-1} and 1055 cm^{-1}. The broad band in the region~ 3050 to 3700 cm^{-1} may be attributed to vibrations of O-H groups belonging to phenols, carboxylic acids and adsorbed water.

Although the spectrum of the GO-PDA product has different absorbance intensities, it also features characteristic vibration values for oxygen functional groups (1720 cm^{-1}, 1575 cm^{-1}, 1390 cm^{-1}, 1210 cm^{-1}, a broad band in the region ~ 3050 to 3450 cm^{-1}). No significant absorbance bands were identified in the spectrum for C-N or N-H bonds which should be present in the broad band (~ 3050 to 3500 cm^{-1}). This is caused by the overlapping of contributions of different functional groups with similar absorption energies. The obtained results are consistent with those published by other teams [19, 20]. Insignificant peaks are visible at 1500 cm^{-1} and 1680 cm^{-1} which correspond to other characteristic bands for N-H and C-N bonds, respectively.

2.2. TGA and DSC analysis

2.2.1. TDA and DSC analysis of GO and GO-PDA

Thermogravimetric analysis (TGA) and differential scanning calorimetry (DSC) were performed on Mettler Toledo STAR^e, temperature range 25 - 750 °C (20 K/min, Air 50 ml/min). The decomposition was examined in air atmosphere. The thermal decomposition curves of GO are shown in **Figure 2**, and of GO-PDA in **Figure 3**. The GO decomposition showed one endothermic effect with the maximum at 77 °C and two continuous exothermic effects [with the maximums at 236 °C and 537 °C]. The overall thermal effect of the composition is 13003 J/g. The thermal decomposition of the GO and DA product shown in **Figure 2** starts from 90 °C with a slight exothermic effect which reaches the maximum at 207 °C. From this maximum to 440 °C there is a steady 9.73 % decline, and the second exothermic effect begins with the maximum at 566 °C. Its shape is a typical Gaussian curve. The weight loss of the sample at the second exothermic effect is 63.22 % and the overall thermal effect is 14590 J/g.

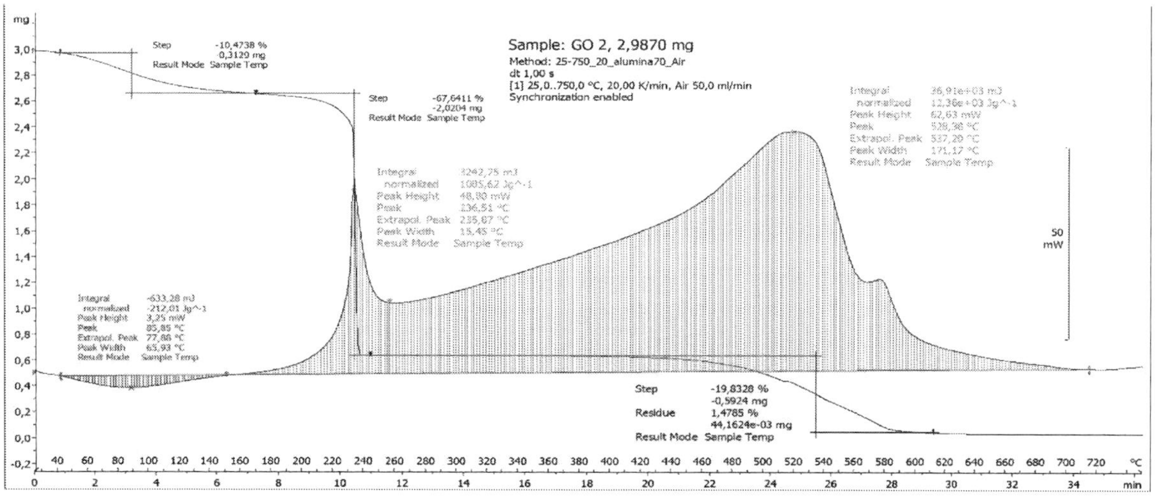

Figure 2 Thermal analysis data measured for GO (TGA/DSC curves)

2.2.2. TDA and DSC analysis of hybrid compounds GO-C_{60} and GO-C_{60} -PDA

In her thesis Roupcová reports that the hybrid compound GO-C_{60} (3:1) decomposes with two exothermic effects. The maximum weight loss (51.1 %) occurs during the first exothermic effect (197 - 205 °C) with the maximum at 205 °C. The second exothermic effect occurs in the range 281-491 °C with the maximum at 390 °C (18 % weight loss) [14]. The thermal decomposition of GO-C_{60} -PDA occurred essentially in two continuous exothermic effects (**Figure 4**). The maximum weight loss (48.1 %) occurred during the first part of the exothermic effect in the temperature range 215-380 °C. The second part of the exothermic effect followed in the 380-561 °C range, with three maximums (430 °C, 456 °C and 470 °C) and with a cumulative weight loss of 32.7 %. The results described above indicate that the PDA coating increased the thermal stability of the original GO-C_{60}.

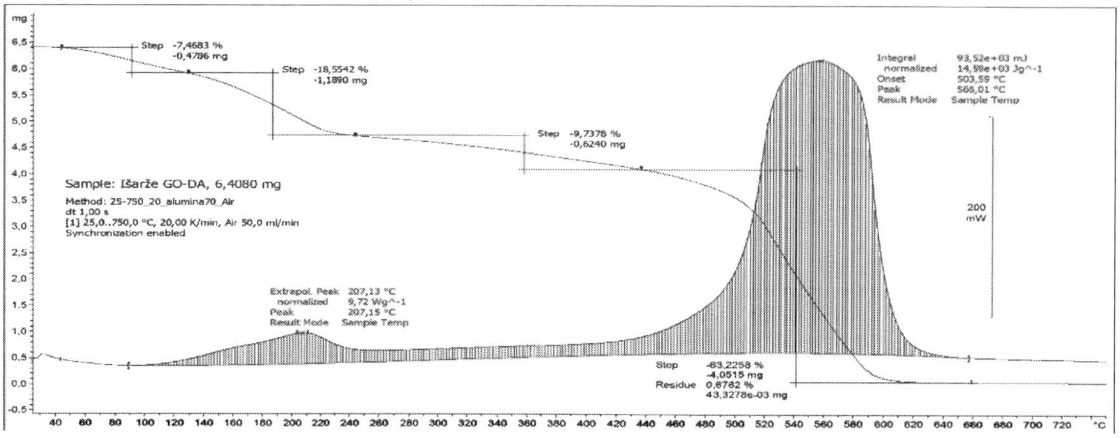

Figure 3 Thermal analysis data measured for GO-PDA (TGA/DSC curves)

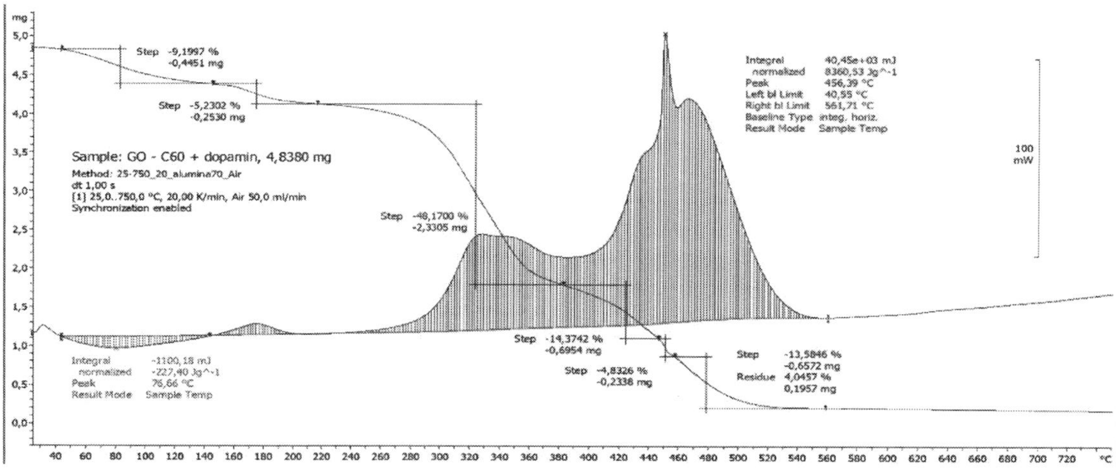

Figure 4 Thermal analysis data measured for GO-C_{60} -PDA (TGA/DSC curves)

2.3. Morphology

Microscopic images (Twist Digital Microscope, Learning Resources) of the dried suspension of GO and GO-PDA on a Petri dish show different surface morphologies of GO and GO after reaction with DA (GO-PDA) even at low magnification (20x) (**Figure 5**). The PDA coating increased the surface roughness. Transmission electron microscopy (TEM) (HT7800, Hitachi, accelerating voltage 100 kV) was used to investigate the GO-PDA and GO-C_{60} -PDA morphology. Nanoparticles present in the filtrate after filtration of prepared suspensions were used for the investigation. TEM image of GO-PDA and GO-C_{60} -PDA are shown in **Figure 6**. It is obvious that nanoparticles of both materials are mostly spherical in shape and form aggregates

Figure 5 Microscopic images of the morphology of the initial GO (left) and the products of its reaction with DA (right) (magnification 20x)

Figure 6 Transmission electron microscopy (TEM) images of GO-PDA (left) and GO-C_{60} -PDA (right)

3. DISCUSSION AND CONCLUSION

The FTIR spectrum of GO contains well-resolved bands, which correspond to vibrations of individual types of bonds in the compound. The spectrum of GO-PDA is quite different. It is formed by broad bands that overlap. These bands correspond to different bond types and oxidation states, different cyclic structures (both heterocyclic and aromatic), and to a variety of bridging bonds and positions. The reason is the PDA structure. The PDA polymer consists of several various monomer units, which are diversely grouped and interconnected and are in different oxidation states [1]. The results of the GO thermal analysis are consistent with those obtained by other research teams [20,21]. The curves suggest that the reaction with DA resulted in a significant reduction of the epoxy bonds (C-O-C) present in both the initial GO and GO-C_{60}. The PDA covering also increased thermal stability of both the initial materials. The PDA covering resulted in a change of the surface morphology. As reported by Bogdan et al. [22], the PDA coating is always made up of two components. The first is a quasi-flat film with intrinsic roughness in the order of 1 nm. The second consists of PDA aggregates lying on top of this film. Larger aggregates can be easily separated by sonication. Therefore sonication may be used to achieve a more even covering and to reduce roughness [22]. Since the GO used in our experiments was not sonicated, we assume that this fact may have influenced the newly formed surface structures. Different monomer units or PDAs could have intercalated between the oxidized carbon layers of the initial graphite. As reported by previous research, thickness of the PDA layer can be influenced by reaction conditions (reaction time, temperature, pH) [1,5,6]. For example, Jia et al. [23] reported that the PDA layer may be up to 50 nm thick. Bogdan et al. [22] therefore suggest that morphological characteristics of PDA coverings should be evaluated based on an agreed standardized protocol. Based on the results of the performed experiments, we assume that the coverings of GO and of the hybrid compound GO-C_{60} changed their sorption properties. The introduction of novel functional groups will allow the formation of new non-covalent bonds with the surface of materials. Therefore the simultaneous use of initial materials and materials with a modified surface could expand the application potential of GO and its reduced forms.

ACKNOWLEDGEMENTS

The publication was written within the project of Norway grants "Innovative carbon-based sorbents as an efficient way of wastewater treatment" project number 3213200008.

REFERENCES

[1] LIEBSCHER, J. Chemistry of Polydopamine - Scope, Variation, and Limitation. *Eur. J. Org. Chem.* 2019, vol. 31-32, pp. 4976-4994. Available from: https://doi.org/10.1002/ejoc.201900445.

[2] BALL, V. Polydopamine Nanomaterials: Recent Advances in Synthesis Methods and Applications. *Front. Bioeng. Biotechnol.* 2018, vol. 6, no. 109. Available from: https://doi.org/10.3389/fbioe.2018.00109.

[3] DELPARASTAN, P., MALOLLARI, K., G., HAESHIN, L., MESSERSMITH, P., B. Direct Evidence for the Polymeric Nature of Polydopamine. *Angew Chem Int Ed Engl.* 2019, vol. 58, no. 4, pp. 1077-1082. Available from: https://doi.org/10.1002/anie.201811763.

[4] PETRONE, L., KUMAR A., SUTANTO C., PATIL N. J., KANNAN S., et al. Mussel adhesion is dictated by time-regulated secretion and molecular conformation of mussel adhesive proteins. *Nature Communications.* 2015, vol. 6, no. 8737. Available from: https://doi.org/10.1038/ncomms9737.

[5] KORD-FOROOSHANI, P., POLEGA, E., THOMSON, K., BHUIYAN, MD. S. A., PINNARATIP, R., TROUGHT, M., KENDRICK, C., GAO, Y., PERRINE, K. A., PAN, L., LEE, B. P. Antibacterial Properties of Mussel-Inspired Polydopamine Coatings Prepared by a Simple Two-Step Shaking-Assisted Method. *Frontiers in Chemistry*, 2019, vol. 7, no. 631. Available from: https://doi.org/10.3389/fchem.2019.00631.

[6] LI, H., XI, J., DONAGHUE, A. G., KEUM, J., ZHAO. Y., AN, K., MCKENZIE, E. R., FEI, R. Synthesis and catalytic performance of polydopamine supported metal nanoparticles. *Scientific Reports.* 2020, vol. 10, 10416. Available from: https://doi.org/10.1038/s41598-020-67458-9.

[7] HUANG, X., Qiao K., LI, L., LIU, G., XU, X., LU, R., GAO, H., XU, D. Preparation of a magnetic graphene/polydopamine nanocomposite for magnetic dispersive solid-phase extraction of benzoylurea insecticides in environmental water samples. *Scientific Reports.* 2019, vol. 9. Available from: https://doi.org/10.1038/s41598-019-45186-z.

[8] YAN, J., WU, R., LIAO, S., JIANG, M., QIAN, Y. Applications of Polydopamine-Modified Scaffolds in the Peripheral Nerve Tissue Engineering. *Front Bioeng Biotechnol.* 2020, vol. 8. Available from: https://doi.org/10.3389/fbioe.2020.590998.

[9] YAN, Z., ZHANG, Y., YANG, H., FAN, G., DING, A., LIANG, H., LI, G., REN, N., VAN DER BRUGGEN, B. Mussel-inspired polydopamine modification of polymeric membranes for the application of water and wastewater treatment: A review. *Chemical Engineering Research and Design.* 2020, vol. 157, no. 1, pp. 195-214. Available from: https://doi.org/10.1016/j.cherd.2020.03.011.

[10] FREEMAN, B. D., PARK, H. B., MCCLOSKEY, B. D. Water purification membranes with improved fouling resistance. EP 2 318 126 B1. 2018. https://patentimages.storage.googleapis.com/33/0c/cc/6961954db14ff0/EP2318126B1.pdf

[11] KONG, F., LIU, Q., YOU, L., LU, P., LIU, T., SUN, G., WANG, Y., CHEN, J. Facile preparation of dopamine mediated graphene oxide composite membranes with enhanced stability for nanofiltration: structure, performance and stability. *Desalination.* 2022, vol. 534. Available from: https://doi.org/10.1016/j.desal.2022.115778.

[12] LEE, J., PATEL, R. Wastewater Treatment by Polymeric Microspheres: A Review. *Polymers.* 2022, vol. 14, no. 9. 1890. Available from: https://doi.org/10.3390/polym14091890.

[13] WEI, D., ZHAO, C., KHAN, A., SUN, L., JI, Y., AI, Y., WANG, X. Sorption mechanism and dynamic behavior of graphene oxide as an effective adsorbent for the removal of chlorophenol-based environmental-hormones: A DFT and MD simulation study. *Chemical Engineering Journal.* 2019, vol. 375. Available from: https://doi.org/10.1016/j.cej.2019.121964.

[14] BUELKE, Ch., ALSHAMI, A., CASLER, J., LEWIS, J., AL-SAYAGHI, M., HICKNER, M. A. Graphene oxide membranes for enhancing water purification in terrestrial and space-born applications: state of the art. *Desalination.* 2018, vol. 448, pp. 113-132. ISSN 00119164. Available from: https://doi.org/10.1016/j.desal.2018.09.008.

[15] PAKULSKI, D., GORCZYŃSKI, A., MARCINKOWSKI, D., et al. High-sorption terpyridine-graphene oxide hybrid for the efficient removal of heavy metal ions from wastewater. *Nanoscale.* 2021, vol. 13, pp. 10490-10499. Available from: https://doi.org/10.1039/D1NR02255E.

[16] ROUPCOVÁ, P., KUBATOVÁ, H., KLOUDA, K., SLIVKOVÁ, S. Preparation and modification of hybrid compounds based on GO-Biochar and verification of their sorption properties. *Chemical Engineering Transaction.* 2021, vol. 84, pp. 61-66. Available from: https://doi.org/10.3303/CET2184011.

[17] HUMMERS, W. S., OFFEMAN, R. E. Preparation of Graphitic Oxide. *Journal of the American Chemical Society.* 1958, vol. 80, no. 6, pp. 1339-1339. Available from: https://doi.org/10.1021/ja01539a017.

[18] ALKHOUZAAM, A., QIBLAWEY, H., KHRAISHEH, M. Polydopamine Functionalized Graphene Oxide as Membrane Nanofiller: Spectral and Structural Studies. *Membranes.* 2021, vol. 11, no. 86. Available from: https://doi.org/10.3390/membranes11020086.

[19] ROUPCOVÁ, Petra. Monitoring of the ecotoxicity of the carbon based nanoparticles. Ostrava, 2018. Disertation Thesis. VSB - Technical university of Ostrava, Faculty of Safety Engineering.

[20] HU, W., HE, G., ZHANG, H., WU, X., LI, J., ZHAO, Z., QIAO, Y., LU, Z., LIU, Y., LI, C. M. Polydopamine-functionalization of graphene oxide to enable dual signal amplification for sensitive surface plasmon resonance imaging detection of biomarker. *American Chemical Society*. 2014, vol. 86, no. 9, pp. 4488-93. Available from: https://doi.org/10.1021/ac500390.5.

[21] SHTEIN, M., PRI-BAR, I., VARENIK, M., REGEV, O. Characterization of Graphene-Nanoplatelets Structure via Thermogravimetry. *Analytical Chemistry*. 2015, vol. 87, no. 8, pp. 4076-4080. Available from: https://doi.org/10.1021/acs.analchem.5b00228.

[22] BOGDAN, D., GROSU, I. G., FILIP C. How thick, uniform and smooth are the polydopamine coating layers obtained under different oxidation conditions? An in-depth AFM study. *Applied Surface Science*. 2022, vol. 597. Available from: https://doi.org/10.1016/j.apsusc.2022.153680.

[23] JIA, L., HAN, F., WANG, H., et al. Polydopamine-assisted surface modification for orthopaedic implants. *Journal of Orthopaedic Translation*. 2019, vol. 17, pp. 82-95. Available from: https://doi.org/10.1016/j.jot.2019.04.001.

MAGNETIC Fe_3O_4 NANOPARTICLES, SYNTHESIS, SURFACE MODIFICATION AND PHYSICOCHEMICAL CHARACTERIZATION

[1]Ondřej SOLNIČKA, [1]Leona KUBÍKOVÁ, [2]Jaromír BAČOVSKÝ, [1]Šárka SOVOVÁ, [1]Petr SEDLÁČEK, [1]Martina KLUČÁKOVÁ, [1]Miloslav PEKAŘ, [1,†]Vojtěch ENEV

[1]*Faculty of Chemistry, Brno University of Technology, Brno, Czech Republic, EU, †enev@fch.vut.cz*

[2]*Delong Instruments a.s., Brno, Czech Republic, EU*

https://doi.org/10.37904/nanocon.2022.4597

Abstract

Surfactants coated magnetic nanoparticles have shown greater application potentials in freshwater source treatment processes for their antibacterial properties.

In this work, we prepared magnetic nanoparticles Fe_3O_4 which surface were modified with Cethyltrimethylammonium bromide (CTAB), tetradecyltrimethylammonium bromide (TTAB) and Carbaethopendecine bromide (Septonex), respectively. Their physicochemical properties and morphology were investigated using thermal analysis (i.e. elemental and thermogravimetric analysis), infrared spectroscopy with Attenuated Total Reflectance (ATR-FTIR), dynamic light scattering (DLS) and transmission electron microscopy (TEM).

The highest surfactant content was obtained for Fe_3O_4 nanoparticles coated with Cethyltrimethylammonium bromide. The presence of $-CH_2-$, $-CH_3$ and ammonium salt functional groups in infrared spectra implied that all surfactants had been successfully anchored onto the surface of native Fe_3O_4 nanoparticles. All modified nanoparticles were surface charged and they not tended to agglomerate in deionize water. The highest value of zeta-potential (approximately 40 mV) was obtained for Fe_3O_4-TTAB. Another characteristic coefficient is polydispersity index (PDI) its relative value was comparable for all modified nanoparticles. The transmission electron microscopy (TEM) analysis showed all modified nanoparticles were approximately spherical morphology with the core size in the range 5-15 nm.

Keywords: Magnetic nanoparticles, modification, surfactants, FTIR, DLS

1. INTRODUCTION

Magnetic nanoparticles such as magnetite (Fe_3O_4) and maghemite (γ-Fe_2O_3), consisting of superparamagnetic core which can be conveniently separated with magnetic field, have been widely used to remove pollutants from water environment [1-3].

With successful and proper functionalization, magnetite and/or maghemite, which are environmentally risk-free and nontoxic inorganic nanomaterials with paramagnetic properties, could be employed to remove pathogenic bacteria (i.e. *Escherichia coli*, *Klebsiella*, *Salmonella*, *Citrobacter*) from freshwater and water environment [4-6]. The unique combination of superparamagnetic nanoparticles and different bioactive substances of natural and anthropogenic origin such as D-mannose, antibiotics and surfactants prompts us to embark on a research work in this scientific field.

CTAB, TTAB and Septonex, cationic surfactants with relative small and long-chain molecules, could be very easily anchored onto the surface of magnetite nanoparticles by a facile and simple two-step their preparation process. These surfactants have been used to specific modification of adsorbent surface to remove negatively charged pollutants from water environment. The positive charged surfactant+ could be interacting with negative charged of pathogenic Gram-negative bacteria.

In this work, we applied modified procedure of preparation magnetic nanoparticles with surfactants using CTAB, TTAB and Septonex as a cationic surfactants and obtained nanomaterials to a complex physicochemical and morphology characterization utilizing methods of spectroscopic, DLS and optic analysis.

2. MATERIALS AND METHODS

2.1. Materials

Cethyltrimethylammonium bromide (CTAB), tetradecyltrimethylammonium bromide (TTAB), Carbaethopendecine bromide (Septonex), $FeCl_2 \cdot 4H_2O$, $FeCl_3 \cdot 6H_2O$, NH_4OH were purchased from Sigma-Aldrich (Sigma-Aldrich, Praha, Czech Republic). All the reagents were analytical grade.

2.2. Preparation Fe_3O_4 nanoparticles

The unedited and surfactants coated Fe_3O_4 magnetic nanoparticles were synthesized with methods modified from literature [7] and [8].

Briefly, 40 mL 1 M $FeCl_2$ and 80 mL 1 M $FeCl_3$ were mixed with a stirring blade in a 300 mL beaker containing 40 mL of degas deionized water (oxygen free Mili-Q water). N_2 gas was used to create an oxygen-free solution during the synthesis process. The solution was heated to 85 °C in software controlled heating circulating water bath. Then NH_4OH (~ 30%) was dropped into the mixture until the pH reached around 11, and black precipitate was formed. The suspension was then stirred at 85 °C for another 30 min. After cooled to room temperature, the black precipitate was collected with magnet and repeatedly washed with Mili-Q water until the pH of suspension was about 7. Then surfactants were anchored onto the Fe_3O_4 nanoparticles surface by explosion the black solid proportion 200 mL of 0.06 M surfactant solution i.e. CTAB, TTAB, Septonex for 30 min under sonication conditions. After that, the black precipitates were collected with magnetic decantation and washed with deionized water five times to wash away the unreacted CTAB, TTAB and Septonex with controlled using ATR-FTIR spectroscopy. The unedited and modified nanoparticles were freeze-drying for 48 h.

2.3. Methods

Elemental and thermogravimetric analysis

The relative content of organic elements in the modified Fe_3O_4 nanoparticles was determined using EA 3000 CHNS/O analyzer (Euro Vector, Pavia, Italy). Approximately 1.0-1.5 mg of the sample was weighted in a tin capsule, the capsule packed and combusted at 980 °C in the analyzer using oxygen as the combustion gas. Calibration of the determination of relative contents of organic elements from the obtained gas chromatograms was provided using sulfanilamide as a reference standard sample.

Thermogravimetry of the nanoparticles was performed using a 5500 Discovery TG analyzer (TA Instruments, New Castel, DE, USA). Approximately 10 mg of sample was weighed into a platinum pan. After inserting the pan into the TG analyzer, the sample was incinerated at a heating rate of 10 °C min^{-1} from laboratory temperature to 1000 °C under nitrogen atmosphere. The weight loss occurring at 100-120 °C was measured and taken as the moisture content. The final residue, after heating to 1000 °C, was measured and taken as the Fe_3O_4 content. During the TG analyses, the relative sample weight was recorded continuously with ± 0.1% mass accuracy.

ATR-FTIR spectrometry

Fourier transform infrared (FTIR) spectra of the modified Fe3O4 nanoparticles were recorded on Nicolet *i*S50 spectrometer (Thermo Fisher Scientific, Waltham, MA, USA) using the Attenuated Total Reflectance (ATR) measuring technique (single reflection built-in diamond ATR crystal and external germanium ATR crystal). MID measurements were taken at 25 °C in the spectral range 4000-400 cm^{-1} at 4 cm^{-1} resolutions as an average of 128 scans. On the other hand, FAR spectra of nanoparticles were measured in the spectral range 1800-200 cm^{-1} at 8 cm^{-1} resolutions as an average of 256 scans. A background spectrum was collected from the clean dry surface of the ATR crystal (diamond and/or Ge) in an ambient atmosphere. Raw absorption infrared spectra with no artificial processing (such as baseline or ATR corrections, atmospheric suppression) are presented and evaluated.

Dynamic light scattering (DLS) and zeta potential measurement

Zeta potentials and average scattered light intensity of modified Fe_3O_4 nanoparticles were measured using Zetasizer Nano ZS (Malvern Panalytical Ltd., UK). All modified Fe_3O_4 dispersions were investigated by the method of electrophoretic light scattering using universal DIP cell (Malvern Panalytical Ltd., UK). The obtained experimental data from DLS and ELS analysis were processed by the Zetasizer software (version 7.11).

TEM analysis

Nanoparticles were sonicated and drop of solution was placed on hexagonal grid with ultrathin carbon supporting layer. Imaging was done by low-voltage transmission electron microscopes LVEM5 and LVEM25 produced by Delong Instruments. These instruments operate with accelerating energy 5kV (LVEM5) and 25kV (LVEM25), which brings a very high image contrast. Micrographs were captured in TEM and also STEM mode.

3. RESULTS AND DISCUSSION

3.1. Elemental and thermogravimetric analysis

The elemental composition, moisture and surfactants content of native and modified Fe_3O_4 nanoparticles are presented in **Table 1**.

Table 1 Surfactant content, moisture content and elemental composition of unedited and modified Fe_3O_4 nanoparticles

sample	surfactant	moisture	C	H	N
	(wt.%)				
Fe_3O_4	-	1.97	0.004	0.002	-
Fe_3O_4-CTAB	3.01	1.21	0.110	0.020	0.070
Fe_3O_4-TTAB	3.89	1.53	0.167	0.032	0.012
Fe_3O_4-Septonex	2.92	1.33	0.082	0.014	0.005

The surfactants content of modified Fe_3O_4 samples ranged from 2.92 % to 3.89 % (relative to the type of surfactant used) varying with the affinity of surfactants to the unedited Fe_3O_4 surface. The highest surfactant content was obtained for Fe_3O_4 nanoparticles modified with TTAB. In contrast, the lowest content of surfactant (2.92 %) was obtained in the case of the sample Fe_3O_4-Septonex modified using Carbaethopendecine bromide. The content of moisture of unedited and modified Fe_3O_4 nanoparticles ranged from 1.21 % to 1.97 %. The lower content of moisture for modified Fe_3O_4 samples is not surprising with respect to surfactants high content of hydrophobic structural constituents such as $-CH_3$ and $-CH_2-$ functional groups.

The organic elements of all modified nanoparticles are presented in **Table 1**. Despite their modification using different surfactants, the elemental compositions of the nanoparticles are comparable: carbon (C), 0.08-0.17 wt. %; hydrogen (H), 0.01-0.03 wt. % and nitrogen (N), ~ 0.01 wt. %. Nitrogen represents a minor component of all modified Fe_3O_4 nanoparticles and its content ranged from 0.005-0.012 wt. %. The highest elemental compositions were found for Fe_3O_4 nanoparticles modified using TTAB surfactant. The presence of organic elements such as carbon, hydrogen and nitrogen implied that surfactants had been successfully anchored onto the surface of Fe_3O_4 nanoparticles.

3.2. ATR-FTIR spectrometry

The ATR-FTIR spectra of unedited and modified Fe_3O_4 nanoparticles are presented in **Figures 1a, b.** Interpretation of their absorption bands has been carried out according to the literature data [4,9].

The broad and intensive absorption band centered at about 3300 cm^{-1} corresponds to O-H stretching of water molecules (moisture), which are connected with an intermolecular hydrogen-bond. This absorption band is well pronounced in FTIR spectrum of unedited magnetic nanoparticles. Presence of the water molecules are usually also manifested by the broad band centered at about 1635 cm^{-1} resulting from the O-H stretching functional groups. The fingerprint zone is characterized by two bands at 1435 cm^{-1} and 1257 cm^{-1} corresponding to the O-H···O stretching in water molecules (i.e. overtone bands of water molecule). The FAR spectrum (marked red in **Figure 1a**) is characterized by a sharp and intensive Fe-O stretching band at 544 cm^{-1} (magnetite).

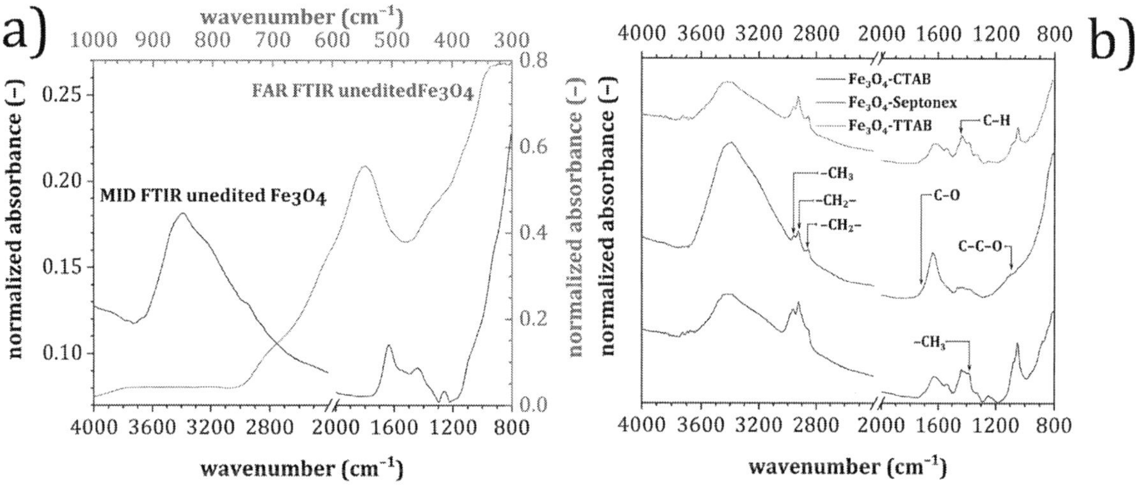

Figure 1a, b ATR-FTIR spectra of unedited and modified Fe_3O_4 nanoparticles

The second set of spectral features that are in common for all the modified Fe_3O_4 samples refer to their content of surfactants such as aliphatic and N-containing molecular moieties. The relative content of aliphatic chains of used surfactants is evaluated primarily in the 3000-2800 cm^{-1} spectral range (see **Figure 1b**). The presence of aliphatic groups is revealed by the bands at 2927 cm^{-1} and 2854 cm^{-1} which are attributed to asymmetric and symmetric C-H stretching in $-CH_2-$ functional groups, respectively. The band at 2962 cm^{-1} is the most intensive in the spectrum of Fe_3O_4-TTAB which is attributed to asymmetric vibration mode of $-CH_3$ groups. The deformation vibrations of the $-CH_2-$ and $-CH_3$ groups at 1431 cm^{-1} and 1384 cm^{-1} respectively occur in the all modified nanoparticles. A band with variable intensity located at around 1253 cm^{-1} is attributed to C-N stretching in ammonium salts. A weak band at 1080 cm^{-1} can be ascribed to C-C-O stretching in alkyl-ester of Septonex.

3.3. Dynamic light scattering (DLS) and zeta-potential

Dynamic light scattering (DLS) is concerned with measurement of particles suspended within in liquid. The size distribution curves of the studied magnetic nanoparticles are shown in presented poster. For the modified Fe_3O_4 nanoparticles, the average size of surfactant-coated Fe_3O_4 nanoparticles ranged from 86 nm to 247 nm. The average sizes of modified particles were found as 85.9 nm Fe_3O_4-CTAB, 108.3 nm Fe_3O_4-TTAB, and

246.6 nm Fe_3O_4-Septonex in dynamic light scattering measurements. For all modified nanoparticles, a polydispersity index (PDI) ranged from 0.12 to 0.23, approaches 1.0 for low PDI values (monodisperse systems), and increases with the system polydispersity. The lowest PDI of samples (0.12) was obtained in the case of the magnetic nanoparticles Fe_3O_4 coated using Septonex.

To verify the effect of surfactant modification on the zeta potential (ζ) and water dispersibility of Fe_3O_4 nanoparticles, samples were diluted to the same concentration and assessed using a Zetasizer Nano ZS particle size analyzer. The zeta potential of modified Fe_3O_4 nanoparticles was found to range between 23.3 ± 1.2 mV and 39.4 ± 4.0 mV, which was in agreement with the findings of previous studies [4,10]. The zeta potential value of unedited Fe_3O_4 was smaller than the three modified nanoparticles, indicating that Fe_3O_4-CTAB, Fe_3O_4-TTAB and Fe_3O_4-Septonex were stable in aqueous suspension.

Table 2 Average size, zeta potential and polydispersity index of modified Fe_3O_4

sample	average size (nm)	zeta potential (mV)	PDI
Fe_3O_4-CTAB	85.9	37.0 ± 0.9	0.23
Fe_3O_4-TTAB	108.3	39.4 ± 4.0	0.13
Fe_3O_4-Septonex	246.6	23.3 ± 1.2	0.12

3.4. TEM images

Figures 2(a, b, c) shows the typical TEM image of modified Fe_3O_4 nanoparticles, from which we can see that the sizes of all samples are practically uniform and considerable part of modified nanoparticles are approximately spherical with the diameter in the range 5-15 nm.

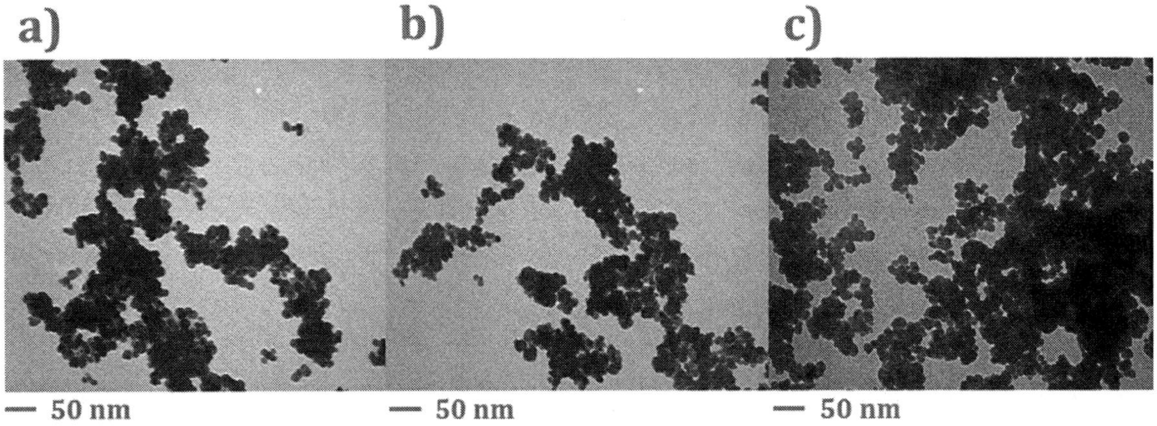

Figure 2 TEM images of modified Fe3O4 nanoparticles: **a)** Fe_3O_4-CTAB; **b)** Fe_3O_4-TTAB; **c)** Fe_3O_4-Septonex

4. CONCLUSION

Magnetic nanoparticles Fe_3O_4 were prepared by co-precipitation method. The unedited Fe_3O_4 nanoparticles were modified using selected cationic surfactants. Unedited and modified magnetic nanoparticles were obtained and subjected to a physicochemical characterization utilizing method of morphological analysis. Advanced instrumental techniques such as Attenuated Total Reflectance (ATR-FTIR), dynamic light scattering (DLS), electrophoretic light scattering (ELS), and TEM analysis were supplemented with thermal analysis.

The nanoparticles have been fully characterized using different methods; in particular TEM images reveal good morphological properties which correspond to the particle size in the range 5-15 nm. Further, it was noticed that there were interactions between the cationic surfactants and the surface of the native nanoparticles as confirmed by thermal analysis and FTIR spectroscopy. The determined zeta potential values and polydispersity index (PDI) of modified nanoparticles show the positive charge of particles (in the range 23-39 mV) and very good PDI values, which supports the expectations that the selected surfactants to stabilize of nanoparticles in water environment.

ACKNOWLEDGEMENTS

The authors would like to thank Delong Instruments a.s. for realization of TEM measurements.

REFERENCES

[1] AKBARZADEH, A., SAMIEI, M., DAVARAN, S. Magnetic nanoparticles: preparation, physical properties, and applications in biomedicine. *Nanoscale Research Letters*. 2012, vol. 7, pp. 1-13. Available from: https://doi.org/10.1186/1556-276X-7-144.

[2] SICILIANO, G., MONTEDURO, A.G., TURCO, A., PRIMICERI, E., RIZZATO, S., DEPALO, N., CURRI, M.L., MARUCCIO, G. Polydopamine-Coated Magnetic Iron Oxide Nanoparticles: From Design to Applications. *Nanomaterials*. 2022, vol. 12, pp. 1-17. Available from: https://doi.org/10.3390/nano12071145.

[3] JIN, Y., LIU, F., TONG, M., HOU, Y. Removal of arsenate by cetyltrimethylammonium bromide modified magnetic nanoparticles. *Journal of hazardous materials*. 2012, vol. 227-228, pp. 461-468. Available from: https://doi.org/10.1016/j.jhazmat.2012.05.004.

[4] JIN, Y., DENG, J., LIANG, J., SHAN, C., TONG, M. Efficient bacteria capture and inactivation by cetyltrimethylammonium bromide modified magnetic nanoparticles. *Colloids and Surfaces B: Biointerfaces*. 2015, 136, pp. 659-665. Available from: https://doi.org/10.1016/j.colsurfb.2015.10.009.

[5] PRABHU, Y.T., RAO, K.V., KUMARI, B.S., KUMAR, V.S.S., PAVANI, T. Synthesis of Fe_3O_4 nanoparticles and its antibacterial application. *International Nano Letters*. 2015, vol. 5, pp. 85-92. Available from: https://doi.org/10.1007/s40089-015-0141-z.

[6] EL-GENDY, N.S., NASSAR, H.N. Biosynthesized magnetite nanoparticles as an environmental opulence and sustainable wastewater treatment. *Science of The Total Environment*. 2021, vol. 774, pp. 145610. Available from: https://doi.org/10.1016/j.scitotenv.2021.145610.

[7] LIU, J.F., ZHAO, Z.S., JIANG, G.B. Coating Fe_3O_4 magnetic nanoparticles with humic acid for high efficient removal of heavy metals in water. *Environmental science & technology*. 2008, vol. 42, pp. 6949-6954. Available from: https://doi.org/10.1021/es800924c.

[8] MAITY, D., AGRAWAL, D.C. Synthesis of iron oxide nanoparticles under oxidizing environment and their stabilization in aqueous and non-aqueous media. *Journal of Magnetism and Magnetic Materials*. 2007, vol. 308, pp. 46-55. Available from: https://doi.org/10.1016/j.jmmm.2006.05.001.

[9] ENEV, V., SEDLÁČEK, P., ŘIHÁK, M., KALINA, M., PEKAŘ, M. IR-Supported Thermogravimetric Analysis of Water in Hydrogels. *Frontiers in Materials*. 2022, vol. 9, pp. 1-8. Available from: https://doi.org/10.3389/fmats.2022.931303.

[10] LI, W., HINTON, C.H., LEE, S.S., WU, J., FORTNER, J.D. Surface engineering superparamagnetic nanoparticles for aqueous applications: design and characterization of tailored organic bilayers. *Environmental Science: Nano*, 2016, vol. 3, pp. 85-93. Available from: https://doi.org/10.1039/C5EN00089K.

SESSION C
BIONANOTECHNOLOGY, NANOMATERIALS IN MEDICINE

Chairmen

Prof. Dr. Anton FOJTIK	Czech Technical University in Prague, Czech Republic, EU
Prof. Dr. Jiri HOMOLA	Institute of Photonics and Electronics of the CAS, Prague, Czech Republic, EU
Prof. Dr. Bohuslav REZEK	Czech Technical University in Prague, Czech Republic, EU

MODELS OF RELEASE OF ACTIVE COMPOUNDS FROM NANOSTRUCTURED MULTI-LAYERED MATERIALS

[1]Jan ŽÍDEK, [1,2]Anna ŠUDÁKOVÁ, [2]Jiří SMILEK, [3]Minh Ha LE, [4]Thi Thu Thuy NGUYEN, [5]Duc Anh NGUYEN

[1]Central European Institute of Technology (CEITEC), Brno, Czech Republic, EU, Anna.Sudakova@ceitec.vutbr.cz, jan.zidek@ceitec.vutbr.cz

[2]Institute of Physical and Applied Chemistry, Faculty of Chemistry, Brno, Czech Republic, EU, smilek@fch.vut.cz

[3]Institute of Natural Products Chemistry (INPC), Vietnam Academy of Science and Technology (VAST), Nghia Do, Cau Giay, Hanoi, Viet Nam, halm2vn@gmail.com

[4]Phenikaa University Nano Institute (PHENA), Phenikaa University, Hanoi, Vietnam, thuy.nguyenthithu@phenikaa-uni.edu.vn

[5]Center for Research and Technology transfer, Vietnam Academy of Science and Technology (VAST), Nghia Do, Cau Giay, Hanoi, Viet Nam, ducanh221100@gmail.com

https://doi.org/10.37904/nanocon.2022.4577

Abstract

Controlled drug release is the transport of an active compound of biologically active agents with a defined time-concentration function. The release of these active compounds from materials is currently achieved by numerical models, based on simple kinetics of the first order, that are used to analyze experimental processes.

In this contribution, we propose a model that enables us to design complex, multi-layered material and define the desired time-concentration function for the release. This approach enables us to tailor the material in such a way that each layer has an individual concentration of the active compound. The result of programming work in models provides instructions for the design of the materials, which are specifically suited to certain situations. The model corresponds to real material structures from nanostructured electrospun material, in which the layers are pasted together with continuous hydrogel layers (for example, collagen).

The final version of the model will enable us to design the patches in the medicine where the release of antibiotics, healing or nutritional compounds will be programmed according to a certain time-release protocol. It is assumed that it will be possible to synchronize the controlled release of two or more compounds, each with different time protocols.

Keywords: Release, layered material, model, medical, electrospun

1. INTRODUCTION

A means of achieving the controlled release of active compounds from polymers and gels has been presented in the literature in recent years and is a widely discussed topic in the field of medicine. The controlled release from various materials is also a suitable method of achieving the fine dosage of compounds to the reaction system, permitting the slow addition of small doses of the compounds to the system. The main strategy in the literature is to find the composition with the optimal physical and chemical structure; this is then optimized to achieve the desired release. The results involve relatively complex physico-chemical systems that control the released amount. The release by such systems is usually described by relatively simple computer models [1].

This article presents the opposite approach; that is, we describe a relatively simple physical and chemical system that is a combination of the diffusion sorption and desorption. We believe that the release behavior of such a system can be designed and controlled with the aid of specific software.

The main aim of this study is to propose a computer program through which a user will define the time profile of the release (that is, how the compound should be released within a certain time-limited interval). The result of the model will be the appropriate design of a substrate with a certain gradient of composition and with the desired release function. A substrate will therefore be proposed that contains a certain gradient of concentration in some compounds (or some other function of concentration). A user will produce a substrate with an initial surface composition, following which a model will be produced for the material, including a programmable release.

Various up-to-date mechanisms of controlled release appear in the literature. The most frequent mechanism is controlled diffusion [2,3]. The next-most frequent mechanism is a release accompanied by the degradation of the substrate material, mostly biodegradable and bioresorbable materials [1,4]. The third mechanism is the release from hydrogel networks, which process involves two steps. The first step is the release of the compound from the network, followed by the desorption of a low molecular compound that is adsorbed on the macromolecules. The second step is slower than the first [5]; it is a gradual release that is intended to be stable for a certain period of time.

There is a special aspect to the release from electrospun material [6,7]: the material is porous and the release from it is relatively slower and more gradual than from solid material.

Our model combines the diffusion, sorption-desorption, and capillary forces. The diffusion is described by Fick's laws: the first law relates to the flux through two arbitrary points in the material, while the second law states that it is possible to derive a relationship between the concentration gradient and changes to the concentration over time. The sorption and desorption were performed by the first law of kinetics, and the capillarity serves to control the flux in one direction. A similar technique is thin-layer chromatography, which has both a mobile and stationary phase.

2. MODELS AND METHODS

The model is connected to a specific type of layered material. The basic patches are geometrically very simple, forming a layer with a uniform concentration of a low molecular compound. The patch can be produced as a single homogenous block of material (**Figure 1.I**). Our new design has a vertical concentration gradient (**Figure 1.II**). The material can be modified by adding thin layers of varying concentrations. The final result will be a patch with a custom concentration profile and a greater amount of released compound (**Figure 1.III**).

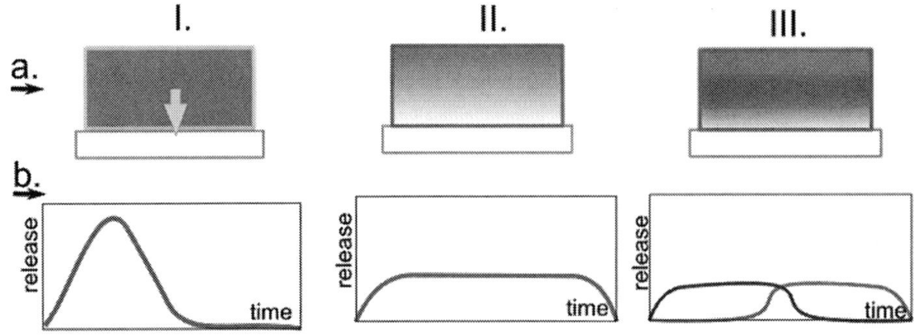

Figure 1 Design of layered materials with I. a homogenous concentration in the substrate; II. a gradient of concentration (single compound); III. a gradient of concentration (more compounds). III.a. design of material; III.b. presumed release curve. Arrow =direction of release.

This article presents the first phase of the model. In the final version of the model, the input is a user-defined time-concentration profile (**Figure 1: II.b.-III.b.**). The output of the model is the concentration function in the substrate (**Figure 1b**), which will be designed in such a way that it can achieve the desired release function.

The input parameter of the model described in this paper is the concentration of layers, and the result is the cumulative time-release function.

The model's mathematical background is analogous to the description of the thin-layer chromatography. The model defines the low molecular compound in both the stationary and mobile phases. Each phase has its own concentration profile. The stationary phase represents the active compound adsorbed on the surface of the substrate. The mobile phase is where the compound is actually desorbed from the substrate and can be transferred by diffusion or capillarity.

2.1. General model

Let us consider a small segment of the volume in the substrate.

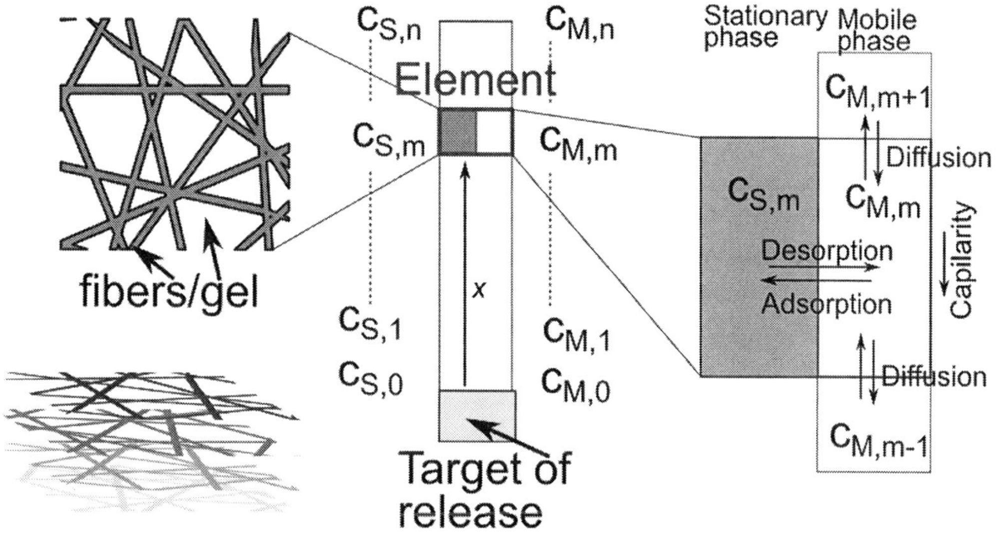

Figure 2 Model of the layered materials: left picture-hydrogel structure; middle picture-model schema of entire multilayer material; right picture-transfer of concentration between volume elements. C_S = stationary phase (concentration of compound adsorbed on surface); C_M = mobile phase (concentration of compound in liquid); x = distance from the release target.

The segment of volume is a certain distance x from the release surface. The concentration is added by desorption from the stationary phase or by transport from neighboring regions (i.e., diffusion vs. capillarity).

1) The desorption and adsorption are solved as kinetics of the first order:

$$-\frac{dC_S}{dt} = \frac{dC_M}{dt} = kC_S. \quad (1)$$

These first-order kinetics mean that the rate of desorption is directly proportional to the concentration of adsorbed compound in the solid material.

2) The capillarity is introduced using the simplest approximation. The front of the solvent during the capillary elevation moves according to the square root of time:

$$x_{fr} = AS\sqrt{t}, \quad (2)$$

where S is the sorptivity coefficient, which depends on the porous structure.

2.2. Model of diffusion

We propose the first simplified version of the complex model shown in **Figure 2**. This equation employs the diffusion equation according to Fick's second law. This equation describes the relationship between differential of the concentration in the time interval and the concentration gradient:

$$\frac{\partial c}{\partial t} = D\frac{\partial^2 c}{\partial x^2}, \tag{3}$$

where c = concentration, t = time and x = distance from the release target. The equation describes the flow of the compound between two neighboring segments. All concentrations in the profile are coded using an array of concentrations:

$$\mathrm{c} = \{c_1, c_2, \dots c_n\} \tag{4}$$

The system of n differential equations describes the flow inside the sample.

Generally speaking, diffusion is a common computing task, and the solution has been described many times. Usually, the task is solved as a diffusion in an extremely thick layer. Mathematically, the boundary approaches ∞, and the concentration on the boundary is 0.

In the case of this material, however, we define specific boundary conditions. The released compound is usually consumed in the release target. For all time intervals, the compound must be removed from the boundary of the model and added to the release profile:

$$\forall t{:}\ c_n = actual; \rightarrow c_n = 0;\ c_r = c_r + actual, \tag{5}$$

where c_r is the cumulative released concentration. The actual removed amount is added to the time profile of the release. In this case, the actual amount of released substance must be subtracted from the substrate.

The consistency of the solution is verified by the condition of mass conservation. The mass of the low molecular weight compound in the substrate and the released fraction must be conserved as a constant during the simulation in each time interval:

$$c_r + \sum_{i=1}^{n} c_i = const. \tag{6}$$

The basic profile occurs when the compound is uniformly distributed in the sample.

The result of the simulation must be in accordance with Fick's first law:

$$J = -D\frac{\partial c}{\partial x}, \tag{7}$$

where $\frac{\partial c}{\partial x}$ is a gradient of concentration. The solution is based on the standard ordinary differential equation (ODE) solvers seen in the MATLAB environment.

3. RESULTS AND DISCUSSIONS

The first modification of the model serves to verify the standard release. The model describes the sample in which the concentration is equally distributed in the substrate, as it is the most frequently seen case in experiments. Here, the release function is an exponential growth function.

The 1D-diffusion problem was solved numerically. The adjustable parameters are the initial concentration of the active compound in the substrate, the diffusion coefficient, and the thickness of the layer. The function of release was modelled from 0.350 mm multi-layer material divided into 350 layers, each 1 μm. The time step was 0.001s, and the diffusion coefficient was 10^{-10} m^2s^{-1}.

The first standard concentration profile is presented in **Figure 3**. Here, the cumulative function of the release profile shows an exponential growth.

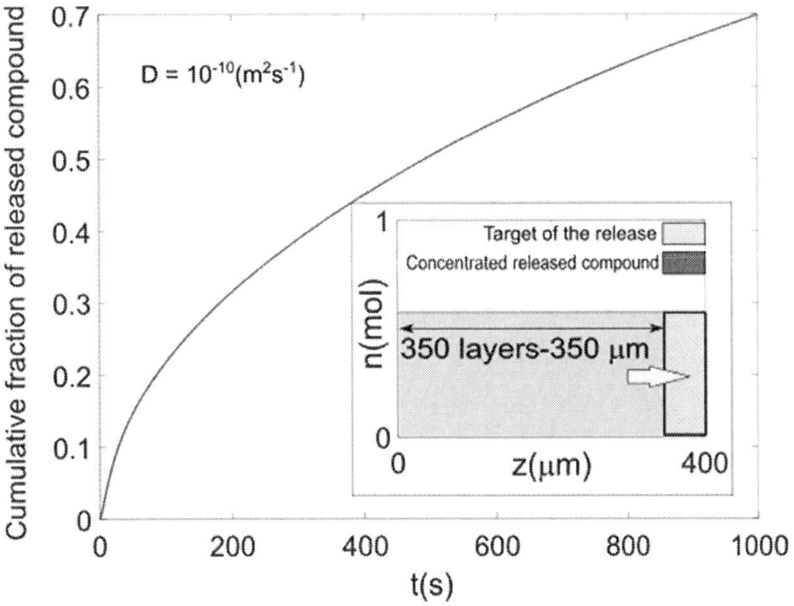

Figure 3 Model's control settings: thin-layer thickness 350 μm

The second version describes a situation in which a more complex release was expected. The total concentration of the compounds is equal to the sample in **Figure 3**. The active compounds are concentrated in two layers, and the concentration profile is defined in the z-direction. As in **Figure 3**, the diffusion coefficient, thickness of the layers, and number of layers are shown.

The results are presented in **Figure 4**. There is an observed bimodal distribution, in which the material runs in two waves. The first release is relatively quick, while the second is slow and gradual.

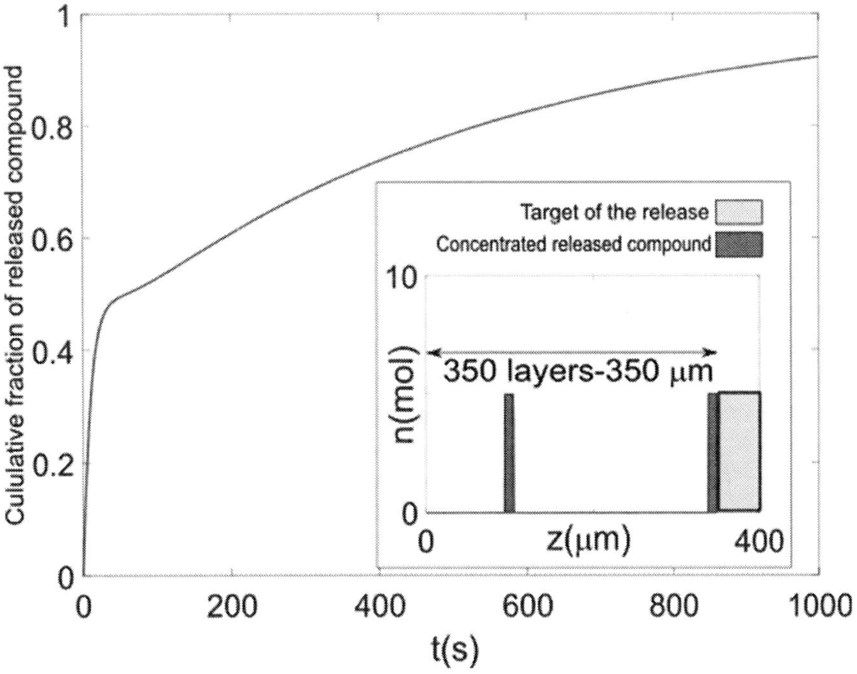

Figure 4 Release of the material's active compound, showing two concentrated layers

The aim of this article has been to confirm the function of the model shown herein. It has been found that, when the model describes the standard release presented in the literature, it outputs the standard cumulative release function. Where a concentration profile is introduced, the more complex release function is observed. It seems that, when mathematical optimization is applied, it is possible to tailor a user-defined release function through the definition of layer concentrations. However, it seems that it is not possible to program any arbitrary release function, and the release function will be always limited by diffusion.

4. CONCLUSION

The paper examines the model-controlled release of active compounds from a complex material. The aim of the model is to effectively predict the release from layered materials, in which each layer can have a different concentration of compound. The model was tested using a standard, uniform sample concentration, which is the most often-used experimental configuration. The release function showed exponential growth, which is similar to the experiments. The second version of the model involved the definition of concentrated layers and the complex release. It was also verified that the model is able to produce both standard and complex release functions.

The next step for our model is to construct layered material from nanostructured layers, with a custom concentration profile for each layer. The custom concentration profile will produce programmable release functions with waves of release at desired times.

ACKNOWLEDGEMENTS

This research was conducted under the project TM03000061 and with financial support from the Technology Agency of the Czech Republic.

REFERENCES

[1] LYSAKOVA, K., HLINAKOVA, K., KUTALKOVA, K., CHALOUPKOVA, R., ZIDEK, J., BRTNIKOVA, J., VOJTOVA, L. A novel approach in control release monitoring of protein-based bioactive substances from injectable PLGA-PEG-PLGA hydrogel. *Express Polymer Letters.* 2022, vol. 16, pp. 798-811.

[2] RITGER, P.L, PEPPAS, N.A. Simple equation for description of solute release I. Fickian and non-Fickian release from non-swellable devices in the form of slabs, spheres, cylinders or discs. *Journal of Controlled Release.* 1987, vol. 5, pp. 23-36.

[3] DANYUO, Y., ANI, C. J., SALIFU, A. A., OBAYEMI, J. D., DOZIE-NWACHUKWU, S., OBANAWU, V. O., AKPAN, U. M., ODUSANYA, O. S., ABADE-ABUGRE, M., MCBAGONLURI, F., SOBOYEJO, W. O. Anomalous release kinetics of prodigiosin from poly-n-isopropyl-acrylamid-based hydrogels for the treatment of triple negative breast cancer. *Scientific Reports.* 2019, vol. 9, no. 3862.

[4] ZILBERMAN, M., MALKA, A. Drug controlled release from structured bioresorbable films used in medical devices. *J Biomed Mater Res B Appl Biomater.* 2009. vol. 9, pp. 155-64.

[5] STREUBEL, A., SIEPMANN, B., PEPPAS, N.A., BODMEIERA, R. Bimodal drug release achieved with multi-layer matrix tablets: transport mechanisms and device design. *Journal of Controlled Release.* 2000, vol. 69, pp. 455-468.

[6] WENG, L., XIE, J. Smart electrospun nanofibers for controlled drug release: recent advances and new perspectives. *Curr Pharm Des.* 2015, vol. 21, pp: 1944-1959.

[7] HU, J., PRABHAKARAN, M.P., TIAN, L., DING, X. RAMAKRISHNA, S. Drug-loaded emulsion electrospun nanofibers: characterization, drug release and in vitro biocompatibility. *RSC Advances.* 2015, vol. 5, pp. 100256-100267.

AIR-JET SPINNING OF BIOCOMPATIBLE POLYMERIC NANOFIBERS FOR WOUND DRESSINGS

[1]Ema CHUDOBOVÁ, [2]Šárka HAUZEROVÁ, [2]Kristýna HAVLÍČKOVÁ, [3]Luboš BĚHÁLEK, [3]Jana MÜLLEROVÁ, [2]Věra JENČOVÁ, [2]David LUKÁŠ, [2]Eva KUŽELOVÁ KOŠŤÁKOVÁ

[1]*Technical University of Liberec, Faculty of Mechatronics, Informatics and Interdisciplinary Studies, Liberec, Czech Republic, EU, ema.chudobova@tul.cz*

[2]*Technical University of Liberec, Faculty of Science, Humanities and Education, Department of Chemistry, Liberec, Czech Republic, EU, eva.kostakova@tul.cz*

[3]*Technical University of Liberec, Institute for Nanomaterials, Advanced Technology and Innovation, Liberec, Czech Republic, EU*

https://doi.org/10.37904/nanocon.2022.4605

Abstract

Air-jet spinning (or solution blowing) is a method of nanofibrous production using high-velocity gas flow of air or gas to transform drops of a polymer solution into fibers. The technology is promising for relatively safe applying nanofibers directly to wounds, in-situ application. In this work, air-jet spinning of several polycaprolactone (PCL) and poly(L-lactic acid) (PLLA) solutions was performed utilizing a commercial airbrush for possible in situ application of skin wound dressings and for the production of flat samples of nanofibrous materials for further in-vitro testing. The average fiber diameters of air-jet spun materials were set 268 nm (6 wt% PCL in dichloromethane (DCM)), 305 nm (8 wt% PCL in DCM), 333 nm (10 wt% PCL in DCM) and 223 nm (6 wt% PLLA in chloroform (CF)). Cytotoxicity of fibrous mats and DNA quantification were investigated. The results showed that air-jet spun mats are nontoxic to mouse fibroblasts and the cell count significantly increased on the mats in comparison to the negative controls. Chemical changes in the biodegradable polymers structure and residual solvents in the fibrous mats were not found using Fourier-transform infrared spectroscopy. The degree of crystallinity was set for fibers from 8 wt% PCL in DCM – 58.9 %; 10 wt% PCL in DCM – 56.4 % and 6 wt% PLLA in CF – 22.0% via differential scanning calorimetry. The results of the tests performed show that air-jet spun PCL materials could be useful for in situ application of skin wound dressings.

Keywords: Air-jet spinning, airbrush, in situ deposition, nanofibrous scaffold, skin wound dressing

1. INTRODUCTION

Tissue engineering is a multidisciplinary field using applied science to provide tissue replacements or reconstruction [1]. There is significant interest in nanofibrous scaffolds mimicking the extracellular matrix. Nowadays, the most widely used way of nanofibrous scaffold fabrication is electrospinning. Nanofibers are produced with devices using DC or AC high voltage sources [2]. Besides this method, the air-jet spinning method (or solution blowing) is increasingly studied for nanofibrous material fabrication. Air-jet spinning uses air or gas flow to produce nanofibers [3]. The hydrodynamic force caused by air or gas friction on the surface of a solution droplet overcomes the surface tension of the polymeric droplet. This transforms the droplets into thin fibers.

It would be a remarkable achievement to fabricate nanofibrous scaffolds directly on a wound (in situ deposition) instead of surgical implantation. Portable electrospinning devices provide nanofibrous production in situ using high voltage, which can be dangerous for the human organism [4]. Electrospinning appears as an unsuitable method for in situ deposition of nanofibrous scaffolds. Air-jet spinning seems relatively safe for applying nanofibers directly to wounds due to the use of gas flow instead of high voltage [4]. One of the devices which can be used for air-jet spinning is a commercial airbrush [3]. A suitable polymeric solution is placed into a reservoir of an airbrush. Then it flows from the reservoir onto a needle in an inner nozzle of the airbrush [5], see **Figure 1**. A solution droplet is formed on the top of the nozzle, and the gas flow elongates the droplet [6]. If the gas flow stretching force overcomes the surface tensions, the jets are made from the solution and taken by aerodynamic force out of the airbrush [6]. Fibers are made by solvent evaporating from the jets [6]. Important air-jet spinning parameters include solution parameters (viscosity, surface tension, molecular weight of the polymer, concentration of the solution, vapor pressure, etc.), process parameters (gas pressure, working distance, solution flow rate, nozzle diameter, and geometry, etc.), and others (temperature, relative air humidity, atmospheric pressure, etc.) [7].

Furthermore, the air-jet spinning method is more versatile than DC electrospinning as it does not require an electrically active collector. It typically has higher productivity as high values of solution flow through the nozzle can be used to form the fibers [4]. Air-jet spinning is useful for coating surfaces of various complex shapes [4]. The method can be easily implemented using a low-cost, user-friendly, portable device [4].

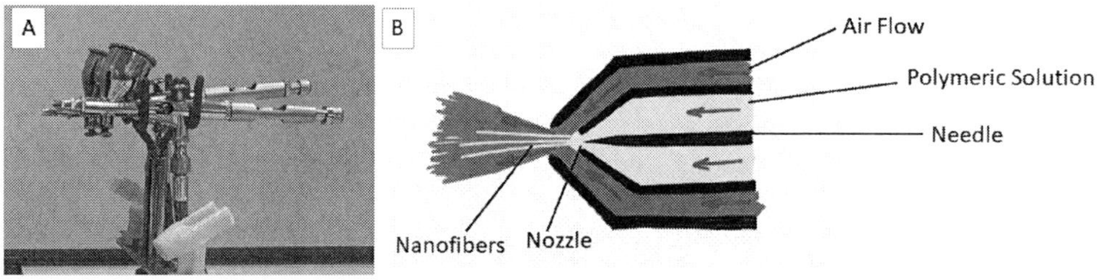

Figure 1 (A) Two airbrushes on a stand. (B) Scheme of air-jet spinning principle.

Air-jet spinning is not a brand-new technology of nanofibrous fabrication. Oliveira et al. fabricated biodegradable polymeric nanofibers using an airbrush and compared them to electrospun nanofibers [5]. Behrens et al. studied air-jet spinning of poly(lactic-co-glycolic acid) nanofibers for in situ deposition of a surgical sealant, hemostatic, and scaffolds [8]. Some water-soluble polymers, such as PVA, are very difficult to air-jet spin. The literature so far reports cases of PVA air-jet spinning with additional device for drying the fibers [9].

In this work, the biodegradable polymeric nanofibers were produced via air-jet spinning using a commercial airbrush. The optimal polymeric solutions for the technology were found. The morphology of the air-jet spun fibers were observed. A cytotoxicity test and DNA quantification were performed to prove suitability for potential medical use. Chemical changes in polymeric nanofibers were set by FTIR, and the degree of crystallinity was evaluated by DSC.

2. MATERIALS AND METHODS

2.1. Materials

The following polymers were used for the preparation of polymeric solutions: polycaprolactone (PCL, Mw 45 000 g·mol^{-1}, Merck, Germany) and poly(L-lactic acid) (PLLA, Mw 45 000–55 000 g·mol^{-1}, PolySciTech, USA). Dichloromethane (DCM, Penta, Czech Republic) and chloroform (CF, Penta, Czech Republic) were used for solvents. For further experiments, the following chemicals were utilized: Dulbecco's Modified Eagle's Medium (DMEM, Merck, Germany), fetal bovine serum (FBS, Biosera, France), Triton X-100 (Merck, Germany), Cell Counting Kit-8 (CCK, Dojindo, Japan), lysis buffer solution (prepared at the Technical University of Liberec, Department of Bioengineering), Quant-iT dsDNA Assay Kit, High Sensitivity (Invitrogen,

USA), phosphate buffer solution (PBS, prepared at the Technical University of Liberec, Department of Bioengineering).

2.2. Solution Preparation

For biocompatible polymeric nanofibers fabrication, the following solutions were prepared: 6 wt%, 8 wt%, and 10 wt% PCL in DCM; 6 wt% and 12 wt% PLLA in CF. The concentration of the solutions was set according to the literature and then other concentration values were adjusted [3], [5].

2.3. Air-jet Spinning and Electrospinning of Biocompatible Polymeric Fibers

Polymeric solutions were placed into a commercial airbrush Fengda BD-180 with 0.3 mm nozzle diameter and 1.8 bar air flow pressure. The air-jet spinning of biocompatible polymeric nanofibers was performed on different days when the temperature ranged between 21.1 and 23.3 °C, and the relative humidity ranged between 19.6 and 51.7%. Firstly, the solutions were air-jet spun as a nanofibrous layer on paper and then on a metal sieve as membranes, which can be taken down and further studied independently. Lastly, the solutions were air-jet spun on a hand inside a nitrile glove as evidence of possible in situ application (see **Figure 2**). Solutions of 10 wt% PCL in DCM and 16 wt% PCL in CF/EtOH (8:2) were fabricated with Nanospider™ (NS 1S500U, Elmarco, Czech Republic) to compare their properties with air-jet spun material of the same weight (50 gsm).

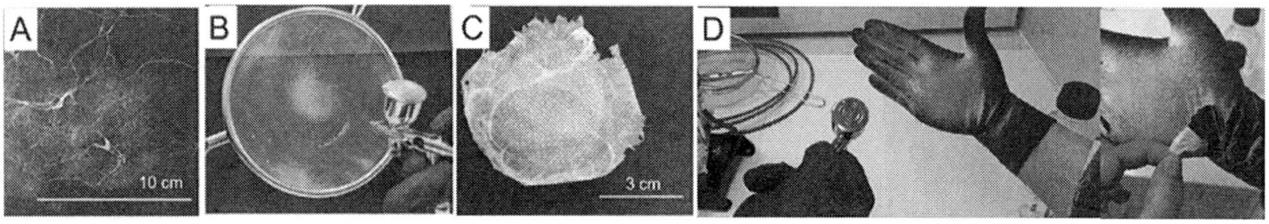

Figure 2 Air-jet spinning of PCL nanofibers onto a black paper (A); onto a metal sieve (B) to produce membrane (C) for further testing; and directly onto a hand inside a black nitrile glove (D).

2.4. Testing Methods of Prepared Fibers

The morphology of fibrous mats was observed with a scanning electron microscope (SEM, Tescan Vega S3B, Tescan Orsay Holding, Czech Republic). Fiber diameters were measured from SEM images using ImageJ software (National Institutes of Health, USA). Over 100 values of diameter were measured for each sample.

The in vitro cytotoxicity test was performed according to the ČSN EN ISO 10993-5 (855220) standard with the use of mouse fibroblasts 3T3-L1. Materials were tested using extracts from the materials. Nanofibrous mats from 8 wt%, 10 wt% PCL in DCM and 6 wt% PLLA in CF were chosen for testing cytotoxicity due to their optimal properties, such as compactness. Each material was air-jet spun for 1.5 minutes on a metal sieve. Immediately after nanofibrous material production, two samples were cut from each material into 1.3x1.3 cm squares and sterilized by UV light in a sterile flow box for 20 minutes from both sides directly following cytotoxicity testing. The cell viability was set from measured data. According to the ČSN EN ISO 10993-5 (855220) standard, the material is considered nontoxic if the value of cell viability is over 70 %. The rapid testing after nanofiber production was carried out to study the effect of any residual solvents in the air-jet spun nanofibrous materials.

After the cytotoxicity test, DNA quantification was studied using solution of Quant-iTTM dsDNA HS reagent and Quant-iTTM dsDNA HS buffer (in ratio 1:200). The excitation wavelength in the spectrophotometer was set to 485 nm, and the measured emitted wavelength was 523 nm.

Chemical changes in polymer chains and the remaining amount of the toxic solvents of air-jet spun mats from 8 wt%, 10 wt% PCL in DCM, and 6 wt% PLLA in CF were measured by Fourier transform infrared spectroscopy

(FTIR, Nicolet iZ10, Thermo Fisher Scientific, Czech Republic) in 4000–400 cm-1 range. The spectra of the nanofibrous mats were compared to the spectra of pure polymers from which the solutions were made. For comparison of air-jet spinning and electrospinning, a spectrum of the electrospun nanofibrous mat from 10 wt% PCL in DCM was measured by FTIR.

The degree of crystallinity of air-jet spun mats from 8 wt%, 10 wt% PCL in DCM, and 6 wt% PLLA in CF was investigated by differential scanning calorimetry (DSC, DSC 1 / 700, Mettler Toledo). Materials were weighted between 2–16 mg and dissolved in an aluminum cup. The temperature interval of heating was set at -20–200 °C in an inert nitrogen atmosphere with a gas flow of 50 ml/min and heating at 10 °C/min. The degree of crystallinity of the nanofibrous mats was compared to the degree of crystallinity of the granular form of polymers from which the solutions were prepared and 10 wt% PCL in DCM (electrospun).

3. RESULTS AND DISCUSSION

Successful nanofiber fabrication was possible from the following solutions: 6 wt%, 8 wt%, and 10 wt% PCL in DCM and 6 wt% PLLA in CF (see **Figure 3**). Average fiber diameter value from mats with confidence interval of 95% probability of: 6 wt% PCL in DCM is 268 ± 27 nm; 8 wt% PCL in DCM is 305 ± 24 nm; 10 wt% PCL in DCM is 333 ± 31 nm and 6 wt% PLLA in CF is 223 ± 19 nm. With a higher concentration of polymer in the solution, the entanglement of the polymer chains increases during air-jet spinning, thereby increasing the diameter of the produced fibers.

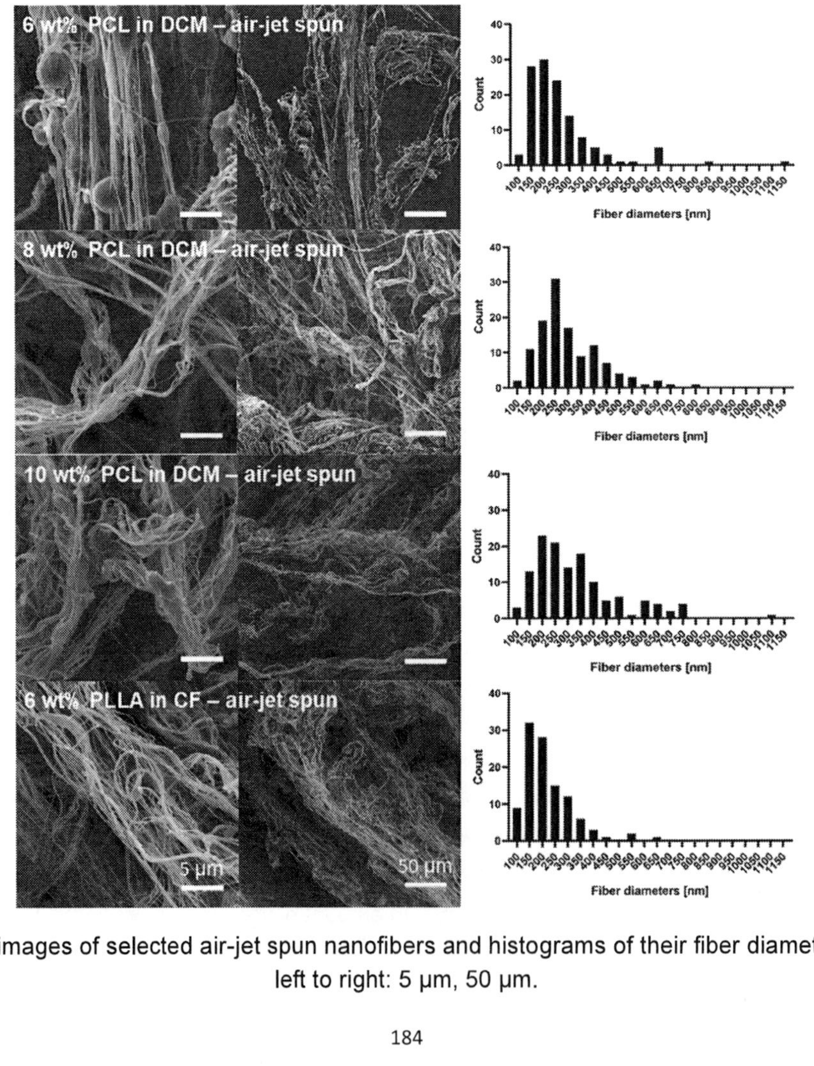

Figure 3 SEM images of selected air-jet spun nanofibers and histograms of their fiber diameters. Scale from left to right: 5 µm, 50 µm.

According to the ČSN EN ISO 10993-5 (855220) standard, the nanofibrous mats are considered nontoxic to the cells (see **Figure 4A**). Value of cell viability of materials from 8 wt% PCL in DCM was 124 ± 7 %, 10 wt% PCL in DCM was 110 ± 11 % and 6 wt% PLLA was 87 ± 12 %. Cell viability values over 100% could have been due to a higher metabolic rate of the cells caused by stress and not their higher cell amount. The viability value is less than 100% for PLLA material. This may be due to the effect of residual CF, which due to its higher boiling point evaporates more slowly during air-jet spinning than DCM. However, this must be a very small amount of residual solvent (CF from PLLA nanofibers) as FTIR showed no measurable residue there. Therefore, it seems that PCL solutions in DCM would be more suitable for in situ application of skin wound dressings by air-jet spinning. DNA quantification was performed to discover if the higher cell viability values are caused by their higher number (see **Figure 4B**). The DNA concentration was calculated from measured data with standard deviation for the wells with the cells from the material extract and negative control: 8 wt% PCL in DCM – 21 ± 4 ng/well, 10 wt% PCL in DCM – 19 ± 5 ng/well, 6 wt% PLLA in CF – 15 ± 4 ng/well and negative control – 13 ± 3 ng/well. For the materials with complete medium, the concentration values per well were higher than for the negative controls. Thus, there may be a higher number of cells per well in the materials than in the negative control wells. The results from cytotoxicity and DNA quantification show that the layers are not toxic and there could be an increase in cell count on these layers. This could be desirable in the use of nanofibrous layers as wound dressings.

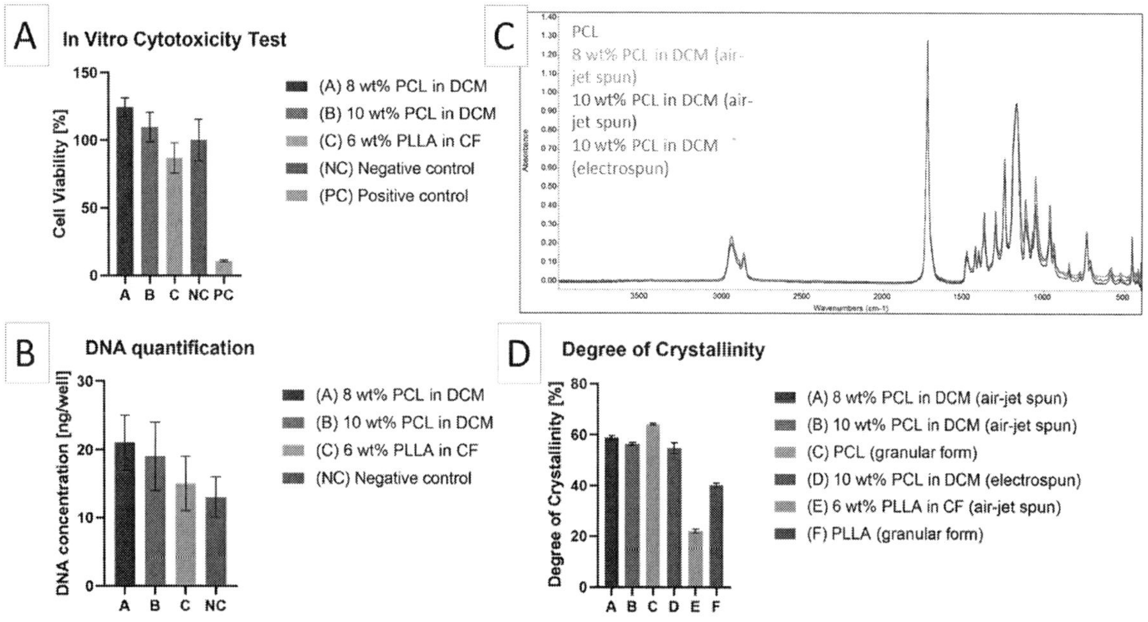

Figure 4 Results from biocompatibility testing (A, B); FTIR testing (C) and DSC testing (D)

No chemical changes were confirmed by FTIR (see **Figure 4C**). The remaining toxic solvents were in undetectable amounts by FTIR which is required. Residual solvents are undesirable in layers for skin wound dressings applied in situ.

The degree of crystallinity was set for following materials: 8 wt% PCL in DCM – 58.9 ± 0.8 %; 10 wt% PCL in DCM (air-jet spun) – 56.4 ± 0.6 %; 10 wt% PCL in DCM (electrospun) – 54.8 ± 2.1 %; PCL (granular form) – 64.2 ± 0.4 %; 6 wt% PLLA in CF – 22.0 ± 0.8 % and PLLA (granular form) – 40.2 ± 0.8 % (see **Figure 4D**). The degree of crystallinity values of the air-jet spun and electrospun materials were lower than those of the original granulates. Furthermore, the possible influence of the technology on the degree of crystallinity of the materials was shown by DSC. Although in electrospinning the fibers fly a longer path through the air and would have more time to form regular crystalline parts, they have a lower crystallinity value than fibers formed by air-jet spinning. In both cases for PCL and PLLA, the granules have a higher degree of crystallinity than the air-jet spun materials. This is clearly an effect of technology i.e., the rate of solvent evaporation or solidification of the materials produced. In larger granules there is certainly more space for longer periods of time for the formation of crystalline regions, but for the relatively very fast liquid-to-solid transformations of the described fiber fabrication technologies there is probably less time for crystal formation. The values of the degree of crystallinity affect biodegradation which is an important parameter for nanofibrous scaffolds for wound dressings. Thus a different course (speed) of degradation can be expected, which was confirmed in further follow-up tests [10].

4. CONCLUSION

Air-jet spinning is being explored as an alternative to electrospinning. The nanofibers are formed by airflow instead of high voltage. This method of nanofiber fabrication allows safe, direct placement of the scaffold on the wound. In this paper, the biocompatible polymeric solutions from PCL and PLLA (6 wt%, 8 wt%, 10 wt% PCL in DCM and 6 wt% PLLA in CF) were successfully air-jet spun for possible in situ deposition of the fibrous scaffolds for wound dressings. The evaluated morphology of the fibers showed that the fiber diameters are according to the nanoscale. The cytotoxicity test proved that all produced nanofibrous materials are noncytotoxic (cytocompatible). The PCL materials showed higher viability compared to the PLLA materials and negative control. Via FTIR, chemical changes or remaining solvents were not present in the nanofibrous mats. The air-jet spun nanofibrous mats had higher degree of crystallinity than electrospun mats but lower than their original granular form according to DSC analysis. The air-jet spinning of 8 wt% PCL in DCM seems to be a great potential candidate for in situ deposition of nanofibrous scaffolds for wound dressings. However, it is necessary to continue studies imitating in situ application, to study possible changes in wound temperature during the application of nanofibers, and to address the issue of sterility of the spinning device.

ACKNOWLEDGEMENTS

Authors thank for the grant "Study of physical networks in polymer materials" of the TUL Grant Program „PURE“, No. PURE-2020-4007 and thank MUDr. Ladislav Dzan, Ph.D. (the head of Oral, maxillofacial and facial surgery) from the Liberec Regional Hospital for expert consultations.

REFERENCES

[1] Y. IKADA. "Challenges in tissue engineering." *Journal of The Royal Society Interface*. 2006, vol. 3, no. 10, pp. 589-601. Available from: https://doi.org/10.1098/rsif.2006.0124.

[2] V. JENČOVÁ, E. KUŽELOVÁ KOŠŤÁKOVÁ, P. MIKEŠ, V. HEDVIČÁKOVÁ, and D. LUKÁŠ. *NANOVLÁKNA Teorie, technologie a použití*. 1. Studentská 1402/2, Liberec: Technická univerzita v Liberci, 2021.

[3] G. C. DADOL *et al.* "Solution blow spinning (SBS) and SBS-spun nanofibers: Materials, methods, and applications." *Materials Today Communications*. 2020, vol. 25, p. 101656. Available from: https://doi.org/10.1016/j.mtcomm.2020.101656.

[4] A. F. HELL, M. M. O. SIMBARA, P. RODRIGUES, D. A. KAKAZU, and S. M. MALMONGE. "Production of fibrous polymer scaffolds for tissue engineering using an automated solution blow spinning systém." *Res. Biomed. Eng.*, Oct. 2018, vol. 34, pp. 273-278. Available from: https://doi.org/10.1590/2446-4740.180039.

[5] J. E. OLIVEIRA, L. H. C. MATTOSO, W. J. ORTS, and E. S. MEDEIROS. "Structural and Morphological Characterization of Micro and Nanofibers Produced by Electrospinning and Solution Blow Spinning: A Comparative Study." *Advances in Materials Science and Engineering*. May 2013, vol. 2013, p. 409572. Available from: https://doi.org/10.1155/2013/409572.

[6] C. JIA, L. LI, J. SONG, Z. LI, and H. WU. "Mass Production of Ultrafine Fibers by a Versatile Solution Blow Spinning Method." *Acc. Mater. Res.* Jun. 2021, vol. 2, no. 6, pp. 432-446. Available from: https://doi.org/10.1021/accountsmr.1c00040.

[7] E. STOJANOVSKA *et al.* "A review on non-electro nanofibre spinning techniques." *RSC Adv.* Sep. 2016, vol. 6, no. 87, pp. 83783-83801. Available from: https://doi.org/10.1039/C6RA16986D.

[8] A. M. BEHRENS *et al.* "In Situ Deposition of PLGA Nanofibers via Solution Blow Spinning." *ACS Macro Letters.* 2014, vol. 3, no. 3, pp. 249-254. Available from: https://doi.org/10.1021/mz500049x.

[9] A. M. C. SANTOS, E. L. G. MEDEIROS, J. J. BLAKER, and E. S. MEDEIROS. "Aqueous solution blow spinning of poly(vinyl alcohol) micro- and nanofibers." *Materials Letters.* 2016, vol. 176, pp. 122-126. Available from: https://doi.org/10.1016/j.matlet.2016.04.101.

[10] E. CHUDOBOVÁ. "*Zvlákňování biodegradabilních materiálů pomocí rozfukování polymerních roztoků.*" Bachelor thesis. Technical University of Liberec, Liberec, Czech Republic, 2022.

STUDY OF THE INTERACTION OF MOUSE FIBROBLASTS WITH BIODEGRADABLE POLYESTERS MICRO/NANOFIBROUS MATERIALS USING BY LIVE CELL IMAGING

[1]Šárka HAUZEROVÁ, [1]Maxim LISNENKO, [1]Juan Pablo Pérez AGUILERA, [2]Andrii SHYNKARENKO, [1]Věra JENČOVÁ

[1]Technical University of Liberec, Faculty of Science, Humanities and Education, Department of Chemistry, Liberec, Czech Republic, EU, sarka.hauzerova@tul.cz

[2]Technical University of Liberec, Faculty of Mechanical Engineering, Department of Manufacturing systems and Automation, Liberec, Czech Republic, EU, andrii.shynkarenko@tul.cz

https://doi.org/10.37904/nanocon.2022.4607

Abstract

Tissue engineering and regenerative medicine are ever-growing multidisciplinary fields. Along with the development of tissue carriers, methods for testing the biocompatibility of these materials are also being developed. This work deals with research, development, and optimisation of the preparation of a fibre system intended for monitoring the direct interaction of the 3T3 mouse fibroblast cell line with polymeric materials in real-time (live cell imaging). The fabricated microfiber model (grid) was seeded with the 3T3 mouse fibroblast cell line, and the interaction between the materials and the cells was monitored for three days by optical microscope. The interaction between the materials and the cells was observed during a picture of cells on the fibres. The advantage of this method is the absence of other chemicals like chemicals for the visualisation of cells (fluorescence staining) or chemicals for measuring cell viability because cells adhering to fibres are visible in transmitted light.

Keywords: Microfibers, drawing, grid model system, cell interaction

1. INTRODUCTION

In recent decades, tissue engineering has been a highly developed multidisciplinary field combining material science, engineering, biology, medicine, etc. The goal of tissue engineering is to create tissue carriers that support, regenerate and restore the function of damaged tissue due to acute or chronic injury [1,2].

With the development of materials intended for tissue engineering and regenerative medicine, methods for testing the biocompatibility of these materials are also being developed. However, monitoring the interaction of cell lines with tissue carriers still has limits. Laboratory methods commonly used in practice are based on monitoring the metabolic activity of cells, or DNA quantification, during the experiment. For morphological analysis, cells adhered to the surface and in the material's structure are fixed and subsequently observed using fluorescence and electron microscopy [3].

One of the methods developed in the last decade is monitoring living cells in real-time using a time-lapse recording. When live cells are observed using optical microscopy in real-time, this method is used with the help of phase contrast to keep the cell population adhered to the bottom of the culture wells and to monitor the influence of different drugs, chemicals or nanoparticles on the cells [4]. However, cells adhered to the surface of the tissue carriers are not detectable using phase contrast. Therefore, cells are used that are affected by fluorescent labelling, which is bound to the cell structure. The cells can thus be observed at specific wavelengths in the emission spectrum using fluorescent microscopes [5]. However, the disadvantage remains the necessity of chemically affecting the cell line, which can cause different behaviour, whether in metabolic activity, adhesion or cell proliferation, compared to an unaffected line.

The geometrically oriented microfiber system prepared by the drawing method allows monitoring of the cellular interaction with polymeric materials over time without using contrast marking. This method could bring a new understanding of cellular interaction with the tested material. Previous studies demonstrated that on oriented fibres prepared by the drawing method, cells which adhered to the fibres were preferentially oriented in the direction of the fibres [6].

2. MATERIALS AND METHODS

2.1. Materials

For fibre preparation was used, poly-ε-caprolactone (PCL, Mw 70000 - 90000, Sigma-Aldrich) was in different solvent systems like chloroform ($CHCl_3$), acetic acid (CH_3COOH) and their combination in ratio 8:2 ($CHCl_3$:CH_3COOH).

2.2. Preparation of fibres using the drawing method

The preparation of fibres by the drawing method uses mechanical forces. A drop of polymer solution is pushed from the tip of the needle onto the pad. Then the hand is moved away from the fall; thanks to the expansive solvent system and the viscoelastic properties of the polymer, fibre is formed (**Figure 1**). A laboratory equipment manipulator based on the robotic arm "Uarm swift pro" (**Figure 1A**) was used for fibre preparation. For handling of polymer solutions was designed as a special extruder for the robotic arm, based on the positive extrusion pipette (Microman pipette) from the Gilson company [7]. The device was constructed by Ing. Andrii Shynkarenko, PhD from the Department of Manufacturing systems and Automation at the Technical University of Liberec (KSA, TUL).

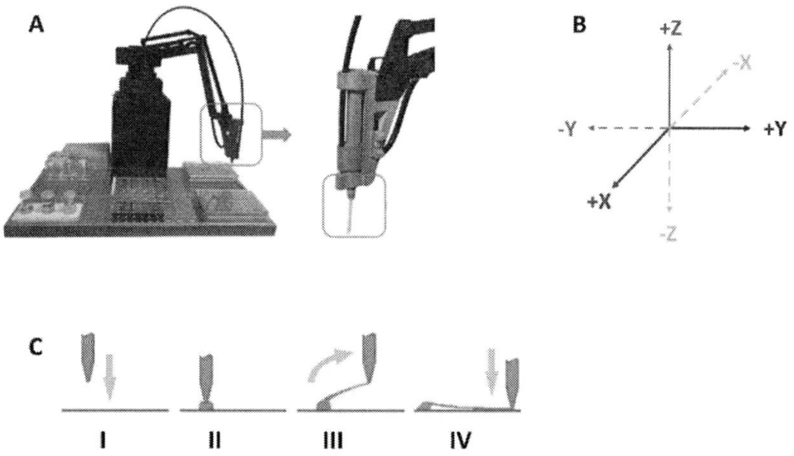

Figure 1 Robotic arm „Uarm swift pro" and detail of extrude with tips for drawing fibres (A), the coordinate system for the movement of the robotic arm (B) and the schematic principle of removing the fibre using the robotic arm (C) [7].

2.3. Optimization of fibre preparation using the drawing method

Optimisation of the fibre preparation by the robotic arm drawing method (Uarm swift pro) based on different solvent systems and PCL polymer solution concentrations. The solutions shown in **Table 1** were used to prepare the fibres. A Tescan Vega 3 scanning electron microscope (Tescan, Czech Republic) was used to evaluate the morphology of the fibres composed of the individual solutions. The morphology of the fibres was assessed from the point of view of fibre diameters. Each polymer solution was prepared with 150 fibres and 100 fibre diameter were measured because some fibre broke during the drawing process.

Table 1 Used concentration of polymer PCL80 and solvent system for drawing fibres

Sample	The concentration of PCL80 (wt%)	Solvent systems
A	10	$CHCl_3$
B	12	$CHCl_3$
C	14	$CHCl_3$
D	10	$CHCl_3$:CH_3COOH (8:2)
E	12	$CHCl_3$:CH_3COOH (8:2)
F	15	$CHCl_3$:CH_3COOH (8:2)
G	10	CH_3COOH
H	15	CH_3COOH
I	20	CH_3COOH

2.4. Microfibre model system

A geometrically oriented microfiber system was prepared to monitor cells' interaction with the polymer material. The microfibers were made from 12% PCL in pure chloroform (sample B, **Table 1**) and arranged perpendicular to each other (grid) (**Figure 2A**). The robotic arm feed speed was 20000 mm/min, the displacement path 100 mm, the arm stroke height 10 mm and the distance between the fibres 0.1 mm. For each direction of the grid, was draw 200 fibres. The resulting grid was fixed in PMMA conical carriers (**Figure 2B**). The carriers were made using injection moulding. These carriers were subsequently sterilised using UV-C for 40 minutes in a sterile flow box; after sterilisation, samples were inserted into a 24-well plate where the microfibers grid was incubated with cells.

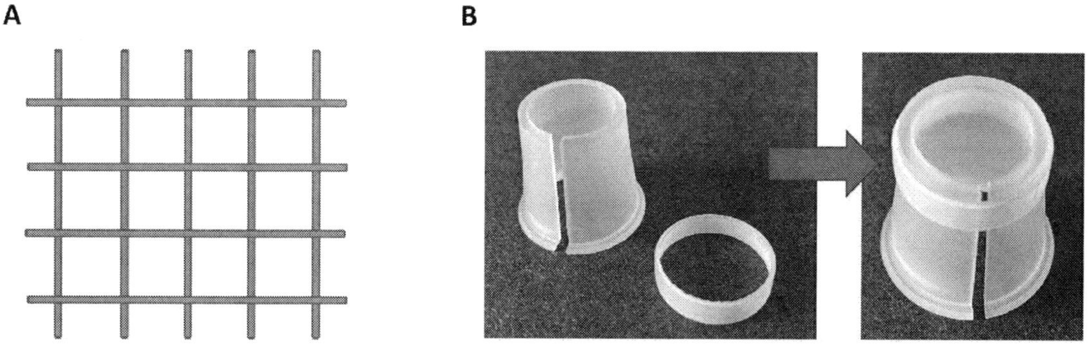

Figure 2 Schema of microfiber grid model system (A) and PMMA conical carrier with fixation ring (B).

2.5. Cell interaction with fibre model system

The cell line for testing was 3T3 mouse fibroblasts. Medium for incubation was DMEM High Glucose (Dulbecco's Modified Engles Medium, Merck, Czech Republic) with 10% fetal bovine serum (Biosera, Czech Republic), 1% glutamine (Biosera, Czech Republic) and 1% antibiotic - Pen / Strep Amphotericin B (Lonza, Czech Republic). Incubation PMMA conical carriers with PCL grid were in 24-well plates. The concentration of cells seeded in the well used was $1.5*10^4$. Samples were incubated with cells for 4 hours in a CO_2 incubator at 37 °C (Heracell Vios 160i, Thermo Fisher Scientific, Czech Republic). Then the PMMA conical carriers with grid and adhered cells on the fibres were removed into the pure well, and add complete medium. The well plate was incubated in a CO_2 incubator at 37 °C (XL incubation model system, Zeiss, Germany). The incubator is part of the Axio Vision optical microscope (Zeiss, Germany). Every 10 minutes, take an image of the fibre grid with cells. Cells were imaged in transmitted light using phase contrast.

3. RESULTS AND DISCUSSIONS

Different concentrations of PCL polymer in other solvent systems were used to optimise the preparation of fibres by the drawing method using the "Uarm" laboratory equipment (**Table 1**). Fibres did not form from solutions D, G, and H. Microfibres' very stable process formation was observed for solutions A, B, C and I. For solutions E and F, fibres frequently had visible breakage during the drawing of the fibre, and the fibres were not evenly arranged. The produced microfibers were morphologically evaluated using scanning electron microscopy (**Figure 3**). For each type of solution, fibre diameters were measured from the images taken and statistically evaluated using a box plot (**Figure 3**).

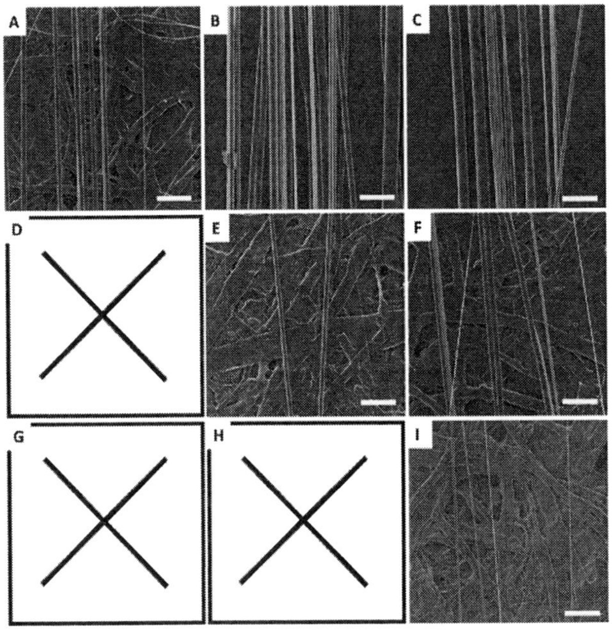

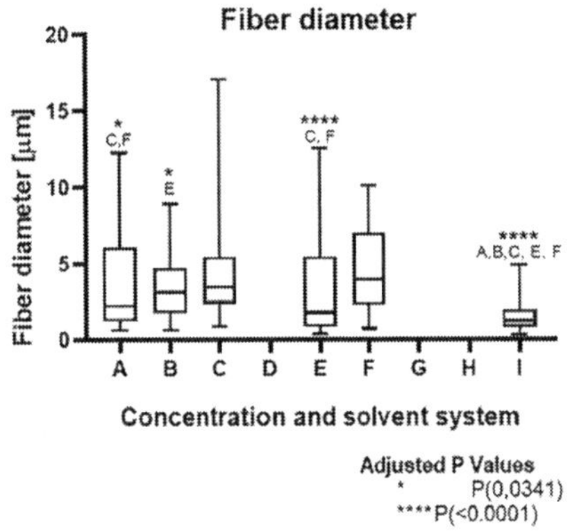

Table with statistic evaluation of fiber diameter		
Sample	**Mean [µm]**	**Standard devition [µm]**
A_10% PCL in $CHCl_3$	3.63	2.85
B_12% PCL in $CHCl_3$	3.46	1.97
C_14% PCL in $CHCl_3$	4.70	3.50
E_12% PCL in $CHCl_3$:CH_3COOH 8:2	3.05	2.68
F_14% PCL in $CHCl_3$:CH_3COOH 8:2	4.94	2.50
I_20% PCL in CH_3COOH	1.44	0.91

Figure 3 Scanning electron microscope images of fibres prepared by drawing method (left, A-I), boxplot of fibre diameters for individual tested solutions (top right), and table of fibre diameter values and standard deviations (bottom left). The scale bar is 50 µm.

The diameter of fibres produced by the drawing method from PCL was in the order of micrometres. The weakest fibres were prepared from the solution I (20% PCL in pure acetic acid). With this solution, approx. 70 % of the diameters of the produced fibres were around 1 µm. The mean value of the fibre diameter is 1.44 µm, and solution I (PCL) showed the lowest fibre variability (standard deviation). Fibres formed from solution B showed the second lowest variability of measured fibre diameters, and the mean value of the fibre diameter is 3.5 µm. The highest fibre diameter was measured for solution F (4.94 µm). Solution C showed the most significant fibre variability.

Based on the results, one selected polymer solution was to prepare a geometrically oriented fibre model (grid), which was subsequently used to monitor the interaction of living cells over time. However, weaker fibres may not be suitable for cell adhesion. There is a reason why was chosen a polymer solution from which formed homogeneous fibres with a diameter of around 3.5 µm, and the variability of fibre diameters was lowest than another solution. Solution B (12% PCL in pure chloroform) was used to prepare the microfibers grid. The created microfiber grid was fixed in a PMMA conical carrier (**Figures 4A and B**) and seeded with mouse fibroblast cells. During 72 hours, cell interaction was monitored by optical microscope Axio Vision (**Figures 4C and D**).

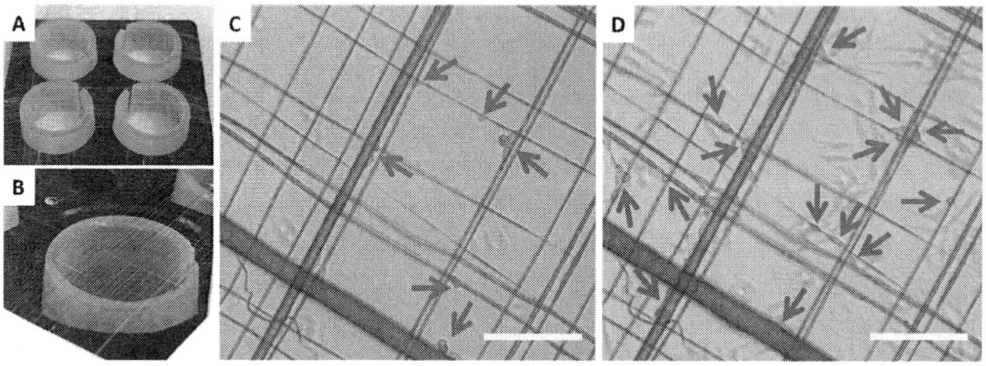

Figure 4 Preparation of a microfiber grid for cell interaction testing (A), detail of the microfiber grid before fixation on to the PMMA conical carrier (B), a picture of cells adhered to the fibres of the model system (grid) after 5 hours from seeded (C) and a picture of cells attached to the fibres after 72 hours from seeded (D). Red arrows indicate adhered cells on fibres. The scale bar is 200 µm.

Figure 4C shows cells that adhered to the PCL fibres of the model system. The first image of the fibre grid with cells was taken approx. 5 hours after deployment. The cells are visibly still round. **Figure 4D** shows cells after 72 hours of interaction with PCL fibres. The number of cells has visibly increased. The shape of the cells here is somewhat elongated, and the cells surround the fibres of the grid.

4. CONCLUSION

As part of the experiment, the production of drawing fibres was optimised. The fibres were prepared on the laboratory equipment "Uarm swift pro" designed by Andrii Shynkarenko (KSA TUL). Different concentrations of PCL (Mn 80000) were tested in other solvent systems. A solution of 12% PCL in pure chloroform (solution B) was selected from tested solutions. The microfibers produced from this solution were homogeneous compared to the other solutions used, and the fibre drawing process was very stable. The diameter of the fibres was around 3.46 µm. This solution subsequently prepared a fibre grid as a model system for monitoring the interaction of living cells over time. The fibrous grid was seeded with 3T3 mouse fibroblasts, and cell interaction was observed for 72 hours. From the images in **Figure 4D**, it can be seen that after 72 hours, the cells adhered to the material (they have an elongated shape), and their number has visibly increased since seeded.

This method is innovative precisely because it is unnecessary to monitor the cell interaction by affecting the cells with any fluorescent or other chemical for its visualisation. Cells are observed on polymer fibres only in transmitted light using phase contrast. The microfiber model system appears suitable for monitoring the interaction of living cells with polymeric microfibers over time. The model system can monitor the movement of cells along fibres. When using a higher concentration of cells and a more prolonged incubation with the materials, the model could be suitable for evaluating the time required for the growth of defined significant defects (distance of individual fibres to each other).

ACKNOWLEDGEMENTS

This work was developed thanks to the financial support of the student grant competition project (SGS-21484) from the Technical University of Liberec.

REFERENCES

[1] R. A. BROWN. *Extreme Tissue Engineering: Concepts and Strategies for Tissue Fabrication.* John Wiley & Sons, 2013.

[2] C. F. GUIMARÃES, L. GASPERINI, A. P. MARQUES, and R. L. REIS, 'The stiffness of living tissues and its implications for tissue engineering, *Nat. Rev. Mater.*2020, vol. 5, no. 5, pp. 351-370. Available from: https://doi.org/10.1038/s41578-019-0169-1.

[3] N. ASADI, A. R. DEL BAKHSHAYESH, S. DAVARAN, and A. AKBARZADEH. 'Common biocompatible polymeric materials for tissue engineering and regenerative medicine. *Mater. Chem. Phys.*, vol. 242, p. 122528, Feb. 2020, Available from: https://doi.org/10.1016/j.matchemphys.2019.122528.

[4] D. DI GIUSEPPE *et al.*, 'Acute cytotoxicity of mineral fibres observed by time-lapse video microscopy. *Toxicology*. Jan. 2022, vol. 466, p. 153081. Available from: https://doi.org/10.1016/j.tox.2021.153081.

[5] L. WANG, M. S. FREI, A. SALIM, and K. JOHNSSON. 'Small-Molecule Fluorescent Probes for Live-Cell Super-Resolution Microscopy'. *J. Am. Chem. Soc.* 2019, vol. 141, no. 7, pp. 2770-2781. Available from: https://doi.org/10.1021/jacs.8b11134.

[6] K. STRNADOVÁ *et al.* 'Drawn aligned polymer microfibres for tissue engineering. *J. Ind. Text.* 2020, vol. 50, no. 3, pp. 263-277. Available from: https://doi.org/10.1177/1528083718825318.

[7] A. SHYNKARENKO *et al.* 'Fabrication of Micro/Nanofibrous Scaffolds using a Robotic Manipulator and Their Application for Tissue Engineering. In Review, preprint. Nov. 2020. Available from: https://doi.org/10.21203/rs.3.rs-112972/v1.

CONTROLLED RELEASE FROM LAYERED NANOSTRUCTURED MATERIALS DRIVEN BY CAPILLARY FORCES

[1,2]Anna SUDAKOVA, [5]Duc Anh NGUYEN, [3]Le MINH HA, [2]Jiri SMILEK, [4]Thi Phuong NGO, [1]Jan ZIDEK

[1]CEITEC - Central European Institute of Technology, Brno, Czech Republic, EU, Anna.Sudakova@ceitec.vutbr.cz, jan.zidek@ceitec.vutbr.cz

[2]Institute of Physical and Applied Chemistry, Faculty of Chemistry, Brno, Czech Republic, EU, smilek@fch.vut.cz

[3]Institute of Natural Products Chemistry (INPC), Vietnam Academy of Science and Technology (VAST), Nghia Do, Cau Giay, Hanoi, Viet Nam, halm2vn@gmail.com

[4]Phenikaa University Nano Institute (PHENA), Phenikaa University, Hanoi, Vietnam, thuy.nguyenthithu@phenikaa-uni.edu.vn

[5]Center for Research and Technology transfer, Vietnam Academy of Science and Technology (VAST), Nghia Do, Cau Giay, Hanoi, Viet Nam, ducanh221100@gmail.com

https://doi.org/10.37904/nanocon.2022.4608

Abstract

Controlled release of active compounds has been investigated in current medical research and there were developed many approaches, how to control it. For instance, complexation of active compound, encapsulation in matrix, or release from gels. However, there are still opportunities to improve its mechanisms.

This contribution presents potentially promising method of controlled release, which is based on two modifications of standard mechanisms: i. release from nanostructured layered material, ii. release forced by capillarity. That mechanism allows us to design systems with gradual release of the active substance.

The release of indigotin dye was performed from materials where layers of non-woven chitosan fabric alternate with gelatine layers. Active compound was incorporated in each layer. Gradual wetting of layers is caused by capillarity effect. The controlled release can be observed visually, e.g. by optical spectroscopy. In the future, the colour compound will be substituted by medical active substances, such as antibiotics etc.

The final systems will be composed of natural and biocompatible materials, such as polysaccharides and proteins. The active compound will be in synergy with the chitosan matrix. Chitosan is a unique natural polymer with positive charge on the surface, it exhibits mucoadhesive and antibacterial properties. Gelatine was chosen because it is a protein with a negative charge on its surface, which forms a polyionic mixture with positively charged chitosan.

Such material will improve the current practice in the field of medical materials. It will allow us to design patches treated with an active substance that could improve the therapeutic effect.

Keywords: Chitosan, nanostructured gel, controlled release, capillary force, layered material

1. INTRODUCTION

The study of controlled release of active particles and drugs is gaining importance nowadays. This is because it can result in an optimal dosage of the drug in the desired location, not only for delivery over longer period of time and one-time application [1]. A major problem here is primarily the burst effect, which is characterized by the rapid release of a certain amount of active substance from the structure in presence of a medium.

This goal is to develop a material composed of nanolayers of non-woven fabric in combination with chitosan hydrogel for the controlled release of an active substance and the polypropylene non-woven fabric will control the release by two methods, the first is multilayer release and the second is by capillary force.

A non-woven polypropylene fabric was used to prepare the layered material, which was glued together by using a collagen solution. Negatively charged indigotin dye was incorporated into this material, which was used to observe the release from the material. The polypropylene non-woven fabric [2] was chosen because it is a material with large pores (approximately10 µm), which exhibits excellent thermal and chemical stability, and at the same time retains its flexibility.

Collagen [3] is a part of larger glycoprotein family, which consist of 27 types of collagen. This linear polymer is a component of extracellular matrices. In its primary structure occurs amino acid glycine in every third position. Higher structure consist of three polypeptide chains folded into triple helices. Gelatine [4] is a product of collagen hydrolysis. In the preparation of the nanolayered material, a gelatine solution is used, it connects the layers of the non-woven fabric by creating a hydrogel, where the individual fibres are connected by covalent bonds.

2. PRINICPLE AND METHOD

2.1. Preparation of layered material

A non-woven polypropylene fabric was chosen for the preparation of the layered material. The initial, squares of non-woven fabric with a size of approximately 4×4 cm were cut out. A concentration series of the blue dye indigotin was prepared with the concentration of 50; 25; 12.5; 6.25; 3.1; 1.6 and 0.8 wt. %. The gelatine solution was prepared at a concentration of 3 wt. %. The layered material was assembled by applying the most concentrated solution of indigotin to one layer of non-woven fabric, followed by a layer of gelatine and two more layers of pure non-woven fabric, which were bonded together by gelatine.

2.2. Release of the indigotin dye

The layered material was treated by indigotin. It was hermetically sealed from water from side in order to prevent the diffusion from side. The diffusion is enabled only from the front. The release was observed to controlled amount of water. The release was observer during 96 hours, the release of the indigotin dye from the layered material into water environment.

2.3. Diffusion through the layered material

The diffusion of the dye through the layered material was also monitored using diffusion cells. Measurement using diffusion cells [5,6] is a very easy method for determining the diffusion coefficient through our layered material. The apparatus consisted of two cells, receiving and source, where one was filled with 60 ml of distilled water and the other with 60 ml of methylene blue solution (0.01 g/l). Prepared layered materials were inserted between these cells. Layered materials with the number of layers 1, 2 and 8 were prepared for this measurement. A methylene blue solution with a concentration of 0,01 g/l was prepared for diffusion through the material in the diffusion cells. The experiment ran for 24 hours, after which samples were taken from the receiving cells for absorbance measurement on a UV-VIS spectrophotometer.

2.4. UV-VIS spectrophotometry

By measuring UV-VIS spectrophotometry, the diffusion coefficient for layered materials with a different number of layers was determined. A Hitachi U-3900H Spectrophotometer was used. The measured wavelength range was from 900 to 300 nm. First, the calibration curve of methylene blue solutions, which had concentrations of 0.01; 0.007; 0.005; 0.003 and 0.001 g·dm^{-3}. From these measurements, the absorbance measured at 665 nm was selected the values were related to the concentration and a graph and linear regression equation were plotted. Using the regression equation and the obtained absorbance values, the concentrations in the receiving and source cells were calculated, as well as the concentration of methylene blue retained in the layered material. The following equation was used to calculate the diffusion coefficient.

$$\mathrm{D} \cdot t = -\frac{1}{\beta} \cdot \ln\left(\frac{(c_{sc} - c_{rc})_t}{(c_{sc} - C_{rc})_0}\right) \tag{1}$$

Where:

D - diffusion coefficient (m^{2}·s^{-1})

β - diffusion cell constant (m^{-2})

c_{sc} - concentration in source cell (g·dm^{-3})

c_{rc} - concentration in receiving cell (g·dm^{-3})

2.5. Atomic Force Microscopy

Atomic force microscopy (AFM) is a method belonging to the group of tunnelling microscopy, which is standardly used to image samples at the nano (micro) level [7]. The scanning principle is based on the measurement of attractive and repulsive forces that are applied between the measuring tip and the sample. In practice, atomic force microscopy is mainly used for imaging the surfaces of samples in aqueous environment, as well as in air. Another possible use is the measurement of the mechanical and adhesive properties of the samples. As a part of this project, this device will be used to investigate the structure of the layered material and at the same time to measure the mechanical and adhesion properties of the material.

3. RESULTS AND DISCUSSION

Subsequently, another layer of non-woven fabric was covered with a dye of a lower concentration. Next, the procedure was repeated up to the layers with the lowest dye concentration (**Figure 1**).

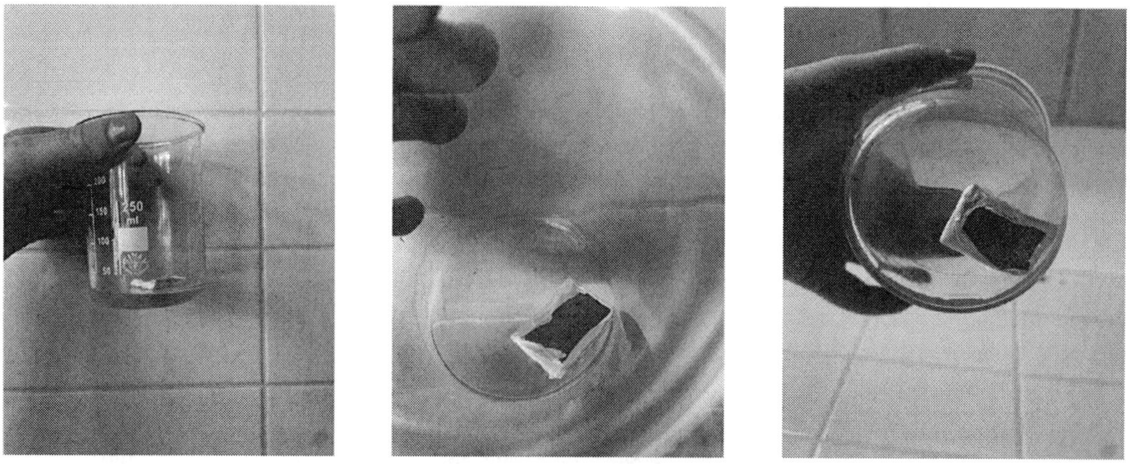

Figure 1 Layered material of a non-woven polypropylene fabric with gradient of concentration of indigotin

3.1. Release of the indigotin dye

During the next 96 hours, the release of the indigotin dye from the layered material into water environment was observed. In the first 25 minutes, a larger amount of dye as observed to be released from the lower layers of the material (**Figure 2**), this is so called burst effect.

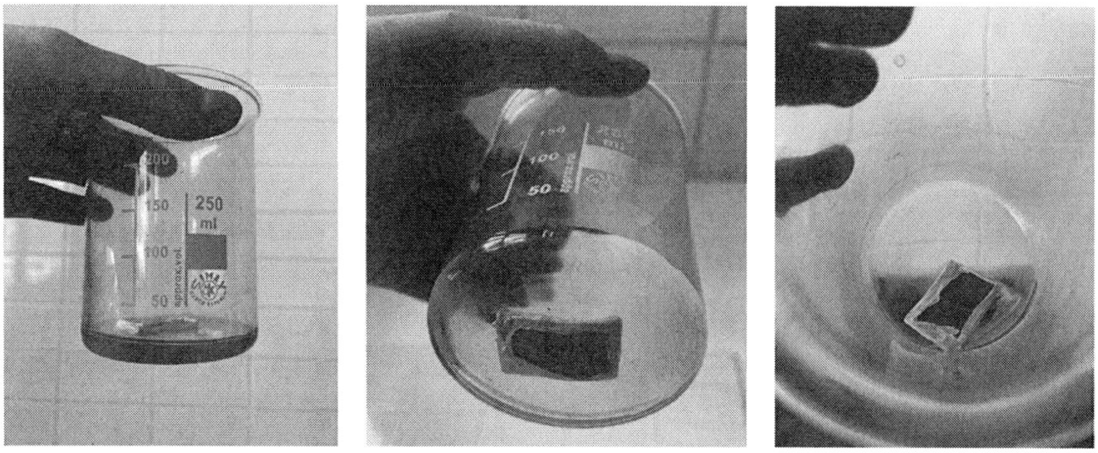

Figure 2 Released indigotin dye from layered material after 25 minutes

Subsequently, the water was replaced with fresh water and the experiment continued. Further release of the dye continued at time intervals. Release of the dye was not as concentrated a in the beginning and each subsequent release was of approximately the same concentration every time (**Figure 3**).

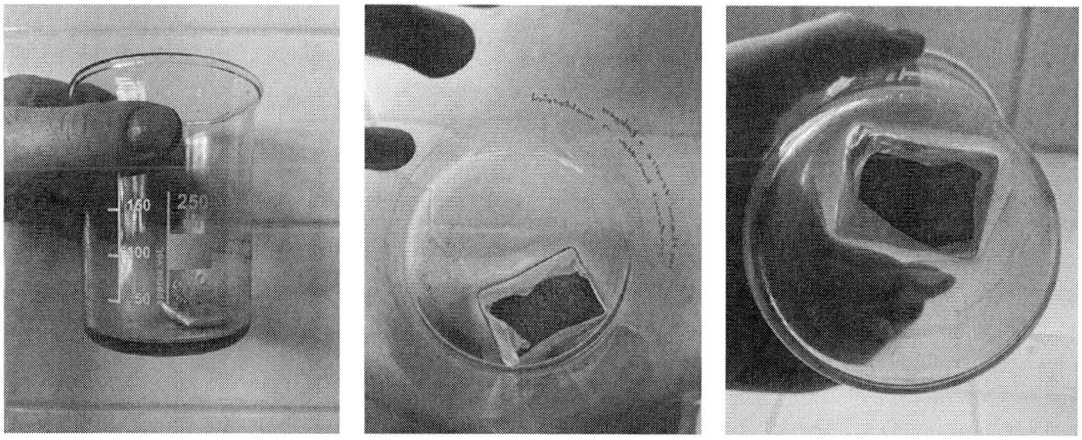

Figure 3 Released indigotin dye from layered material after 24 hours

The total dye release time from the layered non-woven fabric was 96 hours. During this process, the dye was first released into the water from the lowest layers, where the concentration was highest. After a longer period of time, water penetrated into the higher layer of the material, thereby starting the diffusion of the dye through the layers of polypropylene non-woven fabric and chitosan into the lower layers. This caused an equalization of the lower layers of the material, thereby still releasing approximately the same concentration of dye.

During the measurement of the release from the layered material with dye gradient, it was found that after a short time, a larger amount of dye is released from the material. This was caused by the release of the dye from the bottom layer that was not coated with the gelatin layer. The subsequent release was slowed down because water had to penetrate the material first. Only then did diffusion through the material, the dye had to pass through the layers of the non-woven fabric, but the diffusion was influenced the most by the presence of the gelatin, which had much smaller pores.

3.2. UV-VIS spectrophotometry

The measurements revealed not only the diffusion coefficient (D) but also the concentration of methylene blue (c_m), which remained retained in the layered material (**Table 1**).

Table 1 Diffusion coefficient and concentration of methylene blue retained in layered material

number of layers	D ($m^2 \cdot s^{-1}$)	c_m ($g \cdot dm^{-3}$)
1	$6.9 \cdot 10^{-10}$	$2.6 \cdot 10^{-3}$
2	$5.9 \cdot 10^{-10}$	$1.3 \cdot 10^{-3}$
8	$9.9 \cdot 10^{-11}$	$1.2 \cdot 10^{-3}$

It was found that the diffusion coefficient decreases linearly with increasing number of layers. During the calculation of the concentration of methylene blue that remained in the layered material, it is possible to observe that only one layer of polypropylene non-woven fabric captured a larger amount of the used dye. The other two material used between the layers contained gelatin, the retention results were very similar. With the increasing number of layers, it is possible to observe that the concentration of the capture dye in the material decreases and, at the same time, the release time also increases.

4. CONCLUSION

This article deals with the solution of the problem of controlled release of active substances from the layered material of the non-woven polypropylene alternating with gelatin. So far, the release of active substances if mainly controlled by first-order kinetics, i.e. it is dependent on the concentration of the active substance from the material. There is above all a big problem in the so-called burst effect, during which a certain amount of active substance is quickly released from the material. The problem is solved here by releasing the indigotin dye from the layered material into the water environment. Furthermore, the penetration of methylene blue through the prepared materials is measured by diffusion cells and UV-VIS spectrometer. During the measurement, it was concluded that the so-called burst effect can still be observed in the material, but the gelatin in the material can slow down the release enough to last several days. Thanks to measurement on diffusion cells, it was found that the diffusion coefficient decreases with an increasing number of layers. At the same time, we can also observe that there is a absorption of the dye into the layered material. The result show that by creating a layered material with a sufficient number of individual layers, the release time of the active substance will be extended and at the same time the concentration retained in the material will be reduced. The advantage of this method will be material with the active substance that could be applied repeatedly, or two active substances at the same could be incorporated into it with gradual excretion.

ACKNOWLEDGEMENTS

This research was conducted under the project TM03000061 with financial support from the Technology Agency of the Czech Republic.

REFERENCES

[1] HUANG, Xiao a Christopher S BRAZEL. On the importance and mechanisms of burst release in matrix-controlled drug delivery systems. *Journal of Controlled Release*. 2001, vol. 73, no. 2-3, pp. 121-136. ISSN 01683659. Available from: https://doi.org/10.1016/S0168-3659(01)00248-6.

[2] KONG-YIN, Zhao, Chen TIAN, Li DONG-YING, Zhang XIN-XIN, Song HUAN-YU a Wei JUN-FU. Preparation and characterization of mesoporous calcium silicate grafted polypropylene non-woven fabric. *Materials Letters*. 2015, vol. 141, pp. 110-113. ISSN 0167577X. Available from: https://doi.org/10.1016/j.matlet.2014.11.055.

[3] BIRK, David E. a Peter BRUCKNER. Collagen Suprastructures. *Collagen*. Berlin, Heidelberg: Springer Berlin Heidelberg, 2005, 2005-4-12, pp. 185-205. Topics in Current Chemistry. ISBN 978-3-540-23272-8. Available from: https://doi.org/10.1007/b103823.

[4] GÓMEZ-GUILLÉN, M.C., B. GIMÉNEZ, M.E. LÓPEZ-CABALLERO a M.P. MONTERO. Functional and bioactive properties of collagen and gelatin from alternative sources: A review. *Food Hydrocolloids*. 2011, vol. 25, no. 8, pp. 1813-1827. ISSN 0268005X. Available from: https://doi.org/10.1016/j.foodhyd.2011.02.007.

[5] SEDLÁČEK, Petr, Jiří SMILEK a Martina KLUČÁKOVÁ. How the interactions with humic acids affect the mobility of ionic dyes in hydrogels - Results from diffusion cells. *Reactive and Functional Polymers*. [online]. 2013, vol. 73, no. 11, pp. 1500-1509. [Accessed 2022-09-29]. ISSN 13815148. Available from: https://doi.org/10.1016/j.reactfunctpolym.2013.07.008

[6] SMILEK, Jiri, Petr SEDLACEK, Marcela LASTUVKOVA, Michal KALINA a Martina KLUCAKOVA. Transport of Organic Compounds Through Porous Systems Containing Humic Acids. *Bulletin of Environmental Contamination and Toxicology*. [online]. 2017, vol. 98, no. 3, pp. 373-377 [Accessed 2022-09-29]. ISSN 0007-4861. Available from: https://doi.org/10.1007/s00128-016-1926-0.

[7] How AFM Works. *Park Systems* [online]. Korea: Park Systems, 2020.

PREPARATION AND CHARACTERIZATION OF NANOFIBROUS MATERIALS WITH A SHISH-KEBAB STRUCTURE

Kristýna HAVLÍČKOVÁ, Maxim LISNENKO, Šárka HAUZEROVÁ, David LUKÁŠ, Věra JENČOVÁ, Eva KUŽELOVÁ KOŠŤÁKOVÁ

Technical University of Liberec, Faculty of Science, Humanities and Education, Department of Chemistry, Liberec, Czech Republic, EU, kristyna.havlickova@tul.cz

https://doi.org/10.37904/nanocon.2022.4604

Abstract

Biodegradable nanofiber materials are widely used for biomedical applications such as tissue engineering. These materials are characterized by submicron fibre diameter, small pore size, and large specific surface area (surface to volume ratio), and this structure is very similar to the natural extracellular matrix and is a favorable environment for the growth of eukaryotic cells. These days research focuses on other modifications of nanofiber structures for upgrades these structures or to create specific properties e.g. antimicrobial character. Fibre morphology and structure are able modified during their preparation, a suitable method for that is electrospinning technology and post-process modification. Electrospun materials with nanofiber backbone decorated with protruding periodic pattern of polymer crystals are termed nanofiber shish-kebabs. The shish-kebab structure can be formed by post-process recrystallization in suitable liquids (dilute polymer solution, partial solvent, or a mixture of solvent and non-solvent, solvent or partial solvent, or dilute polymer solution evaporation. The crystal period can be controlled to be a few hundreds of nanometres. Partial enzyme-catalyzed degradation of electrospun nanofibers is another method to prepare a similar structure. In this case, amorphous parts of polymer nanofibers are preferentially degraded, and the remaining macromolecules form new structures, mainly crystal structures resembling nanofiber shish-kebab. The aim of this study was preparation, description, and characterization of mentioned structure.

Keywords: Electrospinning, shish-kebab structure, post-process modification, recrystallization, polycaprolactone, enzymatic-degradation

1. INTRODUCTION

Biocompatible polymers are widely used materials for biomedical applications. [1,2] One of the main representatives is the biodegradable polyester polycaprolactone. Mainly due to its properties (biodegradability, cytocompatibility, low glass transition temperature and a melting point of 60 °C, availability, solubility, processability, etc.), it is applied in the areas of tissue engineering in the form of various types of scaffolds. [3] Much attention is focused on nanofibrous scaffolds. Technology suitable for the preparation of nanofibrous scaffolds is electrospinning technology. [4]

Currently, research is focused on modifying the nanofibrous structure. By changing or modifying the structure of nanofibers, new properties can be achieved. For example, the creation of a more attractive environment for the restoration of damaged tissues or the antibacterial character of the material and many others. [5] Modification of fibre structure can be achieved in many ways. [6,7] One of them is post-process recrystallization in a suitable liquid (dilute polymer solution, partial solvent, or a mixture of solvent and non-solvent, solvent or partial solvent, or dilute polymer solution evaporation) Electrostatically spun nanofibers, decorated with periodically protruding patterns of polymer crystals they are called shish-kebab nanofibers. The shish-kebab structure, respectively the distance of the crystalline phases (kebabs) can be controlled. [8,9] Another way to create a shish-kebab structure is the application of partial enzymatic degradation. Short-term exposure of nanofibrous polyester materials to enzymatic action causes hydrolytic cleavage of polymer chains and the restructuring of fibres. [8]

2. MATERIAL AND METHODS

2.1. Materials

Polycaprolactone PCL 80 (Mn 80 kDa; Merck, Germany) was used for the preparation of nanofibrous layers. A mixed solvent was used consisting of chloroform and dimethylformamide (Penta, Czech Republic) in a ratio of 6:4 by weight. For post-process modification of nanofibrous materials was used a solution of Polycaprolactone PCL 45 (Mn 45 kDa; Merck, Germany) dissolved in chloroform (Penta, Czech Republic) and distilled water. Partial enzymatic degradation was performed by Lipase from *Pseudomonas cepacia* (specific activity 30 ≥ $U \cdot mg^{-1}$; Merck, Germany). Sodium azide (NaN_3) was obtained from Merck (Germany).

2.2. Electrospinning

The polymer PCL 80 was added to the mixed solvent and stirred at room temperature until the polymer was completely dissolved to generate 12 wt% solutions. The Nanofibrous layer was prepared using electrospinning technology (Nanospider™ 1WS500U, Elmarco, Czech Republic). The prepared polymer solutions were electrospun at a humidity of 40 % and temperature of 22 °C. The applied voltage was -10 kV (collector) and +30 kV (spinning string). The electrode distance was 180 mm, a carriage was 1 s, and the rewinding speed was 10 mm/min. The diameter of the string was 0.4 mm, and the size of the hole was 0.7 mm. After that, material was stored at room temperature until modification.

2.3. Nanofiber modification

Post-process recrystallization and partial enzymatic degradation were used for nanofibrous materials modification and preparation of the shish-kebab structure.

Post-process recrystallization method (PPR)

Nanofibrous material was cut to the samples (100x100 mm). Diluted PCL 45 solution was used for forming a shish-kebab structure on nanofibers. A mixture of good solvent and non-solvent created a suitable condition for forming a shish-kebab crystal structure. 1.5 wt% solution was prepared by dissolving polymer pellets in acetic acid as a good solvent for PCL on a magnetic stirrer at 70 °C. Distilled water as non-solvent for PCL was added in the next step. Solvents were in the ratio of 77:23 by weight. After that, the solution was cooled down at room temperature and then stored in the fridge at 4 °C until use. A prepared solution was applied to nanofibrous sheets without fixation by a wash bottle (spray). After that, were modified materials were placed on silicon paper and dried at room temperature in a laboratory digester. As the solvent evaporated, the shish-kebab crystal structures formed.

Partial enzymatic degradation method (PED)

Nanofibrous material was cut into small samples (the weight of each sample was 50 ± 5 mg) and placed into 5 ml vials. Enzymatic catalyzed degradation was carried out at 37 °C for 24 hours. As a medium was used phosphate buffer solution (PBS with 0.02 % NaN_3, pH 7.4) with enzyme Lipase (0.5 $U \cdot ml^{-1}$) and non-modified material was incubated only in PBS. After degradation were materials washed twice with distilled water and dried at 25 °C in the laboratory dryer before further investigation.

2.4. Morphological analysis

Morphological analysis was performed by scanning electron microscopy (Vega Tescan 3, Tescan, Czech Republic) after sputter coating 12 nm of gold (Quorum Q50ES, Quorum Technologies, Great Britain). Samples were analyzed at an accelerating voltage of 15 kV. The-as taken images were processed by using of software Image J. The average diameter and the diameter distribution were evaluated from 300 randomly measured values from 4-5 images. Shish-kebab structure was evaluated by a measure of the periodic size of kebabs, size of kebabs, and shish diameters as defined in the schematic image in **Figure 1**. Respectively periodic size distances of kebabs were measured in 1-10 µm sections, and then divided by the number of pitches.

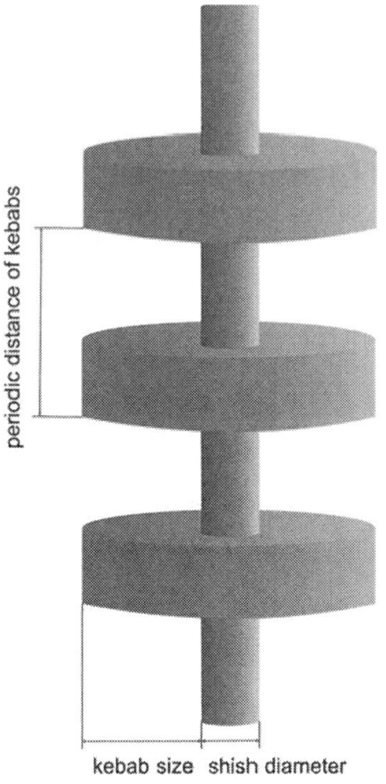

Figure 1 A schematic structure of shish-kebab

3. RESULT AND DISCUSSION

Post-process modification and creation of protruding periodic pattern of polymer crystal called shish-kebab at nanofibers were studied. Post-process recrystallization by diluted solution and partial enzymatic degradation were used as two methods of modifying PCL 80 fibres.

Figure 2 represents the results of SEM analysis for PCL 80 without modification and for PCL80 after individual modifications. It can be seen from the images that the formation of a shish-kebab structure occurs in both types of post-process treatment of nanofibers compared to the original structure of PCL80 nanofibers. However, the PPR method achieves a very regular shish-kebab structure, especially for fibres with a diameter of around 1 µm. Compared to the PPR method, the PED method does not achieve such results. The shish-kebab structure is very irregular and indented; it is observed mainly in fibres with a larger diameter. For fibres with smaller diameters, we observe fibre restructuring, defects and even complete fibre degradation.

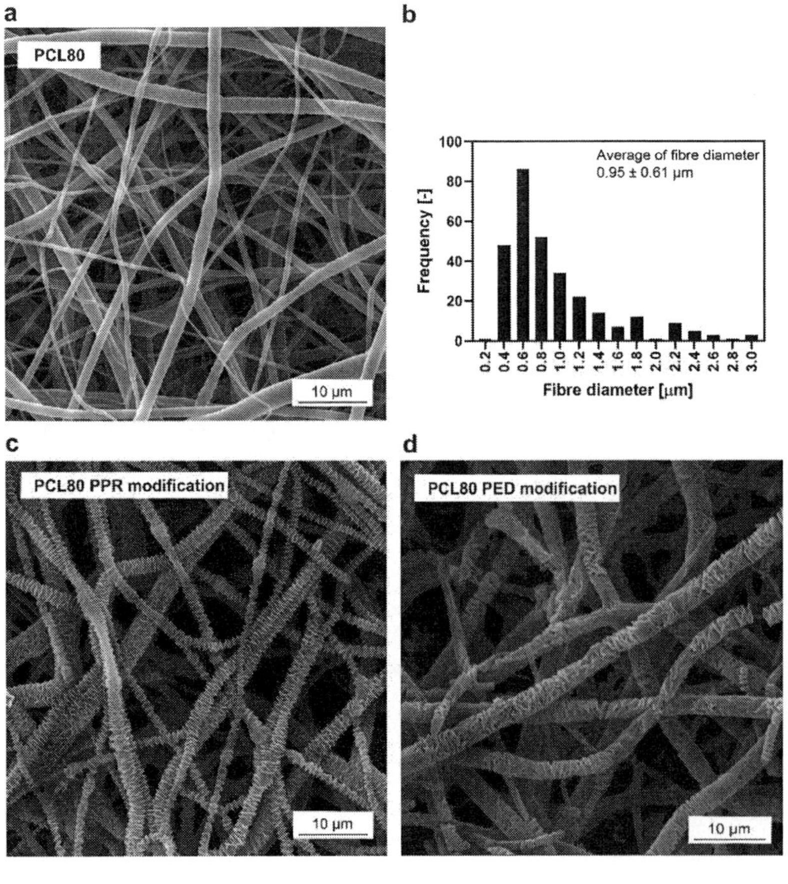

Figure 2 SEM images of PCL 80 (without modification) (a), PCL80 after post-process recrystallization by the diluted solution of PCL45 in AA/dH_2O (77:23, w:w) (c) and PCL80 after partial enzymatic degradation (24 hours) (d). Scale bare 10 □m. The image of PCL80 is appended by the histogram of fibre diameter (b).

The average periodic distance of kebabs for PCL80 after PPR modification was determined by measurement to be 0.46 ± 0.1 µm, and the height of the kebab (kebab size) was determined to be 0.29 ± 0.09 µm. The regularity of the structure observable by SEM analysis can also be supported by histograms of the analyzed characteristics (distances of individual kebabs and kebab size), as shown in **Figure 3**. The average periodic distance of kebabs for PCL80 after PED modification was determined by measurement to be 0.56 ± 0.15 µm. The height of the kebab was not due to the irregularity and fragmentation of the shish-kebab structure measured. An analysis of the diameter of the fibres in terms of the effect of degradation on the formation of the shish-kebab structure or the fragmentation and disintegration of the fibres was attached here. We observe (**Figure 4**) that a shish-kebab structure is formed for fibres of higher diameters (around 3.00 µm ± 0.89 µm), while for fibres around 1.12 ± 0.32 µm, fibre restructuring, fragmentation and fibre degradation already occur. Finer fibres (below 0.9 µm) are observed sporadically. They were probably completely eliminated due to degradation.

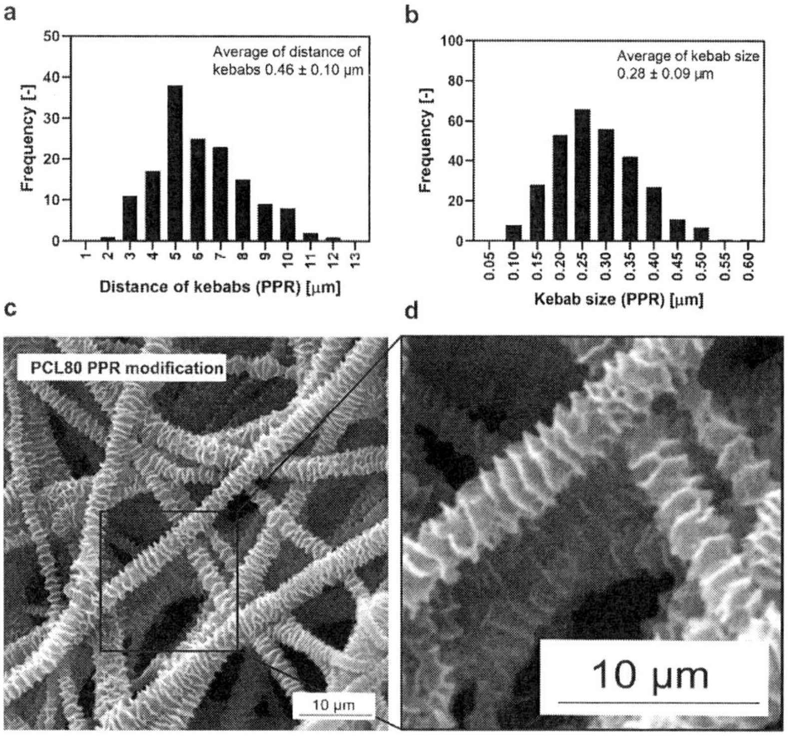

Figure 3 SEM image of the shish-kebab structure of PCL 80 nanofibres after post-process recrystallization (scale bare 10 µm) (c-d); a histogram of observed structure characteristic: a periodic distance of kebabs (a) and size of kebabs (b).

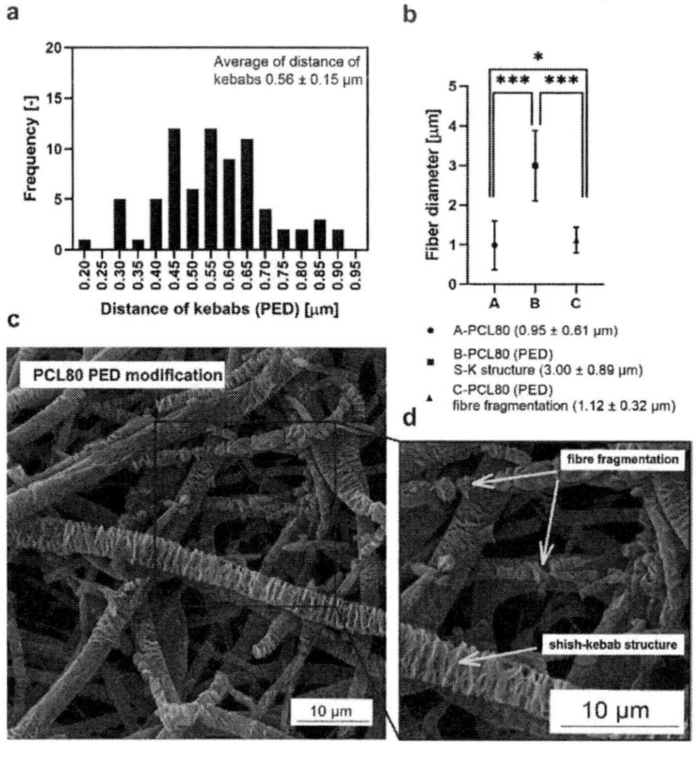

Figure 4 SEM image of the shish-kebab structure of PCL 80 nanofibres after partial enzymatic degradation (c-d). Scale bare 10 µm. A histogram of the periodic distance of kebabs (a) and graph of degradation influence to creation of shish-kebab structure or fibre fragmentation (b).

4. CONCLUSION

In the presented study, two methods of modification of PCL nanofibers were applied with the aim of creating a shish-kebab structure; the post-process recrystallization (PPR) method under the influence of a diluted solution of PCL45 in AA:dH2O (77:23, w:w) and partial enzymatic degradation (PED)) due to lipase (incubation 24 hours). Two main characteristics of the shish-kebab structure of the modified materials, the periodic distance of the kebabs and the height of the kebabs (kebab size), were analyzed. Through morphological analysis, it was observed that a more regular shish-kebab structure was achieved by the PPR method. While the PED method creates a shish-kebab structure irregularly. Probably caused by a different rearrangement of the fibre structure. In PPR, there is a gradual formulation of kebab lamellas in the nuclei of crystalline areas in the fibres. [9], [10] While with the PED method, we observe the hydrolytic cleavage of chains primarily in amorphous regions, then the crystalline phases (remaining macromolecules) form a new structure - individual kebabs on nanofibres. Fibres of finer diameters can be fragmented or even degraded due to hydrolytic cleavage, and no shish-kebab structure is formed. [8]

ACKNOWLEDGEMENTS

Authors thank for the financial support of the grant "Study of physical networks in polymer materials" of the Technical University of Liberec Grant Program „PURE“, No. PURE-2020-4007.

REFERENCES

[1] L. S. NAIR and C. T. LAURENCIN. 'Biodegradable polymers as biomaterials'. *Prog. Polym. Sci.* 2007, vol. 32, no. 8-9, pp. 762-798, Aug. 2007. Available from: https://doi.org/10.1016/j.progpolymsci.2007.05.017.

[2] CSIRO Molecular Science, Bag 10, Clayton South MDC, Vic 3169, Australia and P. Gunatillake, 'Biodegradable synthetic polymers for tissue engineering'. *Eur. Cell. Mater.* 2003, vol. 5, pp. 1-16. Available from: https://doi.org/10.22203/eCM.v005a01.

[3] D. MONDAL, M. GRIFFITH, and S. S. VENKATRAMAN. 'Polycaprolactone-based biomaterials for tissue engineering and drug delivery: Current scenario and challenges'. *Int. J. Polym. Mater. Polym. Biomater.* 2016, vol. 65, no. 5, pp. 255-265. Available from: https://doi.org/10.1080/00914037.2015.1103241.

[4] M. JANMOHAMMADI and M. S. NOURBAKHSH. 'Electrospun polycaprolactone scaffolds for tissue engineering: a review'. *Int. J. Polym. Mater. Polym. Biomater.* Jun. 2019, vol. 68, no. 9, pp. 527-539. Available from: https://doi.org/10.1080/00914037.2018.1466139.

[5] B. KASEMO and J. LAUSMAA. 'Surface properties and processes of the biomaterial-tissue interface'. *Mater. Sci. Eng. C.* 1994, vol. 1, no. 3, pp. 115-119. Available from: https://doi.org/10.1016/0928-4931(94)90041-8.

[6] S. K. NEMANI *et al.* 'Surface Modification of Polymers: Methods and Applications'. *Adv. Mater. Interfaces.* Dec. 2018, vol. 5, no. 24, p. 1801247. Available from: https://doi.org/10.1002/admi.201801247.

[7] K. PARK, Y. M. JU, J. S. SON, K.-D. AHN, and D. K. HAN. 'Surface modification of biodegradable electrospun nanofiber scaffolds and their interaction with fibroblasts'. *J. Biomater. Sci. Polym. Ed.* Jan. 2007, vol. 18, no. 4, pp. 369-382. Available from: https://doi.org/10.1163/156856207780424997.

[8] Y. DONG, S. LIAO, M. NGIAM, C. K. CHAN, and S. RAMAKRISHNA. 'Degradation Behaviors of Electrospun Resorbable Polyester Nanofibers'. *Tissue Eng. Part B Rev.* 2009, vol. 15, no. 3, pp. 333-351. Available from: https://doi.org/10.1089/ten.teb.2008.0619.

[9] X. WANG *et al.* 'A Prerequisite of the Poly(ε-Caprolactone) Self-Induced Nanohybrid Shish-Kebab Structure Formation: An Ordered Crystal Lamellae Orientation Morphology of Fibers'. *Macromol. Chem. Phys.* 2017, vol. 218, no. 24, p. 1700414. Available from: https://doi.org/10.1002/macp.201700414.

[10] H. MURASE, Y. OHTA, and T. HASHIMOTO. 'A New Scenario of Shish-Kebab Formation from Homogeneous Solutions of Entangled Polymers: Visualization of Structure Evolution along the Fiber Spinning Line'. *Macromolecules.* 2011, vol. 44, no. 18, pp. 7335-7350. Available from: https://doi.org/10.1021/ma2008817.

EFFECT OF HUMIDITY DURING ELECTROSPINNING OF BIODEGRADABLE POLYESTER NANOFIBERS

[1]Eva KUZELOVA KOSTAKOVA, [1]Vojtech HOMUTA, [1]Vera JENCOVA, [2]William J. A. HOMER, [1]Lubos BEHALEK, [1]Kristyna HAVLICKOVA, [1]Sarka HAUZEROVA, [1]Maxim LISNENKO, [1]Jan KOVACICIN, [1]Vít NOVOTNY, [1]Jana MULLEROVA, [1]David LUKAS

[1]Technical University of Liberec, Liberec, Czech Republic, EU, eva.kostakova@tul.cz

[2]Aston University, Engineering for Health Research Group, Birmingham, UK

https://doi.org/10.37904/nanocon.2022.4606

Abstract

Among widely electrospun biodegradable polymer materials are aliphatic polyesters: polycaprolactone, polylactide acid or a copolymer of caprolactone and lactic acid (PLCL). Many different modifications of electrospinning can be used to change the fiber structure and, therefore, to set optimal, especially morphological, material effects for application, testing, or treatment with these means. Structural or shape changes include not only fiber diameters, their distribution, but also fiber orientation, their interconnection, their smooth or rough surface, and also the significant arrangement of the internal networks of polymer chains represented, for example, by the degree of crystallinity. The process and material parameters of the production of nanofibrous materials must still be studied to ensure the resulting properties of the materials lead to predictable effects during its application. A general understanding of the effect of five different solvent systems and especially the effect of air humidity (from 20 to 50 % RH) during the needleless electrospinning process on the formation of nanofibrous materials from biodegradable polyesters copolymer (PLCL) is evaluated concerning the actual spinning process and the resulting morphology assessed by SEM. The internal chemical structure or arrangement evaluated using especially DSC, FTIR and GPC is necessary to ensure the reproducibility of the production of materials and thus their application results.

Keywords: Electrospinning, biodegradable nanofibers, ambient conditions, air humidity

1. INTRODUCTION

Nanofibrous materials are widely used in medical applications nowadays and are currently produced mainly by electrospinning [1]. There is a whole range of modifications of electrospinning technology leading to different arrangements of fibers, their shape, and especially surface changes. Morphology change is an essential factor for biomimetic mimicry of the intercellular mass. Mastering the possibilities of structural changes in nano and microfiber materials is important for the repeatability of production and predicting the resulting properties of tissue engineering scaffolds or wound dressings.

The PLCL copolymer belongs, like its homopolymers, to the group of biodegradable aliphatic polyesters. PLCL is characterized by its unique mechanical properties - relatively high flexibility compared to PLA homopolymer due to copolymerization with the much more flexible polycaprolactone. PCL, which has a very low glass transition temperature [2]. Of course, the mechanical properties and also biodegradation development will vary significantly depending on the chosen ratio of the monomers used. It was found that for statistical PLCL, crystallization occurs only in blocks of lactide chain segments, as they are long enough [3]. The glass transition temperatures T_g and melting temperatures T_m increase proportionally with a higher proportion of lactide units up to values similar to PLA alone. For comonomer ratio of 70:30 (L-lactide:ε-caprolactone), which is used in the study, T_g and T_m were around 16-35 °C and 137-158 °C, respectively.

In the research studies devoted to nanofibrous electrospun materials [4-6], the most used PLCL is Purac PLC 7015 with molar ratio L-lactide to ε-caprolactone 70:30 reported by the manufacture Corbion [7]. For the repeatable production of nanofibers using electrospinning technology, it is necessary to obtain a basic polymer source with repeatable properties, but it was found in [6] that the material may have different properties than those defined by the manufacturer (according to ^{1}H NMR, a real molar ratio of 82:18). The type of copolymer is still in the research stage, but it is intended to be used mainly in tissue engineering as biodegradable scaffolds [6], for example, the absorbable tubes for guiding nerves [8], rotator cuff augmentation [9]; the production of vascular bandages [4] or bypasses [6]. Nanofibrous PLCL materials for medical applications are electrospun from polymer solutions with different solvent systems, for example: mixture of chloroform and dimethylformamide [5] or mixture of chloroform, ethanol and acetic acid [6]. Both needle [5] and needle-less [4,6] direct current electrospinning methods and even alternating current electrospinning method [4] can be used for production of PLCL nanofibers. But PLCL is still an understudied material from the point of view of electrospinning and the connection of process and material parameters. This study demonstrates the influence of the used solvent systems and especially of air humidity on the resulting structure and morphology.

2. MATERIALS AND METHODS

A copolymer of L-lactide and ε-caprolactone in a 70/30 (LA 67-73 %; CL 27-33 %.) monomer molar ratio (Purasorb® PLC 7015, Corbion [7]) was used here. The supporting materials was spunbond Pegatex S (PFNonwoven) nonwoven fabric with an area weight of 20 gm^{-2}. The solvent dimethylformamide (DMF) was provided by VWR and formic acid (FA), 99.8% glacial acetic acid (AA), 96% ethanol (ETOH), acetone (Ac), dichlormethane (DCM) and chloroform (CF) were supplied by Penta. Five different solvent systems were used for electrospinning: (i) Ac/AA/FA 1:1:1 w/w (1:1:1); (ii) CF/ETOH/AA 8:1:1 w/w (8:1:1); (iii) CF/ETOH 8:2 w/w (8:2); (iv) CF/DMF 9:1 w/w (9:1); (v) DCM/DMF 7:3 v/v (7:3). Solutions of 10 wt% of PLCL were always freshly prepared immediately before electrospinning.

The electrospinning took place on NS 1S500U & NSAC150 Nanospider™ production line and precisely controlled air conditioning unit (Elmarco). The flow of incoming and outgoing air from the spinning chamber was set to 90 m^3h^{-1} and 110 m^3h^{-1}, respectively. The electrodes were 180 mm apart at an electric potential difference of 50 kV (40 kV on the spinning electrode in the form of a static wire with 0.2mm in diameter and -10 kV on the collector in the form of static wire too) and the withdrawal speed was 30 mm/min. Electrospinning was carried out at 20%; 30%; 40% and 50% (±2%) RH. The velocity of the solution carriage was set to 1 s (approximately 500 mm/s). Metal insert orifice 0.7 mm in diameter was used for control of polymer solution layer on the spinning electrode. All spinning was carried out at a temperature of 22 ± 1 °C. The final nanofibrous layers were removed from the supporting fabric and prepared for further analysis.

Basic characterization of electrospun materials is analysis of the fiber diameters, which represents the morphology and structure. An electrically conductive layer of gold approximately 10 nm thick (Quorum Q150R Plus) was deposited on all samples, and 5 images were taken for each sample at appropriate magnification using a scanning electron microscope (SEM; Vega S3B, Tescan). Fiber diameters were measured by FIJI ImageJ. FTIR analysis (Nicolet iZ10, Thermo Fisher Scientific, ATR technique on a diamond crystal) determined chemical composition and microstructure after the dissolution and electrospinning of the copolymer. GPC analysis (Dionex Ultimate 3000 HPLC) studied the molecular weight of the chains in the fibers and granules. GPC was performed in triplicate dissolved in chloroform and tetrahydrofuran. The mobile phase of tetrahydrofuran flowed through the Phenomenex Phenogel 1E5 column at a temperature of 30 °C at a flow rate of 1 ml/min. Detection of molecular weights was performed using UV scattering. The result of the DSC analysis (Differential Sensing Calorimeter DSC 1/700, Mettler Toledo) is the enthalpies of the individual samples, which makes it possible to compare the arrangement of the copolymer chains. However, due to the absence of an enthalpy value for 100% crystallinity of PLCL, it was not possible to convert the enthalpies to the percentage crystallinity of the samples. Triplets of samples in a weight of 4-6 mg were first left at -20 °C for two minutes and then heated at 10 °C/min to 160 °C. Nuclear magnetic resonance (NMR) analysis was used to measure the actual ratio of monomer units in PLCL copolymer of three different PLCL production batches used during electrospinning tests. Bruker Advance NMR Spectrometer using a high-resolution dual (^1H, ^{13}C) gradients probe was used. Whole monomer signal was calculated according [10].

3. RESULTS AND DISCUSSION

Fiber diameter differences and different complete structure of produced materials can be measured and observed from SEM images, see examples in **Figure 1**. Material produced from solvent system DCM/DMF 7:3(v/v) was not possible to separate from supporting nonwoven, thus it cannot be analysed completely.

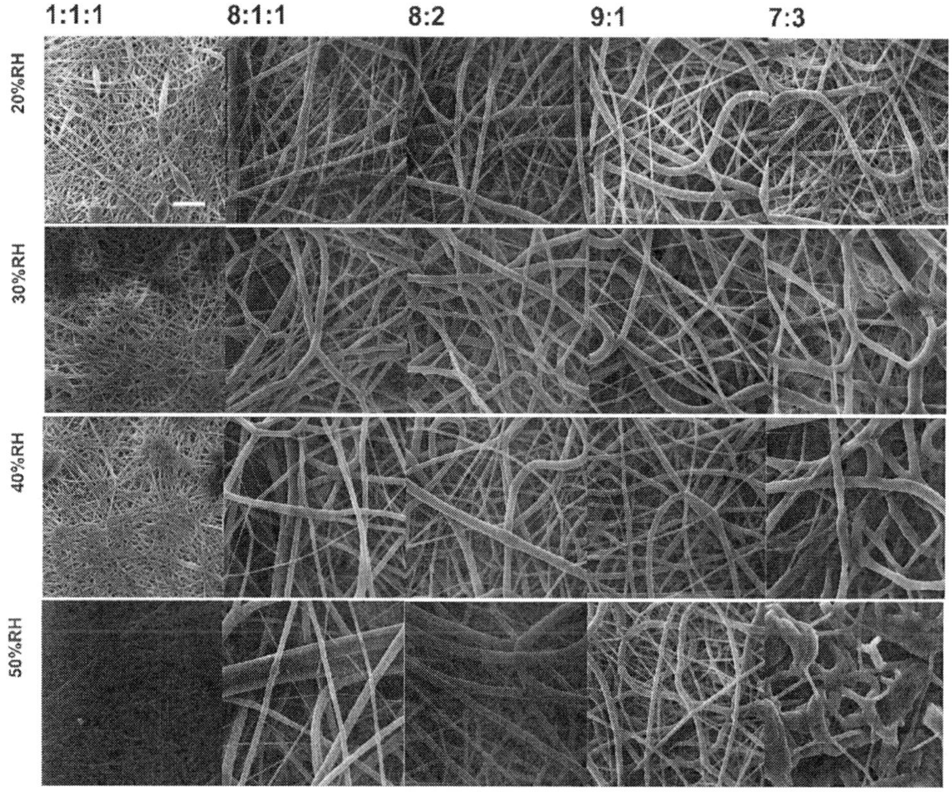

Figure 1 Scanning electrone microscope example images of fibrous materials electrospun from five different solvent systems at four different ambient air humidities. The scale bar represents 10 micrometers.

The complete comparison of average fiber diameters is introduced in **Figure 2a** and of areal weights is shown in **Figure 2b**. FTIR, DSC and GPC analysis was done only for nanofibrous material electrospun under the limit ambient humidity thus at 20 and 50% RH. FTIR analysis of electrospun samples and granules disproved the assumption that the dissolution of the copolymer in individual solvent systems and the subsequent electrospinning of the polymer solutions at different air humidities could significantly affect the internal structure or chemical composition. The results in **Figure 3** show that the chemical composition of the fibers electrospun at 20% and 50% RH was not significantly different from the PLCL granules. If the PLCL copolymer were to degrade in individual solvent mixtures or during electrospinning process, the molecular weight measured by GPC would decrease. Materials spun from the 8:2, 9:1 and 7:3 solvent systems had an average molecular weight M_w approximately the same as the PLCL granules at both relative humidities (**Figure 4a)**. Thus the PLCL chains were not degraded in the given solvents. However, the molecular weight of the PLCL copolymer spun from mixtures 1:1:1 and 8:1:1 decreased significantly (**Figure 4b**).

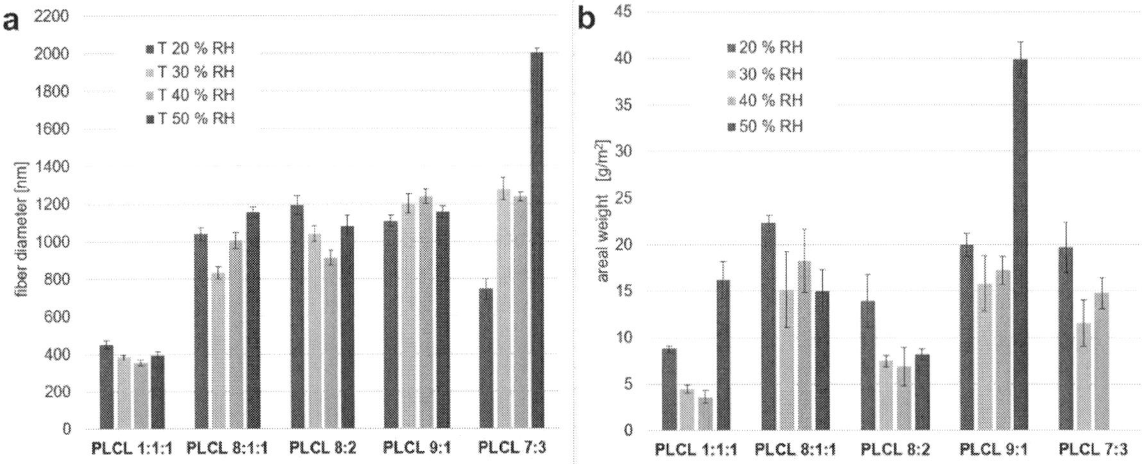

Figure 2 Average fiber diameters (a) and areal weights (b) of PLCL nanofibrous layers electrospun from different solvent systems at different ambient humidity

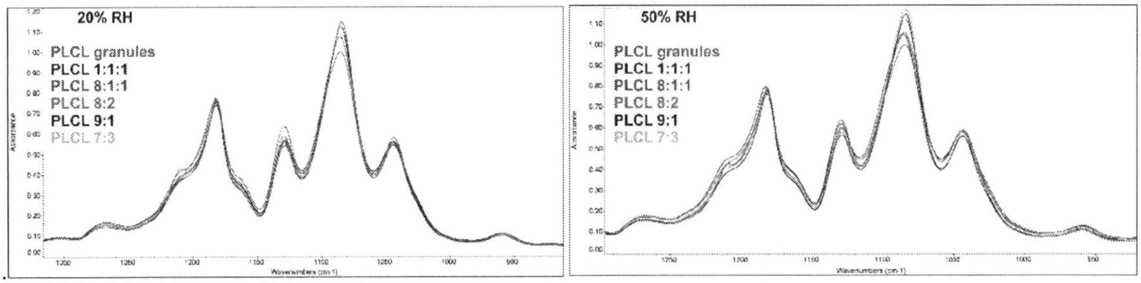

Figure 3 FTIR spectra of electrospun PLCL nanofibrous materials from different solvent systems

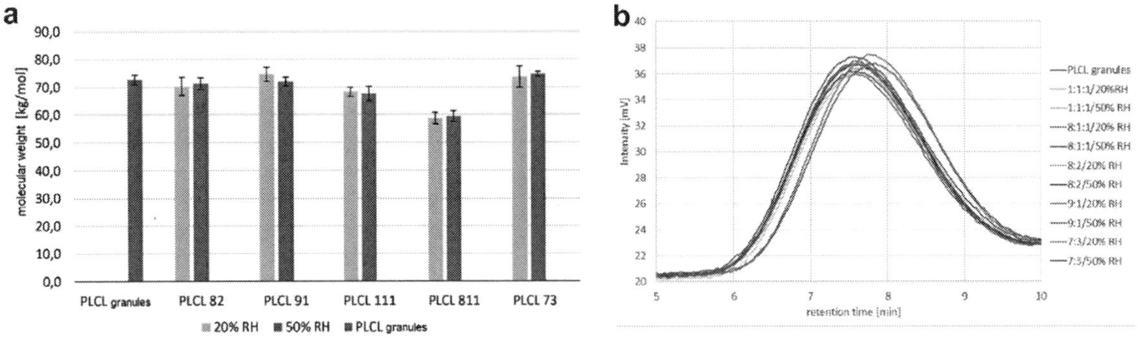

Figure 4 Average molecular weights of PLCL granules and electrospun materials at 50 and 20% RH measured by GPC (a); GPC average spectra for all measured materials (b)

DSC analysis revealed that all the spun samples as well as the initial PLCL granules contain two endothermic peaks (maxima) with different enthalpies in their internal structure, see **Figure 5**. The first peak appeared for all samples around 55 °C and the second around 96 °C. For the PLCL granules, the first peak also formed around 55 °C, but the second maximum only around 111 °C, which is perhaps due to the more rigid chains and the thermal history from the copolymer production. From the DSC results, it could be inferred that both the PLCL samples and granules contain long enough segments of both ε-caprolactone and lactic acid to be able to crystallize. The first peak could belong to the crystalline phase of the ε-caprolactone segments and the second maximum to the crystalline sections from both types of segments, since T_m is between the melting point of PLA and PCL. The PLCL copolymer studied in this work obviously contains sufficiently long stretches of ε-caprolactone that crystallize. Thus, the copolymer is typically not statistical, but rather block. From the enthalpy values, it is possible to estimate the degree of order of the chains in the individual fibers. According the table in **Figure 5**, the most ordered chains in fibers spun at 20% RH were in 1:1:1 and the least in 7:3. At 50% RH, the most ordered chains are in 8:2, followed by 9:1 and 1:1:1, the least in 8:1:1. It should be noted here that the internal ordering in electrospun nanofibers may be related to diameters. It is noteworthy that the order of the ε-caprolactone sections (1st peak) does not necessarily mean the order of the lactic acid sections (2nd peak) and vice versa, which can be seen, when comparing the enthalpies for fibers prepared at 20% RH from solvent systems 9:1 and 8:1:1 For some samples, the fibrous layer is not homogeneous in terms of the arrangement of the internal structure. The higher the relative enthalpy error, the lower the homogeneity. This may be due to the different homogeneity of the distribution of comonomers in the copolymer within one layer caused by their different reactivity in the PLCL synthesis.

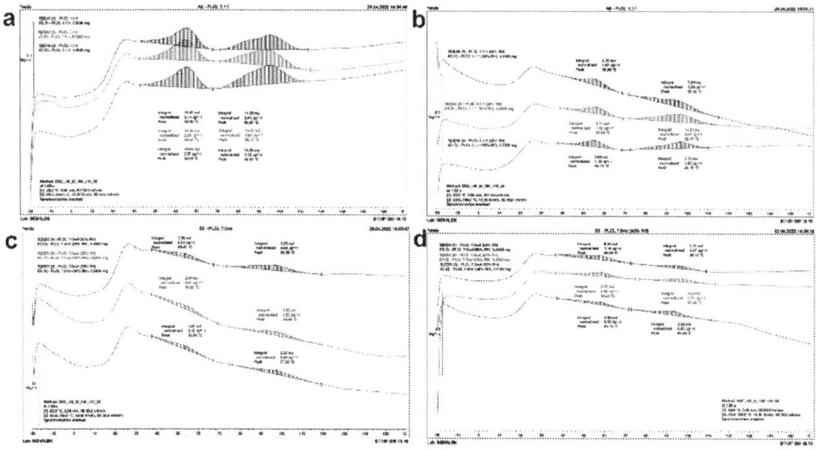

e

	Enthalpy ± SD [J/g]; 20% RH		Enthalpy ± SD [J/g]; 50% RH	
Sample marking	*1st peak*	*2nd peak*	*1st peak*	*2nd peak*
8:2	0.97 ± 0.10	1.27 ± 0.25	1.54 ± 0.17	1.37 ± 0.65
9:1	0.53 ± 0.02	1.30 ± 0.28	1.28 ± 0.21	1.98 ± 0.21
1:1:1	2.35 ± 0.17	2.75 ± 0.31	1.14 ± 0.09	1.90 ± 0.27
8:1:1	0.96 ± 0.31	1.32 ± 0.28	0.67 ± 0.08	0.88 ± 0.08
7:3	0.49 ± 0.05	0.69 ± 0.12	0.89 ± 0.21	0.73 ± 0.14
PLCL granules	1.21 ± 0.24	5.25 ± 0.31		

Figure 5 Selected examples of DSC spectra in triplet measurements for 1:1:1/20%RH (a); 1:1:1/50%RH (b); 7:3/20%RH (c); 7:3/50%RH (d) and table of enthalpies for all electrospun samples and PLCL granules

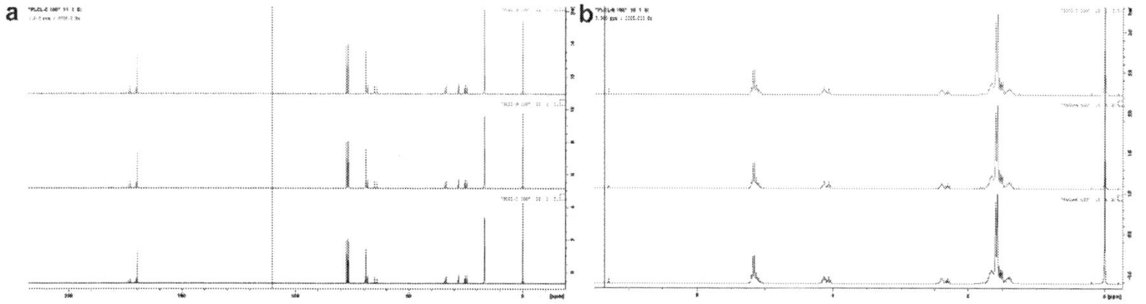

Figure 6 H NMR spectra (a) and C NMR spectra (b) for PLCL granules from three different batches

Side by side comparison of the three samples of PLCL by H NMR (**Figure 6a**) shows near identical spectral for all three materials, indicating a high level of consistency between production runs (batches). These result in a calculated monomer percentage for PLCL granules sample of 69.43 % L-lactide and 30.57 % caprolactone, within manufacturer specified ranges. C NMR (**Figure 6b**) of the 3 granules samples, which have consistent spectra, with the only small variation around chemical shift of 78 ppm, attributed to the chloroform solvent.

4. CONCLUSION

The electrospinning of PLCL is very significantly influenced by the used solvent system and the ambient air humidity. It is obvious that the choice of solvent system, while maintaining constant process parameters, can have an effect on all fiber diameters; the productivity and internal structure of the resulting nanofib. The study of the influence of air humidity clearly indicates that some polymer solutions (especially 7:3) are more sensitive to its changes and some less so (for example 1:1:1). FTIR, DSC and GPC analyzes pointed to very interesting facts that need to be followed up in the development of electrospun materials from PLCL. The rate of degradation in solutions containing carboxylic acids needs to be further studied, but it was clearly demonstrated by GPC analysis that PLCL chains degrade in these solutions. DSC analysis suggests a block copolymer arrangement, which is probably due, among other things, to the different reactivity of the incoming comonomers during production. NMR analysis demonstrated the repeatability of the properties of PLCL Purasorb 7015 input granulate. The PLCL is interesting material mainly in medical applications in the form of electrospun nanofibers. Its use in this area is wide ranging from vascular bandages and replacements to wound coverings and scaffolds for the regeneration of many tissues. The internal polymer structure in the nanofibers, their morphology, representation of defects, porosity and other parameters can significantly influence the resulting unique properties for PLCL, especially mechanical properties, wettability, biocompatibility and biodegradability. These properties give sufficient reason for this copolymer to be further investigated.

ACKNOWLEDGEMENTS

This work was supported by the Czech Health Research Council, project No NU20-02-00368 and the grant project of the "PURE Technical University of Liberec Grant Program", Ref. PURE-2020-4007.

REFERENCES

[1] RAHMATI, M., MILLS, D.K., URBANSKA, A.M., SAEB, M.R., VENUGOPAL, J.R., RAMAKRISHNA, S., MOZAFARI, M. Electrospinning for Tissue Engineering Applications. *Progress in Materials Science.* 2021, vol. 117, 100721, Available from: https://doi.org/10.1016/j.pmatsci.2020.100721.

[2] WOODRUFF, M.A., HUTMACHER, D.W. The Return of a Forgotten Polymer-Polycaprolactone in the 21st Century. *Progress in Polymer Science.* 2010, vol. 35, pp. 1217-1256. Available from: https://doi.org/10.1016/j.progpolymsci.2010.04.002.

[3] FERNÁNDEZ, J., ETXEBERRIA, A., SARASUA, J.-R. Synthesis, Structure and Properties of Poly(L-Lactide-Co--Caprolactone) Statistical Copolymers. *Journal of the Mechanical Behavior of Biomedical Materials.* 2012, vol. 9, pp. 100-112. Available from: https://doi.org/10.1016/j.jmbbm.2012.01.003.

[4] ASATIANI, N., VALTERA, J., LISNENKO, M., HAVLÍČKOVÁ, K., KOVAČIČIN, J., LUKÁŠ, D., KUŽELOVÁ KOŠŤÁKOVÁ, E. Evaluation of Fiber Orientation of AC and DC Electrospun PLCL Nanofibrous Layers. In *Proceedings of the Nanocon 2014, 6th International Conference.* Ostrava: Tanger Ltd, vol. 2021. ISBN 978-80-88365-00-6,

[5] LIN, G., SONG, S.N., WONG, Y.S., TAN, L.P., YONG, K.Y., JASMEN, J., DONG, J.X. Fabrication and Characterization of Electrospun Nano to Microfiber Made of Poly(L-Lactide-Co-ε-Caprolactone). *SSP.* 2012, vol. 185, pp. 122-125. Available from: https://doi.org/10.4028/www.scientific.net/SSP.185.122.

[6] MIKES, P., HORAKOVA, J., SAMAN, A., VEJSADOVA, L., TOPHAM, P., PUNYODOM, W., DUMKLANG, M., JENCOVA, V. Comparison and Characterization of Different Polyester Nano/Micro Fibres for Use in Tissue Engineering Applications. *Journal of Industrial Textiles.* 2021, vol. 50, pp. 870-890. Available from: https://doi.org/10.1177/1528083719848155.

[7] PSDS-PLC-7015.Pdf.

[8] THAPSUKHON, B., THADAVIRUL, N., SUPAPHOL, P., MEEPOWPAN, P., MOLLOY, R., PUNYODOM, W. Effects of Copolymer Microstructure on the Properties of Electrospun Poly(L -Lactide-Co-ε-Caprolactone) Absorbable Nerve Guide Tubes. *J. Appl. Polym. Sci.* 2013, vol. 130, Issue 6, pp. 4357-4366. Available from: https://doi.org/10.1002/app.39675.

[9] REY-VINOLAS, S., CASTAÑO, O., RUIZ-MACARRILLA, L., LLORENS, X., MORA, J.M., ENGEL, E., MATEOS-TIMONEDA, M.A. Development of a Novel Automatable Fabrication Method Based on Electrospinning Co Electrospraying for Rotator Cuff Augmentation Patches. *PLoS ONE.* 2019, vol. 14, e0224661. Available from: https://doi.org/ 10.1371/journal.pone.0224661.

[10] PHETSUK, S., MOLLOY, R., NALAMPANG, K., MEEPOWPAN, P., TOPHAM, P.D., TIGHE, B.J., PUNYODOM, W. Physical and Thermal Properties of L- Lactide/E-caprolactone Copolymers: The Role of Microstructural Design. *Polym Int.* 2020, vol. 69, pp. 248-256. Available from: https://doi.org/ 10.1002/pi.5940.

SURFACE MORPHOLOGY AND MECHANICAL PROPERTIES OF TITANIUM LAYERS MIXED WITH CALCIUM AND PHOSPHORUS

Petr VLCAK, Vojtech SMOLA, Josef SEPITKA, Tomas HORAZDOVSKY

Faculty of Mechanical Engineering, Czech Technical University in Prague, Czech Republic, EU

https://doi.org/10.37904/nanocon.2022.4610

Abstract

Titanium layers doped with biogenic elements attract much attention due to their ability to enhance the biological integration of implant materials. However, the addition of biogenic elements to the coating/layer changes its composition, structure, properties, and surface morphology, which affects their application possibilities. Therefore, the research and surface characterization of such layers is needed. In the present work, the mixed titanium layers with calcium and phosphorus have been synthesized by a dual electron beam evaporation method. The effect of doping in these mixed-titanium layers on their surface morphology and mechanical properties has been investigated. Atomic force microscopy was used for morphological characterization and nanoindentation was used to measure the mechanical properties (indentation hardness and reduced modulus). The surface layers are characterized in connection by tuning the deposition conditions and increasing dopant content.

Keywords: Titanium, doping, morphology, mechanical properties

1. INTRODUCTION

Titanium and its alloys stand out and are used in many engineering applications in the aerospace, chemical, and biomedical industries, due to their excellent strength-to-density ratio, biocompatibility, and high corrosion resistance in many environments [1-3]. Surface properties, not only for titanium and its alloys, depending on the structure, morphology, and chemical composition of their surface. Many methods exist to increase surface properties by changing the morphology and chemical composition of titanium. For example, advanced methods using accelerated ions, such as ion implantation, ion beam-assisted deposition, magnetron sputtering, and plasma-assisted chemical vapour deposition, can be used to improve these surface properties [4-7]. The paper is focused on the dual electron beam evaporation method. Using this method, a layer of titanium with calcium and phosphorus was deposited on commercially pure titanium grade 2. Previously, [8-9] showed an increase in the biocompatibility of titanium by adding phosphorus and calcium. Calcium phosphate and hydroxyapatite help to create chemical bonds with the human bone tissue and optimal bone integration of an implant. Implant fracture due to mechanical failure is related to biomechanical incompatibility. For this reason, biomedical implants require specific mechanical properties, such as hardness, Young´s modulus, elongation, or tensile strength, close to the mechanical properties of human bone.

This work presents the effect of the applied layer with different concentrations of the mixture of titanium, calcium, and phosphorus on the morphology and mechanical properties. Surface morphology was monitored by atomic force microscopy (AFM), and indentation hardness and reduced modulus were investigated by nanoindentation.

2. EXPERIMENTAL

In this study commercially pure titanium grade 2 (CPTi) was used. Samples 5 mm in height and 14 mm in diameter were cut from the CPTi bar and then ground with silicon carbide abrasive papers with grain sizes from 320 to 1000. The samples were mechanically polished using ChemoMet, a soft, porous, chemically resistant synthetic polishing pad, and non-crystallizing amorphous 0.02µm colloidal silica suspension MasterMet2 with a little bit of hydrogen peroxide. The mixed titanium layers with calcium and phosphorus were synthesized by a dual electron beam evaporation method. The scheme of deposition of layers with different concentrations of BTP was the same as in the previous experiment in [10], however, with a BTP target instead of a calcium target and deposition temperature of 500 °C without the assistance of oxygen. The BTP content was controlled by the deposition rate of targets. The TI 950 TriboIndenter® nanomechanical instrument with dynamical mode CMX was used for depth profiling of the deposited layers. Atomic force microscopy was used to characterize the surface morphology. AFM measurements were performed in contact mode.

3. RESULTS AND DISCUSSION

3.1. Surface morphology

The morphology of the deposited layers was measured by atomic force microscopy (AFM). **Figure 1** presents roughness measurements of BTP concentrations of 40, 60, 80, and 100 %. BTP concentrations of 40, 60, and 80 % were measured to have a roughness of around 14 nm. The lowest roughness of approximately 6 nm was achieved at a concentration of 100 % BTP. Morphological changes were also confirmed by AFM images, **Figure 2**. A change in morphology was documented for all BTP concentrations. The formation of a new similar characteristic morphological structure is typical for 40 and 60 % BTP concentrations. As the concentration increases, the structure changes, and the surface becomes smoother.

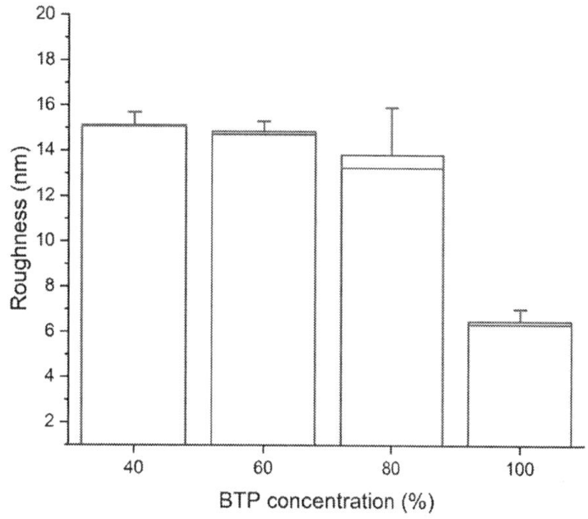

Figure 1 Average roughnesses of different BTP concentrations obtained from AFM topographic measurements

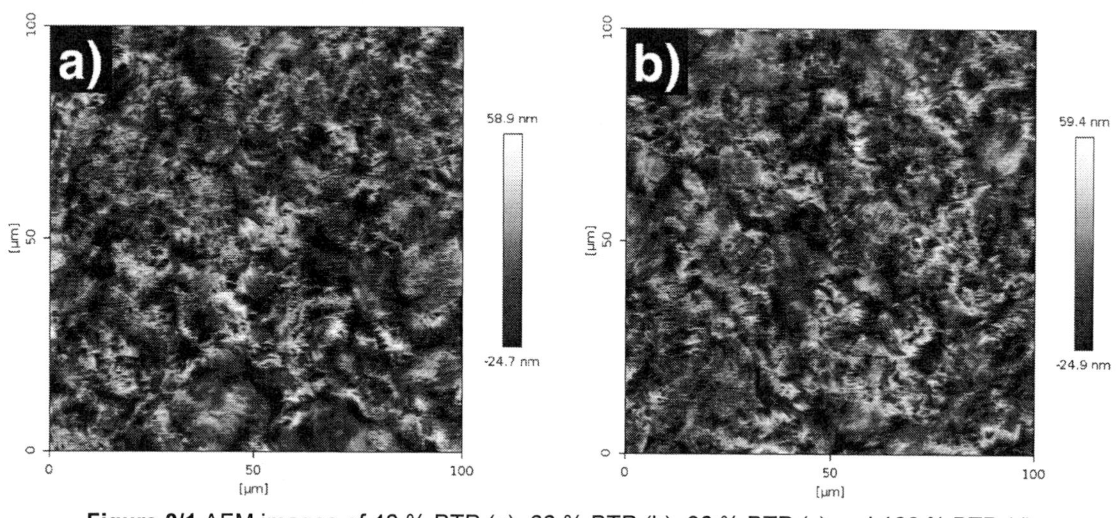

Figure 2/1 AFM images of 40 % BTP (a), 60 % BTP (b), 80 % BTP (c) and 100 % BTP (d)

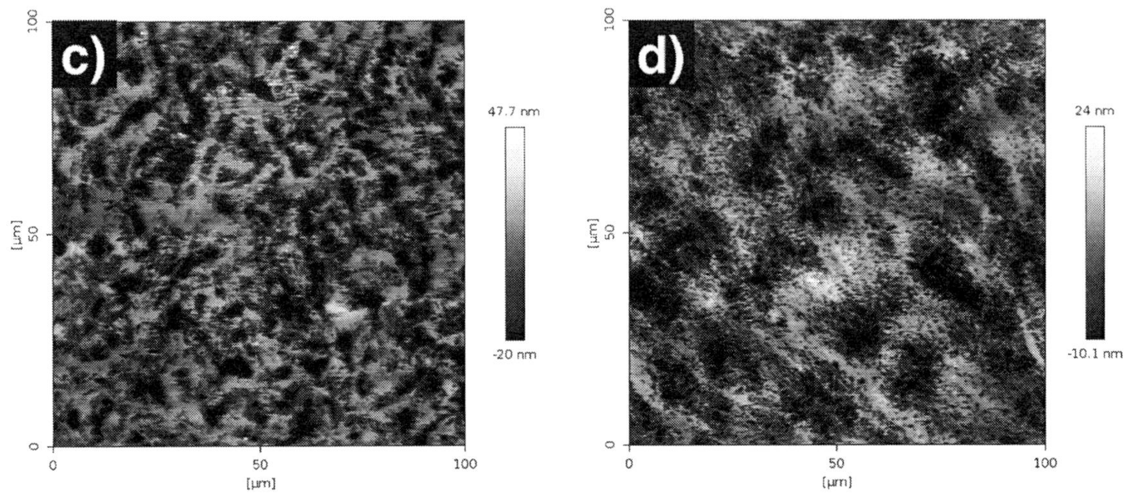

Figure 2/2 AFM images of 40 % BTP (a), 60 % BTP (b), 80 % BTP (c) and 100 % BTP (d)

3.2. Mechanical properties

We measured the mechanical properties (indentation hardness and reduced modulus) of the deposited layers using the nanoindentation method. In both cases, the concentration of 40 % BTP achieved the highest values. The reduced modulus of the 40 % BTP layer was approximately 140 GPa and the indentation hardness was roughly 10.2 GPa. The results of measurements for 60 and 100 % BTP were roughly similar. The reduced modulus was 110 GPa and the hardness was approximately 6 GPa. The 80 % BTP concentration had 105 GPa of reduced modulus and 5.5 GPa of indentation hardness.

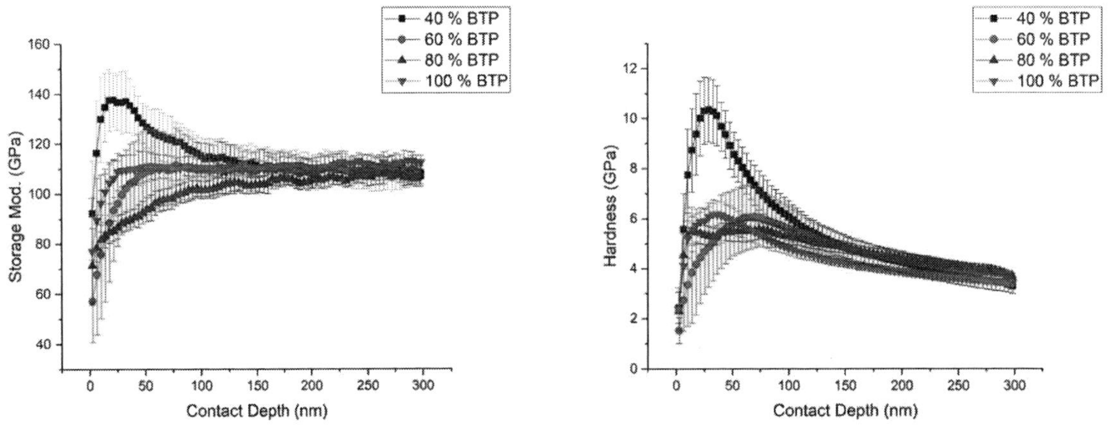

Figure 3 Average roughnesses of different BTP concentrations obtained from AFM topographic measurements

4. CONCLUSION

The influence of the BTP concentration on the surface morphology and mechanical properties has been demonstrated. The deposited mixed titanium layers with calcium and phosphorus create a new characteristic morphology that smooths out with increasing concentration. As the concentration of BTP increases, the surface roughness decreases. The lowest roughness of approximately 6 nm close to the previously polished surface has a concentration of 100 % BTP. A concentration of 40 % BTP conclusively achieves the highest values of hardness (approximately 10.2 GPa) and reduced modulus (approximately 140 GPa).

ACKNOWLEDGEMENTS

This research has been supported by AZV MZ CR project No. NV19-08-00070.

REFERENCES

[1] LEYENS, CH., PETERS, M. *Titanium and Titanium alloys - Fundamentals and Applications*, Köln: Wiley-VCH, 2006.

[2] NIINOMI, M. Recent metallic materials for biomedical applications. *Metallurgical and Materials Transactions A*. 2002, vol. 33, pp. 477-486. Available from: https://doi.org/10.1007/s11661-002-0109-2.

[3] FREESE, H.L., VOLAS, M.G., WOOD, J.R. *Metallury and Technological Properties of Titanium and Titanium alloys*. Berlin: Springer, 2001.

[4] CHODUN, R., NOWAKOWSKA-LANGIER, K., WICHER, B., OKRASA, S., MINIKAYEV, R., ZDUNEK, K. Reactive sputtering of titanium compounds using the magnetron system with a grounded cathode. *Thin Solid Films*. 2017, vol. 640, pp. 73-80.

[5] MOVASSAGH-ALANAGH, F., ABDOLLAH-ZADEH, A., ALIOFKHAZRAEI, M., ABEDI, M. Improving the wear and corrosion resistance of Ti-6Al-4V alloy by deposition of TiSiN nanocomposite coating with pulsed-DC PACVD. *Wear*. 2017, vol. 390-391, pp. 93-103.

[6] WEISS, Z., VLCAK, P., Analysis of shallow depth profiles of titanium nitride and N-implanted titanium by GD-OES: the 'hydrogen effect' after the discharge startup and a correction thereof. *Journal of Analytical Atomic Spectrometry*. 2018, vol. 32, pp. 2476-2484.

[7] VLCAK, P., JIRKA, I. Protective sliding carbon-based nanolayers prepared by argon or nitrogen ion-beam assisted deposition on Ti6Al4V alloy. 2016, Art. No. 1697090, pp. 1-9. Available from: https://doi.org/10.1155/2016/1697090.

[8] MAITZ, M.F., PHAM, M.T., MATZ, W., REUTHER, H., STEINER, G. Promoted calcium-phosphate precipitation from solution on titanium for improved biocompatibility by ion implantation. *Surface and Coatings Technology*. 2002, vol. 158-159, pp. 151-156.

[9] YU, S., Yu, Z., Wang, G., Han, J., Ma, X., DARGUSH, M. Biocompatibility and osteoconduction of active porous calcium-phosphate films on a novel Ti-3Zr-2Sn-3Mo-25Nb biomedical alloy. *Colloids and Surfaces B: Biointerfaces*. 2011, vol. 85, pp. 103-105.

[10] VLCAK, P., NEHASIL, V., KOLLER, J., SEPITKA, J., DRAHOKOUPIL, J. Calcium-doped titanium thin films prepared with the assistance of an oxygen ion beam: The effect of Ca content on microstructure, mechanical properites and adhesion. *Applied Surface Science*. 2022, vol. 573. Available from: https://doi.org/10.1016/j.apsusc.2021.151569

EFFECT OF MICRO- AND NANO-FILLERS ON THE COEFFICIENT OF THERMAL EXPANSION OF COMPOSITE SYSTEMS WITH POTENTIAL APPLICATIONS IN STOMATOLOGY

Rawaa YAHYA, Dora KROISOVÁ, Artur KNAP, Štěpánka DVOŘÁČKOVÁ

TUL - Technical University of Liberec, Liberec, Czech Republic, EU,
rawaa.yahyaovak@tul.cz, dora.kroisova@tul.cz, artur.knap@.cz, stepanka.dvorackova@tul.cz

https://doi.org/10.37904/nanocon.2022.4603

Abstract

This experimental study aims to assess the effect of selected filler types on the longitudinal thermal expansion of composite systems. The studied issue concerns the dimensional stability or durability of white dental fillings, composite systems of photo polymerizing polymer with particulate fillers of micrometer and nanometer dimensions, mainly based on silica and zirconia. ChS Epoxy 455 epoxy resin cured with P11 hardener was used as the matrix of the composite systems on which the effect of selected filler types was evaluated. The fillers were particulate and fibrous materials of different chemical compositions, sizes, and combinations. Synthetic silica nanoparticles, cellulose particles, hybrid fillers derived from modified rice husk, and cellulose nanocrystals were used as fillers. The longitudinal thermal expansion of the prepared composite systems was evaluated in the temperature range of 20 ° to 40 °C, which corresponds to the temperatures to which dental fillings are usually exposed. The selected fillers fracture surfaces of samples were analyzed on a scanning electron microscope, and the longitudinal thermal expansion was measured on a thermomechanical analyzer. Based on the dimensions and chemical compositions of fillers used, the quality of the interphase, and the arrangement of the fillers in the polymer matrix, the contribution of these reinforcing elements to the potential influence of the thermal expansion coefficient was evaluated. In view of results obtained, the following experiments will focus on the study of material combinations of particulate and fibrous fillers, the evaluation of the influence of the environment on materials, and machinability.

Keywords: Composite system, epoxy resin, micro-fillers, nano-fillers, dentistry

1. INTRODUCTION

Composite materials are currently widely used not only in technical applications but we can also find them in medical areas. A typical example is the composite materials used in dentistry as dental fillings. These fillings are a composite or nanocomposite system in which the matrix is a photopolymer, and the filler is a nanoparticle. Most often based on oxides - silicon dioxide, zirconium oxide, barium oxide, and aluminum oxide [1-4]. These fillers ensure the mechanical parameters of the filling, such as hardness, resistance to abrasion, toughness, and color, which is no less critical for these white fillings. Filtek, Ceram and Omnichrom are among the most widely used materials used as dental fillings. However, after a specific time, dental composite material ages. Even if the filling initially has good adhesion to the dentin or tooth enamel, it breaks off, and a micro gap forms between the tooth and the filling. The breaks are followed by the enlargement of the micro gap and the formation of secondary tooth decay. This process is observed over a long time and occurs approximately four to six years after applying a white filling. It is related to dental hygiene, diet, and immunity. The problem lies in the degradation of the composite system used, or instead in the shrinkage of the filling, as a result of which it will detach from the tooth [5-8].

From the theory of composite behavior, it is clear that the use of particle filler cannot provide the matrix with the same mechanical parameters as if fiber fillers were used. The use of fiber fillers in materials for dental fillings is currently not adequately studied. Some undesirable phenomena could be avoided using suitable fiber fillers. Choosing the right fiber filler is also essential, as it is a material that comes into contact with humans [7-10].

The present work aims to analyze samples based on epoxy resin filled with selected particle and fiber fillers and their combinations. In the first step, this analysis is intended to provide insight into the behavior of selected hybrid structures in the epoxy matrix. In the second step, use the chosen fillers or their mixtures, optimize the composition and proportions and transform these fillers into a photo-polymerizing material.

2. EXPERIMENT

The two-component epoxy resin ChS Epoxy 455 cured with hardener P11 (Districhem, s.r.o.) was used as a matrix of composite samples for this work. As fillers, synthetic nanoparticles of fumed amorphous silica Cab-O-Sil LM-150 (CABOT, GmbH), cellulose particles known in practice as abrasive wood pulp (University Pardubice), purified rice husks (Cersum, s.r.o.), fibers from purified rice husks were used as fillers (Cersum, s.r.o.) and cellulose nanocrystals (TUL). Combinations of these materials have also been used. A scanning electron microscope TESCAN MIRA3 (TESCAN ORSAY HOLDING, a. s.) was used to evaluate the fracture surfaces, respectively, the interphase adhesion. A thermomechanical analyzer (Pragolab, s.r.o.) was used to determine the longitudinal thermal expansion.

Due to the existing experience with dental fillings, which are composite systems with particulate fillers, the evaluation of the effect of selected fillers was designed to use particulate, fibrous, and their combination. Synthetically produced silicon dioxide in the form of nanoparticles, used in dental fillings in Filtek, Ceram, and Omnichrom products, were chosen as an inorganic particle filler. The so-called abrasive wood pulp was selected as an organic particle filler based on cellulose. Another type of filler was modified rice husks, provided by Cersum, s.r.o. These are composed of both cellulose fibers and nanoparticles of biogenic silicon dioxide, placing them in the hybrid organic-inorganic fillers group [11]. Fibers separated from treated rice husks were also used, provided by the company Cersum, s.r.o. [12] The last filler was cellulose nanocrystals, characterized by width in units of nanometers and a length in hundreds of nanometers [13].

The filler added to the base epoxy resin was chosen so that the effect of the different behavior of particle filler, fiber filler, and their mixtures, consisting of both fibers and particles, was noticeable. From a chemical point of view, a filler analogous to the typical filler used for dental fillings, i.e., nanoparticles of silicon dioxide, was used. Silica is also contained in rice husks and fibers from treated rice husks. For that reason, combinations of cellulose fibers, particles, nanofibers, and silica were designed and tested, and mixtures were prepared in which fiber and particle fillers were combined **(Table 1).**

Mixtures were prepared by mixing epoxy resin with different fillers, and their combinations were cast into silicone molds and cured. Curing of the samples took place in a pressure vessel to remove inhomogeneities and bubbles at a pressure of 0.7 MPa, for 24 hours at a temperature of 23 °± 2 °C. The prepared samples were not further cured. Before the longitudinal thermal expansion determination, the samples were surface treated according to the requirements for measurement.

3. RESULTS

The results of the experiments were assessed both from the point of view of the measured values of the longitudinal thermal expansion (10 measured samples for every material composition) and from the point of view of the interphase of the designed composite systems **(Table 1). Table 1** shows the composition of hybrid systems and the average values of longitudinal thermal expansion for individual groups of samples for the temperature interval 20 °C to 40 °C. From the fracture surfaces, it can be seen whether the particles and fibers are homogeneously distributed in the polymer matrix and, above all, the quality of the interphase adhesion, which is an essential condition for ensuring the desired behavior in composite systems.

Table 1 Composition of individual samples with the corresponding coefficient of thermal expansion α

	Polymer matrix	Inorganic filler	Organic filler	α (10^{-6}·K^{-1})
1	10 g ChS Epoxy 455	-	-	72.93
2	10 g ChS Epoxy 455	1.0 g silicon dioxide nanoparticles	-	66.45
3	10 g ChS Epoxy 455	-	1.0 g cellulose particles	77.52
4	10 g ChS Epoxy 455	-	1.0 g modified rice husks	71.77
5	10 g ChS Epoxy 455	-	1.0 g fibers from modified rice husks	69.14
6	10 g ChS Epoxy 455	0.5 g silicon dioxide nanoparticles	0.3 g fibers from modified rice husks	67.48
7	10 g ChS Epoxy 455	-	1.0 g cellulose nanocrystals	77.53
8	10 g ChS Epoxy 455	1.0 g silicon dioxide nanoparticles	1.0 g cellulose nanocrystals	72.21

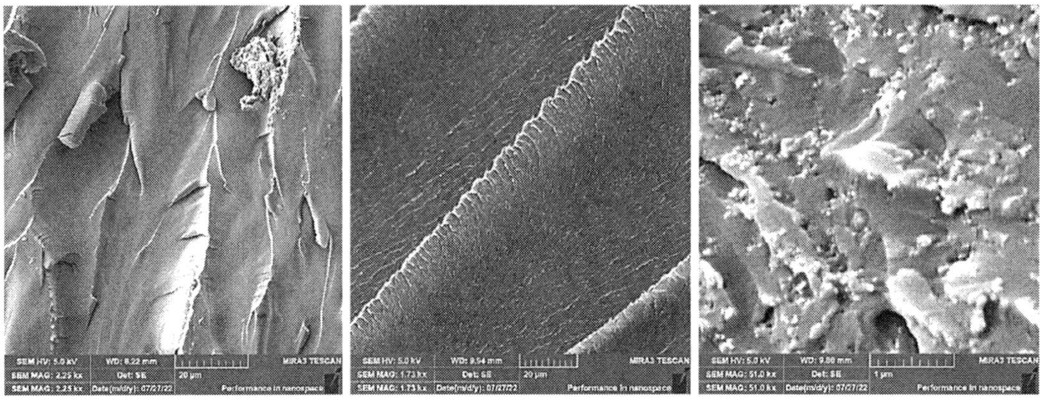

Figure 1 Epoxy resin without fillers (left), epoxy resin with silicon dioxide nanoparticles (middle), and a detailed picture with a visible distribution of silicon dioxide nanoparticles (right)

Figure 1 (left) shows the characteristic brittle fracture of epoxy resin without filler. In the other images of **Figure 1**, it is clear that there is a change after adding silicon dioxide nanoparticles, like the material's fracture. From the overview picture in **Figure 1** (middle), it can be seen that the filler is more or less evenly distributed in the epoxy matrix. When viewing **Figure 1** (right) in detail, it can be seen that the nanoparticles form larger or smaller clusters, and part of the matrix is not filled with them. Adhesion of nanoparticles can be considered reasonable. The measured values of the coefficient of longitudinal thermal expansion α are assigned to the mentioned images, which for a sample of pure epoxy resin reaches a value of $72.93 \cdot 10^{-6} \cdot K^{-1}$, while for a sample with a high content of silicon dioxide nanoparticles, the value is $66.45 \cdot 10^{-6} \cdot K^{-1}$ is the lowest average value measured in the entire series of prepared samples.

Epoxy resin filled with cellulose particles of irregular shape **(Figure 2 left)** does not show good adhesion, the fracture of the surrounding resin is brittle, and the value of the coefficient of longitudinal thermal expansion α is $77.52 \cdot 10^{-6} \cdot K^{-1}$ **(Table 1)**.

Purified rice husks **(Figure 2 middle, right)** are specific for their two-component structure **(Figure 2 right)**, where it can be seen that the outer part of the husk is made up of a few-micrometer layer of silicon dioxide nanoparticles. In contrast, the inner part is built up of organic fibers. Purified rice husks do not show sufficient adhesion in epoxy resin **(Figure 2 middle).** Moreover, these are relatively large particles with a length and width of hundreds of micrometers. As a result, a considerable amount of the epoxy matrix is not affected by their presence, and the fracture is then fragile in these polymer parts **(Figure 2 middle).** The character of the particles and poor adhesion is also matched by the value of the coefficient of longitudinal thermal expansion of the prepared system, which is equal to the value of $71.77 \cdot 10^{-6} \cdot K^{-1}$ and is thus practically at the level of the value of longitudinal thermal expansion of unfilled epoxy resin **(Table 1)**.

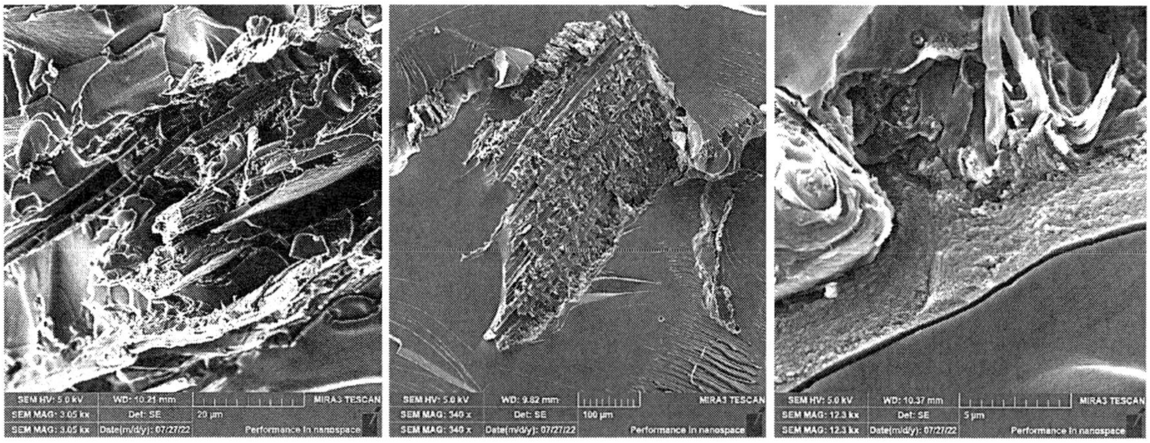

Figure 2 Epoxy resin with cellulose particles (left), epoxy resin with treated rice husk particles (middle), detailed pictures of rice husk particles interfacial adhesion with typical silicone dioxide part and cellulose part (right)

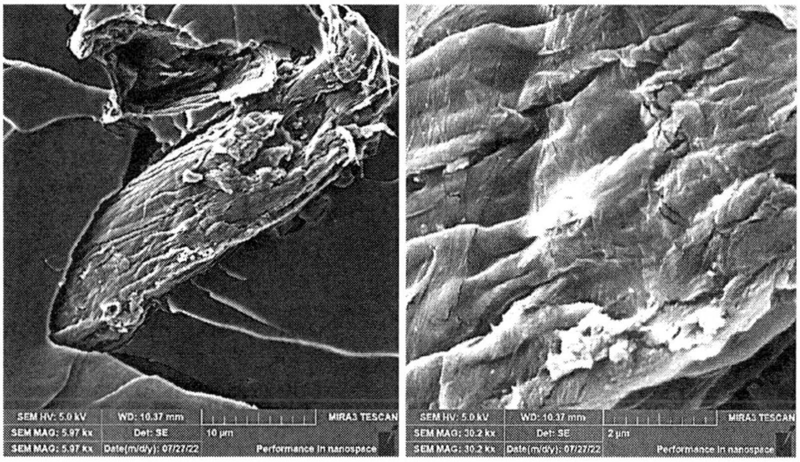

Figure 3 Epoxy resin with treated rice husk fibers in epoxy resin (left), detailed picture of fibers with silicon dioxide nanoparticles on the surface (right)

Cellulose fibers obtained from purified rice husks **(Figure 3)** have a diameter of about 10 micrometers. The length is usually many times greater [12]. On the surface of the fibers, there are residues of nanoparticles of biogenic silica **(Figure 3 right)**, which are primarily formed in rice husks during their growth, and which were not completely removed during their treatment. The surface of the fibers is finely structured. As seen from the picture **(Figure 3 left)**, the adhesion between the fibers and the matrix is not very good. The fracture of the surrounding epoxy resin is brittle. The value of the coefficient of longitudinal thermal expansion α of this composite system is $69.14 \cdot 10^{-6} \cdot K^{-1}$ **(Table 1)**.

Cellulose fibers obtained from purified rice husks and synthetic silica nanoparticles provide good adhesion to the epoxy matrix. The fibers are well coated with resin, the silicon dioxide nanoparticles are more or less homogeneously distributed **(Figure 4 left)**, and the fracture of the composite system is not fragile. Similar to the case of the epoxy resin-silica nanoparticles system, there are visible parts of the resin that is not filled with nanoparticles **(Figure 4 right)**. The value of the coefficient of longitudinal thermal expansion α of this hybrid organic-inorganic composite system is equal to the value of $67.48 \cdot 10^{-6} \cdot K^{-1}$ **(Table 1)**.

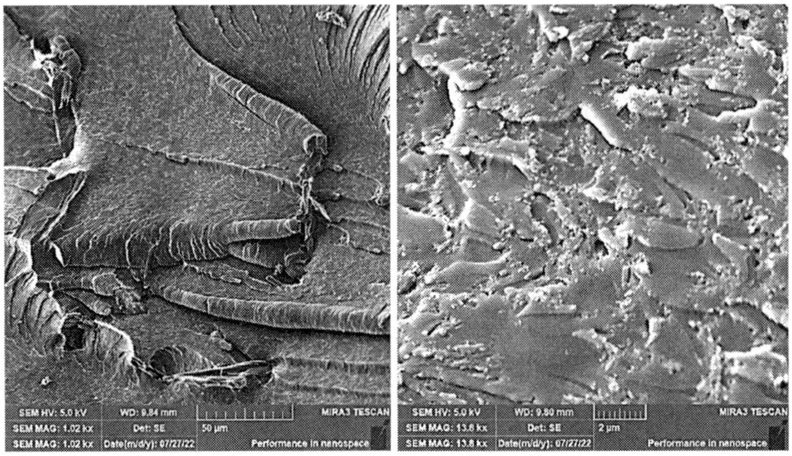

Figure 4 Epoxy resin with treated rice husk fibers and silicon dioxide nanoparticles - adhesion between fibers and epoxy resin (left), distribution of silicon dioxide nanoparticles in epoxy resin (right)

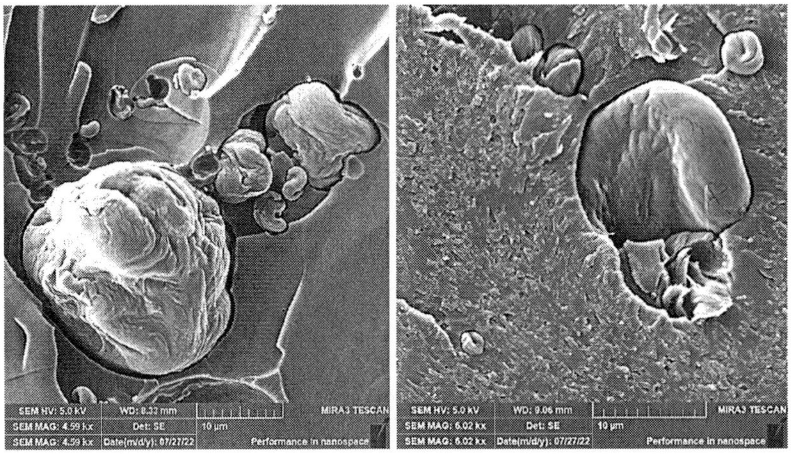

Figure 5 Epoxy resin with cellulose nanocrystals (left), epoxy resin with cellulose nanocrystals and silicon dioxide nanoparticles (right)

Due to their chemical composition, cellulose nanocrystals form spherical structures of larger or smaller dimensions **(Figure 5 left)**. These clusters of nanocrystals do not break down into smaller nanostructures in the epoxy resin, so this filler functions only as a particulate filler. The particles have inferior adhesion to the epoxy matrix, free spaces are visible in the electron microscope images where they are present, and the material shows a brittle fracture **(Figure 5 left)**. Bad adhesion corresponds to the highest ever average measured value of the coefficient of longitudinal thermal expansion α $77.53 \cdot 10^{-6} \cdot K^{-1}$, comparable to the value obtained using cellulose particles or abrasive wood **(Table 1)**. The cause is free spaces between the filler and the binder.

The epoxy resin system filled with cellulose nanocrystals and, at the same time, silicon dioxide nanoparticles **(Figure 5 right)** shows similar behavior as in the previous case, which corresponds to the coefficient of longitudinal thermal expansion α $72.21 \cdot 10^{-6} \cdot K^{-1}$ **(Table 1)**. However, there is a noticeable influence of the present nanofiller based on silicon dioxide, which stands against the non-functional particles of cellulose nanocrystals. As in all previous variants, the silicon dioxide nanoparticles are more or less well distributed in the matrix and change the character of the fracture from initially brittle to tough.

4. CONCLUSION

Based on the measurement of the values of the coefficient of longitudinal thermal expansion and comparison with the character of the fracture surfaces and the quality of interfacial adhesion, it is clear that particle fillers with micrometer dimensions - modified rice husks, cellulose particles, and at the same time also cellulose nanocrystals forming compact units - have a minimal effect on the longitudinal thermal expansion of the system, which has minimal adhesion to the epoxy matrix. The maximum impact leading to a reduction in the coefficient of longitudinal thermal expansion was achieved using nanoparticles of synthetic silicon dioxide. However, it is clear from the available literature and practical experience that particulate fillers will not ensure the functionality of the dental composite system for many years. The experiments showed a significant reduction in the coefficient of longitudinal thermal expansion occurs in the case of a combination of organic and inorganic filler-micrometer cellulose fibers from treated rice husks with residues of biogenic silicon dioxide nanoparticles on their surface and synthetic silicon dioxide nanoparticles. These organic-inorganic fibers are very well incorporated into the polymer matrix, and the value of the coefficient of longitudinal thermal expansion with its value is close to the lowest value achieved when using silicon dioxide nanoparticles. Because of the achieved results, a comprehensive evaluation of the longitudinal thermal expansion coefficients for other material ratios of organic fibers and silica nanoparticles in different polymer matrices will follow.

ACKNOWLEDGEMENTS

This research was funded by the Ministry of Education supported this work, Youth and Sports of the Czech Republic and the European Union - European Structural and Investment Funds in the frames of Operational Programme Research, Development and Education, - project Hybrid Materials for Hierarchical Structures (HyHi, Reg. No. CZ.02.1.01/0.0/0.0/16_019/0000843). Acknowledge belong to Martin Švec for help with microscopic imaging.

REFERENCES

[1] XU, H.H.K., CHENG, L., ZHANG, K., MELO, M.A.S., WEIR, M.D., ANTONUCCI, J.M., LIN, N.J., GIBSON, S.-L., CHOW, L.C., ZHOU, X. Chapter 6 - Nanostructured Dental Composites and Adhesives with Antibacterial and Remineralizing Capabilities for Caries Inhibition. *Nanobiomaterials in Clinical Dentistry*. William Andrew Publishing. 2013, pp. 109-129.

[2] AMINOROAYA, A., NEISIANY, R.E., KHORASANI, S.N., PANAHI, P., DAS, O., MADRY, H., CUCCHIARINI, M., RAMAKRISHNA, S. A review of dental composites: Challenges, chemistry aspects, filler influences, and future insights. *Composites Part B: Engineering*. 2021, vol. 216, 108852. Available from: https://doi.org/10.1016/j.compositesb.2021.108852

[3] ELFAKHRI, F., ALKAHTANI, R., LI, CH., KHALIQ, J. Influence of filler characteristics on the performance of dental composites: A comprehensive review. *Ceramics International*. 2022, vol. 48, issue 19, Part A, pp. 27280-27294. Available from: https://doi.org/10.1016/j.ceramint.2022.06.314.

[4] CHO, K., RAJAN, G., FARRAR, P., PRENTICE, L., PRUSTY, B.G. Dental resin composites: A review on materials to product realizations, *Composites Part B: Engineering*. 2022, vol. 230, 109495. Available from: https://doi.org/10.1016/j.compositesb.2021.109495.

[5] ALBESHIR, E.G., BALHADDAD, A.A., MITWALLI, H., WANG, X., SUN, J., MELO, M.N.S., WEIR, M.D., XU, H.H.K. Minimally-invasive dentistry via dual-function novel bioactive low-shrinkage-stress flowable nanocomposites. *Dental Materials*. 2022, vol. 38, issue 2, pp. 409-420. Available from: https://doi.org/10.1016/j.dental.2021.12.023.

[6] WANG, R., ZHANG, M., LIU, F., BAO, S., WU, T., JIANG, X., ZHANG, Q., ZHU, M. Investigation on the physical-mechanical properties of dental resin composites reinforced with novel bimodal silica nanostructures. *Materials Science and Engineering: C*. 2015, vol. 50, pp. 266-273. https://doi.org/10.1016/j.msec.2015.01.090.

[7] SILVA, R.M., SANTOS, P.H.N., SOUZA, L.B., DUMONT, V.C., SOARES, J.A., SANTOS, M.H. Effects of cellulose fibers on the physical and chemical properties of glass ionomer dental restorative materials. *Materials Research Bulletin*. 2013, vol. 48, issue 1, pp. 118-126. Available from: https://doi.org/10.1016/j.materresbull.2012.10.016.

[8] FUJIHARA, K., TEO, K., GOPAL, R., LOH, P.L., GANESH, V.K., RAMAKRISHNA, S., FOONG, K.W.C., CHEW, C.L. Fibrous composite materials in dentistry and orthopaedics: review and applications. *Composites Science and Technology. 2004, vol. 64,* issue 6, pp. 775-788. Available from: https://doi.org/10.1016/j.compscitech.2003.09.012.

[9] SUNARINTYAS, S., SISWOMIHARDJO, W., TSOI, J.K.H., MATINLINNA, J.P. Biocompatibility and mechanical properties of an experimental E-glass fiber-reinforced composite for dentistry. *Heliyon*. 2022, vol. 8, issue 6, e09552. Available from: https://doi.org/10.1016/j.heliyon.2022.e09552.

[10] UMAPATHY, V.R., NATARAJAN, P.M., JONES, C.S., SWAMIKANNU, B., JOHNSON, W.M.S., ALAGARSAMY, W.M.S., MILON, A.R. Current trends and future perspectives on dental nanomaterials - An overview of nanotechnology strategies in dentistry. *Journal of King Saud University - Science.* 2022, vol. 34, issue 7,102231. Available from: https://doi.org/10.1016/j.jksus.2022.102231.

[11] ADACH, K., KROISOVA, D., FIJALKOWSKI, M. Biogenic silicon dioxide nanoparticles processed from natural sources. *Particulate Science and Technology*. 2021, vol. 39, no. 4, pp. 481-489. Available from: https://doi.org/10.1080/02726351.2020.1758857.

[12] KROISOVÁ, D., DVOŘÁČKOVÁ, Š., YAHYA, R., KŮSA, P. Rice Husks - Potential Source of Cellulose Microfibers/Nanofibers and Biogenic Silicon Dioxide Nanoparticles. *Key Engineering Materials. Trans Tech Publications, Ltd.* 2022. Available from: https://doi.org/10.4028/p-553qz6.

[13] BORŮVKA, M. *Výzkum aplikačních možností bio-kompozitních materiálů plněných nanokrystaly celulózy.* Liberec, 2019. Dissertation. Technical University of Liberec.

INITIAL SCREENING OF SILICA NANOFIBRES AS DRUG DELIVERY SYSTEM FOR ANTIBIOTICS IN WOUND HEALING

[1,2]Miroslava RYSOVÁ, [2]Hana TOMÁNKOVÁ, [2]Markéta SHAABOVÁ, [2]Iva VÍCHOVÁ, [2]Johana KULHÁNKOVÁ, [2,3]Alena ŠEVCŮ

[1]Technical University of Liberec, Faculty of Mechatronics, Informatics and Interdisciplinary studies, Liberec, Czech Republic, EU, miroslava.rysova@tul.cz

[2]Technical university of Liberec, Institute for Nanomaterials, Advanced Technologies and Innovation, Liberec, Czech Republic, EU

[3]Technical university of Liberec, Faculty of Science, Humanities and Education, Liberec, Czech Republic, EU

https://doi.org/10.37904/nanocon.2022.4611

Abstract

Over the last decade, drug delivery systems for wound management have undergone a significant evolution as novel nano-based dosage forms and wound dressing pads for active wound management have been developed. Significance of these materials is based on increasing demand for active wound dressing able to battle infection and inflammation in contaminated and slow-to-heal wounds. Nanofibres represent nanomaterial able to act simultaneously as wound dressing protecting the wound and drug delivery system for *in situ* drug release. Bioavailability of the drug and its release kinetics strongly depend on the drug nature, nanofibrous matrix characteristics and mechanism of drug incorporation. In bacterial infection treatment, fast reaching of minimal inhibitory concentration (MIC) is crucial and can be achieved by mediated surface drug adsorption. In this work, we explored interactions of electrospun silica nanofibres with five model antibiotics from tetracycline family (tetracycline, tetracycline hydrochloride, oxytetracycline hydrochloride, chlortetracycline hydrochloride and doxycycline hydrochloride) in order to evaluate robustness of this novel nanofibrous drug delivery system. The drug adsorption was mediated by surface functionalization with amino groups. Drug functionalized nanofibres were studied in terms of morphology and total drug content. Antibacterial activity against bacteria *E. coli* was tested *in vitro.* Overcoming the MIC value was achieved for all tested antibiotics. Biocompatibility was tested and confirmed on human dermal fibroblasts *in vitro.* Presence of all tested antibiotics led to increased cellular metabolic activity. Based on these results we assumed that silica nanofibres represent a robust system able to bind and release variety of tetracycline antibiotics for contaminated wounds management.

Keywords: Silica, nanofibres, drug delivery, antibiotics, wound healing

1. INTRODUCTION

Development of contaminated non-healing wounds, ulcers and decubitus ulcers represents one of serious issues related to deceases of affluence - diabetes mellitus primarily. According to statistics, the risk of developing foot ulcers is 25 % and leads to one lower limb amputation every 30 seconds among diabetic patients around the world. It is estimated that of all amputations (diabetes related), 85 % are contributed by foot ulceration which further deteriorates to chronic infection and severe forms of gangrene. Overall, the prevalence rate of diabetic foot ulcers is estimated to 5.1 % in Europe with cost demand increasing every year [1]. In 2005, for instance, the average cost of contaminated foot ulcer treatment in Germany was $ 2,957 [2]. This cost doesn´t involve the social impact or any emotional effects on the patient. Under given circumstances, if the ulcer formation can´t be avoided, the most suitable therapy is demanded to prevent the amputation and to minimize the long term impact on patient. The *in situ* therapy seems to be the most effective and the most sensitive treatment and application of a nanofibrous dosage form may bring further benefits in form of wound protection and stimulation [3,4]. Among all materials used for nanofibres production, silica shows unique properties including fast biodegradation [5] and active wound stimulation [6]. Silica nanofibres combine traditional properties of nanofibres such as small fibre diameter and high porosity with unique properties given by their bioactive nature. Silica and organosilane nanofibres formed by sol-gel method and subsequent electrospinning were confirmed as suitable matrix for enzyme conjugation for active wound debridement [7], able to suppress bacterial colonization [8], and stimulate wound healing [9]. Their easy-to-modify surface can be grafted with functional groups introducing surface charge for further drug of choice adsorption. In the era of personalized medicine approach, robustness of such active wound dressing is required in order to meet the needs of individual patients and their specific bacterial profile [10,11].

2. MATERIALS AND METHODS

Tetracycline (TET), tetracycline hydrochloride (TET.HCl), oxytetracycline hydrochloride (OxyTET), chlortetracycline hydrochloride (ChlorTET), doxycycline hydrochloride (DOXY), tetraethyl orthosilicate (TEOS) and (3-aminopropyl)triethoxysilane (APTES) were purchased from Sigma-Aldrich (CZ). The remaining chemicals (acetic acid, ethanol, isopropyl alcohol) were supplied by Petr Švec - Penta (CZ). Cell culture media components and TSA agar were purchased from Fisher Scientific (CZ) and iBiotech (CZ). Biocompatibility testing was performed on adult normal human dermal fibroblasts (aNHDF) purchased from ATCC (USA). *Escherichia coli* (K-12 CCM 7929) was obtained from Czech Collection of Microorganisms (Brno, CZ).

2.1. Nanofibres manufacturing and functionalization

Spinning solution was synthetized by sol-gel method in water/ ethanol mixture from TEOS as a precursor. Silica nanofibres was obtained by subsequent needle-less electrospinning on NanoSpider NS 1WS500U device (Elmarco, CZ)) and thermal stabilization (180 °C). Surface silanization was realized in 3 % APTES solution in water and ethanol mixture with pH adjusted to 4.5 - 5.5. Modification was performed overnight under constant shaking at laboratory temperature. Excess of the silanization solution was washed out and nanofibres were dried at 110 °C for 30 min. Antibiotic immobilization on the surface of APTES treated silica nanofibres was performed in 0.5 % antibiotic solution in ethanol or water for 2 hours in sealed container and under constant agitation (50 rpm). Subsequently, nanofibres were removed from the solution and dried in dark at room temperature.

2.2. Characterization of antibiotic functionalized nanofibres

Morphology of silica nanofibres prior and after the antibiotic immobilization was analysed using scanning electron microscopy (SEM) on device Vega3 (Tescan, CZ). Fibre distribution was performed based on 100 fibre diameter measurements for each sample. Absolute quantity of antibiotic immobilized per sample area was analysed spectrophotometrically using Multi-Mode Microplate Reader (BioTek Instruments, USA). Discs of nanofibres (12 mm diameter) were dissolved in 0.2 M NaOH. Each antibiotic was quantified by its specific absorption maximum values and calculation from calibration curve.

2.3. Biocompatibility and antibacterial activity evaluation

Antibacterial activity of functionalized nanofibres was tested by diffusion test according to the standard AATCC Test Method 147-2004 using Escherichia coli as model bacterial strain. Bacterial suspension (10,000 bacteria per ml) was seeded on trypton soya agar (TSA) plates. Nanofibrous discs (10 mm in diameter) were placed on the agar in triplicates and incubated for 24 hours at 37 °C. After the exposition period, size of diffusion zone inhibiting bacterial growth around samples was measured and photo documented.

Biocompatibility of antibiotic functionalized nanofibres was tested in compliance with ISO 10993-5 on adult NHDF cells. Cell suspension in concentration 10,000 cells per ml was seeded on 24 well plate (10,000 cells per well) and pre-cultured for 24 hours prior the test. After this period, nanofibrous discs (6 mm in diameter) were placed to each well into direct contact with cell population, covered with fresh medium, and incubated for 24 hours (5 % CO_2, 37 °C). After the exposition, cellular viability was tested using MTT assay. 1 mg/ml MTT solution in serum-free medium was placed into each well and incubated for 2 hours in 37 °C. In this widely used assay, MTT (3-(4,5-dimethylthiazol-2-yl)-2,5-diphenyltetrazolium bromide) is converted by active oxidoreductase present in living cells onto its insoluble form MTT formazan. After incubation period, unreacted MTT solution was discarded and produced MTT formazan was solubilized in isopropyl alcohol. Produced MTT formazan was quantified spectrophotometrically by absorbance intensity measuring at 570 nm (and 650 nm for background). Cell viability was calculated as percentage share of living cells exposed to the sample in comparison to living cells present in untreated cell control (CC). DMSO solution (8 %) was used as positive control (PC). Each sample was tested in four repeats.

3. RESULTS AND DISCUSSION

3.1. Nanofibres morphology and immobilized antibiotic quantity

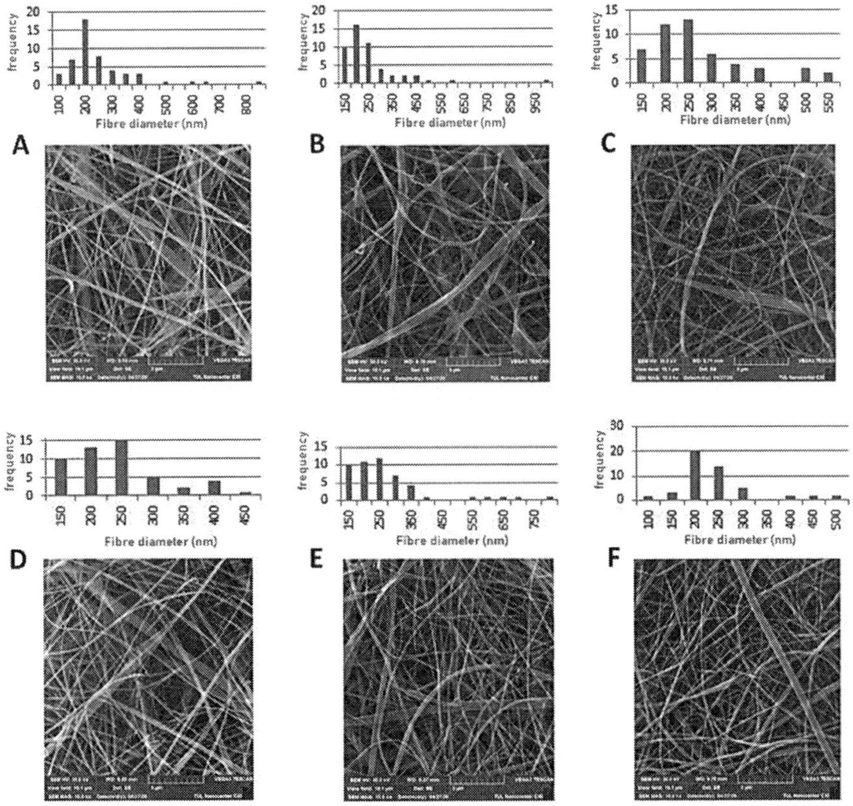

Figure 1 Morphology of SiO_2 nanofibres A. prior the treatment and treated with B. Tetracycline, C. Tetracycline hydrochloride, D. Oxytetracycline hydrochloride, E. Chlortetracycline hydrochloride and F. Doxycycline hydrochloride (Scale 5 μm)

Morphology of silica nanofibres prior and after the silanization procedure and antibiotic immobilization is shown in **Figure 1**. The silica nanofibres show broader fibre diameter distribution in comparison to conventional nanofibres electrospun from commercial synthetic linear polymers. This effect is caused by chain branching and binding at the beginning of sol-gel transition. Comparison of fibre distribution histograms presented in **Figure 1** shows that silanization procedure and subsequent antibiotic immobilization have no significant impact on fibre diameter. For all the samples, the most populated histogram classes lie between 150 nm and 250 nm. All SEM micrographs show opened porous structure without loss of porosity or significant change in fibrous morphology. Functionalization in ethanol or aqueous solution didn´t cause changes in fibres structure, swelling or breaks. This proves good stability of silica nanofibres in variable solvents (except of strong bases), which gives opportunity for functionalization under various conditions and with various molecules. Treatment of nanofibres with 0.5 % antibiotic solution led to homogenous distribution of all model drugs over individual nanofibres.

Quantification of adsorbed antibiotics showed differences in their quantity suggesting differences in interactions between the silanized nanofibres surface and antibiotics molecules. The highest average amount was obtained in the case of doxycycline 244.22 μg/cm². The lowest amount was demonstrated in the case of chlortetracycline (51.36 μg/cm²) and then tetracycline hydrochloride (77.39 μg/cm²). All the obtained results are shown in **Figure 2**.

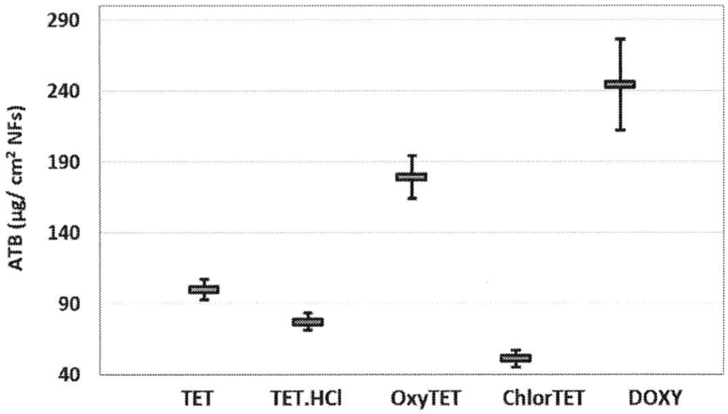

Figure 2 Quantity of tetracycline antibiotics immobilized in silanized silica nanofibres (mean ± S.D.)

3.2. Antimicrobial activity

All the tested tetracycline antibiotics immobilized on silica nanofibres proved antibacterial efficacy against *E. coli* monoculture on TSA substrate. Obtained results are shown in **Figure 3**. Each sample was able to form diffusion zone inhibiting bacterial growth around the nanofibres, The biggest zone was observed in the case of oxytetracycline (33.33 mm) and the smallest zone in the case of chlortetracycline (20.67 mm). These results correlate only partially with the absolute quantities of immobilized antibiotics. This discrepancy can be explained by different specifity and MIC values of tested antibiotics against *E. coli.* This bacterial strain was only used as model bacteria, which is commonly found in contaminated wounds and further investigation against multiple bacterial strains is needed.

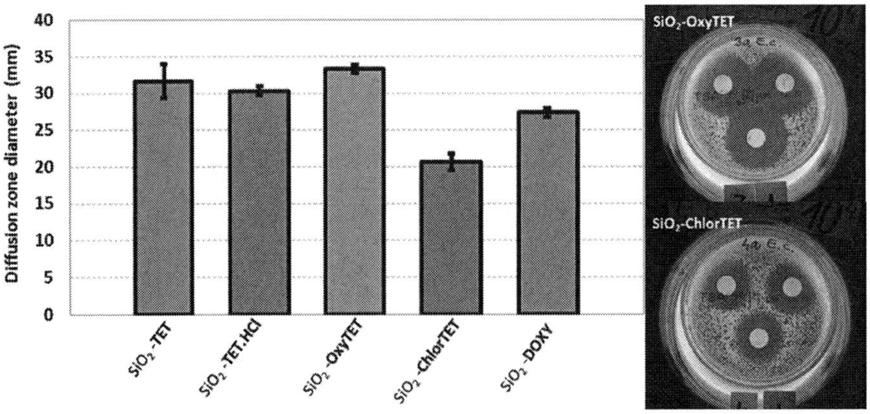

Figure 3 Diffusion zone diameters on *E. coli* monoculture expressed as graph (mean ± S.D.) and photographs of selected samples

3.3. Biocompatibility

Investigation on biocompatibility proved that exposition of human dermal fibroblasts to all the tested samples didn´t result in cell viability drop below 70 %. Therefore, all the tested samples can be declared as biocompatible in compliance to ISO 10993-5. Exposition to silica nanofibres with immobilized antibiotics led to increased metabolic activity compared to cell control (CC). All results are shown in **Figure 4**. Even though the results are presented as cellular viability, it is not possible to determine clearly if presence of antibiotics leads only to increased metabolic activity in NHDF cells or if number of cells increased. Such effect could be very beneficial for wound healing progress after contamination elimination and further instigation is needed.

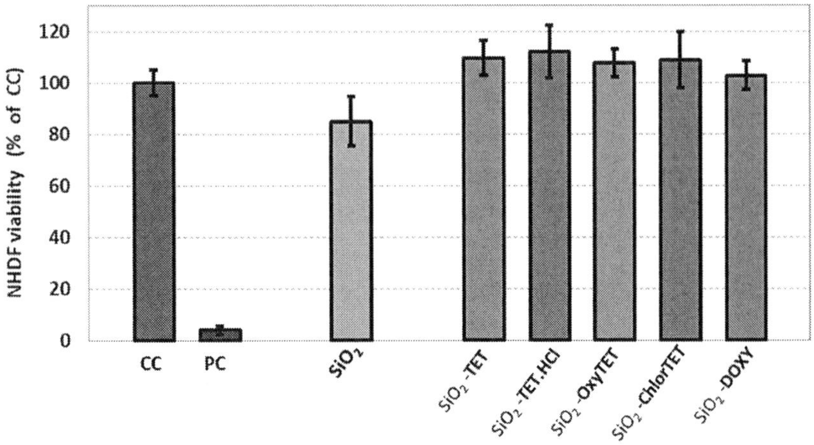

Figure 4 Normal human dermal fibroblasts (NHDF) viability after 24 hours exposure to silica nanofibres with immobilized tetracycline antibiotics *in vitro* (mean ± S.D.)

4. CONCLUSION

In this paper we presented results on initial study of silica nanofibres evaluation as potentially robust system for antibiotics immobilization and *in situ* release for contaminated wounds therapy. Silica nanofibres immobilized with five model antibiotics from tetracycline family were proven be antibacterial against *E. coli* and biocompatible to human dermal fibroblasts *in vitro*. The immobilization procedure in ethanol and aqueous solution didn´t cause changes in the fibrous morphology. Therefore, silanized silica nanofibres represent an interesting candidate for broad range antibiotics immobilization and potential application in personalized wound therapy. To confirm this assumption, further investigation will be needed.

ACKNOWLEDGEMENTS

This research was funded by the European Union-European Structural and Investment Funds in the framework of the Operational Programme Research, Development, and Education project "Hybrid Materials for Hierarchical Structures (HyHi)", Reg. No. CZ.02.1.01/0.0/0.0/16_019/0000843, and by the Student Grant Competition of the TUL under project No. SGS-2022-4059. The authors acknowledge the assistance provided by the Research Infrastructure NanoEnviCz, supported by the Ministry of Education, Youth and Sports of the Czech Republic under Project No. LM2018124.

REFERENCES

[1] RAGHAV, A., KHAN, Z.A., LABALA, R.K., AHMAD, J., NOOR, S., MISHRA, B.K. Financial burden of diabetic foot ulcers to world: a progressive topic to discuss always. *Therapeutic advances in endocrinology and metabolism*. 2018, vol. 9, no. 1, pp. 29-31.

[2] TCHERO, H., KANGAMBEGA, P., LIN, L., MUKISI-MUKAZA, M., BRUNET-HOUDARD, S., BRIATTE, C., RETALI, G.R., RUSCH, E. Cost of diabetic foot in France, Spain, Italy, Germany and United Kingdom: A systematic review. *Annales d´Endocrinologie*. 2018, vol. 79, no. 2, pp. 67-74.

[3] RIEGER, K.A., BIRCH, N.P., SCHIFFMAN, J.D. Designing electrospun nanofiber mats to promote wound healing - a review. *Journal of Materials Chemistry B*. 2013, vol. 1, pp. 4531-4541.

[4] MULHOLLAND, E.J. Electrospun Biomaterials in the Treatment and Prevention of Scars in Skin Wound Healing. *Frontiers in Bioengineering and Biotechnology*. 2020, vol. 8, pp. 1-15.

[5] MÖLLER, K., BEIN, T. Degradable Drug Carriers: Vanishing Mesoporous Silica Nanoparticles. *Chemistry of Materials*. 2019, vol. 31, vol. 12, pp. 4364-4378.

[6] QUIGNARD, S., COADIN, T., POWELL, J.J., JUGDAOHSINGH, R. Silica nanoparticles as sources of silicic acid favoring wound healing in vitro. *Colloids and Surfaces B: Biointerfaces*. 2017, vol. 155, pp. 530-537.

[7] DANILOVÁ, I., LOVĚTÍNSKÁ-ŠLAMBOROVÁ, I., ZAJÍCOVÁ, V., VOLESKÝ, L., RYSOVÁ, M. Immobilization of esterase enzyme onto silica nanofibers for biomedical applications. *Vlakna a Textil*. 2014, vol. 21, no. 2, pp. 3-11.

[8] MÁKOVÁ, V., HOLUBOVÁ, B., TETOUR, D., BRUS, J., ŘEZANKA, M., RYSOVÁ, M., HODAČOVÁ, J. (1S,2S)-cyclohexane-1,2-diamine-based organosilane fibres as a powerful tool against pathogenic bacteria. *Polymers*, 2020, vol. 12, no. 1, pp. 1-12.

[9] GROTHEER, V., GOERGENS, M., FUCHS, P.C., DUNDA, S., PALLUA, N., WINDOLF, J., SUSCHEK, C.V. The performance of an orthosilicic acid-releasing silica gel fiber fleece in wound healing. *Biomaterials*. 2013, vol. 34, no. 30, pp. 7314-7327.

[10] DOWD, S.E., WOLCOTT, R.D., KENNEDY, J., JONES, C., COX, S.B. Molecular diagnostics and personalised medicine in wound care: assessment of outcomes. *Journal of Wound Care*. 2011, vol. 20, no. 5, pp. 232-239.

[11] BESSA, L.J., FAZII, P., DI GIULIO, M., CELLINI, L. Bacterial isolates from infected wounds and their antibiotic susceptibility pattern: some remarks about wound infection. *International Wound Journal*. 2015, vol. 12, pp. 47-52.

SESSION D
MONITORING AND TOXICITY OF NANOMATERIALS

Chairman

Dr. Jan TOPINKA Institute of Experimental Medicine of the CAS, Prague, Czech Republic, EU

TOXIC EFFECT OF NANOTEXTILES AS WASTE OF RESPIRATORS AND FACE MASKS

[1]Petra ROUPCOVÁ, [1]Jan SLANÝ, [1]Jiří PAVLOVSKÝ, [1,2]Karel KLOUDA

[1]VSB - Technical University of Ostrava, Ostrava, Czech Republic, EU,
petra.roupcova@vsb.cz, jan.slany@vsb.cz, jiri.pavlovsk@vsb.cz

[2]Occupational Safety Research Institute, Prague, Czech Republic, EU,
klouda@vubp-praha.cz

https://doi.org/10.37904/nanocon.2022.4613

Abstract

The article deals with the toxic effect of nanotextiles as waste from respirators and face masks. The article is focused on ecotoxicology and basic ecotoxicological tests according to available European standards and Czech legislation. Article describes the production and other characteristics of the nonwovens that were tested. Part of article concludes with a review of foreign studies that have also dealt with the testing of microplastics and nanoplastics on various organisms. The experimental results presents specific experimental characteristics of the textiles and then presents individual tests. Tests on seeds of *Sinapis alba* L. plants growing and *Lemna minor* L. were performed. In all cases, the tested non-woven fabrics had a non-toxic effect on the organisms, but it cannot be confirmed that there is no reaction. Further tests will be carried out to clarify the effects of the nonwovens on the test organisms.

Keywords: Nanotextiles, non-woven fabrics, respirators, ecotoxicity

1. INTRODUCTION

Nonwovens are formations of textile fibres, filaments or yarns which are interconnected by techniques other than weaving, knitting, bobbin-making, plaiting and tufting. However, the production and handling of nanofibres cannot completely eliminate their release into the environment or the working environment. This can occur in four ways, namely during the manufacturing process of the nanofibres that make up the textiles, during the processing of the materials containing the nanofibres (which also results in release to the working environment and exposure of workers), during the use of the nanofibres (as the fibres are mechanically stressed and thus further released to the environment) or during the disposal of the textiles already used, when the nonwovens are discarded freely into the environment and its components, water and soil, where the undesirable effect of adsorbed pollutants trapped during use or absorbed into the layers of the nonwovens may be manifested, which may lead to a change in toxicological properties [1]. Three types of materials were selected to be tested in the tests carried out. These are personal protective equipment used for respiratory protection, which were widely used during the spread of the Covid-19 viral disease. Specifically, they are unlaminated PP-PVDF (polyvinylidene fluoride laminated on one side with a polypropylene layer), laminated PP-PVDF-PP, which is similar to the first material mentioned, but this time laminated on both sides, and the last material is a specific type of Nanovia mask (all of them are shown in **Figure 1**), respectivelly. The ecotoxicity of these three materials was tested on two different organisms - mustard seeds in a range of concentrations of aqueous leachates, following the OEDC methodology, and perch leaves, also in a range of concentrations of aqueous leachates, following the standard EN ISO 20079 (2001) Water quality - Determination of toxic effects of water and wastewater constituents on perch (*Lemna minor* L.) - Perch growth inhibition test. [2].

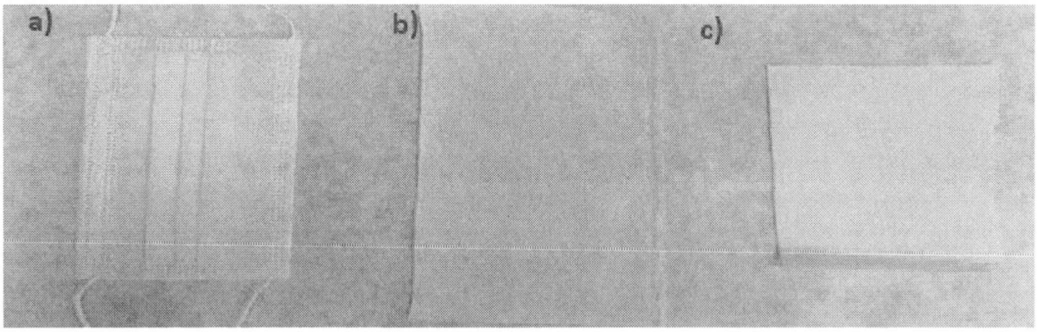

Figure 1 Photographs of materials before preparation of aqueous leachates (from the left Nanovia facemask, unlaminated PP-PVDF, laminated PP-PVDF-PP)

2. EXPERIMENTAL PART AND RESULTS

The basis of the experimental part is the production of polymer nanofibres - these can be prepared in several different ways, some of which can only be discussed in the laboratory for the time being. Others, however, have already been improved to such a level that they can be used in large-scale production - for example, electrospinning or nanospider^{TM}. Other methods include centrifugal force spinning (Force Spinning) or the Melt - Blown method, which is based on the melt blowing principle. The requirements for product properties and production speed are then a decisive factor in the production of the technology. However, the most widely-used production method is undoubtedly electrospinning, the essence of which is the use of the effect of an electrostatic field on a charged viscoelastic liquid (usually a polymer solution), whereby thin filaments of different lengths are formed when given conditions are met [3-6].

2.1. Specification of nonwovens and aqueous leachates

Aqueous leachates were prepared using 100 g of material (nanotextile) and 1000 mL of distilled water. The prepared leachates were shaken on a shaker for 24 hours, then filtered through a 0.40 μm pore membrane. The stock solution of the test substance and dilution solution were also diluted according to ISO 7346 [7,8]. The first two materials used were laminated PP-PVDF and unlaminated PP-PVDF, being multilayered polypropylene fabrics produced either by the spunbond (S) method, where a small volume of ambient temperature air is used, thus achieving coarser textiles with larger diameters and higher tensile and compressive strengths, or by the meltblown (M) method, where high velocity (100-500 m/s) hot air is used to produce smaller fibres with more diameter options. The fibres produced in this way are not very strong, but have excellent filtration properties, which is why both methods are most often combined: a multi-layer structure is used (SM for PP-PVDF, SMS for PP-PVDF-PP). Another option is to combine both methods with nanotextiles to achieve better filtration properties [1,9]. The production itself is then carried out as follows - PVDF nanofibers, which were obtained from a dimethylacetamide solution, are subsequently deposited on the multilayered structure using the electrospinning method. If the aim is to obtain a laminated PP-PVDF-PP material, another non-woven PP fabric (S) is laminated onto the multi-layer structure to form a sandwich trilayer [1,9]. The fabrication of Nanovia Mask 99.97 nanofibres is similar to the two materials mentioned above, i.e. by joining polypropylene fabrics produced by spunbond technology, on top of which lies another layer of fabric produced by the meltblown method, and finally a PVDF nanofibrous structure. Another polypropylene spunbond layer is bonded to the bottom of the PVDF layer using an ethylene vinyl acetate adhesive [9].

2.2. Ecotoxicological test procedure

The first test was to test *Sinapis alba* L. seeds using a concentration range of aqueous leachates of all selected materials to observe the effect of increasing concentrations on the resulting root length compared to the control. The procedure is based on the methodology of the Ministry of Environment for the determination of ecotoxicity of wastes. The basis of this test is to pipette a specified amount of solution (2.5 mL) at different concentrations (10, 20, 50, 100, 200, 500 and 1000 mL/L) and to place 20 seeds (4 rows of 5 seeds = 20 seeds) on a soaked filter paper. The procedure described above, including the formula for calculating root growth inhibition, is then followed. Potassium dichromate at concentrations of 1.25; 2.5; 5; 10; 20; 40; 80; 160 and 320 mL/L was used as a reference substance, which gave results corresponding to the standard results - with increasing concentration, the root length of *Sinapis alba* L. decreased [2]. First, Petri dishes and filter papers are prepared, then the solution is pipetted, always 2.5 mL, (the 1st concentration is always for the blank solution) and a number of seeds (20 in this case) are sown. Finally, the petri dishes are placed in an unlit place where the temperature is constant at (20±1) °C. After 72 hours, a root growth reading is taken and the values obtained are entered into the formula. The percentage of root inhibition can then be calculated. The second test carried out was a semichronic toxicity test on lesser perch, a monocotyledonous plant that often covers the surface of standing water [10,12]. The test procedure is based on the standard EN ISO 20079 (2001) Water quality - Determination of toxic effects of water and wastewater constituents on perch (*Lemna minor* L.) - Perch growth inhibition test and lasts for 7 days, during which the perch grows in a nutrient solution to which different concentrations of the substances whose ecotoxicity is to be determined are added. The result of this semi-chronic toxicity test is the effect of the substance on the growth rate, which varies based on how the number of leaves changes with different concentrations of the test substance. 8 stock solutions of precisely defined concentrations are to be prepared to serve as stock solutions and these 8 solutions are to be mixed in a 2-litre flask. The solution and its pH must be stabilised before the test begins. The procedure for the preparation of the aqueous leachates is given in Section 2.1. The concentrations of the filtrates of each material were then adjusted so that the resulting concentrations of the aqueous leachates were 200 ; 500 and 1000 mL/L (100 g/L), all concentrations being prepared in parallel determinations A and B. However, prior to the actual deployment of the perch, the pH of all filtrates had to be adjusted so that the pH was between 5 and 6. Once the filtrates were adjusted and prepared, four two-leaf and four three-leaf perch colonies were carefully transferred into these beakers. In total, there were therefore 20 leaves. Tweezers, a spoon and possibly a glass rod were used for the transfer. Once the perch had been moved into all the beakers, the entire tray was covered with clear food-grade foil and placed in a culture chamber for 168 hours. During this time, the humidity, temperature, and amount of light were set constant [11].

Total growth inhibition for *Lemna minor* L. was calculated according to the following formulas:

a) growth rate μ

The growth rate indicates the final number of petals per unit time the test was run, calculated for each single concentration and also for blanks according to the following formula, where N_0 is number of petals at the beginning of the test, N_k is number of petals at the end of the test and t_k is duration of the test [11]:

$$\mu = \frac{\ln N_k - \ln N_0}{t_k} \tag{1}$$

b) inhibition (stimulation) of $I_{\mu i}$ growth

The inhibition or stimulation of growth indicates the relative comparison in percent of the growth rate of the blank (number of leaves, per unit time the test was run) to the growth rate of each concentration tested. It is calculated according to the following formula, where μ_c is growth rate of the blank (number of leaves per unit time of test) and μ_i is growth rate of each concentration (number of leaves per unit time of test) [11]:

$$I_{\mu i} = \frac{\mu_c - \mu_i}{\mu_c} \times 100 \tag{2}$$

2.3. Evaluation and discussion of results

A semichronic toxicity test was conducted on *Sinapis alba* L. seeds using a concentration series of aqueous leachates of Nanovia, PP-PVDF-PP and PP-PVDF materials. The calculation of root growth inhibition is in the measurement of root length (root elongation) after the test according to the calculation relationship:

$$IC = \frac{L_c - L_v}{L_c} \times 100 \qquad (3)$$

where *IC* is the root growth inhibition concentration in %, (in case of negative inhibition it is stimulation), L_c is the arithmetic mean of the root length in control in mm, L_v is the arithmetic mean of the root length in the test solution in mm [2,7,8].

Table 1 Resulting inhibition of mustard seeds for a concentration range of aqueous leachates of all materials

Nanovia		PP-PVDF-PP		PP-PVDF	
Leachates concentration [mL/L]	**IC [%]**	**Leachates concentration [mL/L]**	**IC [%]**	**Leachates concentration [mL/L]**	**IC [%]**
10	8.48	10	22.51	10	6.92
20	-4.56	20	8.19	20	1.91
50	-9.96	50	13.78	50	2.89
100	-1.62	100	3.19	100	-16.72
200	-5.74	200	20.26	200	7.41
500	5.74	500	23.31	500	0.15
1000	25.55	1000	-2.01	1000	-3.09

All concentration series results shown in the tables are averages of several parallel determinations. The obtained inhibition results can be easily summarised as they are almost completely different from each other. In all cases there were very low inhibitions or very low stimulations, with none of the inhibitions reaching even 30 % (the highest was 25.55 %). Thus, no EC_{50} (IC_{50}) value can be determined for any of the materials, and it can be reasonably stated that these materials do not appear to be ecotoxic in this type of test, and do not meet the HP 14 property.The results of the semichronic toxicity tests on lesser spotted bass are shown in **Table 2**.

Table 2 Inhibition of periwinkle growth after 168 hours for all materials in parallel determination

Material	**Leachates concentration [mL/L]**	**Number of leaves for parallel determination (A/B)**	**Average number of leaves**	**Inhibition [%]**
	0	28/34/44	35.3	0.0
Nanovia	200	30/32	31.0	23.0
	500	31/34	32.5	14.6
	1000	32/36	34.0	6.8
PP-PVDF-PP	200	38/32	35.0	1.7
	500	41/40	40.5	-24.0
	1000	40/39	39.5	-19.6
PP-PVDF	0	51/51/55/51	52.0	0
	200	74/58	66.0	-24.8
	500	70/56	63.0	-19.7
	1000	55/54	54.5	-4.4

Throughout the experiment, necrosis often appeared on the leaves, and chlorosis was rarely seen. For the first two materials, the growth rate was equal to 0.0813 petals per day, and for the PP-PVDF material was equal to 0.1365 petals per day. It was not possible to trace a trend among the results to confirm that inhibition also increased with the increasing leachate concentration, and in neither case could IC_{80} or IC_{50} be determined. It was only possible to determine IC_{20} for the Nanovia drape sample, where inhibition was exactly 23.0 % for a leachate concentration of 200 mL/L. Negative numbers in the inhibition column then indicate that more than half of the samples had leaf growth stimulation, indicating that the concentrated leachate from the materials used created a more acceptable growth environment for the periwinkle than the standardised nutrient solution alone. The highest stimulation occurred in the unlaminated PP-PVDF sample at a concentration of 200 mL/L, where the value reached -24.8 %, while another high stimulation was observed in the laminated PP-PVDF-PP sample, where the stimulation was -24.0 %.Thus, for the selected concentration range of the selected materials Nanovia, PP-PVDF-PP and PP-PVDF, it can be stated that the leachates were not toxic to lesser celandine leaves, thus not possessing the hazardous property HP 14. However, although the results did not exceed the limit values required to label the tested materials as ecotoxic, organism reactions to the toxic substance did occur, as the leaves were abundantly covered with necrosis and chlorosis. Similar results were obtained in foreign studies [13,14], where, for example, the effects of polyethylene on leaf growth, chlorophyll content, number and length of roots and root cell condition were investigated, with no significant negative or positive effects on any of the above parameters. The second study also tested microplastics and their effect on perch; this time 6 different sources of microplastics were chosen. Attention was focused on the growth rate, root length and chlorophyll a content, which was measured at wavelengths of 662.4 and 648.6 nm. The experimental results showed that growth rate was not negatively affected by microplastics. The same was true for chlorophyll a content. As in the first study, the only parameter affected was root length, which was reduced by 10-40 %. However, this inhibition is explained by the mechanical action of microplastic particles on the test organisms, namely their rough surface, sharp edges and particle size. [13,14]

3. CONCLUSION

Based on the ecotoxicity tests performed on mustard seeds and the aquatic perch plant, it was shown that the test substances (Nanovia drape, laminated PP-PVDF and unlaminated PP-PVDF) do not meet the limit values for being toxic to the selected organisms. However, in no case can the effects of the test materials on the selected organisms be completely excluded. Testing a concentration range of aqueous leachate materials on seeds gave good results, with none of the materials tested acting as an overly large inhibitor or stimulant on mustard seeds. The maximum inhibition achieved was around 25.0 %, while the highest stimulation was 15.0 %. This was the first test on the basis of the results of which it can be stated that aqueous leachates from the materials tested are not ecotoxic, whether the leachate is dilute or maximally concentrated. The semichronic toxicity test on lesser knapweed was aimed at monitoring inhibition (or stimulation) of leaf growth per unit. The results were as follows - the inhibition of leaf growth of the Eurasian watermilfoil occurred essentially only in the Nanovia sample, with the highest inhibition (23 %) occurring when using a 200 mL/L aqueous leachate concentration. From this value, IC_{20} = 276.5 mL/L was subsequently calculated. Which means that at this concentration just 20 % of the test specimens will be inhibited (IC_{50} and IC_{80} could not be determined because such inhibition values were not achieved). The other inhibitions were then only lower (14.6 and 6.8 % for Nanovia, 1.7 % for PP-PVDF-PP). Across the other samples, there was always stimulation, with the highest stimulation achieved in the PP-PVDF sample, when it was close to -25.0 %.The summary of all the experiments performed can be made relatively simply, since in none of the cases were the threshold values reached to make it possible to label the tested materials Nanovia, laminated PP-PVDF-PP and unlaminated PP-PVDF as ecotoxic. At the same time, however, the results show that interactions between organisms and materials are there - both positive and negative. Further tests will be carried out in the future in order to specify and explain this interaction; at present, even the literature outputs do not agree on explanations in all cases.

ACKNOWLEDGEMENTS

The publication was written within the Student grant competition "Release of Nano and Microplastics from Nanotextiles and Determination of Their Effect on the Environment 2" - project number SP2022/119.

REFERENCES

[1] ROUPCOVÁ, P., KUBÁTOVÁ, H., BÁTRLOVÁ, K., and KLOUDA, K. *Nanotextilie - materiál vhodný k ochraně dýchacích cest a zároveň zdroj nano- a mikroplastů pro životní prostředí.* Ostrava: Vysoká škola Báňská - Technická univerzita Ostrava, 2021.

[2] *Metodický návod ke stanovení ekotoxicity odpadů.* Ministerstvo životního prostředí. [online]. 1998, Zpravodaj 12/98. Available from: https://www.mzp.cz/web/edice.nsf/390334EFF595E8B9C1256FC000440F3C/$file/Z_12ekotoxicita.html

[3] ŠVACHOVÁ, V. *Zvlákňování celulóz pomocí elektrospinningu a plazmochemické modifikace celulózových nanovláken* [online]. Brno, 2012 [cit. 2021-04-14]. Graduate thesis. Masarykova univerzita, přírodovědecká fakulta. Thesis supervisor Milan Alberti. Available from: https://is.muni.cz/th/ow2aj/Diplomova_prace-_final_verze_pdf.pdf.pdf.

[4] JIRSÁK, O, SANETRNÍK, F., LUKÁŠ, D., KOTEK, V., MARTINOVÁ, L., CHALOUPEK, J. *Method of nanofibres production from a polymer solution using electrostatic spinning and a device for carrying out the method*, U. S. Pat. 7565437 B2, 2003.

[5] YARIN, A. L.: *Electrospinning of Nanofibres Polymer Solution and Melts.* Institute of Fundamental Technological Research Poland, Warsaw, 2003.

[6] FILATOV, Y., BUDYKA, A. and KIRICHENKO, V. *Electrospinning of micro and nanofibres: fundamentals in separation and filtration processes.* Begell House Inc., Redding, 2007.

[7] *Metodický pokyn*, Laboratoř ekotoxikologie a LCA, Ústrav chemie ochrany prostředí, VŠCHT v Praze, Test semichronické toxicity se semeny hořčice *Sinapis alba*, pp. 1-4.

[8] *ČSN EN 14735: Charakteristika odpadů - příprava vzorků odpadu pro testy ekotoxicity.* Praha: Český normalizační institut, 2007, p. 44.

[9] KLOUDA, K., KUBÁTOVÁ, H., ROUPCOVÁ, P. a BATRLOVÁ, K. *Testování toxicity netkaných textilií využívaných k výrobě ochranných prostředků dýchacích cest.* Ostrava: Vysoká škola Báňská - Technická univerzita Ostrava, 2022.

[10] *Ekotoxikologie.* WikiSkripta [online]. 2018 [cit. 2021-5-2]. Available from: https://www.wikiskripta.eu/w/Ekotoxikologie.

[11] PAVLOVSKÝ, Jiří. *Toxikologie* [online]. Druhé vydání. Ostrava, 2019 [cit. 2021-5-2].

[12] GEBAUEROVÁ, Hana. *Srovnání specifických ekotoxikologických zkoušek pro p-fenylendiamin, včetně jeho sorpce na běžný typ jílovitého materiálu* [online]. Ostrava, 2014 [cit. 2021-11-15]. Diplomová práce. Vysoká škola báňská - Technická univerzita Ostrava

[13] KALČÍKOVÁ, G., ŽGAJNAR GOTVAJN, A., KLADNIK, A., & JEMEC, A. Impact of polyethylene microbeads on the floating freshwater plant duckweed. *Lemna minor. Environmental Pollution.* (Barking, Essex: 1987) 2017, vol. 230, pp. 1108-1115. Available from: https://doi.org/10.1016/j.envpol.2017.07.050.

[14] ROZMAN, U., TURK, T., SKALAR, T., ZUPANČIČ, M., KOROŠIN N. Č., MARINŠEK, M., OLIVERO-VERBEL, J. & KALČÍKOVÁ, G. An extensive characterization of various environmentally relevant microplastics - Material properties, leaching and ecotoxicity testing. *Science of the Total Environment.* 2021, vol. 773, 145576. Available from: https://doi.org/10.1016/j.scitotenv.2021.145576.

PHENOTYPIC RESPONSES OF GREEN ALGA CHLAMYDOMONAS REINHARDTII TO NANOSCALE ZERO VALENT IRON

[1,2]Cheryl S.Y. YEAP, [1]Nhung H.A. NGUYEN, [3]Olga BLIFERNEZ-KLASSEN, [3]Jan H. MUSSGNUG, [4]Tobias BUSCHE, [3]Olaf KRUSE, [1]Alena SEVCU

[1]Technical University of Liberec, Institute for Nanomaterials Advanced Technologies and Innovation, Liberec, Czech Republic, EU, alena.sevcu@tul.cz

[2]Technical University of Liberec, Faculty of Mechatronics, Informatics and Interdisciplinary Studies, Liberec, Czech Republic, EU, cheryl.yeap@tul.cz

[3]Bielefeld University, Algae Biotechnology and Bioenergy, Faculty of Biology, Center for Biotechnology (CeBiTec), Bielefeld, Germany, EU, jan.mussgnug@uni-bielefeld.de

[4]Bielefeld University, Microbial Genomics and Biotechnology, Faculty of Biology, Center for Biotechnology (CeBiTec), Bielefeld, Germany, EU, tbusche@cebitec.uni-bielefeld.de

https://doi.org/10.37904/nanocon.2022.4614

Abstract

The toxicity of two iron-based nanoparticles: nFe_3O_4, and nZVI were assessed on freshwater microalgae, *Chlamydomonas reinhardtii* CC-5325. Microalgae response (total chlorophyll/carotenoids content, photosystem II efficiency, cell shape and total viable cell numbers) to the nanoparticles exposure (100 mg/L) was monitored up to 120 hours. Based on phytochrome and photosystem II analysis performed, almost no significant impact was found. However, microscopic analysis and the total viable cell numbers revealed a certain degree of inhibition effect showing altered cell shape, and higher number of dead cells after the exposure to both nanomaterials. The dead cell numbers increased within one hour after the exposure to nFe_3O_4, while nZVI caused rather slow inhibition effect and persisted until 48 h with the highest dead cell number. In a series of experiments performed, the results may justify that exposure of these two NPs initially slightly inhibited *C. reinhardtii*, however the culture was able to recover towards the end of the study, because of new cell generation and nZVI oxidation.

Keywords: Metal oxide nanoparticles, nZVI, algae, physiological effect

1. INTRODUCTION

Nanoscale zero-valent iron (nZVI) is a strong reducing agent which has been used for more than a decade mainly for environmental remediation to clean up polluted aquifers [1-3] and was also suggested for elimination of harmful cyanobacterial blooms (REF) [4]. Potential application of nanoscale magnetite (nFe_3O_4) could be used as drug delivery system, contrast agent for magnetic resonance imaging, or hyperthermia agent in cancer treatment [5]. To date, the toxicity of iron nanoparticles have been evaluated on numerous cell lines and microorganisms [6-8]. Although different parameters were measured, they were mostly interested in describing the effect of different nanoparticles concentrations rather than in establishing mechanistics links between the measured functions. Thus, in this study, effect of two differernt iron nanoparticles (i) reactive nZVI and (ii) non-reactive nFe_3O_4 was assessed on *Chlamydomonas reinhardtii* which is a model organisms used in many toxicological studies [9]. In particular, several phenotypic effects of *C. reinhardtii* were studied upon these NPs exposure, such as photosystem II efficiency, cell shape and total viable cell count.

2. MATERIALS AND METHODS

2.1. Preparation of nZVI and nFe_3O_4

The supplied NANOFER STAR powder (nZVI) (CAS no: 7439-89-6, size 10-150 nm) and nanoscale magnetite (nFe_3O_4) (CAS No: 1317-61-9, size 50-100 nm) were obtained from NANOIRON s.r.o, Czech Republic, and Sigma Aldrich, Merck, Germany, respectively. Activation of 20% NPs suspension were done according to producer guidelines for 48 h [10].

2.2. Culture condition of *C. reinhardtii* and experimental setup

Chlamydomonas reinhardtii CC-5325, was obtained from the *Chlamydomonas* Resource Center University of Minnesota (USA). The culture was cultivated under a 130 µmol photons·m^{-2} s^{-1} of constant white light illumination and gently bubbled by 1% CO_2 mixed with air at room temperature (24 °C). The cultivations were conducted in a 500 mL flask containing Sueoka's high salt medium (HSM) [11]. A preculture of algae in HSM was started 48 h before addition of nanoparticles (-NPs). At time point t = 48 h in (-NPs), the pre-cultured algae were then harvested from batch culture during mid-log exponential growth (at an initial cell density of 6.026 × 10^6 cells/mL), then added to a flask containing NPs, which is labelled as (+NPs). Three different experimental conditions were set up, namely the: i) unexposed *C. reinhardtii* (control), ii) *C. reinhardtii* with 100 mg/L of nFe_3O_4 and iii) *C. reinhardtii* with 100 mg/L of activated nZVI.

Three replicates for each condition were conducted in 550 mL of HSM, starting with 45 mL of pre-grown algae culture. Total nanoparticle-algae exposure hours were up to 120 h from t = 0 h in (+NPs), under photoautotrophic conditions as mentioned above. Cultures (15 mL) were sampled every 24 h before +NPs exposure (up to 48 h, from -NPs pre-grown culture) and after the start of +NPs exposure at 0, 1, 4 h and thereafter every 24 h up to 120 h for microscopy, viable cell count, chlorophyll and carotenoids and photosystem II analysis. All the measurements were done in three technical replicates per biological sample for each time point and averaged prior to data analysis.

2.3. Phenotypic analysis

2.3.1. Light microscopy, and viable cell count

Algae cell distribution was imaged under optical microscopy (Motic BA310, Motic, China) to assess the substantial physical interaction, NPs aggregation and calculation of cell number. The cell number was determined by manual cell counting using a Neubauer haemocytometer (Superior Marienfeld, Germany).

2.3.2. Chlorophyll quantification and fluorescence analysis

The extraction of photosynthetic pigments (total chlorophyll and carotenoids) were carried out with 90% methanol as described by Lichtenthaler [12] and calculated according to Wellburn [13].

2.3.3. Photosystem II efficiency

Photosystem II efficeincy was measured in a 48-well flat bottom plate using a closed FluorCam FC 800-C Video Imager (Photon Systems Instruments, Czech Republic). Before the measurement, samples were dark-adapted for at least 15 min. The parameters F_v/F_m, non-photochemical quenching (NPQ), PSII photochemistry were calculated as described by Murchie [14].

2.3.4. Statistical analysis

Statistical analysis was performed using Prism (GraphPad, USA). The mean differences of the treatment groups were compared against a *C. reinhardtii* control respectively by two-way ANOVA and Dunnett's multiple comparison test. The differences were considered statistically significant at *$P < 0.05$, **$P < 0.001$, ***$P < 0.0001$.

3. RESULTS AND DISCUSSION

3.1. Morphology changes of algae upon NPs exposure

The light microcopy images (**Figure 1**) show different cell stages and interaction between algae cells and NPs. The cells in the control had intact plasma membrane closely attaching to the cell wall with two flagellar tails (**Figure 1, A1-A3**), and new cells started to divide after 24 h (**Figure 1, A4**). A strong NPs attachment on algae was observed from time zero (**Figure 1, B1, C1**), and cells were slightly deformed or lysed which was observed from 48 h (**Figure 1, B5, C5**) or cell lost their chlorophyll pigments leaving empty cells/cell debris behind (**Figure 1, C2**). Such a cell alteration also appeared in a previous study [7]. At immediate exposure, there was small degree of chlorophyll lost in nFe_3O_4 indicated by the yellow arrows which was not seen in nZVI exposure.

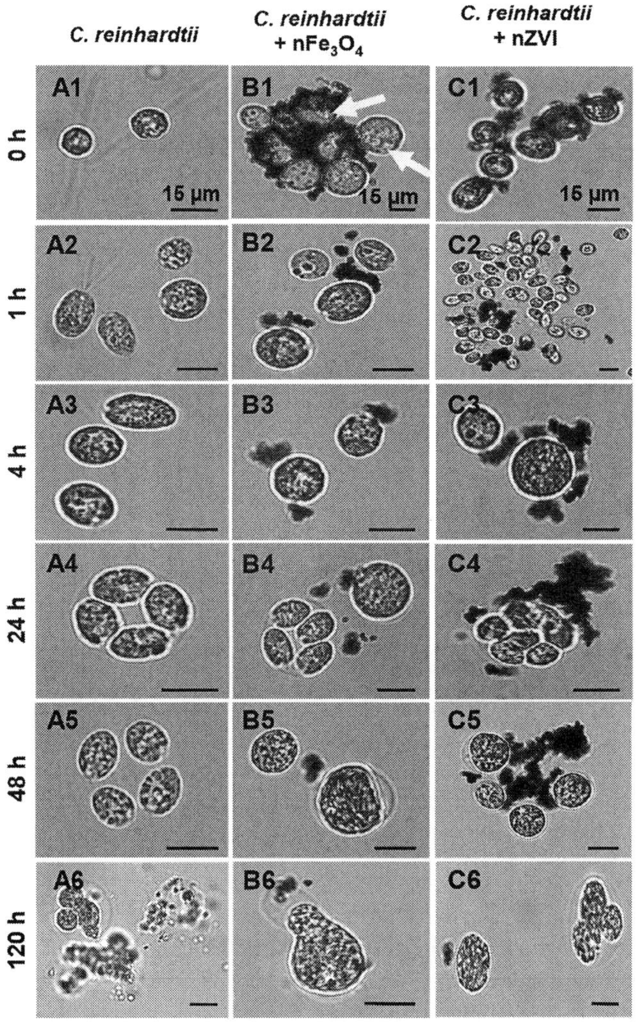

Figure 1 Microscopic imaging of *C. reinhardtii* exposed to nFe_3O_4 and nZVI up to 120 h. (A) Untreated *C. reinhardtii*; (B) *C. reinhardtii* exposed to 100 mg/L nFe_3O_4 (C) *C. reinhardtii* exposed to 100 mg/L nZVI. Scale bar = 15 μm.

3.2. Phenotypic responses of algae upon NPs exposure

The overall phenotypic responses of *C. reinhardtii* was not statistically signicant until 24 h (**Figures 2A-2D**), except NPQ (**Figure 2E**). In the nZVI (indicated by red line), the NPQ value was the lowest comparing three exposure conditions and was significantly affected within 1 h, and these NPQ values continued to decrease until 48 h, and thereafter increased sharply by 120 h, suggesting that cells may be at stress from the initial NPs exposure while were able to recover from the NPs stress after several days. Phenotypic results summarized in (**Figure 2**) display almost no detrimental effect of NPs to *C. reinhardtii*, as their values are similar to control condition (green line). Furthermore, the smaller fluroescence intensity of photosystem II observed in both NPs treatments could be partly due to the shading effect of NPs (**Figure 2F**), whereas *C. reinhardtii* almost regained its photosystem II after 120 h. Exposure to NPs would also caused oxidative stress or DNA damage in cells, due to the oxidation of nZVI and generation of hydroxyl radical [15]. Standard toxicity endpoints are depending mostly on the phenotypic responses of microorganisms such as growth inhibition, or oxidative stress formation. To comprehend the underlying stress beneath the cell membrane on how the cell would defend oxidative insults, more toxicity endpoints are needed for evaluation, such as transcriptomic analysis to better reveal hidden pattern of cell defense [16,17].

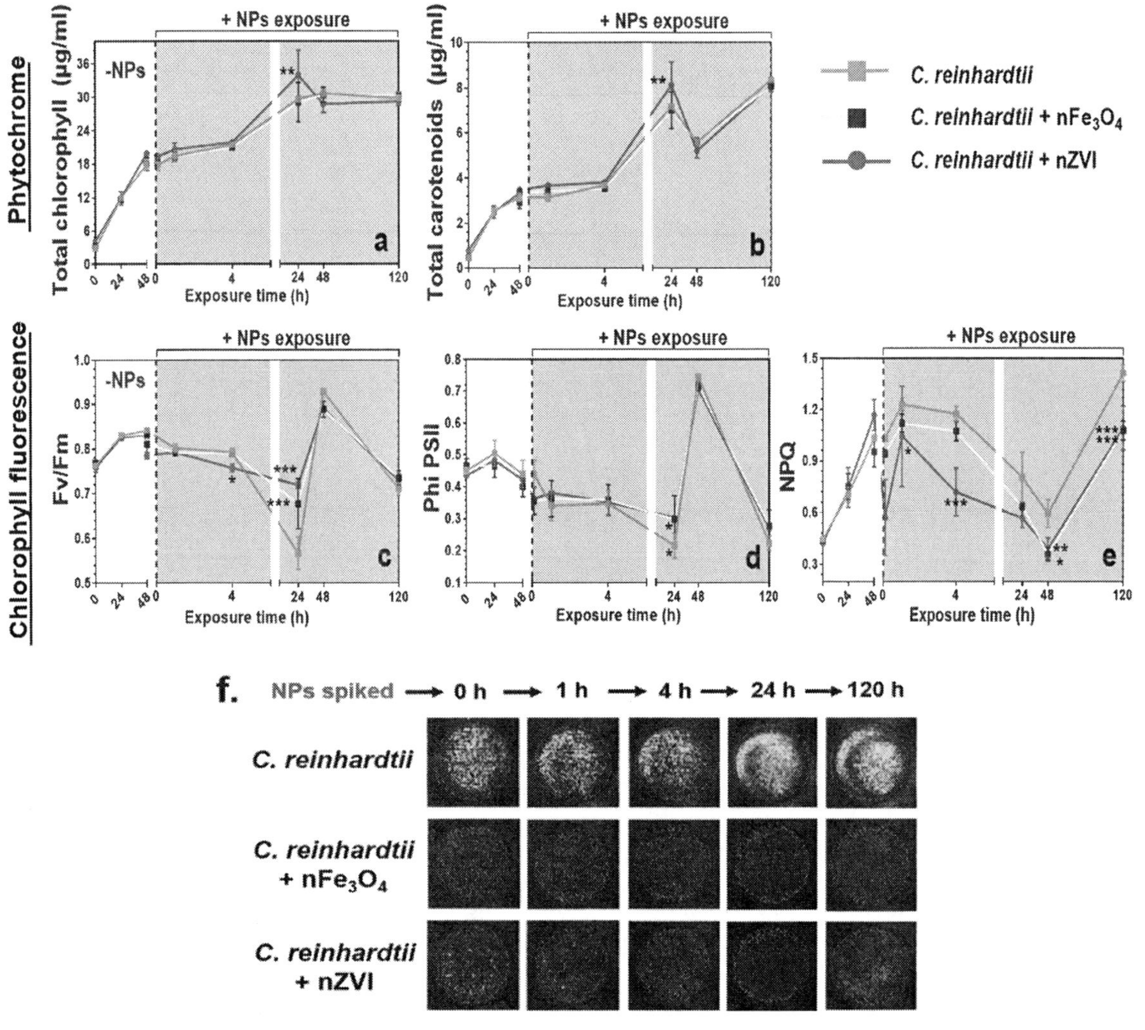

Figure 2 Phenotypic responses of *C. reinhardtii* exposed with nZVI and nFe_3O_4 in HSM. (A,B) Phytochrome content (Total chlorophyll and carotenoid); (C,D,E) Photosystem II efficiency (F_v/F_m, non-photochemical quenching parameters, PSII photochemistry), (mean ± SD; n = 3); (F) Light intensity of chlorophyll fluorescence imaged under FluorCam FC 800-C Video Imager. *= level of significance changes were compared to untreated exposure medium containing only *C. reinhardtii* (*C. reinhardtii,* solid green line), (*P< 0.05, **P < 0.001, ***P < 0.0001). Note: different scales for the y-axis.

3.3. Viable cell numbers of algae upon NPs exposure

The effect of both NPs on *C. reinhardtii* cells was assessed by counting the viable cell (intact cell) and dead cell (cell lost its pigment/plasmolyzed cells) and the results are shown in (**Figure 3**). Both NPs inhibited the algae cell as observed by the reduction of cell viability, however, in a different pace of action. The cell viability in the control was stable with more than 80 % viable cells during the whole experiment (**Figure 3A**). nFe_3O_4 caused a decrease in viability by 10 % comparing to the control - approximately 69 % viable cells were recorded after one hour and this proportion persisted until the end of the study (**Figure 3B**). Effect of nZVI was rather slow and the highest proportion of dead cells occured after 48 h - 30 % decrease in viability comparing to control (**Figure 3C**).

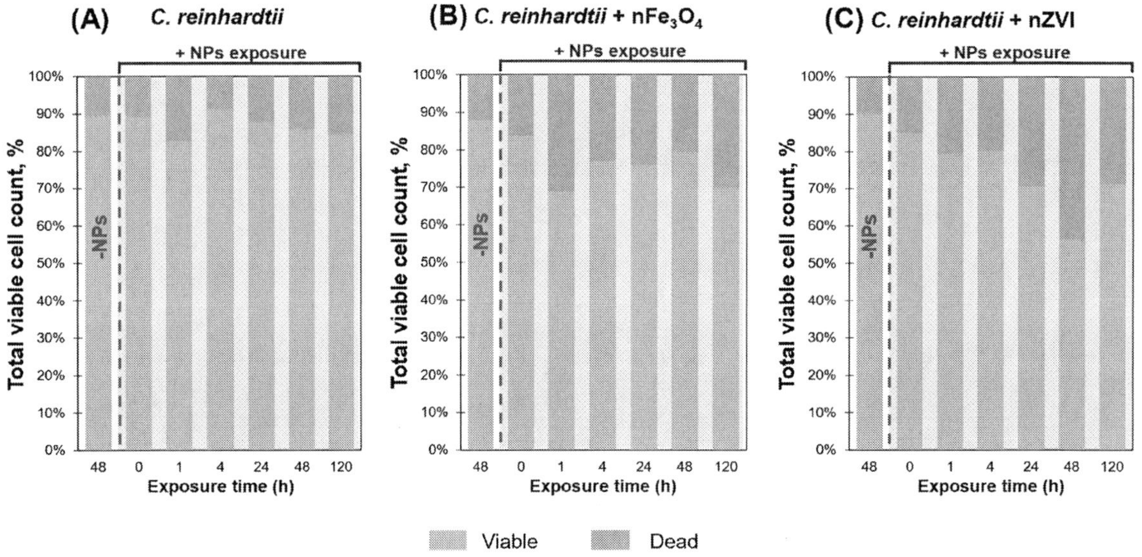

Figure 3 Total viable intact cell and dead cell, determined by manual counting using haemocytometer under light microscope. Values were expressed in percentage, (mean; n = 3).

4. CONCLUSION

Iron NPs caused minor negative effect to *C. reinhardtii* during the initial hours of exposure, while the culture was able to recover towards to the end of the study, probably due to nZVI oxidation, harmless iron oxides and new cell generations. This study amends the current knowledge about the potential risks of nZVI application in the environment.

ACKNOWLEDGEMENTS

This project was financially supported by the Technical University of Liberec through project No. PURE-2019800 (Environmental fate, behaviour and biological effects of engineered nanomaterials) and partly by German Academic Exchange Service short term grants, 2019 (57440917). The authors acknowledge the assistance provided by the Research Infrastructures NanoEnviCz (Project No. LM2018124) supported by the MŠMT of the Czech Republic

REFERENCES

[1] CZINNEROVA, M., N.H.A. NGUYEN, J. NEMECEK, et al. In situ pilot application of nZVI embedded in activated carbon for remediation of chlorinated ethene-contaminated groundwater: effect on microbial communities. *Environmental Sciences Europe*. 2020, vol. 32, no. 1, p. 154.

[2] FILIP, J., T. CAJTHAML, P. NAJMANOVÁ, et al. *Advanced Nano-Bio Technologies for Water and Soil Treatment.* Cham, Switzerland: Springer, 2020.

[3] CZINNEROVÁ, M., O. VOLOŠČUKOVÁ, K. MARKOVÁ, et al. Combining nanoscale zero-valent iron with electrokinetic treatment for remediation of chlorinated ethenes and promoting biodegradation: A long-term field study. *Water Research.* 2020, vol. 175, p. 115692.

[4] FAN, J., Y. HU, and X. LI. Nanoscale Zero-Valent Iron Coated with Magnesium Hydroxide for Effective Removal of Cyanobacteria from Water. *ACS Sustainable Chemistry & Engineering.* 2018, vol. 6, no. 11, pp. 15135-15142.

[5] WU, M. and S. HUANG. Magnetic nanoparticles in cancer diagnosis, drug delivery and treatment. *Molecular and clinical oncology.* 2017, vol. 7, no. 5 .

[6] KARLSSON, H.L., J. GUSTAFSSON, P. CRONHOLM, et al. Size-dependent toxicity of metal oxide particles-A comparison between nano- and micrometer size. *Toxicology Letters.* 2009, vol. 188, no. 2, pp. 112-118.

[7] LEI, C., L. ZHANG, K. YANG, et al. Toxicity of iron-based nanoparticles to green algae: Effects of particle size, crystal phase, oxidation state and environmental aging. *Environmental Pollution.* 2016, vol. 218, pp. 505-512.

[8] HURTADO-GALLEGO, J., G. PULIDO-REYES, M. GONZÁLEZ-PLEITER, et al. Toxicity of superparamagnetic iron oxide nanoparticles to the microalga Chlamydomonas reinhardtii. *Chemosphere.* 2020, vol. 238, p. 124562.

[9] ORGANISATION FOR ECONOMIC CO-OPERATION AND DEVELOPMENT (OECD). *Guidance Document on aquatic and Sediment Toxicological Testing of Nanomaterials* [online]. 2020 [accessed. 15. September 2020]. Retrieved from: www.oecd.org/chemicalsafety/

[10] NANO IRON. *Manual for preparation of an aqueous suspension from dry stabilized iron powder NANOFER STAR* [online]. [accessed. 14. September 2020]. Available from: https://nanoiron.cz/getattachment/7aa3e8c5-5701-4a47-b05c-b91f87279aae/NANOFER-STAR-processing-activation-manual.aspx

[11] SUEOKA, N. Mitotic Replication of Deoxyribonucleic acid in Chlamydomonas reinhardi. *Proceedings of the National Academy of Sciences.* 1960, vol. 46, no. 1, pp. 83-91.

[12] LICHTENTHALER, H.K. Chlorophylls and carotenoids: Pigments of photosynthetic biomembranes. *Methods in Enzymology.* 1987, vol. 148, pp. 350-382.

[13] WELLBURN, A.R. The spectral determination of chlorophylls a and b, as well as total carotenoids, using various solvents with spectrophotometers of different resolution. *Journal of plant physiology.* 1994, vol. 144, no. 3, pp. 307-313.

[14] MURCHIE, E.H. and T. LAWSON. Chlorophyll fluorescence analysis: a guide to good practice and understanding some new applications. *Journal of Experimental Botany.* 2013, vol. 64, no. 13, pp. 3983-3998.

[15] ŠEVCŮ, A., Y.S. EL-TEMSAH, E.J. JONER, et al. Oxidative stress induced in microorganisms by zero-valent iron nanoparticles. *Microbes and environments.* 2011, vol. 26, no. 4, pp. 271-281.

[16] YEAP, C.S.Y., N.H.A. NGUYEN, R. SPANEK, et al. Dissolved iron released from nanoscale zero-valent iron (nZVI) activates the defense system in bacterium Pseudomonas putida, leading to high tolerance to oxidative stress. *Journal of Hazardous Materials.* 2022, vol. 439, p. 129627.

[17] GUO, Z., Y. LUO, P. ZHANG, et al. Deciphering the particle specific effects on metabolism in rat liver and plasma from ZnO nanoparticles versus ionic Zn exposure. *Environment International.* 2020, vol. 136, p. 105437.

DESTRUCTION OF FIBROUS STRUCTURES DURING MACHINING OF CARBON FIBER COMPOSITES

[1]Tomáš KNÁPEK, [1]Dora KROISOVÁ, [1]Štěpánka DVOŘÁČKOVÁ, [1]Artur KNAP

[1]Technical University of Liberec, Liberec, Czech Republic, EU,
tomas.knapek@tul.cz, dora.kroisova@tul.cz, stepanka.dvorackova@tul.cz, artur.knap@tul.cz

https://doi.org/10.37904/nanocon.2022.4612

Abstract

A large part of composite components undergoes a finishing machining process before assembly. Composite systems - carbon fiber/epoxy resin behave differently than metals during machining. The released polymer particles, particularly the carbon fiber segments, can affect human health. The presented study deals with the release of fibers from the composite materials - the formation of chips and their destruction or breaking into shorter segments related to the machining process. Composite samples with recycled carbon fibers and carbon roving were subjected to selected machining processes. The scanning electron microscope was used to characterize the carbon fibers used in both types, especially to study the destruction of fibrous structures formed during machining processes. Based on the study, it was found that at the ends of the carbon fibers, there is a cleavage of microscopic fragments whose dimensions range from 2 to 3 μm, which is the WHO limit size for particles that can be inhaled. Dimensions of separate surface layers are micrometers in terms of length/width and the sub-micrometer level in terms of their thickness. Layers separated from the surface of the carbon fiber were found both in the case of the use of new carbon fibers and recycled carbon fibers.

Keywords: Machining, composite systems, carbon fibers, epoxy resin

1. INTRODUCTION

Composite material components have become commonly used technical elements in recent decades. Advanced composites with carbon fiber reinforcement in roving or fabrics find application from the aerospace and automotive industries to objects of everyday use. The significant expansion of these systems is due to their physical and mechanical parameters, which are responsible for the carbon fibers used, which provide the products with high strength, stiffness, resistance to dynamic stress, low density, corrosion resistance, and maximum dimensional stability [1]. No matter where these products are finally used, it is impossible to do without some of the finishing machining operations associated with the growing production of composite parts. However, unlike metal parts, composite systems behave differently during machining due to their material composition. Chips and dust are formed based on short carbon fibers, most often stored in epoxy or polyester resins [2,3]. The machining method, the type of tool, and the process conditions affect the nature of the resulting chip or the length of the broken carbon fibers, the fiber fracture surface, and whether they remain connected to the matrix or are released as individual fibers or composite segments. The destruction or splitting of the fibers along their axis is related to the type of fibers, their production method, and used material. The potential toxicological hazard is conditioned by the internal structure of carbon fibers, the nature of the fiber structures formed during machining processes, and the number of particles generated in a given volume over a given period selected by the technology [4]. The issue of machining composite systems is currently well described. However, new studies are still emerging that deal with the issue of tool wear, optimization of machining conditions, fiber pulling, and the nature of the machined surface [5-10]. Less work deals exclusively with the issue of chip formation and dust particles released from composite materials based on carbon fiber/polymer matrix, which is formed in connection with the machining process, or other related operations and their conditions and arrangement of fiber reinforcement. Already the work of 1978 states that a small number of particles can be inhaled, of which only less than 1 % are fibrous particles [11]. The particles that can be inhaled have a diameter ranging from 1 to 2.5 µm and a length of up to 15 µm. Further work shows that the mechanical chopping of new fibers produces a significant number of particles or fiber segments with sharp edges, which have the character of particles and the dimensions of inhaled particles [12]. Particles or fragments with a diameter of fewer than 3 micrometers can be inhaled as expressed by WHO [13]. Other published works also addressed the environmental impact and impact on human health in terms of the size of released segments during machining and the possibility of splitting carbon fibers along the axis, which can be very undesirable [14,15,16]. It is necessary to seriously address the toxicological and environmental impacts of working with carbon fibers and recycled carbon fibers and the waste generated by them, especially in terms of the potential risks they may cause in the short and long time.

2. MATERIALS AND METHODS

The two-component low molecular weight epoxy resin ChS-EPOXY 520 (Districhem a. s., Czech Republic) was selected as a matrix for the production of composite samples. Hardener T0492, added in a weight ratio of 100:26 (Districhem a. S., Czech Republic), was used to cure the resin. Carbon fibers were used in two forms as short, recycled fibers Carbiso TM MF (Easy Composites Ltd., Great Britain) with diameters of 7-8 µm and an average length of 100 µm (**Figure 1a**) and carbon roving 3700 tex 50K (HAVEL COMPOSITES CZ, s.r.o.) with fiber diameters 8 µm (**Figure 1b**).

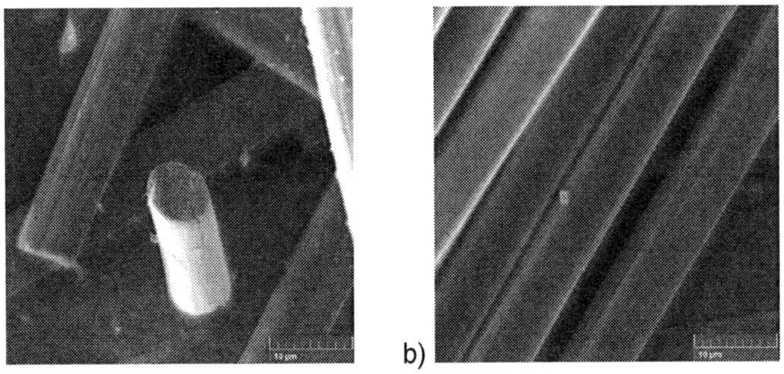

a) b)

Figure 1 (**a**) Recycled carbon fibers, detailed image of the fiber surface and selected fracture area. (**b**) Carbon fibers roving, detailed image of fibers surface.

Both types of composite samples with recycled carbon fibers and carbon roving were identically subjected to selected machining processes, namely milling and grinding. In the case of models with parallel arranged carbon roving, milling and grinding in the direction parallel to the fibers and the direction perpendicular to the fibers were used.

Milling was performed as a face milling machine on a universal cantilever milling machine FNG 32 (TOS Olomouc, Czech Republic). A cylindrical end mill Ø 25 mm with replaceable inserts TNGX 100404SR-F, Grade M6330 (Pramet Šumperk, Czech Republic) was used for milling the samples. The cutting conditions for the selected milling were as follows: speed n = 1000 [min^{-1}], depth of cut a_p = 1 [mm], feed rate v_f = 75 [mm. min^{-1}]. Milling was performed in the absence of process fluid.

Grinding was performed on a BPH 320 A surface grinder (Junker, Germany). A corundum grinding wheel (Tyrolit, Austria) T1 250x32x76 98A 60 K 9 V 40 was used for grinding. Cutting conditions of the selected grinding were as follows: cutting speed v_c = 35 [$m.s^{-1}$], depth of cut a_p = 0.05 [mm], feed rate v_f = 10 [$m.min^{-1}$], grinding was performed in the absence of process fluid.

The TESCAN MIRA3 scanning electron microscope was used to characterize the carbon fibers used in both types, especially to study the destruction of fibrous structures formed during machining processes. The fiber diameter and length, the fiber surface morphology, the fiber surface character, and the fibers' adhesion to the matrix used were evaluated. Samples surfaces were sputtered with a Pt-Pd layer (Quorum Q150R ES) with a 2-4 nm thickness.

3. RESULTS

When milling a recycled carbon fibers sample, fibers are shortened. The average length of fibers is 100 μm but fibers size of 20 to 40 μm were identified. In the images (**Figure 2**), fibers with perpendicular fracture and angles of 40° to 60° fracture were identified. Perpendicular fracture fibers can be released from the polymer matrix in the milling process without destroying fibers. Fracture at an angle of 40° to 60° (**Figure 2b**) is typical for the machining process [10]. **Figure 2b** observed that a thin layer of sub-micrometer size was broken from the fiber surface. **Figure 2c** shows a fiber end jagged break. This type of break was not a unique case, so it may occur in recycled fibers and composite materials. It is evident (**Figure 2c**) that uneven fiber destruction resulting fragments released from the fiber surface. These layers reach dimensions at the sub-micrometer level.

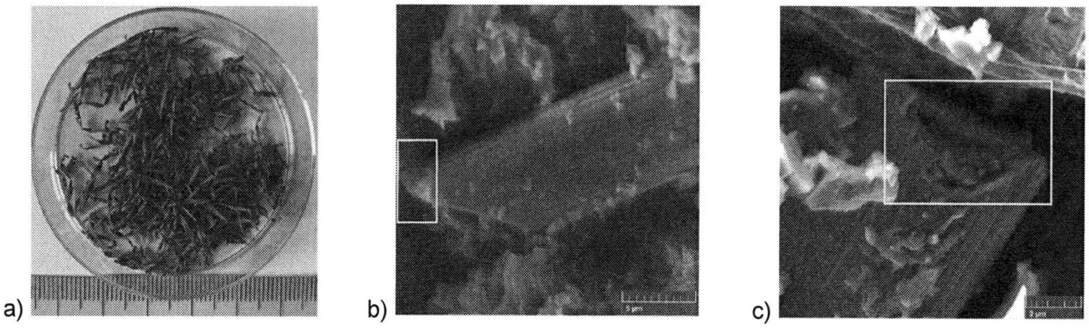

Figure 2 Chip formed after milling a composite sample with recycled carbon fiber. (**a**) Chip overview image.; (**b**) Detailed image of chipped fiber surface layer.; (**c**) Detailed image of the fiber end.

When milling the sample perpendicular to the distribution of carbon fibers in the sample, a coarser chip is formed with a length in units of centimeters and a width in hundreds of micrometers (**Figure 3a**). This chip breaks down into individual segments with dimensions in the tens to hundreds of micrometers (**Figure 3b**), which are made of a heterogeneous material - carbon fibers joined by an epoxy matrix. It is evident (**Figure 3b and Figure 3c**) that the fibers are shortened to tens of micrometers in length, the break at the end of the fibers not being perpendicular but at an angle, similar to the previous case. The surface of the fibers is smooth, without any residual resin. Tiny polymer particles formed during milling adhere to the surface.

Figure 3 Chip formed after milling a composite sample with carbon fiber in the transversal direction. (**a**) Chip overview image.; (**b**) Chip segments with a noticeable connection of fibers and polymer.; (**c**) Detailed image of the fracture at the end of the carbon fiber.

When milling the sample in the longitudinal direction to distribute the carbon fibers in the sample, both more delicate and coarser chips are formed. The coarser and larger parts contain long carbon fibers that come loose during the milling process (**Figure 4a and Figure 4b**). Residues of the polymer matrix are visible on the carbon fibers (**Figure 4b and Figure 4c**). The refraction of the carbon fibers was both perpendicular and at an angle to the direction of the axis fiber.

Figure 4 Chip formed after milling a composite sample with carbon fiber in the longitudinal direction. (**a**) Chip overview image.; (**b**) Carbon fibers are released during the milling process in the longitudinal direction to the fiber-bearing axis.; (**c**) Detailed image of carbon fiber with epoxy resin residues on its surface and the refractive character of the fiber.

A fine chip with a dust character is formed when grinding a sample with recycled carbon fibers (**Figure 5a**). This dust is created by loose carbon fibers connected to the polymer matrix or fibers alone. The average length of used recycled fibers is 100 μm. Fibers with a size of about 20 micrometers were identified in the samples of the collected chips. **Figure 5b** shows a chipped segment at the lower end of the fiber. **Figure 5c** shows the chipping of the surface layer at the left edge of the fiber, the chipped fragment of micrometer dimensions can be seen at the right edge. According to the WHO, carbon fiber fragments are approaching dimensions that are considered potentially dangerous, according to the WHO [13].

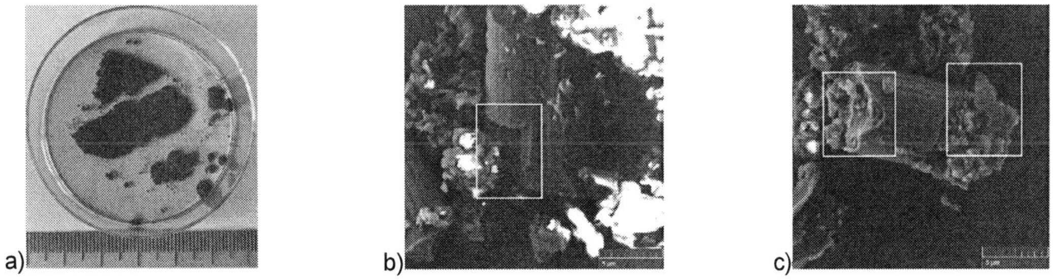

Figure 5 Chip formed during grinding a composite sample with recycled carbon fiber. (**a**) Chip overview image.; (**b**) Snapshot of the fiber with the broken end.; (**c**) Image of a fiber fragment with a chipped surface layer and a fragment.

Figure 6 Chip formed during grinding of a composite sample with carbon fiber in the transverse direction. (**a**) Chip overview image.; (**b**) Overview image of chips with different lengths of carbon fibers.; (**c**) Detailed image of the fracture surface of carbon fiber.

When grinding the sample in the direction perpendicular to the distribution of carbon fibers, a delicate chip with millimeter particle dimensions is formed. The ground resin particles agglomerate and adhere to the surface of the fibers. The chip samples contain fibers with lengths of tens to hundreds of micrometers. The fiber fractures are at an angle of about 40° to 60°.

a) b) c)

Figure 7 Chip formed during grinding of a composite sample with carbon fiber in the longitudinal direction. (**a**) Chip overview.; (**b**) Chip overview with loose long carbon fibers.; (**c**) Detailed image of carbon fiber end with a fracture surface and a chipped surface layer.

When grinding the sample in the longitudinal direction to distribute the carbon fibers in the model, a delicate chip is formed (**Figure 7a**). There is a considerable amount of loose carbon fibers with a length of up to several hundred micrometers (**Figure 7b**). The ground resin adheres to the surface of the carbon fibers (**Figure 7c**). The fracture surface is seen at the end of the carbon fiber. **Figure 7c** shows the destruction of the end of the fiber at an angle at which, however, there is no longitudinal splitting of the fiber. On the surface, the separation of the surface layer is visible as in the previous cases.

4. DISCUSSION

Based on the study, it was found that at the ends of the carbon fibers, there is a cleavage of microscopic fragments (**Figure 5b and Figure 5c**), whose dimensions range from 2 to 3 μm, which is the WHO limit size for particles that can be inhaled [13]. In several cases, it has been observed that part of the surface layer of the fiber separates at the ends of the fracture surfaces of the carbon fibers (**Figure 2b, Figure 2c, Figure 5c, Figure 7c**). Dimensions of separate surface layers are in micrometers in terms of length/width and the sub-micrometer level in terms of their thickness. Separated layers from the ends of the carbon fibers are practically impossible to identify in the surrounding material formed during the machining process. However, by comparing the ends of the fibers before and after machining, it is clear that the layers separate. The explanation of the layer separation in the above description lies in the nature of the internal microstructure of the carbon fibers. The internal microstructure depends on the raw material used and the production process. During the process, fibers with a diameter of 7 to 8 μm are formed, but they are also internally structured. The internal structure consists of graphite layers, which are more or less arranged and parallel to the fiber axis, thus providing the fibers with high values of mechanical parameters [18,19].

The bonds in the graphite layers are strong, in contrast to the bonds between the graphite layers, which are relatively weak. There may be an explanation of the layer separation/splitting above described under mechanical loading by machining processes. Assuming that the used tool hits directly into the carbon fiber, both its destruction in the fiber axis and the chipping of the surface layer will occur (**Figure 7c**). The similar nature of the fragmentation described in the present study has been discussed in the work of other authors [12,16,20].

Identified microscopic fragments analogous to, for example, (**Figure 5c**) with sizes from 2 to 3 micrometers were observed in half of the analyzed samples. Placed sub-micrometer surface layers from the carbon fiber end as in the images (**Figure 2, Figure 5, Figure 7**) were in one-third of all evaluated samples. Due to the large number of small dimension particles generated during the machining of composite samples, it is clear that the number of particles of these dimensions is considerable and should be adequately monitored.

5. CONCLUSION

The experimental study aimed to monitor the fracture surfaces of the carbon fiber ends, reinforcing a composite system with an epoxy matrix. It was found that there are standard types of carbon fiber fractures with the selected machining methods realized under the given conditions. In these processes, small fiber pieces with dimensions of 2 to 3 μm are split simultaneously at the ends of the fibers. Sub-micrometer surface layers separate on the surface of the fiber ends. Comparing the fiber ends before and after machining makes it clear that the separation of layers and segments occurs during the process.

The reason for this study was to assess the possibility of generating micrometer and sub-micrometer fragments due to machining processes, which are currently implemented in many workplaces where finishing operations are performed on composite components. Small parts of the fragmented carbon surface layers may be similar to graphite layers in which the interlayer bonds are not significant, and they are identical to nanocarbon structures. So, it is necessary to address this issue in more depth and assess the potential hazards of these wastes in further studies and steps to eliminate them.

ACKNOWLEDGEMENTS

This research study was funded by the Student Grant Competition of the Technical University of Liberec under the project number SGS-2022-5043. "Research and development in the field of machining of metal and composite materials using new knowledge for industrial practice"

REFERENCES

[1] MALICK, P.K. *Fiber-Reinforced Composites: Materials, Manufacturing and Design*, 3rd ed.; CRC Press: Boca Raton, Florida, USA, 2007. Available from: https://doi.org/10.1201/9781420005981.

[2] UHLMANN, E.; SAMMLER, F.; RICHARZ, S.; HEITMÜLLER, F.; BILZ, M. Machining of carbon fibre reinforced plastics. *Procedia CIRP.* 2014, vol. 24, pp. 19-24. Available from: https://doi.org/10.1016/j.procir.2014.07.135.

[3] KARATAS, M.; GÖKKAYA, H. A review on machinability of carbon fiber reinforced polymer (CFRP) and glass fiber reinforced polymer (GFRP) composite materials. *Defence Technology.* 2018, vol. 14, pp. 318-326. Available from: https://doi.org/10.1016/j.dt.2018.02.001.

[4] RAMULU, M.; KRAMLICH, J. Machining of fiber reinforced composites: Review of environmental and health effects. *Int. J. Environ. Conscious Des. Manuf.* 2004, vol. 11, pp. 1-19.

[5] KOMANDURI, R.; ZHANG, B.; VISSA, C.; Machining of fiber reinforced composites. *Processing and Manufacturing of Composite Materials* 1991, vol. 112, pp. 817-829. Available from: https://doi.org/10.1016/0890-6955(94)00055-O.

[6] AZMI, A. I. Chip formation studies in machining fiber reinforced polymer composites. *International Journal of Materials and Product Technology.* 2013, vol. 112, pp. 1-36. Available from: https://doi.org/10.1504/IJMPT.2013.052790.

[7] UHLMANN, E.; SAMMLER, F.; RICHARZ, S.; REUCHER, G.; HUFSCHMIED, R.; FRANK, A.; PROTZ, F. Machining of carbon and glass fibre reinforced composites. *Procedia CIRP.* 2016, vol. 46, pp. 63-66. Available from: https://doi.org/10.1016/j.procir.2016.03.197.

[8] CAGGIANO, A. Machining of fiber reinforced plastic composite materials. *Materials* 2018, vol. 11, pp. 442. Available from: https://doi.org/10.3390/ma11030442.

[9] KODAMA, H.; OKAZAKI, S.; JIANG, Y.; YODEN, H.; OHASHI, K. Thermal influence on surface layer of carbon fiber reinforced plastic (CFRP) in grinding. *Precision Engineering.* 2020, vol. 65, pp. 53-63. Available from: https://doi.org/10.1016/j.precisioneng.2020.04.005

[10] TENG, G.; CHANGE, L.; MIN, Y.; YANBIN, Z.; DONGZHOU, J.; WENFENG D.; SUJAN, D.; TIANBIAO Y.; ZAFAR, S.; JUN, W. Mechanics analysis and predictive force models for the single-diamond grain grinding of carbon fiber reinforced polymers using CNT nano lubricant. *Journal of Materials Processing.* 2021, vol. 290. Available from: https://doi.org/10.1016/j.jmatprotec.2020.116976.

[11] HOLT, P. F.; HORNE, M. Dust from carbon fiber. *Environmental Research.* 1978, vol. 17, pp. 276-283. Available from: https://doi.org/10.1016/0013-9351(78)90030-0.

[12] MAZUMDER, M. K.; CHANG, R. J.; BOND, R. L. Aerodynamic and morphological properties of carbon-fiber aerosols. *Aerosol Science and Technology.* 1982, vol. 1, pp. 427-440. Available from: https://doi.org/10.1080/02786828208958606.

[13] The WHO/EURO man-made mineral fiber reference scheme, WHO/EURO Technical Committee for Monitoring and Evaluating MMMF. *Scand J Work Environ Health.* 1985, vol. 11, no. 2, pp. 123-9. Available from: https://doi.org/10.5271/sjweh.2251, PMID: 3890161.

[14] BOATMAN, E. S.; COVERT, D.; KALMAN, D.; LUCHTEL, D.; OMENN, G. S. Physical, morphological, and chemical studies of dust derived from the machining of composite-epoxy materials. *Environmental research.* 1988, vol. 45, pp. 242-255. Available from: https://doi.org/10.1016/S0013-9351(89)80070-2.

[15] WANG, J.; SCHLAGENHAUF, L; SETYAN, A. Transformation of the released asbestos, carbon fibers and carbon nanotubes from composite materials and the changes of their potential health impacts. *Journal of nanobiotechnology.* 2017, vol. 15, pp. 1-16. Available from: https://doi.org/10.1186/s12951-017-0248-7.

[16] KEHREN, D.; SIMONOV, B.; BÄGER, D.; DZIUROWITZ, N.; WENZLAFF, D.; THIM, C.; PLITZKO S. Release of respirable fibrous dust from carbon fibers due to splitting along the fiber axis. *Aerosol and air quality research.* 2019, vol. 19, pp. 2185-2195. Available from: https://doi.org/10.4209/aaqr.2019.03.0149.

[17] DVOŘÁČKOVÁ, Š.; KROISOVÁ, D. Thermal Expansion of Composite System Epoxy Resin/Recycled Carbon Fibers. *Materials Science Forum.* 2020, vol. 994, pp. 162-169.

[18] DIEFENDORF, R. J.; TOKARSKY, E. High-performace carbon fibers. *Polymer Engineering and Science.* 1975, vol. 15, pp. 150-159. Available from: https://doi.org/10.1002/pen.760150306.

[19] DRESSELHAUS, M. S. Fifty years in studying carbon-based materials. *Physica Scripta.* 2012. Available from: https://doi.org/10.1088/0031-8949/2012/T146/014002.

[20] SCHLAGENHAUF, L.; KUO, Y. Y.; MICHEL, S.; TERRASI, G.; WANG, J. Exposure assessment of high-energy tensile test with large carbon fiber reinforced polymer cables. *Journal of occupational and environmental hygiene.* 2015, vol. 12, pp. 178-183. Available from: https://doi.org/10.1080/15459624.2015.1029614.

NANOPARTICLES FROM BRAKE PADS WEAR

[1]Eva ŠVÁBENSKÁ, [1,2]Pavla ROUPCOVÁ, [1]Martin FRIÁK, [1]Oldřich SCHNEEWEISS

[1]*Institute of Physics of Materials, AS CR, Brno, Czech Republic, EU,*
svabenska@ipm.cz, schneew@ipm.cz, friak@ipm.cz

[2]*CEITEC Brno university of Technology, Brno, Czech Republic, EU,*
pavla.roupcova@ceitec.vutbr.cz

https://doi.org/10.37904/nanocon.2022.4616

Abstract

Exposure to particulate air pollution has been associated with a variety of health problems. One of the main sources of metal-rich airborne particulate pollution in roadside environments are brake-wear emissions. The scanning electron microscopy (SEM) and transmission electron microscopy (TEM), Mössbauer spectroscopy (MS), X-ray powder diffraction (XRD) was used to a make quantitative evaluation of wear particles in two samples. Sizes, morphologies, oxidation state, and trace element compositions of wear particles were investigated using combination of several methods revealed high concentrations of Fe species and phases, often associated with carbonaceous material. SEM and TEM observations show that brake wear particles with a size of tens to hundreds of nm mainly form particle agglomerates. However, ultrafine (<100 nm) particles are likely to pose a health threat after inhalation and/or ingestion.

Keywords: Wear particles, analysis, TEM, iron oxides, metallic particles

1. INTRODUCTION

Exposure to airborne particulate matter (PM) has been related to some health problems. In particular, exposure to fine (2.5-0.1 μm) and ultrafine (<0.1 μm) particles is associated with deterioration of cardiovascular and neurological health or impaired of pulmonary [1,2]. The airborne PM suspended in air can be of both natural and anthropogenic origin. Anthropogenic PM reaches to the atmosphere due to industrial activity, heating and automobile traffic. The brake emissions account for one of the major sources of airborne ultrafine particles (UFP) in the roadside environment [3,4]. Some works [5,6] reported that brake wear PM contributes to the total PM10 mass by approximately 11 % - 21 %.

The iron is dominant metal in brake-wear particles. Iron chips and/or Fe-containing compounds are one of the many components that brake pads commonly contain [7,8]. However, wear particles are released from the brake pads and at the same time from the brake disc made from cast iron. This is the reason, why Fe-bearing particles often constitute about 50 % of the total mass of particles emitted from brake systems [9]. The various forms of iron inclucing oxides or hydroxides, carbides and or metallic Fe (α-Fe) can be found in Fe-bearing particles in brake emissions. The knowledge of the oxidation state of Fe, particle size distribution and concentration of each Fe-phases may be particularly important [10] in the context of human health hazard as well as a potential association with neurodegeneration and Alzheimer's disease.

From the point of view of studies dealing with braking emissions and the composition of wear particles, we can find two main directions of research. Part of the works is focused on studying the effects of braking conditions on PM production [11]. Others are focusing on particles in the air in the context of selected localities or population groups [12,13]. In our study we use a combination of scanning (SEM) and transmission (TEM) electron microscopy together with X-ray powder diffraction and Mössbauer spectroscopy analysis to obtain chemical and phase composition and morphology of wear particles from automotive brake systems.

2. MATERIALS AND METHODS

Wear particles (debris) were collected by sweeping brake pads obtained from passenger car services. Two samples of wear debris, labelled as F and R, were analysed using combinations of analytical and microscopic techniques.

A TESCAN LYRA 3XMU FEG/SEM scanning electron microscope equipped with an XMax80 Oxford Instruments detector for energy dispersive X-ray analysis (EDX) was used to follow the chemical composition. For the SEM study, the wear particles were captured to an aluminum stub by a double-sided sticky carbon tape. The secondary electron (SE) images and energy dispersive X-ray spectra (EDS) were obtained at 12 kV and 15 kV accelerating voltage, respectively.

A Jeol JEM 2100F HRTEM with a Schottky cathode operating at 200 kV was applied for study the finest fraction of wear particles. For the TEM study, part of samples was placed in a 1.5 ml Eppendorf tube filled with 1 ml pure ethanol, and mix for 2 minutes. Then a drop of the finest fraction was placed onto a copper TEM grid coated with a lacey carbon film and left to dry. The conventional TEM mode were used for bright-field (BF) images, scanning transmission (STEM) mode were applied for high-angle annular dark-field (HAADF) images and EDS elemental maps.

The phase compositions studies were carried out with Empyrean diffractometer (Panalytical) in the Bragg-Brentano geometry with CoKα radiation (λ = 0.17902 nm) at room temperature (RT). The analysis of patterns was realized by HighScore® software and the ICDD (PanAnalytical) database.

The ^{57}Fe Mössbauer measurements were carried out at RT using a $^{57}Co(Rh)$ source in transmission geometry with detection of 14.4 keV gamma radiation (MS) using a standard Mössbauer spectrometer. A pure α-Fe foil was used as a calibration standard for the velocity scale. The computer processing of the spectra was done using CONFIT program package [14].

3. RESULTS AND DISSCUSION

The scanning (SEM) and transmission (TEM) electron microscopy were used to obtain independent information on the composition, structure, morphology and particle size of the brake-wear particles. Overview of prevalent size of brake-wear particles shows SEM image on **Figure 1**. The main fraction of wear particles was around 1 μm in size. The larger, sharp-edged particles were also observed in both samples. Agglomerates composed of different sizes particles were noted.

Figure 1 SEM images of brake wear particles obtained from sample F(left) and R (right)

All particles, agglomerates but also larger grains, have similar elemental composition typical of brake emissions. The dominant elements are Fe, C and O, and Cu, Al, Si, S, Mg, Mn, Zn, and Ca are present to a lesser extent. According to EDX analysis is possible to say that the wear particles predominantly contain iron.

The TEM images show that the agglomerates comprise prolific numbers of particles ~10 nm in size (**Figure 2**, right). The elemental analysis using EDX in STEM mode showed the presence of Fe, O and C, with lower concentration of other elements analogous to result SEM/EDX analysis. However, the distribution of these minor elements is not uniform throughout the agglomerates. The influence of the original material of the brake pads is also evident, when the composition of samples F and R were different. The finest fraction of brake products from sample F contain carbon structures with some degree of crystallographic order (**Figure 2**, left). Interplanar distances of this carbon structures are approximately 3.3 Å, similar structures observed Gonet [15] in agglomerates of ultrafine brake-wear particles. Nanoparticles were trapped in the carbon structure.

Compared to that only agglomerates of nanoparticles without any carbon matrix were observed on sample R (**Figure 2**, right). High concentration of carbon in the form carbon matrix (sample F) can be probably connected with type of resin used for production of this brake pads.

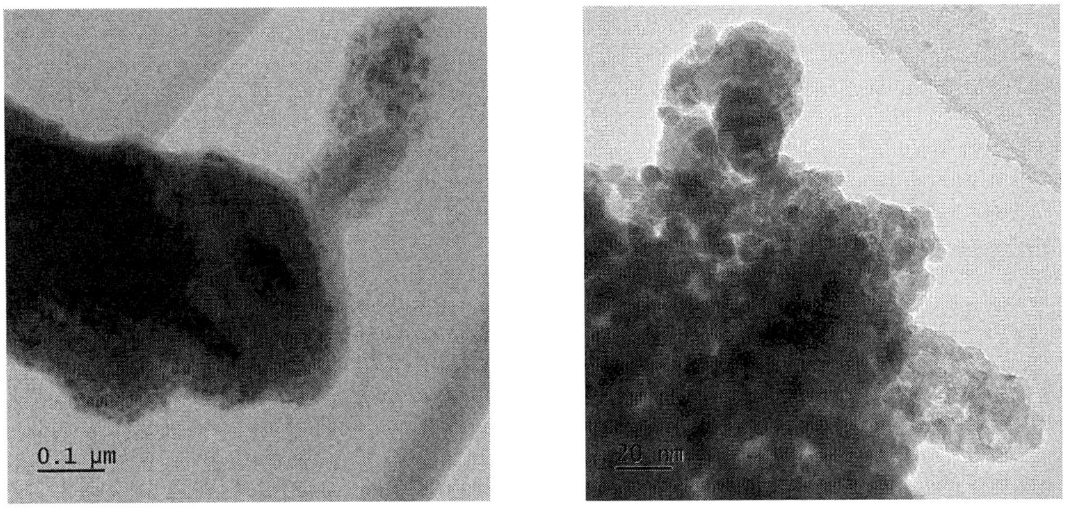

Figure 2 TEM images of brake wear particles. Matrix with trapped nanoparticles in sample F (left) and detail of agglomerate found in sample R (right).

Table 1 The phases composition of wear particles (weight percent) according XRD measurements.

Phase	Formula	Sample	
		R	F
Hematite	Fe_2O_3	46	52
Goethite	FeO.(OH)		22
Magnesium diiron (III) diodixe	$(MgFe_2)O_4$	42	
Iron / ferite	α-Fe		11
Manganese iron carbide	$(Mn_4Fe_3)C_3$	8	
Iron (II) Iron (III) aluminium hydroxide sulfate hydrite	$Fe_2(Fe_{0.76}Al_{0.24})).(OH)_6(SO_4)_{0.5}(H_2O)_{2.4}$	2	
Chalcopyrite	$CuFeS_2$		1
Zhanghengite	CuZn		1
Quartz low	SiO_2		13
Calcite	$CaCO_3$	1	
Carbon / graphite	C	1	

The phase composition of wear particles samples determined by XRD shows that the wear particles contain 80 or more weight percentage of iron-containing compounds. Various iron oxides and iron hydroxides had been noted. The phase composition of wear particles samples is present in **Table 1**. Our previous study [7] was targeted the possible pollution caused by brake pads. The brake disc was determined as a source of the most of particles. The iron oxides, hydroxides and carbides are product of friction and thermal interaction of brake pads and mainly disks. On the contrary, the other minerals, sulphides and carbon, come from brake pads.

Due to high iron content the Mössbauer spectroscopy (MS) was used for more detailed identification of wear particles. Spectra analysis yields the relative contents of individual Fe-containing phases in atomic fraction of iron (A) and corresponding hyperfine parameters of phases: isomer shift (IS), quadrupole shift (Q_{Sh}) and quadrupole splitting (QS_{pl}), and hyperfine induction (B_{hf}). In view of the fact the similar content of individual iron phases, spectra of the same appearance were expected. As the main component of the spectra, a doublet representing the paramagnetic phase was present in both samples (**Figure 3**) [16]. In the sample F comprises was contents of paramagnetic phase in atomic fraction of iron atom equal 1. The spectrum was constituted only by doublet. Compared to that in the spectra of R sample minor sextets can be observed, except dominant doublet.

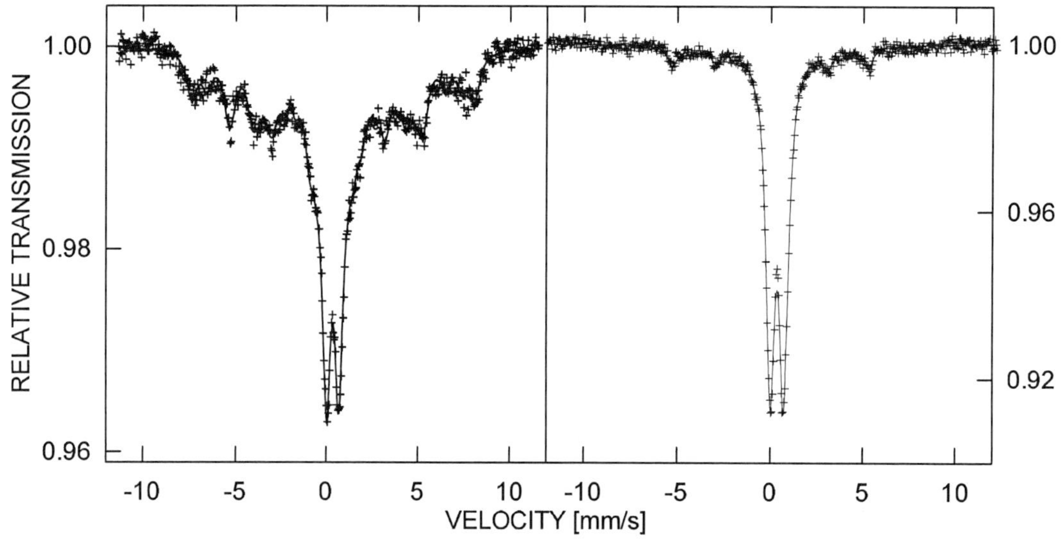

Figure 3 Mössbauer spectrum of the sample R (left) and sample F (right).

The several groups were identified on the basis of corresponding hyperfine parameters for components and attributed to the following phases: (i) Fe based oxides particles with magnetic splitting, (ii) metallic particles of iron, (iii) iron carbon Fe_3C, and (iv) paramagnetic particles. Their contents are reported in **Table 2.**

Table 2 Mössbauer spectroscopy results; Contents of phase groups in atomic fraction of iron atoms for samples measured at RT.

Sample	Fe based oxides	Fe	Iron carbon	Paramagnetic particles
R	0.37	0.16	0.10	0.37
F				1.0

A combination of MS, XRD and TEM results demonstrated that the paramagnetic fraction can be attributed to the presence of fine nanocrystalline particles (≤ 10 nm). The composition of paramagnetic particles of sample F according to the next MS experiments roughly corresponds with phases observed by MS in the sample R.

The iron-bearing particles containing metallic iron and iron carbides that originate from interaction brake pads with brake disc made from cast iron.

4. CONCLUSION

The wear particles originating from brake pads have been studied using a combination of scanning electron microscopy (SEM) and transmission electron microscopy (TEM), Mössbauer spectroscopy (MS), X-ray powder diffraction (XRD). Wear particles are mainly formed by various iron oxides and hydroxides. Based on MS experiments, the presence of the following phases can be defined: (i) Fe-based oxide particles with magnetic cleavage, (ii) metallic iron particles, (iii) iron carbon Fe_3C and (iv) paramagnetic particles. The paramagnetic fraction observed during MS experiments at RT corresponding with presence of fine nanocrystalline particles (≤ 10 nm). The expected particle size in the nm range was confirmed by the TEM results.

Based on current knowledge, it is possible to say that emissions of particles generated by the wear of brake pads/discs can be probably influenced by adjusting the composition of the brake pads. However, great care is needed so that these modifications do not lead to an even greater threat to health and environment.

ACKNOWLEDGEMENTS

We thank Dr. Naděžda Pizúrová and Ivana Podstranská for their help with TEM. We acknowledge CzechNanoLab Research Infrastructure supported by MEYS CR (LM2018110). We appreciate the thoughtful comments provided by the reviewers.

REFERENCES

[1] POPE III, C.A., DOCKERY D.W. Health effects of fine particulate air pollution: lines that connect. *J. Air Waste Manag. Assoc.* 2006, vol. 56, no. 6, pp. 709-742. Available from: https://doi.org/10.1080/10473289.2006.10464485.

[2] MAHER, B.A., GONZÁLEZ-MACIEL, A., REYNOSO-ROBLES, R., TORRES-JARDÓN, R., CALDERÓN-GARCIDUEÑAS, L. Iron-rich air pollution nanoparticles: An unrecognised environmental risk factor for myocardial mitochondrial dysfunction and cardiac oxidative stress. *Environmental Research.* 2020, vol.188, article no. 109816. Available from: https://doi.org/10.1016/j.envres.2020.109816.

[3] GONET, T., MAHER, B.A. Airborne, Vehicle-Derived Fe-Bearing Nanoparticles in the Urban Environment: A Review. *Environ. Sci. Technol.* 2019, vol. 53, pp. 9970−9991. Available from: https://doi.org/10.1021/acs.est.9b01505.

[4] GRIGORATOS, T., MARTINI, G. Brake wear particle emissions: a review. *Environ Sci Pollut Res.* 2015, vol. 22, pp. 2491-2504. Available from: https://doi.org/10.1007/s11356-014-3696-8.

[5] BUKOWIECKI, N., GEHRIG, R., LIENEMANN, P., HILL, M., FIGI, R., BUCHMANN, B., FURGER, M., RICHARD, A., MOHR, C., WEIMER, S., PREVOT, A., BALTENSPERGER, U. *PM10 Emission Factors of Abrasion Particles from Road Traffic.* [online]. 2009. [viewed: 2020-03-24]. Available from: https://trimis.ec.europa.eu/sites/default/files/project/documents/20150710_141622_66365_priloha_radek_1052.pdf.

[6] LAWRENCE, S., SOKHI, R., RAVINDRA, K., MAO, H., PRAIN, H.D., BULL, I.D. Source apportionment of traffic emissions of particulate matter using tunnel measurements. *Atmos. Environ.* 2013, vol. 77, pp. 548-557. Available from: https://doi.org/10.1016/j.atmosenv.2013.03.040.

[7] ŠVÁBENSKÁ, E., ROUPCOVÁ, P. Skryté nebezpečí otěrových částic. *Chem. Listy.* 2022, vol. 116, no.4, pp. 228-234. Available from: https://doi.org/10.54779/chl20220228.

[8] CHAN, D., STACHOWIAK, G.W. Review of automotive brake friction materials. *In: Proceedings of the Institution of Mechanical Engineers, Part D: Journal of Automobile Engineering.* 2004, vol. 218, no. 9, pp. 953-966. Available from: https://doi.org/10.1243/0954407041856773.

[9] SANDERS, P.G., XU, N., DALKA, T.M., MARICQ M.M. Airborne brake wear debris: size distributions, composition, and a comparison of dynamometer and vehicle tests. *Environ. Sci. Technol.* 2003, vol. 37, pp. 4060-4069. Available from: https://doi.org/10.1021/es034145s.

[10] MAHER, B.A. Airborne magnetite- and iron-rich pollution nanoparticles: potential neurotoxicants and environmental risk factors for neurodegenerative disease. *J. Alzheim. Dis.* 2019, vol. 71, no. 2, pp. 361-375. Available from: https://doi.org/10.3233/JAD-190204.

[11] VOJTÍŠEK, M., VACULÍK, M., PECHOUT, M., HOPAN, F., RAJ, A. F. A., PERUMARTI S., HORÁK J. S., POPOVICHEVA, O., ONDRÁČEK, J., DOUŠOVÁ, B. Effects of braking conditions on nanoparticle emissions from passenger car friction brakes. *Science of The Total Environment*. 2021, vol 788, article no. 147779. Available from: https://doi.org/10.1016/j.scitotenv.2021.147779.

[12] MILER, M. Airborne particles in city bus: concentrations, sources and simulated pulmonary solubility. *Environ Geochem Health*. 2021, vol. 43, pp. 2757-2780. Available from: https://doi.org/10.1007/s10653-020-00770-5.

[13] MUXWORTHY, A. R., LAM, C., GREEN, D., COWAN A., MAHER, B. A., GONET, T. Magnetic characterisation of London's airborne nanoparticulate matter. *Atmospheric Environment*. 2022, vol. 287, article no. 119292. Available from: https://doi.org/10.1016/j.atmosenv.2022.119292.

[14] ŽÁK, T, JIRÁSKOVÁ, Y. Confit: mossbauer spectra fitting program. *Surf. Interface Anal.* 2006, vol. 38, pp. 710-714. Available from: https://doi.org/10.1002/sia.2285.

[15] GONET, T., MAHER, B.A., NYIRÖ-KÓSA, I., PÓSFAI, M., VACULÍK, M., KUKUTSCHOVÁ, J. Size-resolved, quantitative evaluation of the magnetic mineralogy of airborne brake-wear particulate emissions. *Environmental Pollution*. 2021, vol. 288, article no. 117808. Available from: https://doi.org/10.1016/j.envpol.2021.117808.

[16] GONSER, U. *Mossbauer Spectroscopy,* Berlin, Springer, 1975.

INTERACTION OF DRUGS WITH SOIL ORGANIC MATTER STUDIED BY MICROCALORIMETRY

[1]Jitka KROUSKÁ, [1]Nadezhda PAPKO, [1]Martina KLUČÁKOVÁ

[1]*Brno University of Technology, Faculty of Chemistry, Brno, Czech Republic, EU,*
krouska@fch.vut.cz

https://doi.org/10.37904/nanocon.2022.4615

Abstract

The usage of pharmaceuticals increases worldwide every year. The residues of these substances are excreted by the human or animal bodies and thus come into direct contact with the environment, mainly soils and ground water. The accumulation of pharmaceuticals in natural ecosystems may be problematic in the future therefore the knowledge of their behaviour, transport and interactions in the ecosystems is essential for their treatment. The most reactive substances in soils are humic as well as fulvic substances. Two fulvic acids stand as the model reactants with different pharmaceuticals in this study. These reactive compounds contain several kinds of active functional groups, such as hydrophilic hydroxylic or carboxylic groups, and also diverse hydrophobic structures. The interactions between fulvic acids and solutions of pure drugs were studied by isothermal titration calorimetry. The aim of the study was to observe the heat effects of the interactions and to describe the intensity of the binding process.

Keywords: Humic substances, drug, interaction, microcalorimetry

1. INTRODUCTION

Soils can be contaminated by pharmaceuticals [1-3]. It has been documented [4-6] that pharmaceuticals are not entirely removed from wastewater in wastewater treatment plants. As a result, pharmaceuticals contaminate surface and ground waters [4,7]. Municipal and hospital wastewater is the main sink of human pharmaceuticals, which are in the end predominantly degraded through processing in sewage treatment plants. However, balances of the input and output of pharmaceuticals into sewage sludge after technological treatment usually reveal that many pharmaceuticals are recalcitrant to biological and abiotic degradation and are therefore not removed quantitatively, thus entering the environment where they represent a risk for both aquatic and terrestrial ecosystems [8-10]. Introduction of drugs to the environment via irrigation is a highly relevant exposure route in semiarid zones where recycled wastewater is an important source of irrigation water [10,11]. Highly mobile drugs have the potential to leach to the groundwater, whereas strongly sorbing pharmaceuticals can accumulate in the top-soil layer. These compounds can subsequently affect the soil microbial community and may be taken up by plants [10,12]. Dissolved organic matter can affect transport and sorption of organic compounds in soils [10,13]. Dissolved organic matter can increase the solute's apparent solubility and therefore enhance its mobility. Alternatively, dissolved organic matter can reduce mobility due to sorption to the soil solid phases. These processes are controlled by the binding affinity of the contaminant to organic matter [10,13].

The most important constituents of organic matter are humic substances [14,15]. There is a growing interest in the investigation of their interactions with different types of pollutants [1-3,14-16]. Limited research has been conducted to determine the occurrence and direct health risks of primary drugs, environmental degradation, transformation, and the fate of pharmaceuticals are poorly understood [17]. Recently published studies are focused mainly on adsorption processes and determination of parameters of different adsorption isotherms, e.g. [2,8,17]. Some authors calculated thermodynamics parameters (as $\Delta G°$, $\Delta S°$ etc.) on the basis of experimentally determined adsorption characteristics [18-20]. Thermodynamic parameters reflecting interactions of drugs in soils have been determined indirectly, usually batch sorption techniques performed at various temperatures. There are no direct, calorimetric studies. On general level, there is a very little knowledge on the thermodynamic consistency and true relevance of indirectly determined parameters. We hope that the thermodynamic parameters have high impact for the behaviour of drugs in nature, their mobility and immobilization ability in soils. However, our approach is different. This contribution is focused on direct determination of the thermodynamic parameters by means of microcalorimetry (similarly as in ref. [21]). The directly determined changes in Gibbs energy, enthalpy, and entropy can provide true values characterized interactions between humic substances and used pharmaceuticals.

2. MATERIALS AND METHODS

Standards of fulvic acids (FA1 - Fulvic acid Nordic Aquatic 1R105F, FA2 - Fulvic acid Pahokee Peat 2S103F) were purchased from the International Humic Substances Society, USA. The pharmaceuticals diclofenac and ibuprofen were purchased from Sigma-Aldrich, Germany. Stock solutions of fulvic acids were prepared by dissolving the precise amount of its powder in pure water (ELGA PURELAB flex), molar concentrations of both acids were 8.56 mM. The solutions of pharmaceuticals were prepared in pure water (diclofenac $6.75 \cdot 10^{-5}$ M, ibuprofen 24.2 mM) as well and all the solutions were stirred overnight.

The calorimetric experiments were performed using the microcalorimeter PEAQ-ITC, Malvern Instruments, at 25 °C. The method is based on the titration of one solution with higher concentration from the syringe (40 μl) into the second solution of lower concentration filled in the measuring cell of the instrument (200 μl) under constant temperature. The heat effects (exothermic or endothermic) occurred during the experiment after each injection are observed as separated peaks on the raw signal of the heat flow detected as the temperature difference between the sample and reference cell of the instrument. The solution of FA was filled in the measuring cell in the case of ibuprofen and filled in the syringe in the case of diclofenac. Experimental setup: 19 aliquots, titration interval 150 s.

3. RESULTS AND DISCUSSION

This study was based on the determination of the thermodynamic parameters and on the comparison of two different fulvic acids (they have different origin) from the calorimetric point of view. The obtained results were evaluated in the instrument software according to the standard procedure. The signal of the blank experiments was subtracted from the real experiment data. Then, in the first step of the data evaluation from the enthalpogram (**Figure 1**) the peaks were integrated to obtain the enthalpies for every single injection.

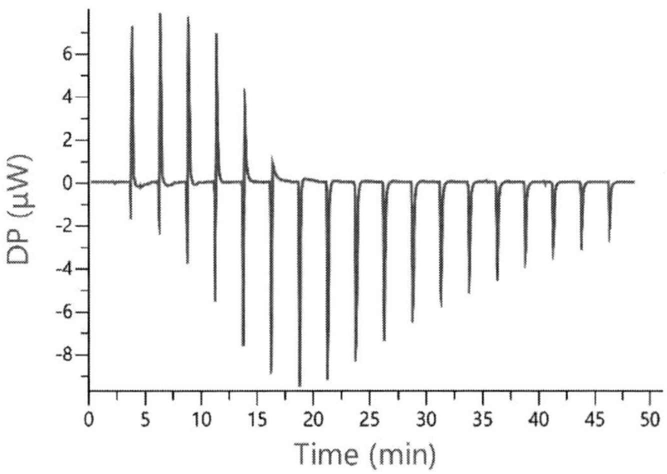

Figure 1 Raw titration data for the interaction of ibuprofen and FA2. DP stands for differential power (heat flow), after the baseline correction, at 25 °C. Initial concentration of ibuprofen in the syringe was 24.2 mM and FA2 in the measuring cell 8.56 mM.

Further, the dependence of the enthalpy change was plotted versus molar ratio of the reactants or concentration as presented in **Figure 2** and **Figure 3**. The experimental conditions should be set as good as to get the S-shape curve suitable for fitting in the software. Without the satisfactory shape of the fitting curve there are no reliable parameters determined.

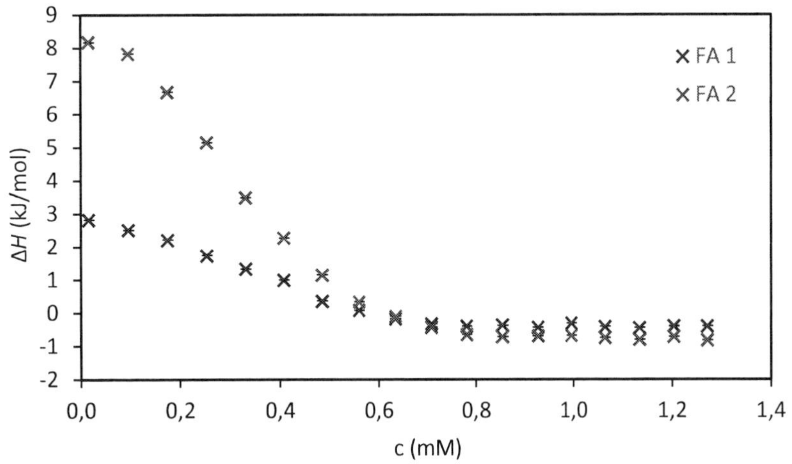

Figure 2 Calorimetric titration curves at 25 °C obtained after the raw data integration, titration of fulvic acid (initial conc. 8.56 mM) into diclofenac (initial conc. 0.0675 mM). Exothermic binding process, higher enthalpy change observed for interaction between diclofenac and FA2.

The main aim of this study was to observe and compare the behaviour of the soil organic matter with the pharmaceuticals and describe the basic parameters of these interactions, mainly the intensity of the binding process. As shown in **Figure 2** and **Figure 3**, there are significant differences between the intensities of the binding process when compared two selected pharmaceuticals.

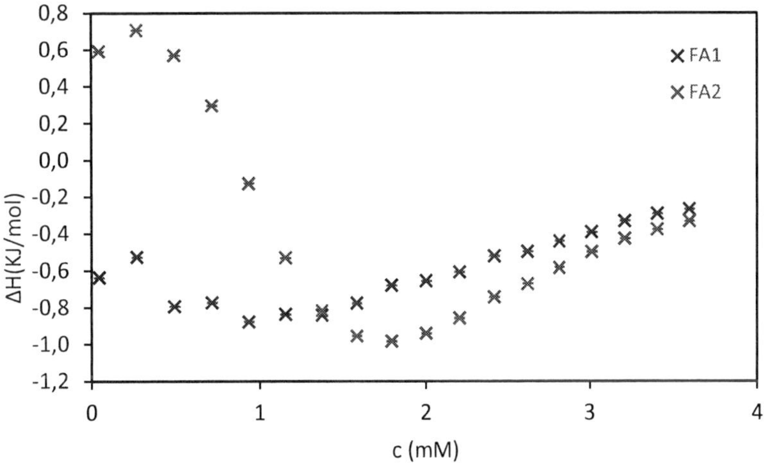

Figure 3 Calorimetric titration curves at 25 °C obtained after the raw data integration, titration of ibuprofen (initial conc. 24.2 mM) into fulvic acid (initial conc. 8.56 mM). Extremely low heat effects observed, different shapes of the curves occurred in comparison to behaviour with diclofenac.

The obtained thermodynamic parameters for diclofenac are summarized in **Table 1**, they were calculated automatically in the software after the data fitting process.

Table 1 Thermodynamic parameters of interactions between both fulvic acids and diclofenac

	Δ*H* (kJ/mol)	**Δ*G* (kJ/mol)**
FA 1	-16	-23
FA 2	-15	-22

The results of the titration of ibuprofen into the solution of fulvic acids are presented in **Figure 3**. It was quite complicated to fit the data for ibuprofen because of the presence of the minimum around 1.5 mM. This unusual shape can be evaluated using another fitting model (two sets of sites).

We observe different shapes of the curves in both figures which is caused by the variety of functional groups of the pharmaceutical molecules. Diclofenac shows more intensive interaction with both fulvic acids. Moreover, the enthalpy of binding FA2 with diclofenac is almost ten times higher than with ibuprofen which might be caused by better reactivity of FA2 (higher amount of reactive functional groups in FA structure) or better combination of those functional groups with diclofenac.

4. CONCLUSION

This paper describes the interactions between widely used pharmaceuticals ibuprofen and diclofenac with soil organic matter, standards of two different fulvic acids. The interactions were studied by a microcalorimetry technique isothermal titration calorimetry to obtain main thermodynamics parameters such as enthalpy or Gibbs energy of the binding process. The concentrations of pharmaceuticals used in this study was set according to the previous preliminary results of our research team as well as the knowledge of the possible concentration in the soils or wastewater. This study provided the basis for the subsequent research in this topic.

REFERENCES

[1] CHUANGAB, Y-A., LIU, C.-H., SALLACH, J.B., HAMMERSCHMIDT, R., ZHANG, W., BOYD, S.A., LI, H. Mechanistic study on uptake and transport of pharmaceuticals in lettuce from water. *Environment International*. 2019, vol. 531, pp. 104976. Available from: https://doi.org/10.1016/j.envint.2019.104976.

[2] PARK, J., CHO, K.H., LEE, E., LEE, S., CHO, J. Sorption of pharmaceuticals to soil organic matter in a constructed wetland by electrostatic interaction. *Science of The Total Environment*. 2018, vol. 635, pp. 1345-1350. Available from: https://doi.org/10.1016/j.scitotenv.2018.04.212.

[3] DING, Y., TEPPEN, B.J., BOYD, S.A., LI, H. Measurement of associations of pharmaceuticals with dissolved humic substances using solid phase extraction. *Chemosphere*. 2013, vol. 91, pp. 314-319. Available from: https://doi.org/10.1016/j.chemosphere.2012.11.039.

[4] KODEŠOVÁ, R., KLEMENT, A., GOLOVKO, O., FÉR, M, NIKODEM, A., KOČÁREK, M, GRABIC, R. Root uptake of atenolol, sulfamethoxazole and carbamazepine, and their transformation in three soils and four plants. *Environmental Science and Pollution Research*. 2019, vol. 26, pp. 9876-9891. Available from: https://doi.org/10.1007/s11356-019-04333-9.

[5] GOLOVKO, O., KUMAR, V., FEDOROVA, G., RANDAK, T., GRABIC, R. Seasonal changes in antibiotics, antidepressants/psychiatric drugs, antihistamines and lipid regulators in a wastewater treatment plant. *Chemosphere*. 2014, vol.111, pp. 418-426. Available from: https://doi.org/10.1016/j.chemosphere.2014.03.132.

[6] LOOS, R., CARVALHO, R, ANTONIO, D.C., COMERO, S. LOCORO, G., TAVAZZI, S., PARACCHINI, B., GHIANI, M., LETTIERI, T., BLAHA, L., JAROŠOVÁ, B., VOORPOELS, S., SERVAES, K., HAGLUND, P., FICK, J., LINDBERG, E.H., SCHWESING, D., GAWLIK, B.M. EU-wide monitoring survey on emerging polar organic contaminants in wastewater treatment plant effluents. *Water Research*. 2013, vol. 47, pp. 6475-6487. Available from: https://doi.org/10.1016/j.watres.2013.08.024.

[7] FEDOROVA, G., RANDAK, T., GOLOVKO, O., KODEŠ, V., GRABICOVA, K., GRABIC, R. A passive sampling method for detecting analgesics, psycholeptics, antidepressants and illicit drugs in aquatic environments in the Czech Republic. *Science of The Total Environment*. 2014, vol. 487, pp. 681-687. Available from: https://doi.org/10.1016/j.scitotenv.2013.12.091.

[8] PASTRELLO, A., MOSETTI, D., CANTONE, P., LEITA, L. Interaction between diclofenac and soil humic acids. *Soil and Sediment Contamination: An International Journal*. 2009, vol. 18, pp. 489-496. Available from: https://doi.org/10.1080/15320380902962353.

[9] TER LAAK, T.L., GEBBINK, W.A., TOLLS, J. 2006. Estimation of soil sorption coefficients of veterinary pharmaceuticals from soil properties. *Environmental Toxicology and Chemistry. 2006*, vol. 25, pp. 933-941. Available from: https://doi.org/10.1897/05-229r.1.

[10] CHEFETZ, B., MUALEM, T., BEN-ARI, J. Sorption and mobility of pharmaceutical compounds in soil irrigated with reclaimed wastewater. *Chemosphere*. 2008, vol. 73, pp. 1335-1343. Available from: https://doi.org/10.1016/j.chemosphere.2008.06.070.

[11] KUMAR, K., GUPTA, S., BAIDOO, S.K., CHANDER, Y., ROSEN, C.J. Antibiotic uptake by plants from soil fertilized with animal manure. *Journal of Environmental Quality. 2005, vol.* 34, pp. 2082-2085. Available from: https://doi.org/10.2134/jeq2005.0026.

[12] THIELE-BRUHN, S. Pharmaceutical antibiotic compounds in soils - a review. *Journal* of *Plant Nutrition and Soil Science*. 2003, vol. 166, pp. 145-167. Available from: https://doi.org/10.1002/jpln.200390023.

[13] TOTSCHE, K.U., KÖGEL-KNABNER, I. Mobile organic sorbent affected contaminant transport in soil: numerical case studies for enhanced and reduced mobility. *Vadose Zone Journal*. 2004, vol. 3, pp. 352-367. Available from: https://doi.org/10.2136/vzj2004.0352.

[14] KLUČÁKOVÁ, M., PEKAŘ, M. Behaviour of partially soluble humic acids in aqueous suspension. *Colloids and Surfaces, A: Physiochemical and Engineering Aspects*. 2008, vol. 318, pp. 106-110. Available from: https://doi.org/10.1016/j.colsurfa.2007.12.023.

[15] SEDLÁČEK, P., SMILEK, J., KLUČÁKOVÁ, M. How interactions with polyelectrolytes affect mobility of low molecular ions - Results from diffusion cells. *Reactive and Functional Polymers*. 2013. vol. 73, pp. 1500-1509. Available from: https://doi.org/10.1016/j.reactfunctpolym.2013.07.008.

[16] KLUČÁKOVÁ, M. Complexation of metal ions with solid humic acids, humic colloidal solutions, and humic hydrogel. *Environmental Engineering Science*. 2014, vol. 31, pp. 612-620. Available from: https://doi.org/10.1089/ees.2013.0487.

[17] EBELE, A.J., ABDALLAH, M.A.E., HARRAD, S. Pharmaceuticals and personal care products (PPCPs) in the freshwater aquatic environment. *Emerging Contaminants*. 2017, vol. 3, pp.1-16. Available from: https://doi.org/10.1016/j.emcon.2016.12.004.

[18] DE ANDRADE, J.R., QLIVIERA, M.F., DA SILVA, M.G.C., VIERA, M.G.A. Adsorption of pharmaceuticals from water and wastewater using nonconventional low-cost materials: A review. *Industrial & Engineering Chemistry Research. 2018, vol. 57, pp. 3103-3127. Available from:* https://doi.org/10.1021/acs.iecr.7b05137.

[19] *KHAZRI, H., GHORBEL-ABID, I., KALFAT, R., TRABELSI-ABADI, M. Removal of ibuprofen, naproxen and carbamazepine in aqueous solution onto natural clay: equilibrium, kinetics, and thermodynamic study. Applied Water Science. 2017, vol. 7, pp. 3031-3040. Available from:* https://doi.org/10.1007/s13201-016-0414-3.

[20] *MIHCIORUKUR, H., OGUZ, M. Removal of oxytetracycline and determining its biosorption properties on aerobic granular sludge. Environmental Toxicology and Pharmacology. 2016, vol. 46, pp. 174-182. Available from:* https://doi.org/10.1016/j.etap.2016.07.017.

[21] *TARABA, B., VESELÁ, P.* Sorption of lead(II) ions on natural coals and activated carbon: Mechanistic, kinetic, and thermodynamic aspects. *Energy Fuels*. 2016, vol. 30, pp. 5846-5853. Available from: https://doi.org/10.1021/acs.energyfuels.6b00976.

SESSION E
ADVANCED METHODS OF PREPARATION AND CHARACTERIZATION OF NANOMATERIALS

Chairmen

Dr. Petr KLAPETEK Czech Metrology Institute, Brno, Czech Republic, EU

Dr. Oldrich SCHNEEWEISS Institute of Physics of the CAS, Prague, Czech Republic, EU

PLASMA-ASSISTED VAPOUR THERMAL DEPOSITION WITH CONTINUOUS MATERIAL FEED

[1]Jaroslav KOUSAL, [1]Zdeněk KRTOUŠ, [1]Pavel SOLAŘ, [1]Ivo KŘIVKA, [1]Ivan KRAKOVSKÝ

[1]*Charles University, Faculty of Mathematics and Physics, Prague, Czech Republic, EU, jaroslav.kousal@mff.cuni.cz*

https://doi.org/10.37904/nanocon.2022.4579

Abstract

The Plasma-Assisted Vapour Thermal Deposition (PAVTD) is thin-film deposition technique utilizing a classical polymer as a source of material. The polymer is heated to fragmentation/evaporation under low pressure. The released fragments (effective a "monomer") with molar mass of 10^2 to 10^3 g/mol are then repolymerized in the rf plasma. The high molar mass of the film building blocks offers an opportunity to tune the structure and properties of the films in an exceptionally broad range for a plasma polymer, effectively bridging the gap between classical polymers and PECVD films.

Currently, the PAVTD method has several drawbacks compared with PECVD. The deposition must be done as a batch process, governed by the capacity of the crucible. The thermal release of the precursor fragments is highly temperature-and history- sensitive. The resulting fluctuations in deposition rate make retaining good reproducibility of the process rather tricky. To overcome these technical limitations, a modification of the setup utilizing a filament for FDM 3D-printing fed into a modified filament heater/extruder was made.

In this paper, overview of the possibilities of PAVTD using PLA as the source material will is given along with the first results obtained using an improved deposition setup with continuous material feed. Significant improvements in deposition rate and control of stability of the deposition are presented.

Keywords: Plasma polymer, plasma assisted vapour thermal deposition, continuous process, polylactic acid

1. INTRODUCTION

Plasma polymerization (PECVD) is well-established method for the preparation of organic highly crosslinked thin films [1,2] for applications such as protective or barrier coatings [3,4], in packaging, wettability control [5,6] or in the biomedical field [7-9].

Typical plasma polymerization methods utilize volatile precursor, which is fragmented and repolymerized in the plasma zone. Necessity of volatile precursors limits PECVD to use only low molar mass monomers. As the result of fragmentation and subsequent repolymerization are semi-random structures, the final molecular structure if the film is typically quite different from the original precursor [10]. This significantly limits the ability of PECVD to prepare chemically complex materials. To overcome this issue, plasma-assisted vapour thermal deposition (PAVTD) has been developed [11].

Since heavy molecules have low volatility, PAVTD uses high molar mass oligomers obtained in-situ from thermal degradation/evaporation from a solid polymer as a precursor. The subsequent process of plasma polymerization is the same as for classical PECVD. As the evaporated fragments are typically much larger than single monomeric unit of the original precursor the resulting product of plasma polymerization preserves much better the original chemical structure of the precursor, while still being crosslinked. Recently, as model case of the molecular structure, polylactic acid (PLA) was chosen, since it exhibits interesting properties dependent on its molecular structure, like biodegradability and hydrolyzability [12-14]. Using PAVTD, it was clearly demonstrated that the PLA structure can be retained in a plasma polymer thin film and controllably modified, allowing for control of the film properties in an usually broad range [15,16].

The main advantage of PAVTD (usage of high molecular precursor) is also the main disadvantage of the process as it brings the complexity of polymer evaporation into consideration. The macromolecules do not have well-defined boiling point and vapour pressure. The evaporation process of polymers is rather driven by the thermal decomposition of the large macromolecules into shorter fragments that are volatile. Due to the changing parameters inside of the evaporation crucible (temperature, actual amount material, process history, etc.), it is quite challenging to stabilize the deposition process. Unstable deposition rate leads to low reproducibility [17] as the product of the process is mostly dependent on power to precursor mass flow ratio i.e. Yasuda scaling law [18,19].

To overcome this issue, we propose a modification of the original batch PAVTD process into a process with a continuous feed of the material into the evaporation crucible. Having all the time only a small amount of material in the crucible is expected to improve the control over the process.

2. EXPERIMENTAL

The continuous feed version of PAVTD setup was strongly influenced by the Fused Deposition Modelling (FDM) 3D printing technology, both in used components and polymer material form. Since PLA is one of the most common 3D printing materials, it was possible to compare the films prepared using older batch (powder-filled) and a new continuous feed (filament-fed) setup.

2.1. Deposition setup

The original batch PAVTD setup was already described elsewhere [11,15]. The difference between batch and continuous feed setups is schematically shown in the **Figure 1**. The main difference is the feed system itself. It utilizes mostly off-the-shelf FDM 3D printer components with zero or minimal modifications (heater/heating block, heat break, extruder, stepper motor), the water-cooling being the main difference from the common 3D printing parts. The system is controlled by a proprietary Arduino-based circuitry.

The crucible had diameter of 20 mm and height of 30 mm. The centre of the circular (diameter 60 mm) rf electrode was 40 mm and the sample plane (with vacuum load-lock system) was 120 mm above the top of the crucible. The whole PAVTD setup was placed in a steel vacuum chamber with 500 mm height and 330 mm diameter with diffusion pump backed by a rotary pump.

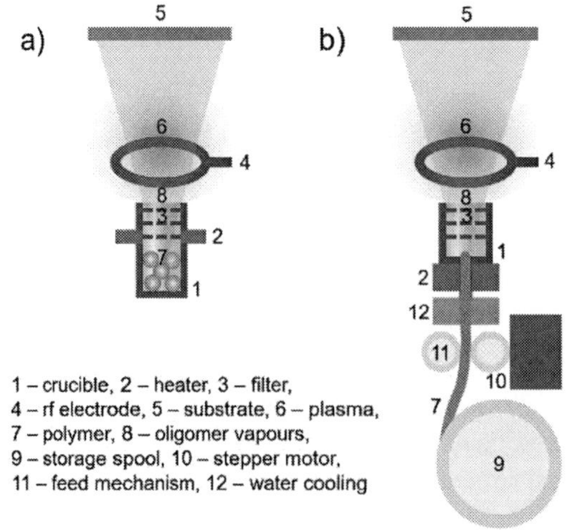

Figure 1 Scheme of the experimental setup for PAVTD a) batch setup b) continuous feed setup

2.2. Materials

As a source of the material, PLA printing filament ("natural", Gembird) with standard 1.75 mm diameter was used. As substrates, glass, one-side polished silicon and gold-coated glass were used. Argon was used as a carrier gas.

2.3. Deposition conditions

The heater (65 W) temperature for PLA filament was kept in the range of 320-350 °C, where the deposition rate can be controlled directly by the feed rate from the storage spool (10-50 g of the PLA filament). During testing of the setup, the filament material feed rate was varied between 0.3 g/h and 15 g/h. The sample series prepared to demonstrate the discharge power dependence of the film composition was prepared at the filament feed rate of 7 g/h. The chamber was fed with argon gas flow of 4.5 cm^3_{STP}/min to maintain working pressure 0.35 Pa. The discharge power (rf, 13.56 MHz) was varied from 0 W (pure thermal deposition) to 16 W.

2.4. Characterisation of the process and thin films

The deposition rate of the film was monitored by quartz crystal microbalance (QCM) sensor placed 60 mm above the substrate. In-situ QCM measurements in Hz/s were calibrated to nm/s by ex-situ film thickness obtained with spectroscopic ellipsometry (J. A. Woollam M-2000 DI). Chemical composition of the films was characterized using infrared spectroscopy (FTIR, Bruker Equinox 55 in reflection-absorption mode on gold-coated glass substrates) and X-ray photoelectron spectroscopy (XPS, Specs Phoibos 100).

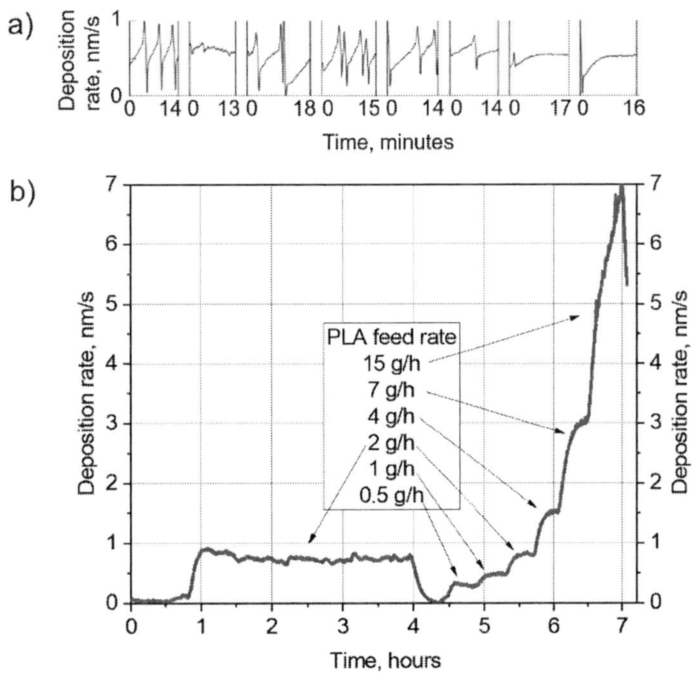

Figure 2 Comparison of deposition stability and duration of PAVTD process measured by QCM in a) batch in high-deposition rate mode (data: Z. Krtous, thesis, 2020) b) continuous feed variant.

3. RESULTS AND DISCUSSION

3.1. Stability of the deposition

The stability of deposition was studied both in terms of the short- and long-term fluctuations.

Repeated experiments did not reveal any fundamental limitation in the deposition length. Several experiments with run time of over 6 hours were performed. The deposition time was limited only by the amount of filament stored inside the vacuum chamber, which can be easily upgraded. Since the PAVTD operates at about 100 °C temperature higher than a 3D printer for the same material, limits of the deposition rate were found to be mostly caused by blockages of the feed system both at low and high end of deposition rate ranges (caused by heat creep and material overflow, respectively). The deposition rate inside these limits was found to be proportional to the feed rate. This makes the control much simpler than in the case of the batch process, where the fluctuations of over 50 % of the mean value were observed at high deposition rates (**Figure 2a**), so in most experiments, deposition rates had to be kept below 0.08 nm/s. The continuous feed setup was possible to operate at nearly 10 times higher value (**Figure 2b**). The stability at deposition rate levels comparable to those of the batch setup is also significantly better.

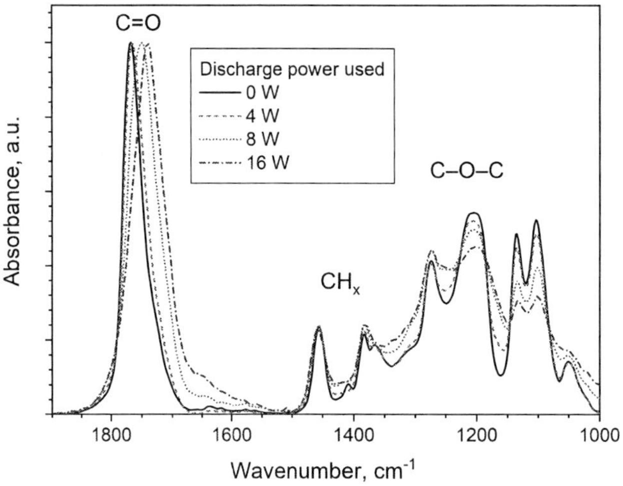

Figure 3 FTIR spectra (normalized to C=O peak) of PLA-like films prepared using PAVTD at the deposition rate of 7 g/h at varying discharge power

3.2. Chemical composition of the deposited thin films

The chemical composition of the film was found to be very similar to the films prepared previously using batch setup [15-17], as expected.

The infrared spectra of the films correspond well to the PLA structure. With increasing discharge power, the spectra show signs of slightly more disordered/fragmented structure of the films, exhibited by broadened peaks. The shift and shape change of the carbonyl peak (~1750 cm^{-1}) can be explained by the presence of more carbonaceous moieties (like C=C, ~1650 cm^{-1}). This is supported by the XPS data, where a significant decrease of the O/C ratio was observed in film prepared at higher discharge powers. This effect was also observed previously in the films prepared in the batch setup.

Table 1 Changes in the elemental composition of PAVTD PLA-like films with the discharge power (note: O/C ratio of original PLA is 0.66)

Discharge power (W)	C (XPS at. %)	O (XPS at. %)	O/C ratio
0	62.5	37.5	0.60
4	64.6	35.4	0.55
8	67.7	32.3	0.48
16	71.1	28.9	0.41

4. CONCLUSION

The plasma-assisted vapour thermal deposition setup with a continuous material was developed. The chemical properties of the PLA-like films prepared in this setup correspond well to the films prepared in the earlier batch variant of the method. The good control of the properties of the films is reproduced in the continuous process.

The continuous feed setup allowed to increase the deposition rate more then by one order of magnitude. At the same time, the total running time of over 7 hours was demonstrated (limited only by the filament storage) while maintaining much better deposition rate stability. This makes the continuous-feed PAVTD both an interesting tool to study classical polymer - plasma polymer transition and a practical deposition method.

ACKNOWLEDGEMENTS

This contribution was supported by the grant GA22-21007S of the Czech Science Foundation.

REFERENCES

[1] FRIEDRICH, J. Mechanisms of Plasma Polymerization-Reviewed from a Chemical Point of View. *Plasma Processes Polym.* 2011, vol. 8, pp.783-802.

[2] DENES, F. Macromolecular plasma-chemistry: an emerging field of polymer science. *Prog. Polym. Sci.* 2004, vol. 29, pp. 815-885.

[3] BEWILOGUA, K., BRAUER, G., DIETZ, A., GABLER, J., GOCH, G., KARPUCHEWSKI, B., SZYSZKA, B. Surface technology for automotive engineering. *CIRP Ann. - Manuf. Technol.* 2009, vol. 58, pp. 608-627.

[4] GRUMDEIER, G., THIEMANN, P., CARPENTIER, J., BARRANCO V. Tailored thin plasma polymers for the corrosion protection of metals. *Surf. Coatings Technol.* 2003, vol. 174-175, pp. 996-1001.

[5] CVERBAL, U., WALSH, J.L., CERNAK, M., DE VRIES, H.W., REUTER, S., BELMONTE, T., CORBELLA, C., MIRON, C., HOJNIK, N., JUROV, A., et al. White paper on the future of plasma science and technology in plastics and textiles. *Plasma Process. Polym.* 2019, vol. 16.

[6] MILELLA, A., DI MUNDO, R., PALUMBO, F., FAVIA, P., FRACASSI, F., D'AGOSTINO, R. Plasma nanostructuring of polymers: Different routes to superhydrophobicity. *Plasma Process. Polym.* 2009, vol. 6, pp. 460-466

[7] BEKESCHUS, S., FAVIA, P., ROBERT, E., VON WOEDTKE, T. White paper on plasma for medicine and hygiene: Future in plasma health sciences. *Plasma Process. Polym.* 2019, vol. 16.

[8] TRUICA-MARASESCU, F., WERTHEIMER, M.R., Nitrogen-Rich Plasma-Polymer Films for Biomedical Applications. *Plasma Process. Polym.* 2008, vol. 5, pp. 44-57.

[9] KUMAR, V., PULPYTEL, J., RAUSCHER, I.H., MANNELLI, I., ROSSI, F., AREFI-KHONSARI, F. Fluorocarbon Coatings Via Plasma Enhanced Chemical Vapor Deposition of 1H, 1H, 2H, 2H- perfluorodecyl Acrylate - 2, Morphology, Wettability and Antifouling Characterization. *Plasma Process. Polym.* 2010, vol 7, pp. 926-938.

[10] NISOL, B., WATSON, S., LEROUGE, S., WERTHEIMER, M.R. Energetics of reactions in a dielectric barrier discharge with argon carrier gas: IV ethyl lactate. *Plasma Process. Polym.* 2016, vol. 13, pp. 965-969.

[11] CHOUKOUROV, A., HANUS, J., KOUSAL, J., GRINEVICH, A., PIHOSH, Y., SLAVINSKA, D., BIEDERMAN, H. Thin polymer films from polyimide vacuum thermal degradation with and without a glow discharge. *Vacuum.* 2006, vol. 80, pp. 923-929.

[12] LENDLEIN, A., SISSON, A. Handbook of Biodegradable Polymers: Isolation, Synthesis, Characterization and Applications. Weinhem: John Wiley & Sons, 2011.

[13] KUCHARCZYK, P., HNATKOVA, E., DVORAK, Z., SEDLARIK, V. Novel aspects of the degradation process of PLA based bulky samples under conditions of high partial pressure of water vapour. *Polymer Degradation and Stability.* 2013, vol. 98, pp. 150-157.

[14] DESTEFANO, V., KHAN, S., TABADA, A. Applications of PLA in modern medicine. *Eng. Regen.* 2020, vol. 1, pp. 76-87.

[15] KRTOUŠ, Z., HANYKOVÁ, L., KRAKOVSKÝ, I., NIKITIN, D., PLESKUNOV, P., KYLIÁN, O., SEDLAŘÍKOVÁ, J., KOUSAL, J. Structure of Plasma (re)Polymerized Polylactic Acid Films Fabricated by Plasma-Assisted Vapour Thermal Deposition. *Materials.* 2021, vol. 14, pp. 459.

[16] KRTOUŠ, Z., KOUSAL, J., SEDLAŘÍKOVÁ, J., KOLÁŘOVÁ RAŠKOVÁ, Z., KUČEROVÁ, L., KRAKOVSKÝ, I., ALI-OGLY, S., PLESKUNOV, P., CHOUKOUROV, A. Thin films of cross-linked polylactic acid as tailored platforms for controlled drug release. *Surface and Coatings Technology.* 2021, vol. 421.

[17] KOUSAL, J., KRTOUŠ, Z., KOLÁŘOVÁ RAŠKOVÁ, Z., SEDLAŘÍKOVÁ, J., SCHAFER, J., KUČEROVÁ, L., SOLAŘ, P., HURAJOVÁ, A., BIEDERMAN, H., LEHOCKÝ, M. Degradable plasma polymer films with tailored hydrolysis behavior. *Vacuum.* 2020, vol. 173.

[18] HEGEMANN, D., NISOL, B., WATSON, S., WERTHEIMER, M.R. Energy Conversion Efficiency in Plasma Polymerization - A Comparison of Low- and Atmospheric-Pressure Processes. *Plasma Process. Polym.* 2016, vol. 13, pp. 834-842.

[19] SHELEMIN, A., ZABEIDA, O., KLEMBERG-SAPIEHA, J-E., MARTINU, L. Ion beam assisted chemical vapor deposition of hybrid coatings-Process diagnostics and mechanisms, *Jourmal of Vacuum Science & Technology A.* 2021, vol. 39.

LASER-ASSISTED CONTROL OF BISMUTH DOPING IN MAGNETIC NANOPARTICLES

[1,2]Ondrej HAVELKA, [1]Jan BRAUN, [1]Sabrin ABDALLAH, [1]Rafael TORRES-MENDIETA

[1]*Institute for Nanomaterials, Advanced Technologies and Innovation, Technical University of Liberec, Liberec, Czech Republic, EU,* ondrej.havelka@tul.cz, rafael.torres@tul.cz

[2]*Faculty of Mechatronics, Informatics and Interdisciplinary Studies, Technical University of Liberec, Liberec, Czech Republic, EU*

https://doi.org/10.37904/nanocon.2022.4578

Abstract

The fine control of mono- or bi-metallic nanoparticles' (NPs) chemical composition is one of the most critical concerns in nanomaterial manufacturing; however, when metals holding dissimilar characteristics form bi-metallic NPs, this becomes an unresolved matter. Herein we present the successful control of doping two different magnetic elements - nickel and iron- into bismuth to produce magnetic NPs with optical characteristics of semiconductors. For this, we have employed reactive laser ablation in liquids (RLAL) employing a magnetic foil (either nickel or iron) as the ablation target, which was immersed in an acetone solution with a hard-to-dissolve $Bi(NO_3)_3$ salt.

The employment of various $Bi(NO_3)_3$ concentrations showed that values between 0.01 and 1 mM are optimal for manipulating the doping degree on the NPs, which was verified by scanning electron microscopy. Moreover, their magnetic manipulation testing and inspection by ultraviolet-visible spectroscopy allowed assessing that the magnetic NPs formed by Bi doping of Fe display the most appropriate optical properties towards the material exploitation in photocatalysis. Furthermore, the minimal Bi salt employment resulted in chemical waste suppression verified by inductively coupled plasma spectroscopy. Since room conditions and no hazardous reducing agents were required for this synthetic procedure, the introduced alternative represents an up-and-coming practice for producing magnetic NPs doped by Bi.

Keywords: Nanoalloys, magnetic nanoparticles, bismuth compound, laser ablation in liquids, acetone

1. INTRODUCTION

By the end of the 20th century, nanotechnology had become a rapidly expanding field, continuing unabated over the years [1]. As a result, newly developed nanomaterials started to be employed in real applications outperforming the bulk materials from which they originated. This trend, moreover, did not end there; when nanomaterials formed by multiple elements started to be explored, it became evident that the arising new properties would be able to face the most demanding challenges our society faces. Therefore, a global effort began toward creating multi-element nanomaterials, which, by connecting multiple useful properties, gave rise to precious materials [2]. A clear example of this is NPs made up of two different metals, each providing a different type of particle behavior, like photonic and magnetic, which can be helpful in critical fields, such as photocatalysis. This widespread field has long been searching for recyclable materials that could be activated by the ubiquitous source of energy on our planet, sunlight radiation. In the past, magnetic Fe_3O_4/TiO_2 core-shell NPs were most often used for this purpose because they showed a suitable band gap and the possibility of being magnetophoretically recovered [3]. However, the main drawback of these magnetic NPs with a TiO_2 shell was, as in the case of pure TiO_2, the rapid recombination of photo-induced charge carriers and the promotion of catalytic reactions only by the ultraviolet (UV) part of the solar spectrum [4], representing only 4 % of the solar irradiance spectra. For this reason, the search for these recyclable materials with a bandgap capable of being activated by radiation falling in the solar irradiance visible (Vis) region, which is up to 10 times more abundant than the UV region, is crucial [5]. The state-of-the-art materials address such an issue by doping the TiO_2 side by plasmonic materials, which can transfer photo-induced charge carriers to TiO_2, expanding the material's ability to get activated by a broader part of the solar irradiance spectrum [6].

Although this is by itself an up-and-coming solution to overcome the TiO_2 optical limitations, TiO_2 is not the only photocatalyst under exploration. Other photocatalysts, like bismuth oxides also display wide band gaps, yet their slower charge recombination rate makes them even more interesting than TiO_2 [5]. Besides, unlike the TiO_2 case [3], doping bismuth-based photocatalysts' surfaces by transition metals as Ni or Fe can be of great benefit [7]. Specifically, by reducing the recombination rate of electron-hole pairs, thus, the photons used in photocatalysis remain available for the catalytic process longer.

Herein, we present a facile methodology to dope Ni or Fe by Bi oxide, which besides improving its electronic effects, expands its optical properties by narrowing its band gap, thus allowing it to utilize the sunlight irradiation spectrum more efficiently. Moreover, the ferromagnetic nature of the doped NPs provides it with magnetophoretic motility, which can serve for their reusability in photocatalytic reactions, preventing it from becoming waste after the first catalyst saturation, thus, making its use more sustainable [7]. Besides, to ensure complete materials sustainability, this work explores the synthesis of these doped NPs by RLAL, an eco-friendly laser-mediated route for multielement NPs preparation. In contrast to current multielement NPs synthesis techniques, RLAL enables their production without ligands and avoids hazardous chemicals usage, while ambient synthesis conditions are employed [8]. Thus, considering that this process enables the eco-friendly creation of NPs combining great optical and magnetic performance, we believe that the presented study will be of great importance in stimulating the generation of recyclable and sustainable photocatalysts with tailored optical properties.

2. EXPERIMENTAL

The NPs production consisted of irradiating a magnetic foil (nickel with > 99.98 % Ni, iron with > 99.99 % Fe, both from Sigma-Aldrich) immersed in a 40 mL acetone solution by pulsed laser radiation (Onefive Origami XP-S; output power 5.1 W, pulse duration < 400 fs, central wavelength 1030 nm). To get different elements' stoichiometries between Bi and the different magnetic elements within NPs, the $Bi(NO_3)_3 \cdot 5H_2O$ salt (> 99.999 %, Sigma-Aldrich) was dissolved in acetone at various concentrations (0, 0.01, 0.1, 1 mM), which range was based on a previous study conducted in aqueous media [9]. Note that larger concentration values were disregarded due to a massive laser beam scattering, which disabled production of magnetic NPs. The irradiation process followed the scheme displayed in **Figure 1**. In brief, a galvanometric mirror system (intelliscan 14; moving at a 2 m/s speed) was employed to move the focused laser beam by an F-Theta Lens (160 mm focal length) over the entire surface of the metallic foil while the Bi solution was magnetically stirred at 100 rpm. The employed laser and irradiation parameters were selected to maximize the NPs production rate (maximal average power: 5.1 W, optimized repetition rate: 0.7 MHz, optimized irradiation time: 20 min, scanning pattern: spiral). Note that the acetone volume was decreased on average by about 10 % to 36.1 mL due to the solvent heating-evaporation process caused by 20 min irradiation, which slightly influenced Bi salt concentration in the acetone during the NPs synthesis.

After the NPs preparation, the samples were cleaned of residual salt using a centrifuge (Centrifuge MiniSpin plus) for 15 minutes at 14500 rpm. The sediments obtained from the process were re-dispersed in Eppendorfs with fresh acetone by an ultrasonic cleaner (SONOREX DIGITEC DT 510 H, 35 kHz, 9.7 L). After the homogenization of samples, the magnetic NPs were collected on one side of Eppendorf using a neodymium magnet. The remaining non-magnetic material was removed. Then, the centrifugation followed by magnetic separation processes were done twice to ensure proper NPs cleaning.

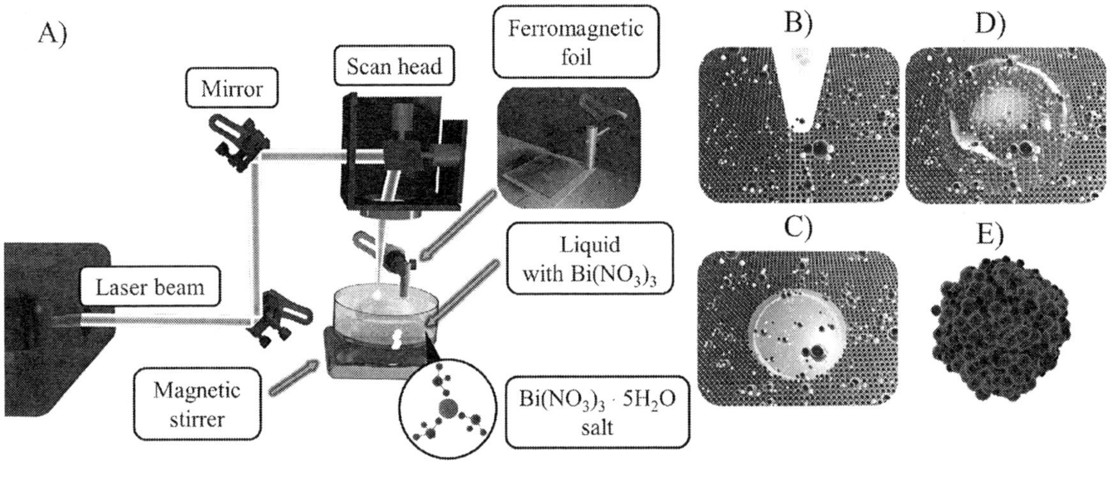

Figure 1 A) RLAL of an acetone-based solution of $Bi(NO_3)_3 \cdot 5H_2O$ and magnetic foil consisting of B) laser irradiation of a foil immersed in a solution, C) creation of the plasma from metal species and dissociated liquid molecules, D) cavitation bubble evolution, from which are E) formed bimetallic NPs released after bubble's collapse.

After the cleaning process, the samples' element composition was analyzed by an inductive-coupled plasma mass spectroscope (ICP-MS; NexION 3000D, Perkin Elmer) and optical emission spectroscope (ICP-OES; Optima 2100Dv, Perkin Elmer). Besides, the samples' morphology, size distribution, and chemical composition were then determined using a scanning electron microscope (FE-SEM; UHR Carl Zeiss Ultra Plus, Jena) equipped with an energy dispersive X-ray detector (EDX; Oxford X-Max 20). Finally, the optical features of samples were assessed by a UV-Vis spectroscope (PerkinElmer Lambda 35, Perkin Elmer) in a wavelength range from 230 to 800 nm and at a wavelength resolution of 1 nm.

3. RESULTS AND DISCUSSION

3.1. ICP-MS and ICP-OES

The NPs element compositions obtained after simple purification by centrifugation and after the entire cleaning process, including the magnet, can be found in **Table 1**. For simplicity, the samples are marked as S1-S8 according to the table. The results showed that, in general, the mass fraction of each element in the synthesized particles is proportional to the concentration of the Bi salt solution, while all or most of the salt is consumed in the synthesis. At the same time, the magnetically cleaned samples displayed a slightly higher Bi content. It is, at first sight, non-intuitive due to the possible formation of NPs with no or low magnetic element content or to the incomplete separation of redundant Bi salt during the centrifugation process. Since the location (not just the amount) of bismuth in the structure of NPs can play a crucial role in their magnetic response, further analysis of the NPs' internal structure is needed. Moreover, although there is a material loss, after magnet-assisted purification, minimally 93 % of NPs were recovered, indicating most of the material holds the desired magnetic properties. This result is especially significant in the case of Fe-based samples, in which >99 % NPs recovery rate indicates a better Fe doping by Bi. Although, according to the Hume-Rothery rules [10], neither Bi-Fe nor Bi-Ni are element combinations able to produce substitutional alloys, Bi and Fe share a property that may help them form a more stable system; both elements can be arranged in a cubic crystal centered in the body. This can enable a more efficient Bi allocation in the crystal formed by Fe, thus, leading to better element doping without compromising the inherent Fe magnetophoretic motility properties, which can be beneficial, for instance, for the magnetic recovery of NPs after their use in catalysis.

Table 1 Chemical composition of laser-prepared colloids. Average weight percentage of Bi, Ni, Fe in the no-magnetically purified and magnetically purified samples determined by ICP-MS, ICP-OES.

Sample No	Foil	No-magnetically purified			Magnetically purified			Magnetically recovered NPs (%)
		Salt (mM)	Bi (wt%)	Ni/Fe (wt%)	Salt (mM)	Bi (wt%)	Ni/Fe (wt%)	
S1	Ni	0	0	100	0	0	100	-
S2		0.01	2.7	97.3	0.01	3.3	96.7	97.5
S3		0.1	13.8	86.2	0.1	19.4	80.6	98.0
S4		1	88.6	11.4	1	94.2	5.8	92.9
S5	Fe	0	0	100	0	0	100	-
S6		0.01	2.8	97.2	0.01	4.2	95.8	99.0
S7		0.1	18.6	81.4	0.1	19.8	80.2	99.3
S8		1	66.5	33.5	1	72.2	27.8	99.1

3.2. SEM

The SEM micrographs displayed in **Figure 2** show a significant morphological difference in the Fe- and Ni-based NPs (samples S4 and S8 were selected as their difference is most evident). As anticipated by ICP data, there is a lower mixing of Ni and Bi, where, as indicated by EDS mapping, Bi is in fact found in the form of separated NPs clusters accompanied by residues of the Bi salt. In contrast, the element mapping of the Fe-based sample exhibited a homogeneous distribution of both elements, indicating an excellent mixture. Although its micrograph reveals an agglomeration of NPs with similar average size, there are large-looking NPs decorated by smaller ones as well. This is a result that, even when it can be further corroborated by TEM analysis, largely agrees with the previous observation that in nanoalloys with segregated phases, the element more affine to the solvent used in the synthesis tends to migrate to the outermost part of the NPs [11,12]. In our case, the acetone represents a solvent, which can enable the separation of Bi to the desired outermost part due to the $Bi(NO_3)_3$ affinity to acetone [13].

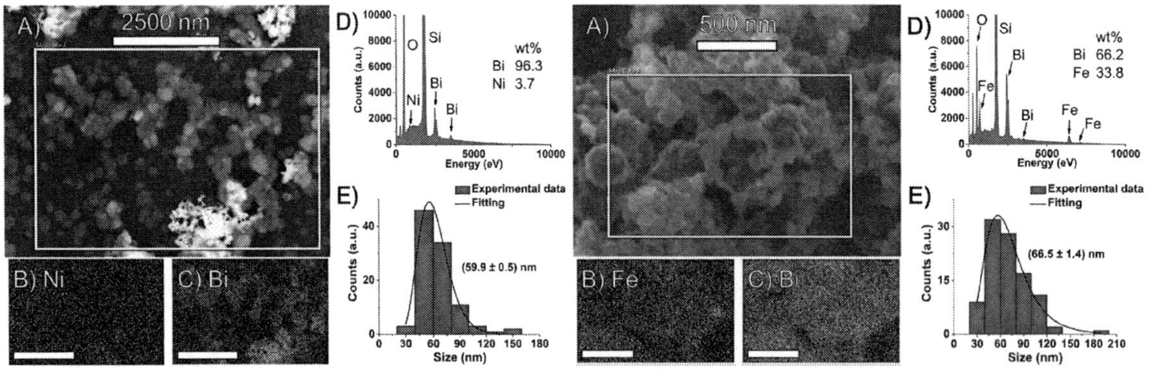

Figure 2 Representative SEM (left side - S4, right side - S8), A) micrograph, B) magnetic element and C) Bi EDX mapping with D) corresponding spectrum, and E) NPs size histogram fitted with the log-normal distribution.

3.3. Optical properties

Based on the solar irradiance spectra comparison with the samples' band gap, which was calculated according to the Tauc plot [14,15], the most promising sample was S8. This sample exhibited a unique band gap position at 2.6 eV (**Figure 3**), approaching the blue (Vis) part of the solar irradiance, which is the most intensive part of spectra during sunny days and the noon daylight illumination [16].

Moreover, the sample displays a secondary band gap at 3.7 eV, which is typical for Bi oxide forms used in photocatalysis [17]. According to the NPs morphology revealed by SEM analysis, this double band gap behavior can be a consequence of the NPs surface alternating between Fe and Bi elements. Thus, providing the NPs with the possibility to be activated by either visible or UV light, expanding the capabilities of Bi-based photocatalysts for using the sunlight radiation more efficiently.

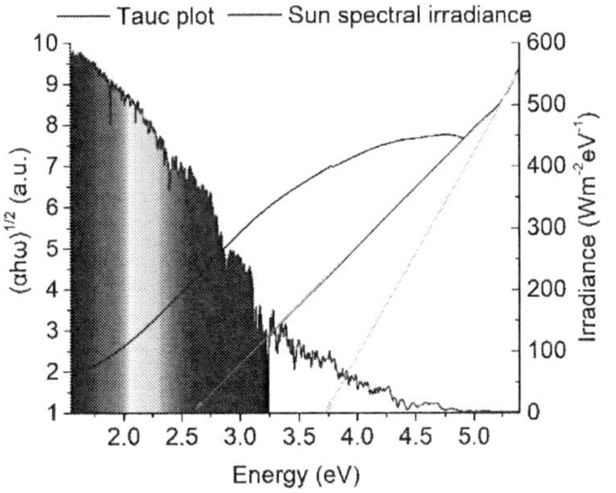

Figure 3 Comparison of Sun spectral irradiance during periods of moderate solar activity [18] with Tauc plot of S8 sample showing the proximity of Sun irradiance maximum and band gap position of synthesized NPs (red line), while the remaining Bi-related band gap (orange line) present in S8 meets the UV part of the solar spectra.

4. CONCLUSION

The current study reveals that RLAL methodology enables the production of Fe- and Ni-based magnetic NPs doped by Bi. Moreover, the control of Bi precursor concentration directly affects the doping degree found in the NPs, where Fe shows the best affinity for the allocation of Bi. Since these two elements forming a phase segregated nanoalloy are also found as alternating in the NP's surface, the produced NPs are capable of displaying two band gaps, one at the Vis region and one at the UV region of the sunlight radiation spectrum. This implies that the doping of Fe by Bi not only leads to the production of a semiconductor with the ability to be activated by visible and UV radiation, but its magnetophoretic motility inherited from Fe turns it into a recyclable, and because of this, a very valuable alternative towards photocatalysis. Finally, since the current synthesis approach largely minimizes chemical waste production, it is considered that the current results will be of fundamental importance for the future development of potent and magnetically recyclable photocatalysts.

ACKNOWLEDGEMENTS

The research presented in this article was supported by the Internal Grant of the Technical University of Liberec (SGS-2022-3008) and the Ministry of Education, Youth and Sports in the Czech Republic under the Research Infrastructures NanoEnviCz (Project No. LM2018124).

REFERENCES

[1] ROCO, Mihail C. The long view of nanotechnology development: the National Nanotechnology Initiative at 10 years. In: *Nanotechnology research directions for societal needs in 2020.* Springer, Dordrecht, 2011, pp. 1-28.

[2] FERRANDO, Riccardo; JELLINEK, Julius; JOHNSTON, Roy L. Nanoalloys: from theory to applications of alloy clusters and nanoparticles. *Chemical reviews*, 2008, vol. 108, no. 3, pp. 845-910.

[3] ÁLVAREZ, Pedro M., et al. Preparation and characterization of magnetic TiO2 nanoparticles and their utilization for the degradation of emerging pollutants in water. *Applied Catalysis B: Environmental*. 2010, vol. 100, no.1-2, pp. 338-345.

[4] WEI, Zhidong; LIU, Junying; SHANGGUAN, Wenfeng. A review on photocatalysis in antibiotic wastewater: Pollutant degradation and hydrogen production. *Chinese Journal of Catalysis*. 2020, vol. 41, no. 10, pp. 1440-1450.

[5] PELAEZ, Miguel, et al. A review on the visible light active titanium dioxide photocatalysts for environmental applications. *Applied Catalysis B: Environmental*. 2012, vol. 125, pp. 331-349.

[6] BIELAN, Zuzanna, et al. Defective TiO2 core-shell magnetic photocatalyst modified with plasmonic nanoparticles for visible light-induced photocatalytic activity. *Catalysts*. 2020, vol. 10, no. 6, pp. 672.

[7] REN, Ao, et al. Enhanced visible-light-driven photocatalytic activity for antibiotic degradation using magnetic NiFe2O4/Bi2O3 heterostructures. *Chemical engineering journal*. 2014, vol. 258, pp. 301-308.

[8] FRIAS BATISTA, Laysa M., et al. Generation of nanomaterials by reactive laser-synthesis in liquid. *Science China Physics, Mechanics & Astronomy*, 2022, vol. 65, no. 7, pp. 1-45.

[9] ETTEL, David, et al. Laser-synthesized Ag/TiO nanoparticles to integrate catalytic pollutant degradation and antifouling enhancement in nanofibrous membranes for oil-water separation. *Applied Surface Science*. 2021, vol. 564, pp. 150471.

[10] MIZUTANI, Uichiro. Hume-Rothery rules for structurally complex alloy phases. *MRS Bulletin*. 2012, vol. 37, no. 2, pp. 169-169.

[11] ETTEL, David, et al. Expansion of Optical Properties in Tiox Nanoparticles by their Laser-Mediated Alloying with Ag. *Education*. 2020, pp. 109-115.

[12] JOHNY, Jacob, et al. Multidimensional thermally-induced transformation of nest-structured complex Au-Fe nanoalloys towards equilibrium. *Nano Research*. 2022, vol. 15, no.1, pp. 581-592.

[13] OLLEVIER, Thierry, JALBA, Angela and KEIPOUR, Hoda. Bismuth(III) Nitrate Pentahydrate. *In Encyclopedia of Reagents for Organic Synthesis*, 2016.

[14] TAUC, Jan. Optical properties and electronic structure of amorphous Ge and Si. *Materials research bulletin*. 1968, vol. 3, no. 1, pp. 37-46.

[15] MAKUŁA, Patrycja; PACIA, Michał; MACYK, Wojciech. How to correctly determine the band gap energy of modified semiconductor photocatalysts based on UV-Vis spectra. *The Journal of Physical Chemistry Letters*. 2018, vol. 9, no. 23, pp. 6814-6817.

[16] TAYLOR, A. H.; KERR, G. P. The distribution of energy in the visible spectrum of daylight. *JOSA*. 1941, vol. 31, no.1, pp. 3-8.

[17] GBASHI, Kadhim R., et al. Structural, morphological, and optical properties of nanocrystalline (Bi2O3) 1− x:(TiO2) x thin films for transparent electronics. *Plasmonics*. 2019, vol. 14, no. 3, pp. 623-630.

[18] GUEYMARD, Christian A. The sun's total and spectral irradiance for solar energy applications and solar radiation models. *Solar energy*. 2004, vol. 76, no. 4, pp. 423-453.

SIMULATION OF LIFETIME CHARACTERIZATION OF TEXTURED SILICON AND COMPARISON WITH EXPERIMENTAL RESULTS

[1]Ruhani DADASHOV, [1]Jan VOVES, [2]Deep Pakki ABHINAV

[1]*Czech Technical University in Prague, Department of Microelectronics, Faculty of Electrical Engineering, Czech Republic, EU, dadasruh@fel.cvut.cz*

[2]*Czech Technical University in Prague, Department of Electrotechnology, Faculty of Electrical Engineering, Czech Republic, EU, pakkiabh@fel.cvut.cz*

https://doi.org/10.37904/nanocon.2022.4617

Abstract

We present the surface lifetime characterization of the textured silicon surface for the application in solar cells. The measurement by the Sinton WCT-120 PL lifetime tester is compared with the simulation by the Silvaco TCAD software to get possible values of the surface recombination velocity. The n-type silicon textured surface wafers were used as the substrate. In the first experiment, the carrier lifetime was measured and simulated of the pure textured surface. The second experiment was done after deposition of aluminum oxide by the Atomic Layer Deposition on the textured surface. The thermal method was used with TMA as a precursor. The two-dimensional TCAD calculation was realized on the silicon structure with textured surface. The surface profile was determined by the AFM measurement. Two electrodes were placed on both sides of the structure to calculate the current flow generated by the light pulse. The real Sinton optical source spectrum was approximated by the five beams in the range 690 - 830 nm with different intensities. The current decrease connected with the decay of the optical intensity was simulated. The comparison of experimental results with the TCAD simulation gives the possibility to distinguish between the bulk carrier lifetime and surface recombination velocity. The coincidence of the measured and simulated graphs by the fitting the simulation parameters result in the realistic value of the surface recombination velocity. It can be used for the classification of the silicon surface passivation influence on the solar cells' efficiency.

Keywords: Surface recombination velocity, carrier lifetime, textured silicon

1. INTRODUCTION

The knowledge about the surface recombination velocity (SRV) and charge carrier lifetime of silicon is necessary for an analysis of the semiconductor material quality and the efficiency of the work power of solar cells. [1]. Charge-carrier lifetime can be described as the characteristic time τ describing how long an excess charge carrier remains in a conductive state before becoming immobilized due to trapping or recombination. It is clear that τ is of essential importance for electronic and optoelectronic devices because the performance of these devices is mainly determined by the efficiency of the transport of excess charge carriers injected by light or contacts [2]. Surface recombination can affect devices in two ways: by increasing saturation currents (decreasing output voltage) and by decreasing short circuit currents [3]. The surface recombination velocity S_r is defined as the number of carriers recombining on the surface per unit area per unit time of excess carriers at the surface or interface [4]. For extremely pure Si samples, the bulk lifetime τ_b is so high that surface recombination dominates the effective lifetime. However, for lower quality, i.e., low resistivity or solar grade silicon wafers, the lifetime is lower and both τ_b and S_r must be determined. Photoconductance decay, surface photovoltage, and quasi-steady state photoconductance are widely used for characterizing the minority carrier lifetime and diffusion length. Separation of bulk and surface recombination is possible by varying the sample thickness or by evaluating the initial decay modes of photoconductance decay [5]. The overall goal of this work was to find reasonable efficiency values for effective carrier lifetime and surface recombination velocity by comparison of Silvaco TCAD simulation results and compare them with the experimental part measurement of carrier lifetime by Sinton WCT 120 photo-conductance lifetime tester. All simulations and experiments are carried out on randomly textured silicon wafers with a thickness of 140 µm and approximately 1 ms bulk carrier lifetime. We measured wafers with a 9.6 nm deposited passivation layer of Al_2O_3 [6] applied via the Atomic Layer Deposition machine Sentech Si 500 and without a passivation layer for comparison.

2. EXPERIMENTAL

The first step of the experiment started with characterising the pure wafer's surface with 2.5 Ω·cm base resistivity via Atomic Force Microscopy (AFM) and using these pictures to observe the roughness of the textured wafer surface and how to is the surface deposited by Al_2O_3 affected as shown in **Figure 1**. These images were used to determine the dimensions of pyramids and to get the average size for one shape to apply to the simulation for the next steps. Deposition of Al_2O_3 was executed by thermal ALD using TMA precursor and 120 deposition cycles. The layer thickness per cycle is 0.8 Å. The aim of the deposition was to create an additional thin sheet which has a passivation role on the surface. The effective carrier lifetime was measured by the Sinton lifetime tester immediately before and after the deposition of the passivation layer. The results are in **Table 1**.

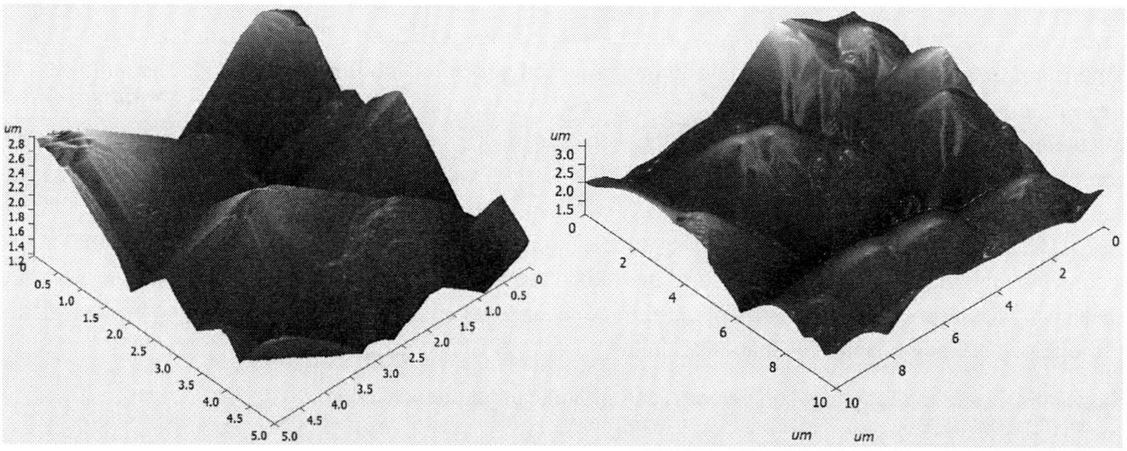

Figure 1 AFM picture of pure silicon surface (left), and silicon surface with passivation layer (right)

Table 1 Comparison between Sinton WCT-120 PL lifetime tester results before and after deposition of the passivation layer to the surface

	Maximum value of photoconductance (siemens)	**Minority carrier lifetime (µm)**	**Minimum minority carrier density (cm^{-3})**	**Maximum minority carrier density (cm^{-3})**
Before	0.0026	1.16	8.88e+12	6.82e+14
After	0.0082	3.26	6.68e+13	2.14e+15

3. MODELING

The simulation was based on creating a part of the silicon model (20000 µm x 140 µm) and adding two electrodes for both sides of it to apply a small voltage to see the current drop produced by carrier recombination during the illumination drop. We measured the optical spectrum of the Sinton optical source by means of the spectrometer Ocean Optics USB 2000 (**Figure 2**) to acquire the wavelengths and intensities of five optical beams (**Table 2**) to approximate the real spectrum. These beams in the range 690-830 nm at different intensities we applied in a perpendicular direction to the wafer surface. Turning off the light with an exponential decrease of light power during 0.015 ms time was simulated on the pure and passivated wafer surface. We calculated the relative beam intensities and for the final beam intensity we used the definition that 1 sun is equal to 0.1 W/cm^2 [7]. The constant voltage of 0.1 V was applied between side electrodes during transient simulation to obtain the resulting photocurrent decay. We kept a constant bulk lifetime for both electrons and holes (10^{-3} s) and mainly focused on calibrating for the near value of SRV to get an accurate graph for both minority and majority carriers.

Table 2 Used intensities of beams with different wavelength according to **Figure 2**

Wavelength (nm)	690	714	735	763	830
Intensity (a.u.)	250	350	380	450	700
Relative intensity (%)	0.12	0.16	0.18	0.21	0.33

The simulation and experiment give us only the effective lifetime:

$$\frac{1}{\tau_{eff}} = \frac{1}{\tau_b} + \frac{1}{\tau_s} \tag{1}$$

where the surface carrier lifetime τ_s can be replaced by the surface recombination velocity S_r where W is the wafer thickness and D is the diffusion constant of the minority carriers (in $\mathrm{cm^2 s^{-1}}$) [8]:

$$S = \frac{W}{2\left(\tau_s - \frac{1}{D}\left(\frac{W}{\pi}\right)^2\right)} \tag{2}$$

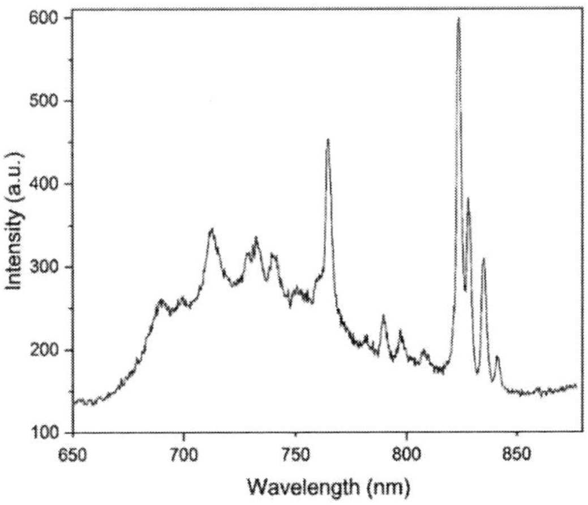

Figure 2 The spectrum of Sinton optical source measured via spectrometer Ocean Optics USB 2000

Our other goal was to optimize the SRV value and to obtain the simulated photocurrent curves most close to the experimental ones.

4. RESULTS

The experimental results (**Figure 3)** show approximately 3.1 times increase in the maximum value of photoconductance between the clean surface and the surface with deposited passivation layer. The optimized simulation results have almost the same amount of increase as shown in **Figure 4**.

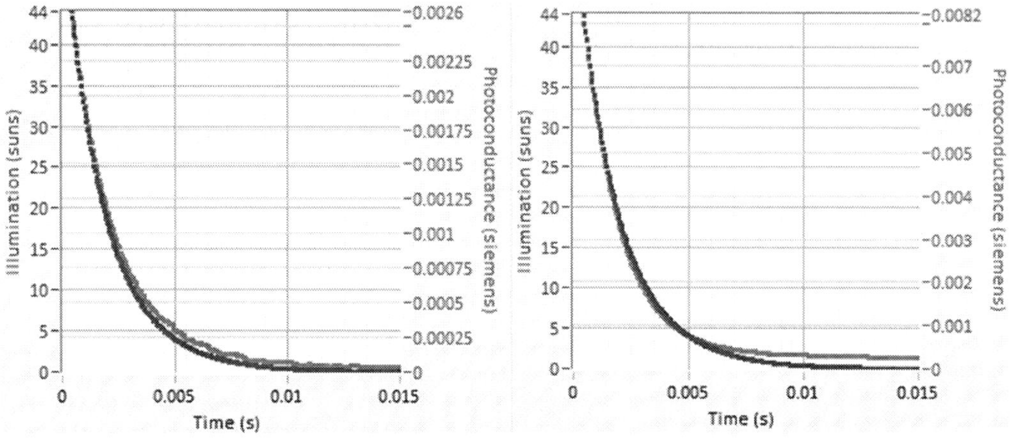

Figure 3 Photocurrent dependence during the turning off the light according to results of Sinton WCT 120 photo-conductance lifetime tester. The surface without Al_2O_3 (left), with Al_2O_3 (right)

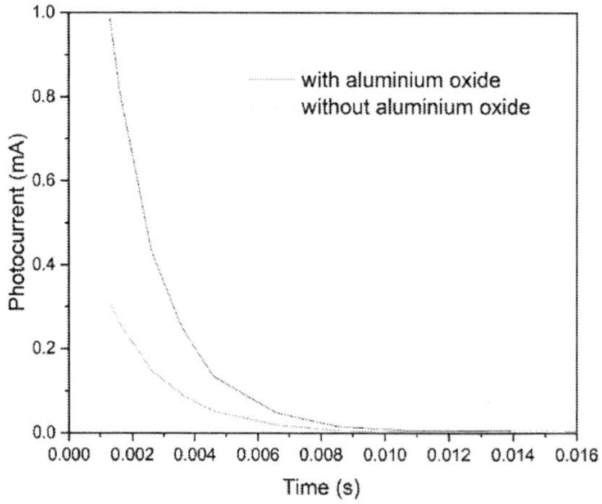

Figure 4 Photocurrent dependence during the turning off the light by means of Silvaco TCAD

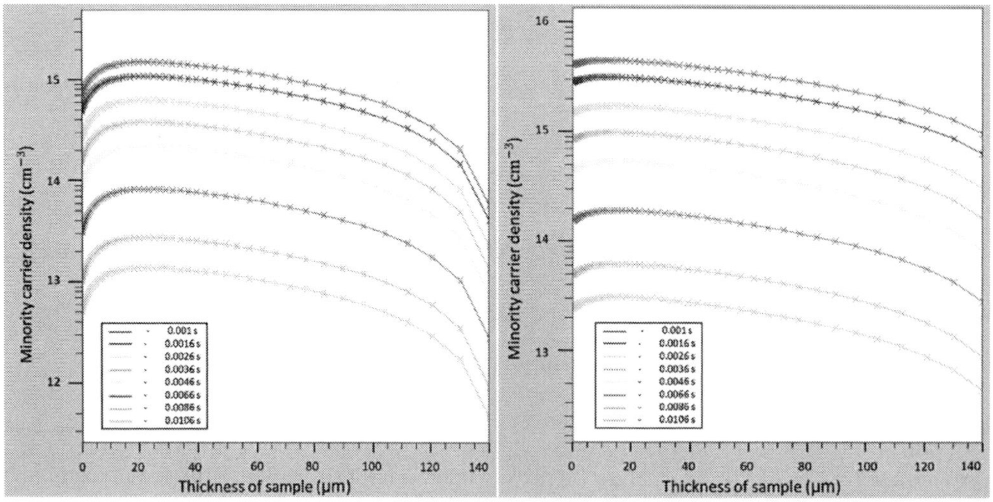

Figure 5 Minority carrier density (holes) during the turning off the light. The surface without aluminum oxide (left), with aluminum oxide (right)

During photocurrent change, we tried to keep the same values for minority carrier densities compared with experimental results as shown in **Table 1** by calibrating surface recombination velocity for both minority and majority carriers for passivated and pure surfaces. We found the near-optimal values for surface recombination velocities for electrons and holes is 3.45 .10^4 cm/s for the pure surface, and 6.79 .10^3 cm/s for surface with the passivation layer. The results are shown in **Figure 5**.

5. CONCLUSION

The comparison of the Sinton WCT lifetime measurement with the Silvaco TCAD simulation allowed us to obtain unknown values for surface recombination velocities for both electrons and holes. The ability to distinguish between the bulk lifetimes and surface recombination velocities would help to improve the efficiency of solar cells. It can be used for optimizing the choice of the different materials for the surface passivation and for the shape of the surface.

ACKNOWLEDGEMENTS

The research has been supported by the student grant of CTU No. SGS20/176/OHK3/3T/13 and by the project Centre of the Advanced Applied Natural Sciences No. CZ.02.1.01/0.0/0.0/16_019/0000778 supported by the Operation Programme Research, Development and Education co-financed by European Community and by Ministry of Education Czech Republic.

REFERENCES

[1] JOHANNES, A. Giesecke, RONALD, A. Sinton, MARTIN, C. Schubert, STEPHAN, Riepe, WILHELM, Warta. Determination of bulk lifetime and surface recombination velocity of silicon ingots from dynamic photoluminescence. *IEEE journal of photovoltaics.* 2013, vol. 3, pp. 1311-1318.

[2] KLEIN, D., WUENSCH, F., KUNST, M. The determination of charge-carrier lifetime in silicon. *Basic solid state physics.* 2008, vol. 245, issue 9, pp. 1865-1876.

[3] MATTHEW, Justin Stocks. *High efficiency multicrystalline silicon solar cells.* Canberra, 1998. PhD thesis. Australian National University.

[4] HARI, Singh Nalwa. *Handbook of Surfaces and Interfaces of Materials.* San Diego: A Harcourt Science and Technology Company. 2001.

[5] DAEYUL, Baek, SERGEI, Rouvimov, CHANG-BAE, Kim, et al. Surface recombination velocity of silicon wafers by photoluminescence. *Applied Physics Letters.* 2005, vol. 86, pp 1-4.

[6] FLORIAN Werner. *Atomic layer deposition of aluminum oxide on crystalline silicon: Fundamental interface properties and application to solar cells.* Leibniz. 2014. Dissertation. Gottfried Wilhelm Leibniz Universität Hannover.

[7] PETER, R. Michael, DANVERS, E. Johnston, WILFRIDO, A conversion guide: solar irradiance and lux illuminance. *Journal of Measurements in Engineering.* 2020, vol. 8, pp. 153-166. Available from: https://doi.org/10.21595/jme.2020.21667.

[8] ANGÈLE, Reinders, PIERR, Verlinden, WILFRIED, van Sark, ALEXANDRE, Freundlich. *Photovoltaic Solar Energy: From Fundamentals to Applications.* Chichester: Wiley, 2017.

SHAPE VARIABILITY OF PHYTOSYNTHESIZED GOLD NANOPARTICLES

[1]Ruslan MARIYCHUK, [1]Romana SMOLKOVA, [2]Liudmyla M. GRISHCHENKO, [2]Vladyslav V.LISNYAK

[1]Department of Ecology, Faculty of Humanities and Natural Sciences, University of Presov, Presov, Slovakia, EU, ruslan.mariychuk@unipo.sk

[2]Department of Analytical Chemistry, Faculty of Chemistry, Taras Shevchenko National University of Kyiv, Kyiv, Ukraine

https://doi.org/10.37904/nanocon.2022.4621

Abstract

Biomedical applications of nanomaterials depend on the size and morphology of nanoparticles and their biocompatibility. However, the armory of trusted protocols usually includes reducing agents, including sodium borohydride, methoxy polyethylene glycol, potassium tartrate, etc., and capping agents, such as sodium dodecyl benzyl sulfate, polyvinylpyrrolidone, etc., which are usually toxic. The unavoidable presence of toxic compounds in nanocolloid systems limits the use of nanoparticles for biological systems. This study aims to synthesize non-spherical gold nanoparticles using plant extracts and characterize their properties for biomedical applications. Gold nanoparticles were obtained through the direct interaction of Au^{3+} cations with aqueous extracts of selected plants (elderberry and peppermint) under different synthetic conditions. Ultraviolet and visible spectroscopic measurements confirmed the response of the obtained Au nanocolloid solutions to near-infrared radiation. This response opens the prospect of using the obtained nanoparticles in nanomedicine for antitumor therapy, targeted drug transport, diagnostics, or drug delivery systems. The position of the maximum near-infrared radiation can be adjusted by the synthetic conditions (extract composition, concentration, reagent ratio, pH, and temperature). Transmission electron microscopy and atomic force microscopy showed the formation of a mixture of circum-spherical Au nanoparticles and non-spherical Au nanoparticles: nanotriangles with sizes from 20 nm and nanohexagons with sizes > 100 nm.

Keywords: Plasmonic nanoparticles, phytosynthesis, non-spherical nanoparticles, biocompatible nanoparticles

1. INTRODUCTION

Nanomaterials have become an important branch of modern science due to their unique chemical and physical properties. Due to their unusual optical, electric, magnetic, and catalytic properties, nanomaterials were recognized as new and efficient components that can be used for the creation of novel advanced materials.

Biocompatibility and optical properties (surface plasmon resonance, SPR) of noble metal nanoparticles (NPs) open the possibility of their applications in modern nanomedicine, which is currently being actively developed. Typically, the NPs of noble metals, especially gold (Au NPs), are employed to deliver therapeutics and mediate heat and light to specific types of tissues [1]. The structure, composition, size, and shape of metal NPs define the optical and electronic properties which can be adjusted for utilization for advanced photothermal therapy [2] and controlled drug release systems [3].

Au NPs with their excellent chemical stability and biocompatibility are suitable for biomedical applications. However, Au NPs with reasonably small size (< 60 nm) exhibit their surface plasmon resonance in the range of 510-560 nm, far beyond the biological transparency window of 650-1350 nm. This region is divided into two optical near-infrared ranges, at λ = 650-850 nm and 950-1350 nm, respectively. The effective utilization of Au NPs for photothermal therapy due to the deeper penetration of long-wave radiation requires NPs with a response in the second near-infrared range [4]. This is the reason, why the significant interest of scientists is focused on the synthesis of Au NPs with different shapes - spheres [5,6], nanorods [7], nanoshells [8], nanostars [9], nanoprisms [10], nanotriangles [11], and nanohexagons [12,13].

Various methods for synthesizing metal NPs were developed in recent decades: hydrothermal synthesis, co-precipitation, microemulsion, inert gas condensation, ion sputtering scattering, microwave, pulse laser ablation, sol-gel, spark discharge, sonochemical, and biological synthesis. Physical methods usually require the utilization of highly expensive equipment and do not provide the control over parameters of final nanoparticles (size and shape). From this point of view, traditional wet chemistry methods are preferable. However, the most efficient protocols involve the use of toxic reducing and capping agents which remain in the resulting nanocolloid solutions. As result, such as Au NPs show high cytotoxicity towards living organisms and are unsuitable for application in biomedicine. Biological synthetic methods, including the application of biological materials, e.g., plants, fungi, and bacteria, can be considered to be a prospective alternative. Such methods are also economical and time-effective and environmentally friendly [14,15]. One of the routes for obtaining biocompatible non-spherical Au NPs in the greenest manner is phytosynthesis - the use of plant extracts and natural surfactants [16]. Despite the extensive studies, the wide application of photosynthetic protocols is limited because the plant extracts are complex systems and the knowledge of the formation mechanism of non-spherical Au NPs is limited [17].

This study aims to consider the current state in the development of phytosynthesis protocols for the preparation of Au NPs with optical activity in the near-infrared region.

2. EXPERIMENTAL

2.1. Preparation of nanoparticles

Au NPs were synthesized using aqueous extracts of elderberry (*Sambucus nigra*) and peppermint (*Mentha piperita*) according to protocols [6] and [18], respectively. Syntheses were performed at an ambient laboratory temperature (20-25 °C) by the direct interaction of the plant extracts with 1 mM of $HAuCl_4$ aqueous solution under continuous stirring.

2.2. Characterization of nanoparticles

The obtained nanocolloid solutions were characterized by UV-Vis spectroscopy using Shimadzu UV-1800 spectrophotometer with matched 1-cm quartz cells. Transmission electron microscopy (TEM) imaging was conducted with a JEOL JEM-2100F microscope. TEM specimens were prepared by dropping a sonicated aqueous suspension of Au nanoparticles on a carbon-coated copper grid and followed by drying it under the infrared lamp. TEM images of different magnifications were captured at a maximum acceleration voltage of 200 kV.

3. RESULTS AND DISCUSSION

UV-Vis spectroscopy was applied to follow the formation of the Au NPs through the observation of the SPR peak position and shape (**Figure 1**). The collected UV-Vis spectra of Au NPs prepared using different extracts are different in principle. To the best of our knowledge, the presence of the SPR peak maximum can be seen in the spectral range from 530 to 570 nm. This SPR peak is true evidence of the formation of spherical Au NPs, as confirmed by the literature [5, 6]. Besides, we should notice that in rare cases, the second maximum can be seen in the near-infrared range of the recorded spectra.

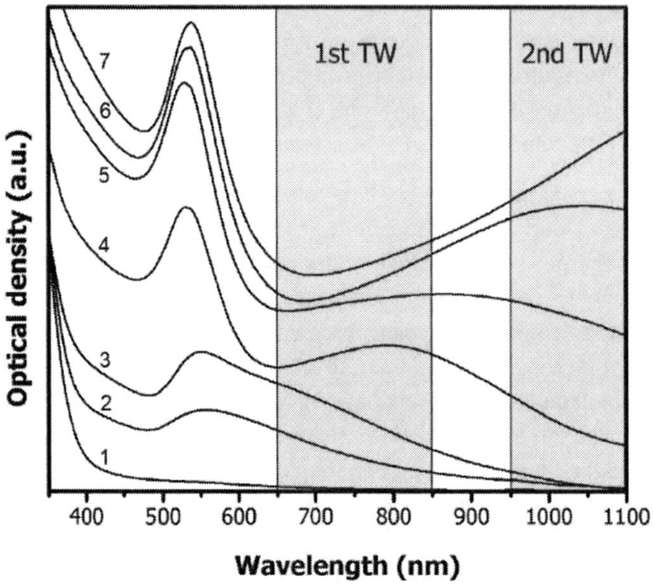

Figure 1 UV-Vis spectra of *Mentha piperita* mediated Au NPs prepared with different concentrations of Au^{3+}: 1 - 0.0250; 2 - 0.1250; 3 - 0.3125; 4 - 0.3750; 5 - 0.5000; 6 - 0.5625 and 7 - 0.7500 mM compared with the first and second therapeutic windows (TW). Spectra in scale but shifted along *y*-axis for better view

The size and shape of Au NPs have a real influence on the position of SPR absorption maximum. In many cases, the growth of the spherical Au NPs leads to a small shift of absorption maximum in the infrared range, however, this dependence for non-spherical Au NPs is very different [19].

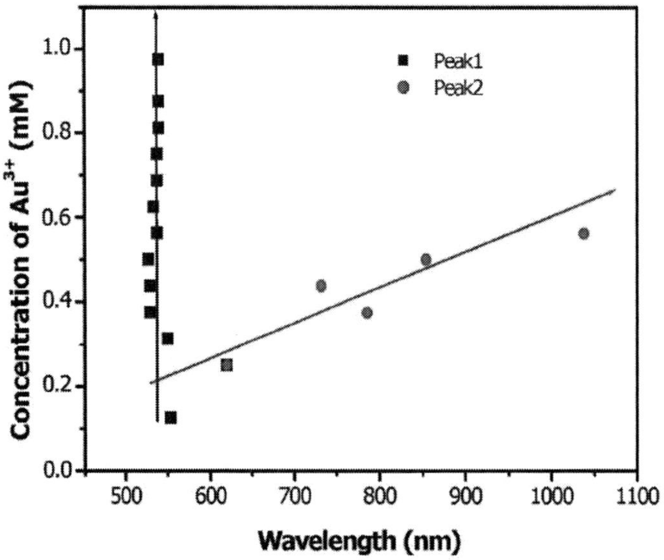

Figure 2 Dependencies of the positions of first and second absorption maxima for Au nanocolloids prepared with different initial concentrations of Au^{3+}

Here slight changes in the size or in the aspect ratio, which is a geometrical presentation of shape deviation from an ideal sphere, results in a considerable shift of the SPR absorption maximum in the near-infrared spectral range [20]. On the basis of UV-Vis spectra, show that, depending on the composition, almost perfect spherical Au NPs (*Sambucus nigra* extract) [6] and non-spherical Au NPs (*Mentha piperita*) can be obtained from the extracts. And although in the second case, a mixture is formed, it is possible to adjust the adsorption and position to a maximum in the NIR.

Figure 2 shows that two surface plasmon resonance (SPR) peaks form in all samples. The first SPR peak, at 528 nm, corresponds to the contribution of the spherical Au NPs, definitely, its position is unchanged. The second SPR peak is formed because of the presence of non-spherical Au NPs in nanocolloid solutions. When increasing the concentration of Au^{3+} in the reaction mixture, Au NPs can increase in size, which is manifested a rapid shift of the maximum in the IR range, and the concentration of larger newly formed Au NPs increases.

Figure 3a shows the synthesized spherical Au NPs with an average diameter of 6.6 ± 1.2 nm. A TEM image shows a solitary spherical Au nanoparticle with size of 7.7 nm (**Figure 3b**), and one can see here the morphology of the seed, with an interplanar spacing characteristic of the {111} plane in face-center cubic Au structure. Based on TEM images, one can suggest that the composition of extracts impacts the morphology of the resulting Au NPs.

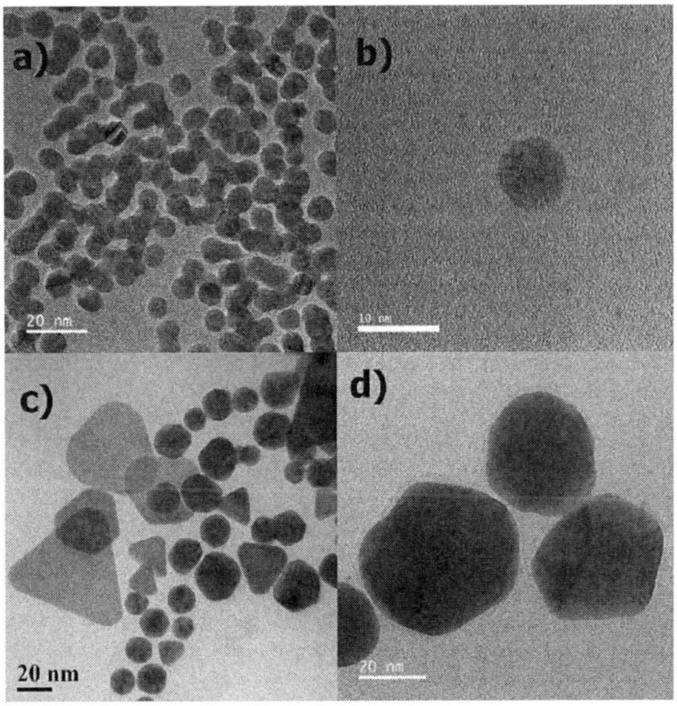

Figure 3 TEM images of Au NPs prepared using polyphenolic fraction of *Sambucus nigra* extract (a, b) and aqueous extract of *Mentha piperita* (c, d)

Whereas in some cases almost perfect Au nanospheres are produced (extracts of *Sambucus nigra*), in other cases, a mixture of spherical and non-spherical Au NPs (extracts of *Mentha piperita*) is formed. Moreover, depending on the amount of Au^{3+} ions in the reaction mixture, the formation of thin Au nanotriangles and also thin Au nanohexagons are registered. The contribution of the latter shifts the absorption maximum of SPR further into the near infrared spectral range. Besides, it should be noticed that the spherical Au NPs are not quite spherical but with ribs, which indicates a presence of a special organic substance, which probably, under optimal conditions, could increase the yield of non-spherical Au NPs. Separation by centrifugation or electrophoresis could be principal methods for the preparation of mono-disperse irregular shape Au NPs from the multicomponent colloidal solutions of gold. Spherical Au NPs are among the well-studied nanoobjects considering their synthesis and characterization; their surface usually is modified extensively with ligands. The affinities of extract components to the surface of spherical Au NPs cause the translation of the affinity into the final morphology of particles when different phyto-organics were used. Typically, the extract components with low binding affinity can support stabilization of large spherical Au NPs in colloidal solutions. In such conditions, the polyphenolic fraction of *Sambucus nigra* extract enables just a growth of gold seeds to bigger spherical Au NPs, without influencing their shape. We should point out that when using the extract of *Mentha piperita*, a mixture of crystallographic habits is usually obtained, including single-crystalline, mono-twinned, and pentatwinned populations of pseudospherical Au NPs. In fact, seed-mediated growth of non-spherical Au NPs is strongly dependent on the crystal structure of the seeds, which are mono-twinned or multiplies twinned. The pseudo-spherical single-crystalline seeds may produce triangles or hexagons (**Figure 3c**), in good yields. Increasing the negative charge on the Au seeds at the presence of excess amount of reducing agent can be a reason for the migration of Au atoms to low-energy facets, for example, to {111}. Conversely, at low concentrations of reducing agents, a slower process of Au^{3+} reduction takes place, resulting in penta-twinned gold intermediates which are appropriate for the growth of the six-branched Au NPs. These results strongly suggest that while crystalline defects may be necessary for such branching anisotropic growth, such twinning can be both induced and specifically tuned by dosing one of reactants. So, it is not surprising, that we can also capture the Au NPs with truncated tetrahedral morphology, as well as the Au NPs decahedral with additional tetrahedral growth and with icosahedral morphologies (**Figure 3d**).

4. CONCLUSION

Based on the results of our studies, we have considered the optical characteristics in the region of SPR and the morphological characteristics of Au NPs, both spherical and specific irregularly shaped Au NPs, prepared by using green protocols. It was found that the use of extracts of some plants (*Sambucus nigra*) leads to the formation of almost perfectly spherical Au NPs. The details of the types of defects observed by the TEM method are presented. According to the obtained data, we do not rule out the possibility of obtaining the required non-spherical Au NPs with a response in the near-infrared range using extracts of selected plants, for example, *Mentha Piperita*. The synthesis of Au NPs makes it possible to obtain biocompatible metal NPs for biomedical purposes; however, there are still quite complex issues related to the mechanisms of growth of non-spherical particles. Herein, we have proposed a view of the growth of such particles, considering the formation of twinning and multi-twinning planes of Au seeds.

ACKNOWLEDGEMENTS

This work was supported by the Scientific Grant Agency of the Ministry of Education, science, research and sport of the Slovak Republic and the Slovak Academy of Sciences, No. 1/0882/21.

REFERENCES

[1] QUINTANA, C., CIFUENTES, M.P., HUMPHREY, M.G. Transition metal complex/gold nanoparticle hybrid materials. *Chemical Society Reviews*. 2020, vol. 49, no. 8, pp. 2316-2341. Available from: https://doi.org/10.1039/c9cs00651f.

[2] YANG, W., LIANG, H., MA, S., WANG, D., HUANG, J. Gold nanoparticle based photothermal therapy: Development and application for effective cancer treatment. *Sustainable Materials and Technologies*. 2019, vol. 22, AN e00109. Available from: https://doi.org/10.1016/j.susmat.2019.e00109.

[3] SANOJ REJINOLD, N., CHOI, G., CHOY, J.-H. Recent trends in nano photo-chemotherapy approaches and future scopes. *Coordination Chemistry Reviews*, 2020, vol. 411, AN 213252. Available from: https://doi.org/10.1016/j.ccr.2020.213252.

[4] SMITH, A.M., MANCINI, M.C., NIE, S. Bioimaging: Second window for in vivo imaging. *Nature Nanotechnology*. 2009, vol. 4, no. 11, pp. 710-711. Available from: https://doi.org/10.1038/nnano.2009.326.

[5] FIZER, M.M., MARIYCHUK, R.T., FIZER, O.I. Gold nanoparticles green synthesis with clove oil: spectroscopic and theoretical study. *Applied Nanoscience*. 2022, vol. 12, no. 3, pp. 611-620. Available from: https://doi.org/10.1007/s13204-021-01726-6.

[6] MARIYCHUK, R., PORUBSKÁ, J., OSTAFIN, M., ČAPLOVIČOVÁ, M., ELIAŠOVÁ, A. Green synthesis of stable nanocolloids of monodisperse silver and gold nanoparticles using natural polyphenols from fruits of *Sambucus nigra* L. *Applied Nanoscience*. 2020, vol. 10, no. 12, pp. 4545-4558. Available from: https://doi.org/10.1007/s13204-020-01324-y.

[7] LOHSE, S.E., MURPHY, C.J. The quest for shape control: A history of gold nanorod synthesis. *Chemistry of Materials*. 2013, vol. 25, no. 8, pp. 1250-1261. Available from: https://doi.org/10.1021/cm303708p.

[8] LOPATYNSKYI, A.M., MALYMON, Y.O., LYTVYN, V.K., MOGYLNYI, I.V., RACHKOV, A.E., SOLDATKIN, A.P., CHEGEL, V.I. Solid and Hollow Gold Nanostructures for Nanomedicine: Comparison of Photothermal Properties. *Plasmonics*. 2018, vol. 13, no. 5, pp. 1659-1669. Available from: https://doi.org/10.1007/s11468-017-0675-1.

[9] DUAN, H., WANG, D., LI, Y. Green chemistry for nanoparticle synthesis. *Chemical Society Reviews*. 2015, vol. 44, no. 16, pp. 5778-5792. Available from: https://doi.org/10.1039/c4cs00363b.

[10] BAO, C., BEZIERE, N., DEL PINO, P., PELAZ, B., ESTRADA, G., TIAN, F., NTZIACHRISTOS, V., DE LA FUENTE, J.M., CUI, D. Gold nanoprisms as optoacoustic signal nanoamplifiers for in vivo bioimaging of gastrointestinal cancers. *Small*. 2013, vol. 9, no. 1, pp. 68-74. Available from: https://doi.org/10.1002/smll.201201779

[11] JAKHMOLA, A., KRISHNAN, S., ONESTO, V., GENTILE, F., PROFETA, M., MANIKAS, A., BATTISTA, E., VECCHIONE, R., NETTI, P.A. Sustainable synthesis and theoretical studies of polyhedral gold nanoparticles displaying high SERS activity, NIR absorption, and cellular uptake. *Materials Today Chemistry*. 2022, vol. 26, AN 101016. Available from: https://doi.org/10.1016/j.mtchem.2022.101016.

[12] HAZARIKA, M., KALITA, G.D., PRAMANIK, S., BORAH, D., DAS, P. Bio-functionalized anisotropic gold nanoparticles as efficient catalyst for nitrile hydration and hydrogenation of nitrophenol. *Current Research in Green and Sustainable Chemistry*. 2020, vol. 3, AN 100018. Available from: https://doi.org/10.1016/j.crgsc.2020.100018.

[13] MARIYCHUK, R., GRULOVA, D., GRISHCHENKO, L.M., LINNIK, R.P., LISNYAK, V.V. Green synthesis of non-spherical gold nanoparticles using *Solidago canadensis* L. extract. *Applied Nanoscience*. 2020, vol. 10, no. 12, pp. 4817-4826. Available from: https://doi.org/10.1007/s13204-020-01406-x.

[14] NADAF, S.J., JADHAV, N.R., NAIKWADI, H.S., SAVEKAR, P.L., SAPKAL, I.D., KAMBLI, M.M., DESAI, I.A. Green synthesis of gold and silver nanoparticles: Updates on research, patents, and future prospects. *OpenNano*. 2022, vol. 8, AN 100076. Available from: https://doi.org/10.1016/j.onano.2022.100076.

[15] JANNATHUL FIRDHOUSE, M., LALITHA, P. Biogenic green synthesis of gold nanoparticles and their applications - A review of promising properties. *Inorganic Chemistry Communications*. 2022, vol. 143, AN 109800. Available from: https://doi.org/10.1016/j.inoche.2022.109800.

[16] BAO, Z., LAN, C.Q. Advances in biosynthesis of noble metal nanoparticles mediated by photosynthetic organisms - A review. *Colloids and Surfaces B: Biointerfaces*. 2019, vol. 184, AN 110519. Available from: https://doi.org/10.1016/j.colsurfb.2019.110519.

[17] LUCENA-SERRANO C., CONTRERAS-CÁCERES R., SÁNCHEZ-MOLINA M., CASADO-RODRÍGUEZ M.A., CLOAREC J.M., MAINETTI E., LÓPEZ-ROMERO M. Preparation and application of non-spherical metal nanoparticles: reality and perspectives. *Current Organic Chemistry*. 2017, vol. 21, pp. 1-22. Available from: https://doi.org/10.2174/1385272821666161107160143.

[18] MARIYCHUK, R., SMOLKOVÁ, R., BARTOŠOVÁ, V., ELIAŠOVÁ, A., GRISHCHENKO, L.M., DIYUK, V.E., LISNYAK, V.V. The regularities of the *Mentha piperita* L. extract mediated synthesis of gold nanoparticles with a response in the infrared range. *Applied Nanoscience*. 2022, vol. 12, no. 4, pp. 1071-1083. Available from: https://doi.org/10.1007/s13204-021-01740-8.

[19] MOCK, J.J., BARBIC, M., SMITH, D.R., SCHULTZ, D.A., SCHULTZ, S. Shape effects in plasmon resonance of individual colloidal silver nanoparticles. *Journal of Chemical Physics*. 2002, vol. 116, no. 15, pp. 6755-6759. Available from: https://doi.org/10.1063/1.1462610.

[20] RODRÍGUEZ-OLIVEROS, R., SÁNCHEZ-GIL, J.A. Gold nanostars as thermoplasmonic nanoparticles for optical heating. *Optics Express*. 2012, vol. 20, no. 1, pp. 621-626. Available from: https://doi.org/10.1364/OE.20.000621.

EVAULATION OF THE APPLICABILITY OF VARIOUS POROSIMETRY METHODS IN INVESTIGATION OF THE ULTRASTRUCTURE OF HYDROGELS

[1]Monika TRUDIČOVÁ, [1]Jan ZAHRÁDKA, [1]Jiří SMILEK, [2]Kamila HRUBANOVÁ, [2]Kateřina MRÁZOVÁ, [1]Petr SEDLÁČEK, [1]Miloslav PEKAŘ

[1]Brno University of Technology, Faculty of Chemistry, Brno, Czech Republic, EU, xctrudicova@fch.vut.cz

[2]Institute of Scientific Instruments of the Czech Academy of Sciences, Brno, Czech Republic, EU.

https://doi.org/10.37904/nanocon.2022.4619

Abstract

Properties of hydrogels that define their application potential (e.g., stiffness, mechanical strength, transport properties) are primarily based on the morphology of the internal structure. Accurate analysis of the ultrastructure of hydrogels is therefore important for understanding the parameters affecting the functionality of hydrogels. Accordingly, this work was focused on the optimization and testing of methods suitable for the analysis of the nanostructure of hydrogel materials with a focus on determining the pore size. Conventional porosimetry methods (DSC thermoporometry), direct visualization methods (SEM, cryo-SEM, and AFM) as well as rheology, turbidimetry, and DLS microrheology were chosen as methods for structural analysis. Hydrogels based on polyvinyl alcohol (PVA) were chosen as representative hydrogel systems for testing these methods. The results obtained by the individual methods were compared.

Keywords: Hydrogel, polyvinyl alcohol, structural analysis

1. INTRODUCTION

Hydrogels are natural or synthetic polymeric materials with a three-dimensional internal structure that evinces versatile chemical, physical, and biological properties. Their structure is formed by physically or chemically connected polymers creating a spatial network. The properties of these materials are unimaginably diverse and applicable in many practical directions. Their great advantage is the ability to absorb a large amount of water into their internal structure for a long period. In this way, they resemble human tissues to a large extent. Obtaining detailed information on the morphology of the internal structure of gels, focusing on the pores, and understanding how they can be modified depending on the desired properties is crucial for preparing these materials for specific applications.

The main goal was to find a method or a set of methods that would be able to precisely define the internal and surface structure with a focus on porosity in materials with a high-water content i.e., hydrogels. Determining the internal structure of hydrogels is crucial to understanding the relationship between the structure and properties of these systems. A significant problem with most of the used methods consists in their sample dehydration prior to the analysis. When converting hydrogels into xerogel form, the microstructure inevitably collapses. Based on the experience with these materials, it could be said that PVA hydrogels show very good mechanical properties from the point of their viscoelastic properties, and therefore should resist growing ice crystals at higher molecular weights and should even limit their growth during the lyophilization process. This assumption is partially true; however, it is necessary to consider that even a small change in experimental conditions can affect the process of ice crystal growth to such an extent that it can disrupt the resulting structure of the hydrogel.

2. MATERIALS AND METHODS

In this work, PVA polymers with different molecular weights but the similar degree of hydrolysis were selected to restrict the variation of the structure and material properties among the tested gel solely to the PVA chain length. All polymers used in this work were purchased from Sigma-Aldrich (polyvinyl alcohol (PVA) CAS: 9002-89-5). Concentration series of stock PVA solutions were prepared. The exact amount of PVA polymer was transferred together with the required amount of deionized water into a reagent bottle. First, these mixtures were mixed for 30 minutes at laboratory temperature, and then this suspension was transferred to a water bath in which it was mixed at 95 °C for 3 hours. Subsequently, the polymer dispersion prepared in this way was poured into prepared molds, in which it was subjected to a cross-linking process under different conditions. All samples of PVA hydrogels were prepared by the freeze-thaw method at two freezing speeds. **Table 1** refers to the parameters of the various freezing conditions.

Table 1 Conditions of the freeze/thaw method during the preparation procedure of hydrogels

	Temperature (°C)	**Freezing time (min)**	**Melting time (min)**	**The average number of freeze/thaw cycles**
Laboratory freezer	−20 to −15	~135	~15 to 30	5
Salt ice bath	−2 to 0	~315		9

A wide portfolio of methods was selected to determine the internal porous structure, which should cover the most frequently used methods that are used for these purposes.

3. RESULTS AND DISCUSSION

3.1. DSC thermoporometry

The literature [1,2] provides a high-quality mathematical apparatus for many porosimetric methods. Among them, determination of the pore size via monitoring a decrease in the melting temperature of intraporous ice by DSC (differential scanning calorimetry) represents a simple and widely available experimental option [1]. For any DCS experiment, there is a certain correlation between the heating rate of the experiment and the resolution of the signal, which provides information on the pore size. During the evaluation, it is necessary to determine the decrease in the melting temperature that corresponds to the intrapore water. In the case of microporous solids, the signal is often represented by the separated melting peak that occurs at the subzero temperature. The measured PVA hydrogels (using TA Instruments, Q 2000) do not have even a hint of a separate peak that would define the porous phase and therefore it is not possible to characterize the pore size in the quantitative manner. Despite this, it is possible to extract somewhat interesting qualitative information from the measured data. With the additional TGA measurement (using TA Instruments, TGA 5500), it is possible to determine the amount of freezable bound water. Based on the detected amount of water and the values measured by DSC, it is possible to evaluate that the onset temperature of the hydrogels shifts to lower temperatures as the molecular weight of the polymer increases. It indicates that the greater amount of freezing water, the more the normal and stable form of water is disturbed. Therefore, the higher the freezing water content, the larger the pores in the hydrogel.

3.2. Direct visualization methods

In the evaluation of methods focused on direct visualization of the gel ultrastructure, atomic force microscopy (AFM; Bruker Corporation, NanoWizard 4 XP BioScience), scanning electron microscopy (SEM; Carl Zeiss AG, EVO LS 10), and cryogenic scanning electron microscopy (CryoSEM; Thermo Fisher Scientific, Magellan 400) were used.

AFM is fundamentally different from other types of microscopes since the resulting image is not created by focusing light or electrons on the surface of the sample. Using AFM, the sample is spatially mapped based on the atomic forces between the sample and a sharp probe consisting of a tip (probe), which is located at the end of the beam (cantilever). Therefore, the samples could be measured in their native swollen state and the structure of the sample could not be destroyed by artifacts created during sample preparation for measurement (e.g., freezing). A big advantage of AFM images (**Figure 2A**) compared to SEM images is their homogenous illumination and contrast. The images taken of PVA hydrogels did not have as well discernable polymer network structure as we have previously found for macroporous gels, e.g., agarose (data not shown). Most likely this was caused by the fact that we are two to three orders of magnitude lower than the agarose hydrogels. The resulting pore sizes confirm the trend (**Figure 1**) of decreasing pore size with increasing PVA concentration, as described in [3]. Even these values for the lower molecular weight sample match the values reported in this paper.

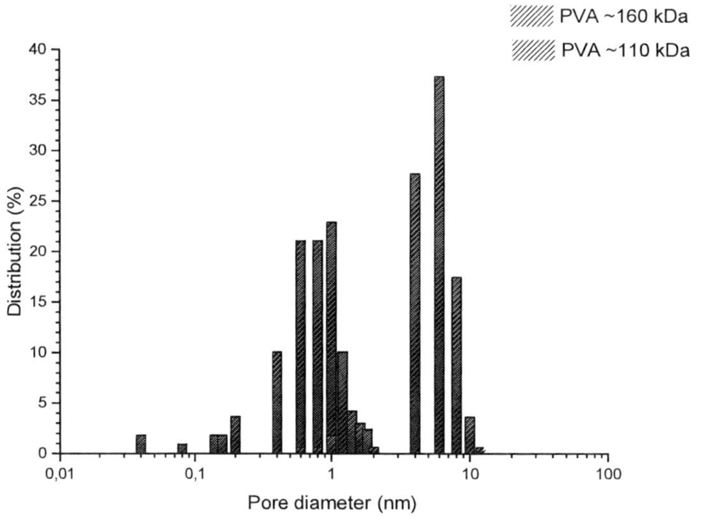

Figure 1 Histogram of the distribution of pore sizes obtained from AFM measurements and image analysis for PVA hydrogels (results of image analysis provided by Particle analysis tool in ImageJ as described previously [4])

Compared to AFM, samples for SEM measurements were converted into xerogel form. This modification of the sample is likely to create artifacts that can disrupt the internal structure of the samples. No measurement using this method produced adequate results and these results showed (**Figure 2B**) a distorted structure. A possible solution to this problem could be measurement under cryogenic conditions using the CryoSEM technique.

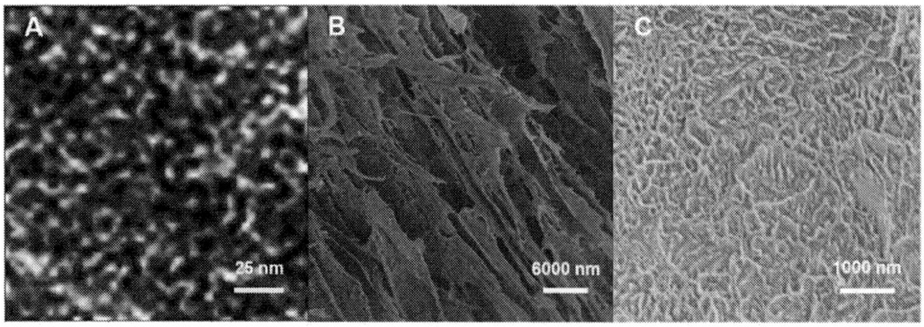

Figure 2 PVA sample (110 kDa and 10 wt.% of polymer) visualized by AFM (A), SEM (B), and CryoSEM (C)

The results obtained with CryoSEM show similar nanostructures (**Figure 2C**) as with AFM, except that the AFM images are displayed at a scale almost 10 times larger. CryoSEM shows a lower contrast, which could increase with further modification of the samples, however, these results cannot yet be used for image analysis. Compared to classic SEM a swollen structure can be analyzed, this structure could be considered as real, however, it is necessary to take into account all modifications that could alter the sample structure during sample preparation for analysis.

3.3. Rheology

The mechanical properties depending on the internal structure of the fully swollen hydrogels were characterized using a rotational rheometer (TA Instruments, Discovery Hybrid Rheometer (HR-2)). Since hydrogels are viscoelastic materials (i.e., they exhibit both viscous and elastic behavior), rheology (the measure of flow and deformation behavior of liquids and solids) is an appropriate method for characterizing hydrogel mechanical properties. The sample is placed between the plates with a defined gap, and the upper plate performs an oscillating movement. All hydrogels showed a common plateau in the frequency sweep, i.e., the elastic moduli were frequency independent in the range from about 0.1 to 10 rad/s. Therefore, these hydrogels exhibit a soft rubbery behavior. Polymers with lower concentrations of PVA and molecular weights show lower values of elastic moduli, and their plateau has a relatively narrow range. In contrast, hydrogels with higher mass concentrations and molecular weights have the opposite trend. The change in toughness or stiffness of these hydrogel networks can therefore be correlated with their composition and internal structure.

The mesh size of the hydrogels was calculated based on the rheological data using the rubber elasticity theory [2]. In hydrogels, the crosslinking density increases with increasing polymer concentration. This leads to the increase in the elastic modulus and at the same time to the decrease in the molecular weight of the chain between the two nodes. Hydrogel networks with a higher concentration and molecular weight of the polymer show a smaller average mesh size. In a wide range of concentrations, the trend of decreasing porosity with increasing concentration and molecular weight of PVA was well reproducible, as you can see in **Figure 3**.

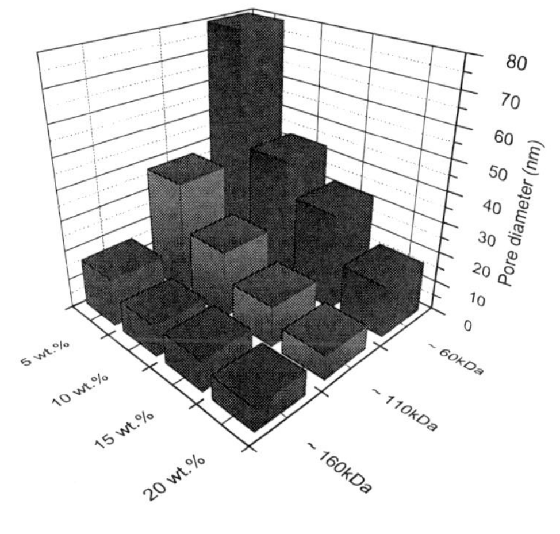

Figure3 Dependence of the input parameters of the PVA polymer on the resulting pore sizes; hydrogels prepared in a laboratory freezer

3.4. DLS microrheology

The DLS (dynamic light scattering) measurement is based on observing the movement of tracer particles with defined particle size in the sample via monitoring the intensity of the light scattered by these particles. The selection of suitable particles for microrheological measurements can critically affect the measurement results. Based on theoretical knowledge [3] of the framework pore size in the selected PVA hydrogels, polystyrene (PS) particles with an average size of 50 nm, which are larger than the predicted meshes of the polymer network, were selected. If the selected particles were smaller, their free passage through the mesh would occur and the method would not provide an accurate rheological response. Furthermore, if the particles were much larger, there would be complex interactions of the particles with the entire network. Micrometer-sized particles also begin to exhibit non-Brownian behavior with thermal fluctuations reduced below realistically measurable levels. The measured data (using Malvern Panalytical, ZetaSizer Nano ZS) of the MSD curves were transformed into the form of viscoelastic moduli using complex mathematical apparatus. The subsequent evaluation was based only on the middle part of the average elastic moduli, the so-called plateau. To evaluate this part of the modules, we decided to use the mathematical apparatus of inverse derivation. The results of these measurements provide data that are almost correlated with the values discussed in the literature [3].

3.5. Turbidimetry

Polymer networks creating pores in hydrogels are responsible for the turbidity of final hydrogels. Measuring the loss of intensity of transmitted light caused by the light scattering due to this turbidity can offer information about the internal structure. The turbidimetric measurement (performed using UV-VIS spectrophotometer: Hitachi High-Technologies Corporation, U-3900H) of PVA hydrogels was carried out based on an article [5] in which the author dealt with agarose hydrogels. To evaluate the measured data, the part of the spectrum that corresponds to the shape of the scattering spectrum without radiation interference was selected. The part of the spectrum in which the interference occurred was fundamental to the final calculation and the limitation of the correlation curve was based on the wave exponents associated with this part. Therefore, it was not possible to calculate the exact pore sizes. Nonetheless, it was possible to determine the predicted trends regarding the porosity of these hydrogel materials. According to the directions of the individual curves, a decreasing trend can be confirmed depending on the increasing molecular weight, e.i., the trend that was observed for all previous methods. For ordinary comparison of set of samples, this method is suitable and simple to perform.

4. CONCLUSION

This work focused on the optimization and testing of methods suitable for the analysis of the microstructure of hydrogel materials with a particular focus on the determination of pore size. Each method tested has specific advantages and disadvantages (summarized in **Table 2**).

Table 2 Comparison of discussed methods

	Advantages	**Disadvantages**	**Economic demands of operation**
DSC thermoporometry	small amount of material; both dried and hydrated samples can be measured	poor results for aqueous solvents, better results with organic solvents	$$$
SEM	topography measurement	impossible to measure hydrated samples; deformation of the sample during preparation	$$$$
CryoSEM	measurement of the samples in the hydrated state	possible sample deformation during cryofixation or sublimation	$$$$$
AFM	universal method; allows obtaining qualitative and quantitative information about physical parameters, 3D	low scan speed; both the sample and the cantilever can be damaged during the measurement; prolonged optimization of new materials	$$
Rheology	quick and easy method; samples are measured in the hydrated state	it is not always possible to prepare a sample of the same size	$$
DLS microrheology	small amount of material; fast method with higher measurement ranges	complex sample preparation; selection of a suitable probe for measurement	$
Turbidimetry	the simplest of the commented techniques	complex sample preparation; there is no mathematical apparatus connecting the issues for a wider range of materials	$

The AFM method shows admirable and, in our opinion, the most reasonable results, which, however, are redeemed by relatively complex measurements and optimization for each measured sample. Therefore, it would be advisable to combine this technique with the simpler technique for a faster determination of e.g., rheology, microrheology, turbidimetry, or DSC. The optimization of the combination of these two chosen methods should be carried out on a wide range of different materials and should try to find a connection between the results, thanks to which it should be possible to subsequently use a simpler and faster variant in the routine porosity-mapping assays. However, the limitations of individual methods and the dynamic nature of hydrogel polymer networks must still be considered.

REFERENCES

[1] LANDRY, Michael R. Thermoporometry by differential scanning calorimetry: experimental considerations and applications. *Thermochimica Acta*. 2005, vol. 433, pp. 27-50. Available from: https://doi.org/10.1016/j.tca.2005.02.015.

[2] PESCOSOLIDO, Laura, FERUGLIO, Luigi, FARRA, Rossella, FIORENTINO, Simona, COLOMBO, Italo, COVIELLO, Tommasina, MATRICARDI, Pietro, HENNINK, Wim E., VERMONDEN, Tina, GRASSI, Mario. Mesh size distribution determination of interpenetrating polymer network hydrogels. *Soft Matter*. 2012, vol. 8, pp. 7708-7715. Available from: https://doi.org/10.1039/C2SM25677K.

[3] HICKEY, Alla S., PEPPAS, Nikolaos A. Mesh size and diffusive characteristics of semicrystalline poly(vinyl alcohol) membranes prepared by freezing/thawing techniques. *Journal of Membrane Science*. 1995, vol. 107, pp. 229-237. Available from: https://doi.org/10.1016/0376-7388(95)00119-0.

[4] TRUDICOVA, Monika, SMILEK, Jiri, KALINA, Michal, SMILKOVA, Marcela, ADAMKOVA, Katerina, HRUBANOVA, Kamila, KRZYZANEK, Vladislav, SEDLACEK, Petr. Multiscale Experimental Evaluation of Agarose-Based Semi-Interpenetrating Polymer Network Hydrogels as Materials with Tunable Rheological and Transport Performance. *Polymers*. 2020, vol. 12, pp. 1-25. Available from: https://doi.org/10.3390/polym12112561.

[5] AYMARD, Pierre, MARTIN, David R., PLUCKNETT, Kevin P., FOSTER, Tim J., CLARK, Allan H., NORTON, Ian T. Influence of Thermal History on the Structural and Mechanical Properties of Agarose Gels. *Biopolymers*. 2001, vol. 59, pp. 131-144. Available from: https://doi.org/10.1002/1097-0282(200109)59:3<131::AID-BIP1013>3.0.CO;2-8.

THE EFFECT OF RICE HUSKS AND $mTiO_2$ ON THE THERMAL PROPERTIES OF LLDPE/RH/$mTiO_2$ BIOCOMPOSITES

Emi GOVORČIN BAJSIĆ, Mario MEHEŠ, Ljerka KRATOFIL KREHULA, Ana PERŠIĆ, Veljko FILIPAN, Igor SUTLOVIĆ

University of Zagreb, Faculty of Chemical Engineering and Technology, Zagreb, Croatia, EU,
egovor@fkit.hr, mmehes@fkit.hr, krehula@fkit.hr, apersic@fkit.hr, vfilipan@fkit.hr, isutlo@fkit.hr

https://doi.org/10.37904/nanocon.2022.4620

Abstract

Petroleum-based products as Linear low-density polyethylene (LLDPE) are most commonly used in packaging of products like food, pharmaceuticals, cosmetics and detergents because of his low cost, good mechanical properties, easy processing and flexibility. However the lack of LLDPE as non-degradable packaging material has a major impact on environmental pollution. During the last few decades, petroleum-based products because of improper management and disposal practice cause serious environmental pollution problems that could persist for centuries. This has raised growing concern about degradable polymers and promoted develop of new synthetic plastics with plants (biopolymers) which can be biodegradable and may replace currently used plastics at least in some of the fields. Rice husk (RH) is one of the major agricultural residues produced as a byproduct during the rice milling process and it can be widely used due to its properties. The incorporation of RH into polymer matrices provides advantageous characteristics, such as biodegradability, low cost, light weight, toughness, and resistance to weathering. The aim of this study was to examine the influence of the RH content and $mTiO_2$ on the thermal properties of LLDPE. In addition to RH, rutile ($mTiO_2$) was added due to its antibacterial and photocatalityc properties as well as resistance to high temperatures. LLDPE/RH/$mTiO_2$ biocomposites were prepared by mixing RH in content of 10, 20, 30, 40 and 50 wt % and 5 wt % of $mTiO_2$ in the neat LLDPE in a Brabender mixer. By DSC techniques phase transitions, TGA technique the thermal stability while by DMA technique viscoelastic properties of the LLDPE/RH/$mTiO_2$ biocomposites were obtained with the respect of RH content and addition of mTiO2. According to DSC results for LLDPE/RH/$mTiO_2$ biocomposites changes in the melting (T_m) and crystallisation (T_c) temperature indicated that there was some degree of interaction between LLDPE, RH and $mTiO_2$. The TGA results show that the LLDPE is more thermally stable with addition of $mTiO_2$. The viscoelastic properties were improved with the addition of $mTiO_2$.

Keywords: Linear low density polyethylene, rice husks, $mTiO_2$, biocomposites, thermal properties

1. INTRODUCTION

During the last few decades, polymer composites because of their superior properties such as high strength to weight ratio, good electrical insulation, ability to transfer load, and easy and inexpensive processing have been of interest to industry and academia. However, petroleum-based products creates ecological and enviromental problems due to non-biodegradable polymers which release harmful chemicals and toxic gases. This results in synthesis of polymers for the production of biodegradable polymers by inexpensive and renewable resources such as cellulose, starch, lignin, proteins and vegetable oils has attracted the attention of scientists around the world and has shown a promising path to the sustainable development of new products. Biodegradable polymers/fibers are biocompatible and degrade in the suroundiing enviroments without causing any environmental pollution. Natural fibers such as banana, rice husks, jute, sisal and hemp have advantages such as low cost, less weight, easy to process and bio-degradable compared to the synthetic fibers.[1] Because of that natural fibers can be used an alternative material to conventional synthetic fibers such as glass, carbon, boron and kevlar fibers. Most natural fibers consist of cellulose, hemicelluloses, lignin and other low molecular weight compounds. Over the last years, remarkable attention has been paid to eco-friendly, sustainable materials for various applications. In this context, rice husks (RH) was used as reinforcing element in green polymer composites. RH is characterized by its abundance, renewability, low-price, low density and high strength to weight ratio. Polymer composites incorporating fillers can result in materials with improved the different properties such as mechanical, thermal, flame retardancy and water absorption behaviour [2]. The improvement of mechanical and thermal properties can be achieved by the addition of inorganic filler namely titanium dioxide (TiO_2). TiO_2 is a versatile material with interesting characteristics biocompatible, chemical stability, high reactivity, antibacterial properties, electrochemical properties, low cost, and safe production. Thermal analysis (TA) is one of the group of techniques, which is used in the characterization of polymeric materials to give information, which are often unavailable via using other techniques. Therefore, thermal analysis is a perfect and versatile tool for characterizing composite materials in a wide temperature range. In this work thermogravimetry (TG), differential scanning calorimetry (DSC) and dynamic-mechanical analysis (DMA) were used to characterize the influence of RH and microsized TiO_2 on the thermal and dynamic-mechanical properties.In the aim to develop biocomposites with better biodegradation properties RH particles were added in the LLDPE polymer. For expect is that the RH particles will stimulate biodegradation process of synthetic LLDPE polymer. This study is the preliminary for the further investigations which will be directed on the determination of photocatalytic effect of $mTiO_2$ in the LLDPE/RH biocomposites and biodegradation process.

2. EXPERIMENTAL

2.1. Materials

Linear low density polyethylene (LLDPE) (LLDPE-EFDA 7047), with a MFR of 1. 0 g/10 min at 190 °C and density of 0.918 g/cm^3 was supplied from Equate Pertrochemical Company, Kuwait. The rice husk (RH) was donated from rice factory in Kocani, Macedonia, and the rice cultivar was 'Sant Andrea' (Oryza sativa L. 'Sant Andrea'), Italy. Microsized TiO_2 (Aeroxide P25, Titania, Titanium dioxide) in the form of white powder, molecular weight 79.87 gmol^{-1} and density of 3.9 g cm^{-3} was purchased from Sigma-Aldrich Chemistry.

2.2. Preparation of biocomposites

In this work, two series of biocomposites formulation of LLDPE reinforced rice husk without and with TiO_2 incorporation were prepared. LLDPE biocomposites containing RH (content 10-50 wt%) and 5 wt% of microTiO_2 were prepared by the melt-mixing in the Brabender kneading chamber at 155 °C for 5 minutes (screw rotation speed was 60 rpm). The RH were chopped to small size in IKA-WERKE M20 mixer and dried in an air-circulating oven at 105 °C for 24 h to remove any residual water prior to mixing with LLDPE. The biocomposites were pressed in a hydraulic press Fontune, Holland (SRB 140, EC 320x320NB) at the temperature of 160 °C and a pressure of 25 kPa for 5 min.

2.3. Characterization

The viscoelastic properties (storage modulus E' and loss modulus E") of the LLDPE and LLDPE/RH biocomposites without and with $mTiO_2$ were measured using a dynamic mechanical analyser (DMA 983, TA Instruments). The measurements were performed at a frequency of 1 Hz and at a oscillation amplitude of 0.2 mm. The temperature range was from -100 °C to 120 °C at a heating rate of 3 °C/min. The specimen dimensions were 20 x 13 x1 mm. The samples were cooled to -100 °C using liquid nitrogen.

DSC analysis was carried out using a differential scanning calorimeter DSC Mettler Toledo 822e. The measurements were carried out in the temperature range from -100 to 150 °C at a heating rate of 10 C/min, in two heating/cooling cycles and inert atmosphere of nitrogen. Samples of 9-10 mg were placed into aluminium pans with a lid and were hermetically-sealed. Thermal parameters such as melting temperature (T_m), crystallization temperature (T_c), heat of melting (ΔH_m) and the percentage of crystallinity (X_c), of the studied biocomposites were analyzed.

Thermogravimetric (TG) and Differential Thermogravimetric (DTG) measurements were performed with TA Instruments Q500 system analyzer to obtained the thermal stability of the neat LLDPE and LLDPE/RH biocomposites without and with $mTiO_2$. Samples of approximately 10 mg was heated from 25 °C to 600 °C at a heating rate of 10 °C /min under a nitrogen atmosphere (60 ml/min).

3. RESULTS AND DISCUSSION

3.1. Dynamic Mechanical Analysis (DMA)

Dynamic mechanical analysis (DMA) has become a widely used technique for the determination of the loss modulus, E", storage modulus,E', and tanδ. Generally, the DMA properties of a composite material depend presence and content of the additives like filler,compatibiliser, and the mode of testing.The storage modulus E' of the neat LLDPE and LLDPE/RH biocomposites without and with TiO_2, shown in **Figure 1.**

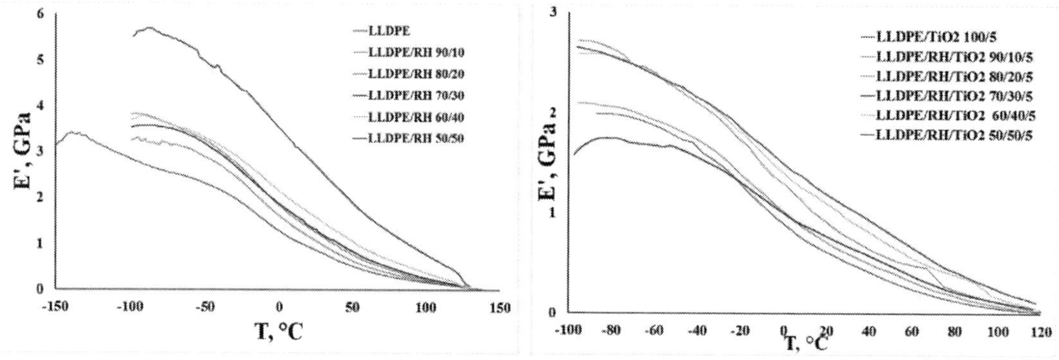

Figure 1 Storage modulus (E') as a function of temperature of the neat LLDPE and LLDPE/RH biocomposites without and with TiO_2

The values of (E') at 25 °C for the neat LLDPE and LLDPE/RH biocomposites without and with $mTiO_2$ are reported in **Table 1** and **Table 2.** As can be observed, the E' values increased with the increase of the RH content in the biocomposites without and with $mTiO_2$, due to the reinforcement imparted by the fillers that allowed greater stress transfer at the interface.

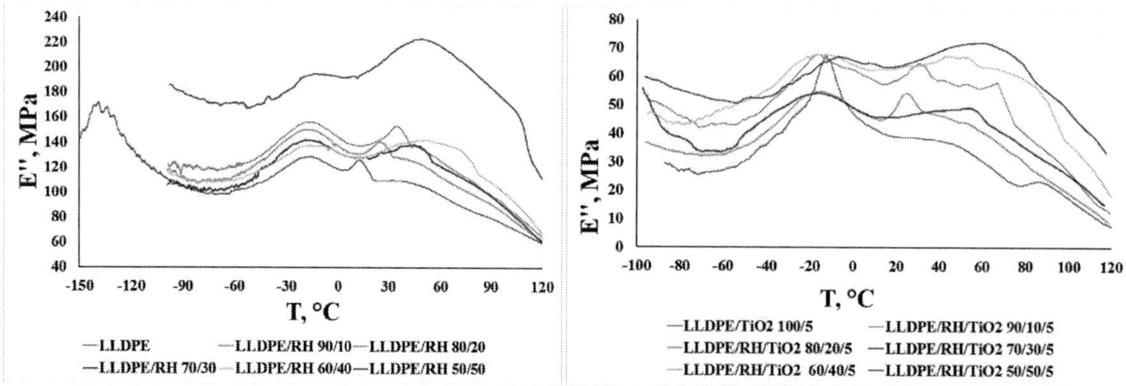

Figure 2 Loss modulus (E") as a function of temperature of the neat LLDPE and LLDPE/RH biocomposites without and with $mTiO_2$

Figure 2 shows the temperature dependence of the loss modulus, *E"* of the neat LLDPE and LLDPE/RH biocomposites without and with $mTiO_2$. For all biocomposites two relaxations are clearly observed in this temperature range. The low-temperature peak is related to the glass transition (T_g) of the amorphous LLDPE fractions. The high-temperature peak corresponds to the α_h transition related to the LLDPE crystalline fractions [3]. The E" values corresponding to the T_g increased with incorporation of the $mTiO_2$ (**Table 2**). It may be attributed to the presence of $mTiO_2$ can enhance interaction between LLDPE matrix and RH in the biocomposites. A good interfacial interaction can be evaluated based on capability of $mTiO_2$ particles in restricting the mobility of polymer chain. As it is seen, that the α_h transition peak is significantly shifted to higher temperatures by the addition of RH (**Table 1**) and $mTiO_2$ (**Table 2**). Sirotkin et al. [4] suggest that in polyethylene α_h relaxation temperature increases with the lamellar thickness, irrespective of grade or crystallinity and is associated with c-shear within the crystalline lamellae.

Table 1 DMA results of the neat LLDPE and LLDPE/RH biocomposites without $mTiO_2$

Sample	LLDPE100/0	LLDPE/RH 90/10	LLDPE/RH 80/20	LLDPE/RH 70/30	LLDPE/RH 60/40	LLDPE/RH 50/50
T_g(°C)	-16.4	-18.9	-16.3	-16.8	-16.5	-10.0
α_h (°C)	12.5	24.9	34.0	44.2	51.5	48.5
$E'_{25°C}$(GPa)	0.876	1.086	1.264	1.333	1.600	2.725

Table 2 DMA results of the neat LLDPE and LLDPE/RH biocomposites with $mTiO_2$

Sample	LLDPE/$TiO_2$10 0/5	LLDPE/RH/TiO $_2$ 90/10/5	LLDPE/RH/TiO $_2$ 80/20/5	LLDPE/RH/TiO $_2$ 70/30/5	LLDPE/RH/TiO $_2$ 60/40/5	LLDPE/RH/TiO $_2$ 50/50/5
T_g(°C)	-12.70	-15.38	-16.20	-14.90	-14.00	-7.08
α_h (°C)	25.75	24.31	30.42	53.92	46.76	56.87
$E'_{25°C}$(GPa)	0.5600	0.6500	0.8325	0.7210	1.0270	1.1250

3.2. Thermal behaviour of biocomposites

Differential scanning calorimetry (DSC) was used to investigate the effect of the RH and $mTiO_2$ filler on the thermal properties of biocomposites. The DSC curves of 2nd heating and cooling cycles of the neat LLDPE and LLDPE/RH biocomposites without and with $mTiO_2$ are shown in **Figure 3** and **Figure 4,** respectively.

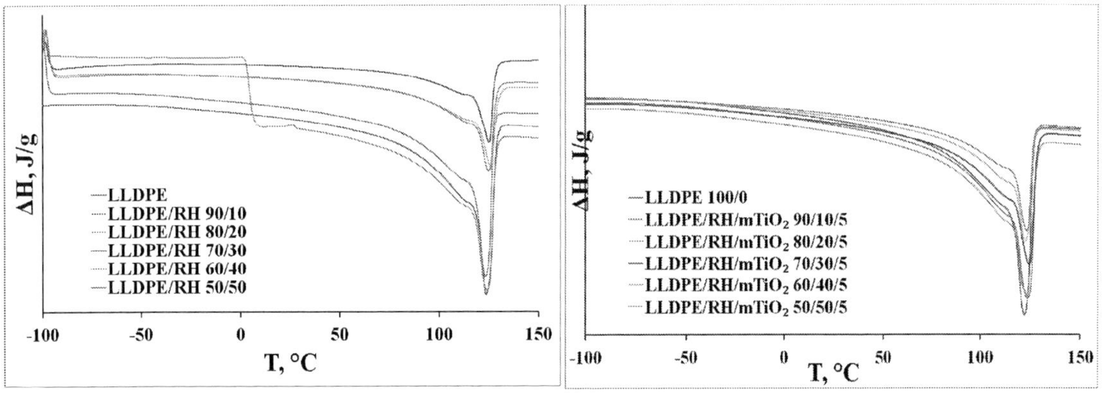

Figure 3 DSC curves of 2nd heating cycle for the neat LLDPE and LLDPE/RH biocomposites without and with $mTiO_2$

The effect of RH content and introduction of $mTiO_2$ on T_m for the LLDPE/RH biocomposites is presented in **Table 3** and shows no significant changed in T_m for the biocomposites. The DSC curves of the neat LLDPE and LDPE/RH biocomposites without and with $mTiO_2$ during the 2nd cooling cycle are shown in **Figure 4**. The determined crystallization temperature, T_c, values are listed in **Table 3** and **Table 4**. The T_c of the biocomposites had slightly shifted to a higher temperature compared to the neat LLDPE with the incorporation of RH and $mTiO_2$. A change in crystallization temperature in the presence of fillers reflects changes in the nature of relaxation processes occurring in thin layers at the polymer-phase interface. A low shoulder observed in the low point of temperature side of the cooling peak of the LLDPE indicating paracrystalline structure.

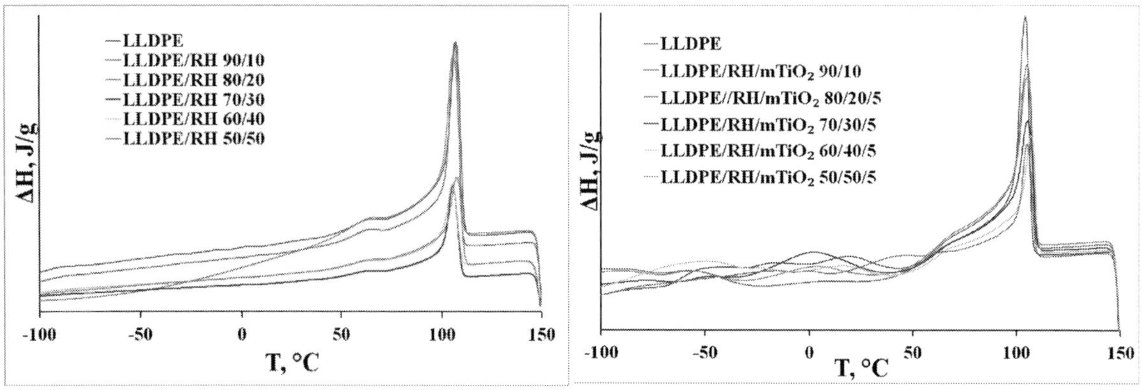

Figure 4 DSC curves of 2nd cooling cycle for the neat LLDPE and LLDPE/RH biocomposites without and with $mTiO_2$

Table 3 DSC results after 2nd heating/cooling cycle of the neat LLDPE and LLDPE/RH biocomposites without $mTiO_2$

Sample	LLDPE100/0	LLDPE/RH 90/10	LLDPE/RH 80/20	LLDPE/RH 70/30	LLDPE/RH 60/40	LLDPE/RH 50/50
T_m (°C)	120.8	122.5	124.0	123.4	123.6	123.7
T_c (°C)	104.2	105.5	104.3	105.1	105.5	106.1
ΔH_m (Jg^{-1})	118.41	113.80	105.01	92.99	75.02	63.36
χ_c (%)	40.4	43.2	44.8	45.3	42.7	43.2

Table 4 DSC results after 2nd heating/cooling cycle of the neat LLDPE and LLDPE/RH biocomposites with $mTiO_2$

Sample	$LLDPE/TiO_2$ 100/5	$LLDPE/RH/TiO_2$ 90/10/5	$LLDPE/RH/TiO_2$ 80/20/5	$LLDPE/RH/TiO_2$ 70/30/5	$LLDPE/RH/TiO_2$ 60/40/5	$LLDPE/RH/TiO_2$ 50/50/5
T_m (°C)	121.3	122.1	124.0	123.6	121.6	122.6
T_c (°C)	104.9	105.8	104.3	105.8	107.6	105.4
ΔH_m (Jg^{-1})	125.4	120.6	119.94	94.6	76.16	60.9
χ_c (%)	42,8	48.1	53.9	48.6	45.6	44.0

The crystallinity of LLDPE and LLDPE/RH biocomposite without and with $mTiO_2$ (χ_c %) was calculated according to Equation (1). A increase of the crystallinity can be observed when the $mTiO_2$ are incorporated in the LLDPE/RH biocomposites (**Table 4**). These results can be explained by the strong nucleation ability of the $mTiO_2$ on LLDPE.

$$\chi c = \left(\frac{\Delta H_m^0}{\Delta H_m^{100} x \left(1 - \frac{\%RH}{100}\right)} \right) x100 \qquad (1)$$

where:

ΔH_m^0 - the value of melting enthalpy read from the surface of melting endotherm (Jg^{-1})

ΔH_m^{100} - the melting enthalpy of 100 % crystalline LLDPE (293,0 Jg^{-1}).

3.3. Results of Thermogravimetric Analysis (TGA)

A TG analysis was very useful tehnique to determine quantitatively the degradation behavior, the composition of the materials and thermal stability. In this study the thermal stability of the neat LLDPE and LLDPE/RH biocomposites without and with $mTiO_2$ was investigated. The TG and DTG curves of LLDPE/RH biocomposites without and with $mTiO_2$ are represented in **Figure 5** and **Figure 6**, respectively.

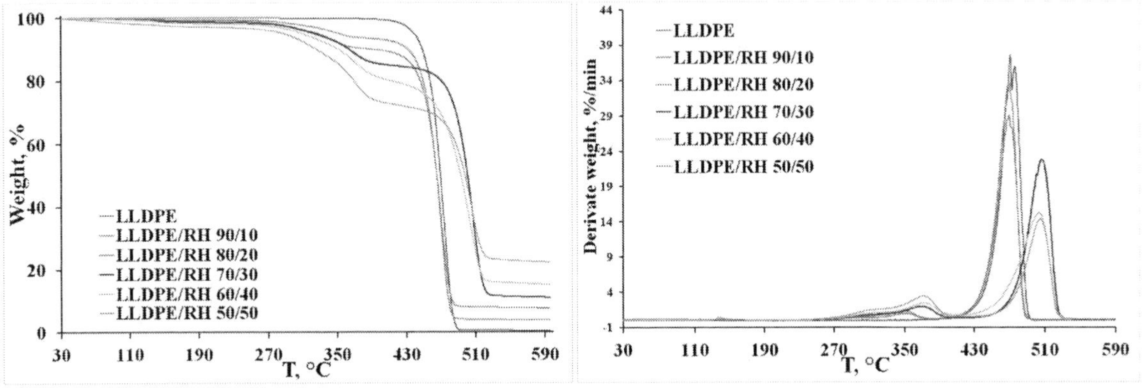

Figure 5 TG and DTG curves of the neat LLDPE and LLDPE/RH biocomposites without $mTiO_2$

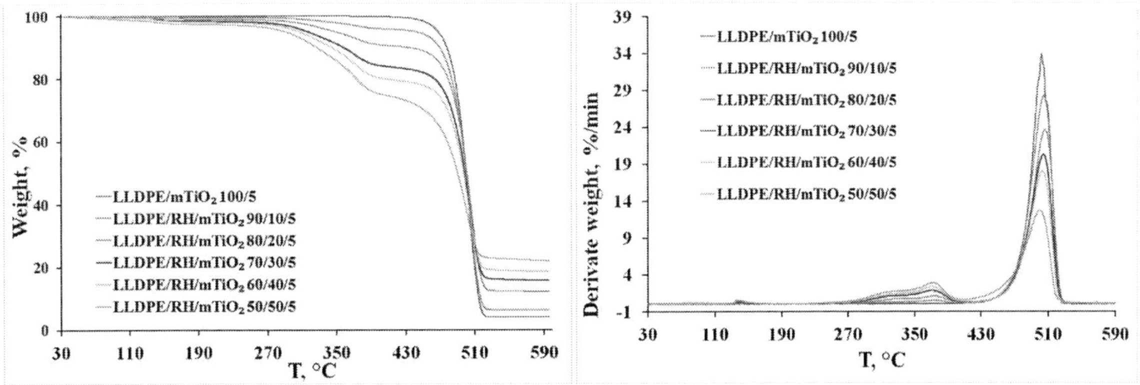

Figure 6 TG and DTG curves of the neat LLDPE and LLDPE/RH biocomposites with $mTiO_2$

Table 2 shows the initial degradation temperature ($T_{ini.}$), the temperature corresponding to final temperature of thermal degradation (T_{end}) and the residual mass of the biocomposites at 600 °C. As can be seen from the DTG curve LLDPE degrades in one degradation stage while the both LLDPE/RH biocomposites series degrades up to three degradation stages. The first two stages corresponding to those of RH (due to the loss of water in RH and to the thermal decomposition of hemicellulose, cellulose and lignin in RH) and the third one was due to thermal degradation of LLDPE. As expected, the area of the first two peaks in the DTG curves increases with the RH content, whereas the area of the third peak decreases with it. The increase in RH content in the LLDPE/RH biocomposites decreased the thermal stability of biocomposites. This is due to the lower thermal stability of the RH. In addition, the weight loss starts at the higher temperature for biocomposites with $mTiO_2$. It shows that thermal stability of LLDPE/RH biocomposites was significantly improved by the incorporation of $mTiO_2$. The value of the residual at 600°C of LDPE/RH biocomposites increase with increasing

RH content and it was mainly related to the silica content of RH (96 %). It can be seen that the residue of LLDPE/RH/mTiO_2 biocomposites is a little higher than that of LLDPE/RH biocomposites.

Table 5 TGA results of the neat LLDPE and LLDPE/RH biocomposites without mTiO_2

Sample	LLDPE100/0	LLDPE/RH 90/10	LLDPE/RH 80/20	LLDPE/RH 70/30	LLDPE/RH 60/40	LLDPE/RH 50/50
$T_{ini.}$ (°C)	439.8	352.9	329.4	324.4	310.6	308.1
$T_{fin.}$ (°C)	484.1	480.7	481.3	520.2	516.5	517.6
Residue at 600°C	0.515	3.8	7.7	11.1	15.2	22.3

Table 6 TGA results of the neat LLDPE and LLDPE/RH biocomposites with mTiO_2

Sample	LLDPE/TiO_2 100/5	LLDPE/RH/$TiO_2$90/10/5	LLDPE/RH/TiO_2 80/20/5	LLDPE/RH/TiO_2 70/30/5	LLDPE/RH/TiO_2 60/40/5	LLDPE/RH/TiO_2 50/50/5
$T_{ini.}$ (°C)	467.6	467.9	467.8	468.2	475.1	457.5
$T_{fin.}$ (°C)	513.7	515.7	467.8	522.4	525.0	512.6
Residue at 600°C	4.1	6.2	12.2	15.7	18.5	22.2

4. CONCLUSION

In this study we prepared the LLDPE/RH biocomposites with different content of RH particles and withouth and with the addition of TiO_2 with the aim to investigate the effect of RH and mTiO_2 on the phase transitions, viscoelastic properties and thermal stability. For expect is that the RH particles and mTiO_2 will improve thermal properties important in the application of this biocomposites as packing materials. By adding and increasing the content of RH the storage modulus (*E'*) for the LLDPE/RH biocomposites was higher than that of the neat LLDPE, because of the increasing stiffness of the biocomposites. The incorporation of mTiO_2 as filler to LLDPE/RH biocomposites, the T_g increases due to the reduced segmental mobility of LLDPE. On the other side the presence of the RH and mTiO_2 does not affect T_m of the LLDPE/RH biocomposites. DSC results indicate that the addition of mTiO_2 favors the nucleation process; thus, the *Xc* values of the LLDPE/RH/mTiO_2 biocomposites are higher than those of LLDPE/RH biocomposites. On the bases of TGA measurements the incorporation of the mTiO_2 gives rise to a considerable increase of the thermal stability of LLDPE/RH biocomposites.

REFERENCES

[1] VAISANEN T., DAS O., TOMPPO L. A review on new bio-based constituents for natural fiber-polymer composites, *Journal of Cleaner Production*. 2017, vol. 149, pp. 582-596. Available from: https://doi.org/10.1016/j.jclepro.2017.02.132.

[2] PRAVEENKUMARA J., GIRIJAPPA Y. G., Thyavihalli, PUTTEGOWDA M., RANGAPPA S. M., SIENGCHIN S. Effect of natural filler materials on fiber reinforced hybrid polymer composites: An Overview, *Journal of Natural Fibers*. 2020, vol. 19, pp.1-16. Available from: https://doi.org/10.1080/15440478.2020.1854145.

[3] TURI, E. A. "Thermal Characterization of Polymeric Materials". 2nd Edition. Academic Press, 1997, pp 2420.

[4] SIROTKIN, R. O., BROOKS, N. W. The Dynamic Mechanical Relaxation Behavior of Polyethylene Copolymers Cast from Solution. *Polymer*. 2001, vol.42, pp. 9801-9808. Available from: https://doi.org/10.1016/S0032-3861(01)00535-3.

SERS OF ORGANIC ANALYTE ON THE FILTER PAPER WITH TiO_2 ALD COATING AND AG NANOPARTICLES

[1]Jan VOVES, [1]David MARES, [1]Ruhani DADASHOV, [2]Wei-Nien SU, [2]Steven SUWITO, [2]Agaje Bedemo BEYENE

[1]Department of Microelectronics, Faculty of Electrical Engineering, Czech Technical University in Prague (CTU), Prague, Czech Republic, EU, voves@fel.cvut.cz

[2]Division/Department Graduate Institute of Applied Science and Technology, National Taiwan University of Science and Technology (TAIWAN TECH), Taiwan, wsu@mail.ntust.edu.tw

https://doi.org/10.37904/nanocon.2022.4618

Abstract

Surface Enhanced Raman Spectroscopy (SERS) is an important method for characterizing different materials and chemical analytes at low concentrations. The Raman signal plasmonic amplification is strongly dependent on the surface nanostructure. The filter paper could be used as the substrate for the chemical analysis of organic materials. We investigated the influence of different TiO2 thicknesses deposited on the ashless filter paper surface on the signal efficiency. TiO2 is hydrophilic and biocompatible. High-surface area TiO2 on the filter paper will help create 3D heterostructure to collect more analyte molecules. We used Sentech SI 500 thermal Atomic Layer Deposition (ALD) System with 10, 20, and 50 growth cycles and the TTIP precursor. Surface structure was investigated by Atomic Force Microscopy (AFM). The silver (Ag) colloid solution was dropped on the filter paper before standard R6G biomarker solutions were deposited. The Raman signal measured by In Via Renishaw spectrometer was compared for the different samples. TiO2 ALD-20 cycle gave the highest enhancement factor (~2.3 X 105) compared to other ALD-coated samples. Interestingly, the TiO2 ALD coating generally results in an enhancement factor more than 20 times higher than the reference sample (namely filter paper without the ALD coating). SERS mapping results suggest the uniformity and reliability of SERS measurements across a large area (40 x 40 um2). The finding paves a new direction for preparing simple, flexible, and biocompatible SERS substrates, which could be readily applied in detecting various biomolecular or organic analytes in our coming works.

Keywords: Surface enhanced Raman spectroscopy, filter paper, titanium oxide, atomic layer deposition

1. INTRODUCTION

Surface-enhanced Raman scattering (SERS) is a non-destructive sensing technique in which inelastic light scattering by molecules is enhanced in comparison with clasical Raman scattering by the orders of magnitude when the molecules are adsorbed onto corrugated metal surfaces such as silver or gold nanoparticles (NPs). The enhancement is produced by localized field of the plasmonic oscilations between the nanoparticles. SERS enables the examination of small numbers of molecules. SERS has increasing number of applications in several different directions, including electrochemistry, catalysis, biology, medicine, art conservation, materials science, and others [1-3]. SERS substrates are produced mainly by metallic nanostructures on glass substrates or silicon wafers. However, three-dimensional (3D) structures are more suitable because they enlarge both the interaction volumes and total interaction surface areas between the localized field and the molecules to be detected. But the complexity of the fabrication and low-flexibility of the substrates, limit the practical application of these SERS devices. The new methods of the preparation of 3D SERS substrates use polymer nano-fibers or paper fiber networks as the frame for the plasmonic nanostructures [4]. Particular advantages attributed to filter paper-based SERS are the ability to create greater hotspot sites and to concentrate analyte. Filter paper based SERS substrates have recently shown strong capability in detecting trace amount of small molecules [5]. The crystalline TiO2 with Ag NP or silica-titania fibers obtained by sol-gel and electrospinning processes were also successfuly used for SERS substrate preparation [6,7]. Rhodamine 6G (R6G) is often used as a model molecule in the SERS spectroscopy with resulting EF up to 10^8 on roughened nano-Au films or detection limit of 10^{-13} M R6G on urea and formaldehyde (UF) microsphere with Ag NP [8 - 11]. Here we report a study of the properties of the Ag NP covered ashless filter paper with different TiO_2 coverage and the resultant surface enhanced Raman scattering (SERS) activity using R6G as a analyte.

2. EXPERIMENTAL

Several samples were prepared for the Raman and SERS measurement. The reference samples were placed on the silicon substrates. SERS samples were prepared on the ashless filter paper by two methods: The first with the synthetized rutile TiO_2 and Ag NP, the second with the thin TiO_2 layers prepared by 10, 20 and 50 cycles of the thermal Atomic Layer Deposition (ALD) SI 500 Sentech covered again by the Ag NP. The rutile TiO_2 was synthetized by mixing of 0.75 M of urea (99.5 %) and 1 M of TiCl4 (99.9 %) in an autoclave by 200 °C for 2 hours. TiO2 nanoparticles were obtained by centrifugation, washing, and drying in the vacuum oven at 80 °C overnight. The filter paper surface has been characterized by the optical and Atomic Force Microscopy (AFM) as shown in **Figure 1**. The paper fiber diameter is approximately 10 µm and the filter paper structure creates gaps with dimensions of several tens of µm where the analyte could be accumulated. Resulting samples were inspected by the Scanning Electron Microscopy (SEM). In the **Figure 2** the R6G molecule and the surface of the sample with synthetized rutile TiO_2 and Ag NP are shown. The ALD covered sample if shown in the **Figure 3** with different scales. The diameters of the Ag NP are in the range 40 - 120 nm.

Figure 1 The AFM and optical microscope (insert) picture of the filter paper surface

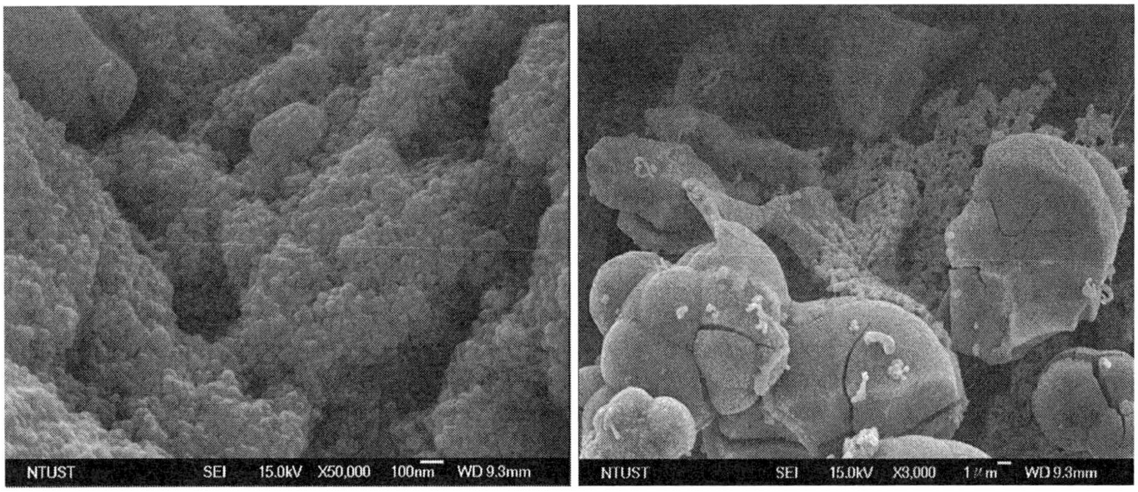

Figure 2 SEM picture of R6G molecule (left) and rutile/Ag NC surface (right)

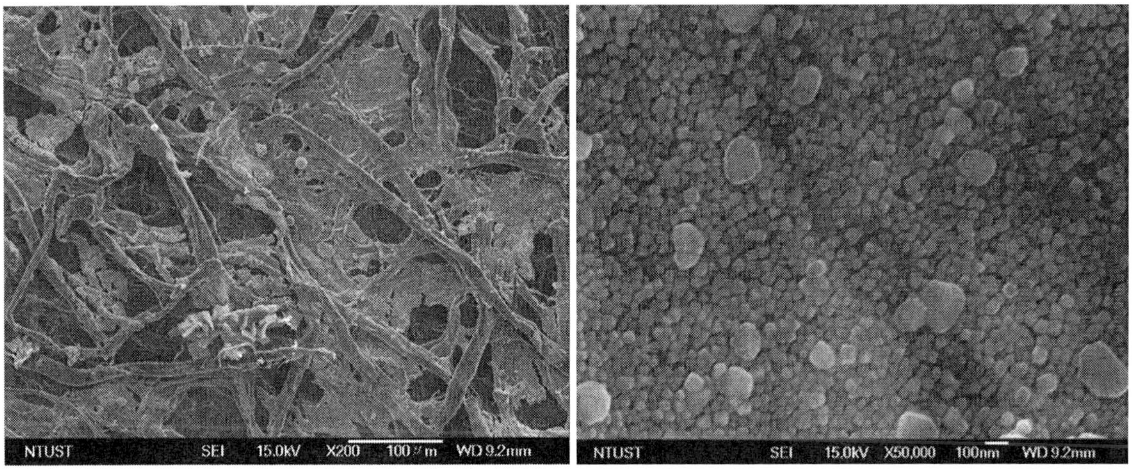

Figure 3 SEM picture of ALD TiO_2/Ag NC surface in microscale (left) and in nanoscale (right)

3. RESULTS

First, the Raman spectra of 10 mM and 1 mM R6G on silicon were performed (**Figure 4**). All measurements were done on Renishaw In Via Raman spectrometer. We used the 633 nm laser with 10 % intensity and we compared the signal at Raman shift on the 1364 cm^{-1} peak. SERS enhancement factor (EF) is defined as the ratio of SERS intensity contributed by each surface molecule to the ordinary Raman intensity contributed by each free molecule:

$$EF = \frac{(I_{SERS} / C_{SERS})}{(I_{Raman} / C_{Raman})}$$

where I_{SERS} and I_{Raman} are the SERS and ordinary Raman intensities, respectively. C_{SERS} and C_{Raman} are the concentrations of molecules probed by SERS and ordinary Raman. For the samples with synthesized rutile TiO_2/Ag NC layer we obtained the height for 1 mM R6G 452286 a.u as shown in **Figure 5**. In comparison with the height of reference peak for 10 mM R6G on the silicon wafer (451 a.u.) we obtained EF 10^4. For the 1 μM of R6G on the ALD TiO_2/Ag NC surface we measured the samples with different thicknesses of ALD TiO_2 layer using 10, 20 and 50 growth cycles on Sentech SI 500 thermal ALD with the TTIP precursor. Resulting SERS spectra are shown in **Figure 6**. The highest peak 10591 a.u. of 1 µM R6G at1364 cm^{-1} was observed for the 20 cycles of ALD (TiO_2 layer thickness 0.4 nm). Resulting EF is 2.3 x 10^5.

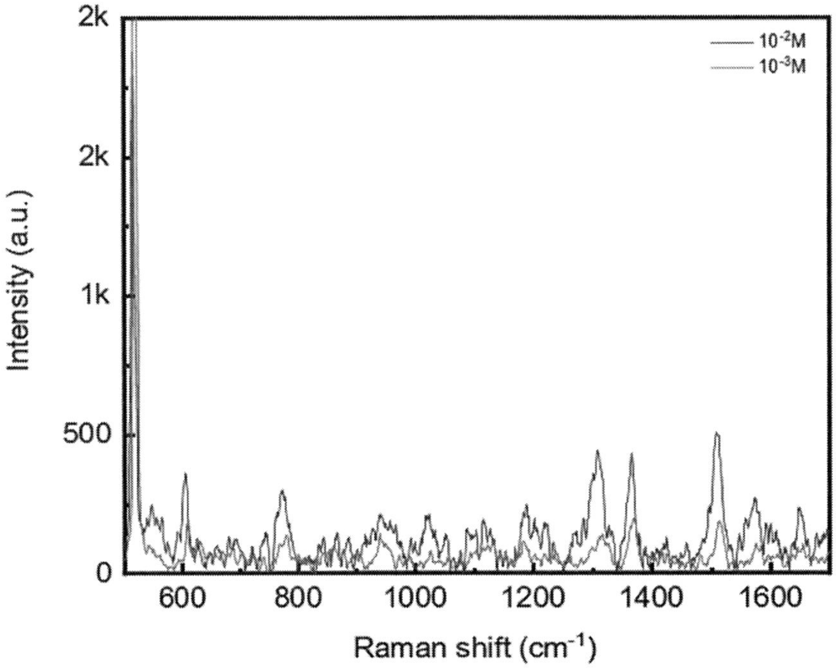

Figure 4 Raman spectra of 10 mM (black) and 1 mM (red) R6G on silicon with 10 % power of 633 nm laser

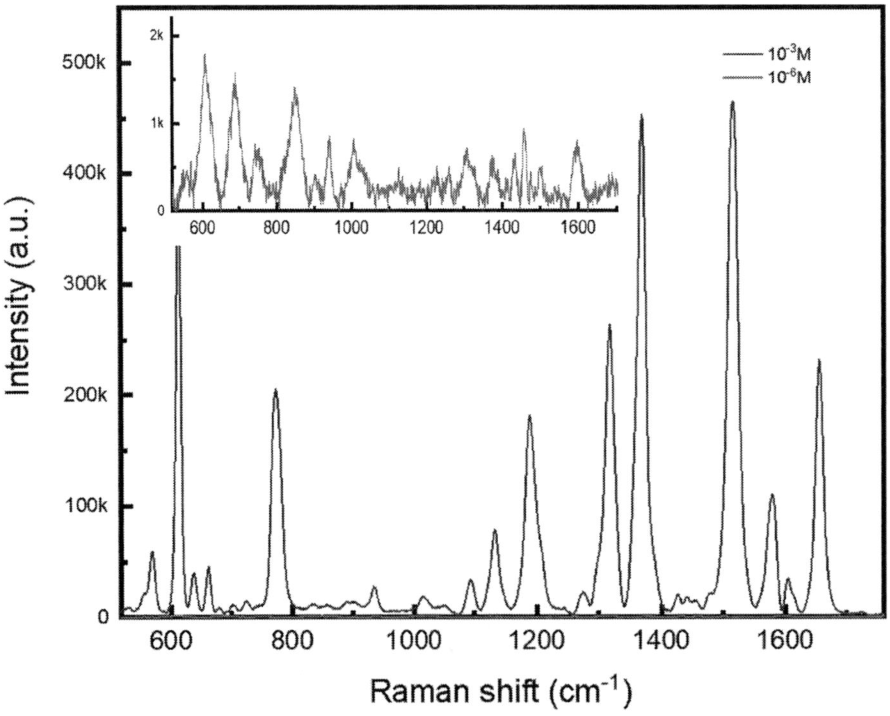

Figure 5 SERS spectra of 1 mM and 1 µM (insert) R6G (insert) on rutile TiO_2/Ag NP

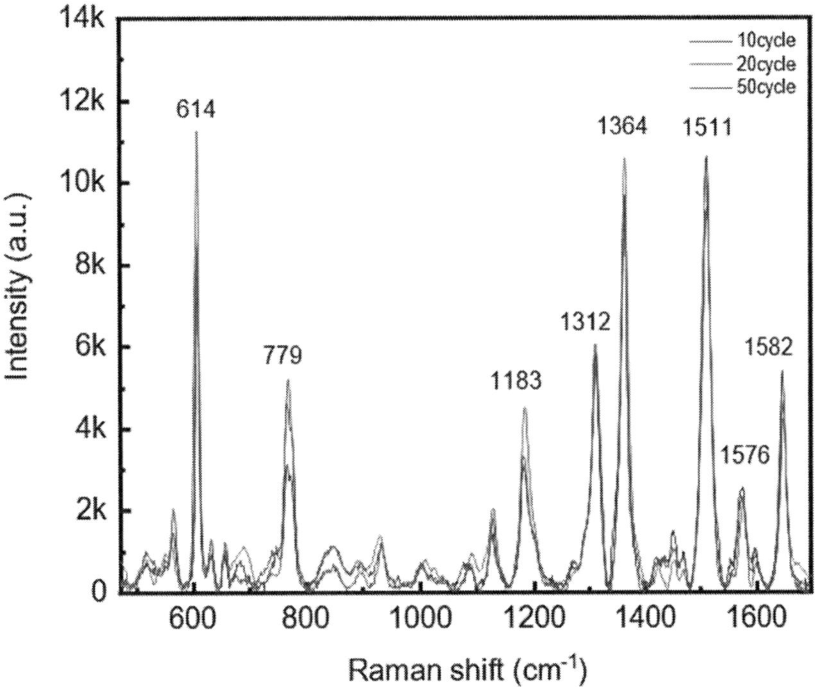

Figure 6 SERS spectra on ALD TiO_2/Ag NP

The SERS signal homogeneity was measured as well by the Raman mapping (**Figure 7**). The 1364 cm^{-1} peak showed good homogeneity in the area of several tenths of µm.

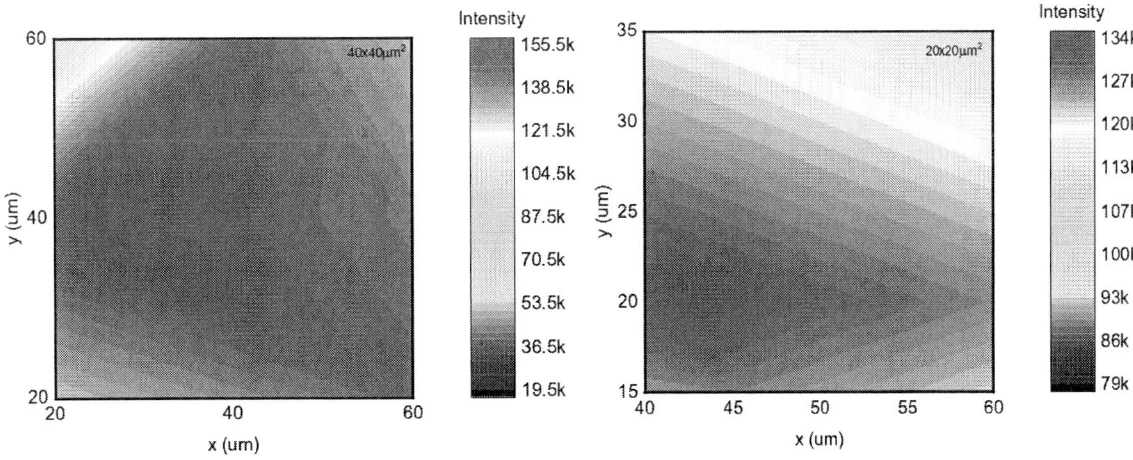

Figure 7 SERS mapping on droplet 50 ALD cycle for 1364 cm^{-1} (left) and for 1511 cm-1 (right).

4. CONCLUSION

The Raman spectra of the R6G analyte on the pure silicon substrate in comparison with SERS spectra on TiO_2 coated filter paper with Ag NP show the enhancement factor larger than 10^5. Low costs, thin, and flexible devices are the main advantages of the presented SERS substrate over the conventional 3D designs. The samples with thin ALD TiO_2 coverage have 10 times higher enhancement factor than the samples with sythetized pristine TiO_2. It probably follows from the retained filter paper 3D structure due to very thin oxide coverage.

ACKNOWLEDGEMENTS

The research has been supported by the grant CTU-NTUST-110-02 of the Joint Research Program between Czech Technical University in Prague and National Taiwan University of Science and Technology, by the by the student grant of CTU No. SGS20/176/OHK3/3T/13 and by the project Centre of the Advanced Applied Natural Sciences No. CZ.02.1.01/0.0/0.0/16_019/0000778 supported by the Operation Programme Research, Development and Education co-financed by European Community and by Ministry of Education Czech Republic.

REFERENCES

[1] LANGER Judith et al. Present and Future of Surface-Enhanced Raman Scatteri. *ACS Nano*. 2020, vol. 14, pp. 28−117.

[2] PEREZ-JIMENEZ, Ana Isabel, LYU, Danya, LU, Zhixuan, LIU, Guokun, REN, Bin. Surface-enhanced Raman spectroscopy: benefits, trade-offs and future developments. *Chem. Sci.* 2020, vol. 11, 4563.

[3] KIHARA, Shinji, CHAN, Andrew, IN, Eugene, TALEB, Nargiss, TOLLEMACHE, Cherie, YICKAB, Samuel, McGILLIVRAY, Duncan J. Detecting polystyrene nanoplastics using filter paper-based surface-enhanced Raman spectroscopy. *RSC Adv.* 2022, vol. 12, 20519.

[4] XU, Tianwen, WANG, Xueqin, ZHANG, Xin, BAI, Zhongchen, Compact Ag nanoparticles anchored on the surface of glass fber flter paper for SERS applications. *Applied Physics A.* 2022, pp. 128-317.

[5] MU, Yunyun, ZHANG, Xinping, A Paper-Fiber-Supported 3D SERS Substrate. *Plasmonics.* 2020, vol. 15, pp. 889-896.

[6] MA, Lingwei, HUANG, Yu, HOU, Mengjing, XIE, Zheng, ZHANG, Zhengjun. Ag Nanorods Coated with Ultrathin TiO2 Shells as Stable and Recyclable SERS Substrates. *Scientific Reports.* 2015, vol. 5, 15442.

[7] ROQUE-RUIZ, José Hafid, MARTÍNEZ-MÁYNEZ, Héctor, ZALAPA-GARIBAY, Manuela Alejandra, MORAQUECHO, ARIZMENDI, Ana, FARIAS, Rurik, REYES-LÓPEZ, Simón Yobanny: Surface enhanced Raman spectroscopy in nanofibers mats of SiO_2-TiO_2-Ag. In: *11th IEEE International Conference on Nanotechnology, Results in Physics.* 2017, vol. 7, pp. 2520-2527.

[8] KAHRAMAN, Mehmet, DAGGUMATI, Pallavi, KURTULUS, Ozge, SEKER, Erkin, WACHSMANN-HOGIU, Sebastian. Fabrication and Characterization of Flexible and Tunable Plasmonic Nanostructures. *Scientific Reports.* 2013, vol. 3, 3396.

[9] WANG, Jiangcai, QIU, Cuicui, MU, Xijiao, PANG, Hua, CHEN, Xinchun, LIU, Dameng, Ultrasensitive SERS detection of rhodamine 6G and p-nitrophenol based on electrochemically roughened nano-Au film. *Talanta.* 2020, vol. 210, 120631.

[10] HAO, Zhixian, MANSUER, GUO, Mulati, ZHU, Yuqing, Zhirong, WANG Xiaogang, Ag-nanoparticles on UF-microsphere as an ultrasensitive SERS substrate with unique features for rhodamine 6G detection. *Talanta.* 2016, vol. 146, pp. 533-539.

[11] PRISTINSKI, Denis, TAN Siliu, EROL, Melek, DU, Henry, SUKHISHVILI, Svetlana. In situ SERS study of Rhodamine 6G adsorbed on individually immobilized Ag nanoparticles. *J. Raman Spectrosc.* 2006, vol. 37, pp. 762-770.